METHODS IN MOLECULAR BIOLOGY

For further volumes:
http://www.springer.com/series/7651

For over 35 years, biological scientists have come to rely on the research protocols and methodologies in the critically acclaimed *Methods in Molecular Biology* series. The series was the first to introduce the step-by-step protocols approach that has become the standard in all biomedical protocol publishing. Each protocol is provided in readily-reproducible step-by-step fashion, opening with an introductory overview, a list of the materials and reagents needed to complete the experiment, and followed by a detailed procedure that is supported with a helpful notes section offering tips and tricks of the trade as well as troubleshooting advice. These hallmark features were introduced by series editor Dr. John Walker and constitute the key ingredient in each and every volume of the *Methods in Molecular Biology* series. Tested and trusted, comprehensive and reliable, all protocols from the series are indexed in PubMed.

Myofibroblasts

Methods and Protocols

Edited by

Boris Hinz

Laboratory of Tissue Repair and Regeneration, Faculty of Dentistry, University of Toronto, Toronto, ON, Canada

David Lagares

Matrix and Mechanobiology Program, Massachusetts General Hospital, Charlestown, MA, USA; Laboratory of Tissue Regeneration, Fibrosis and Tumor Microenvironment, Massachusetts General Hospital, Harvard Medical School, Boston, MA, USA

Humana Press

Editors
Boris Hinz
Laboratory of Tissue Repair and Regeneration
Faculty of Dentistry
University of Toronto
Toronto, ON, Canada

David Lagares
Matrix and Mechanobiology Program
Massachusetts General Hospital
Charlestown, MA, USA

Laboratory of Tissue Regeneration
Fibrosis and Tumor Microenvironment
Massachusetts General Hospital
Harvard Medical School
Boston, MA, USA

ISSN 1064-3745 ISSN 1940-6029 (electronic)
Methods in Molecular Biology
ISBN 978-1-0716-1381-8 ISBN 978-1-0716-1382-5 (eBook)
https://doi.org/10.1007/978-1-0716-1382-5

This Humana imprint is published by the registered company Springer Science+Business Media, LLC part of Springer Nature.
The registered company address is: 1 New York Plaza, New York, NY 10004, U.S.A.

Preface

Myofibroblasts are activated mesenchymal cells that orchestrate wound healing responses to tissue damage by producing growth factors and extracellular matrix (ECM) and contracting the neo-ECM into mechanically stabilizing scar tissue. The outcome of these activities in physiologically controlled conditions is promoting tissue regenerative processes and repair. However, persistent and excessive myofibroblast activities drive physiological tissue repair into the development of pathological fibrosis. Fibrosis can affect virtually every organ in our body and is a terminal condition of various human diseases, including idiopathic pulmonary fibrosis, systemic sclerosis, cardiac fibrosis, liver cirrhosis, progressive kidney disease, and desmoplastic tumors.

Fibrotic scars disfigure and immobilize skin and fascia of patients with burns, suffering from severe trauma, or being affected by Dupuytren's disease. On the contrary, diminished myofibroblast activities underlie impaired healing of skin wounds of elderly and/or diabetic individuals, resulting in chronically open wounds. Given their central effector role, myofibroblasts have become prime targets for therapeutic treatment of organ fibrosis or chronic wounds. In this volume, leading academic scientists discuss the fundamentals of myofibroblast biology in tissue repair, fibrosis, and tumors as well as cutting-edge laboratory methods used to investigate myofibroblast functions in physiological and pathological settings in vitro and in vivo. This volume is divided into three sections. Part I focuses on fundamental methods to study myofibroblast biology and covers topics such as methods for detecting myofibroblasts and senescent myofibroblasts in cell culture and histology, single-cell RNA sequencing to identify myofibroblast subsets in fibrotic tissues, and functional assays to assess TGF-β activation, myofibroblast apoptosis, or matrix deposition and cross-linking. Part II discusses methods to investigate the mechanobiology of myofibroblasts in vitro, including the fabrication of functional hydrogels with tunable stiffness, the use of atomic force microscopy to characterize matrix and cellular stiffness, as well as molecular assays to assess fibroblast mechanotransduction pathways and durotaxis. Part III describes multiple animal models to investigate myofibroblast functions across organs in vivo as well as human organoid systems, precision tissue slices, and decellularized 3D tissue scaffolds to assess myofibroblast functions in relevant human ex vivo models.

Toronto, ON, Canada — *Boris Hinz*
Boston, MA, USA — *David Lagares*

Contents

Contributors

SOUMEYA ABED • *Department of Medicine, Firestone Institute for Respiratory Health, McMaster University and The Research Institute of St. Joe's Hamilton, Hamilton, ON, Canada*

TASLIM A. AL-HILAL • *Department of Pharmaceutical Sciences, The University of Texas at El Paso, El Paso, TX, USA*

KJETIL ASK • *Department of Medicine, McMaster University and The Research Institute of St. Joe's Hamilton, Firestone Institute for Respiratory Health, Hamilton, ON, Canada*

MOHAMMADNABI ASMANI • *Department of Biomedical Engineering, University at Buffalo, The State University of New York, Buffalo, NY, USA*

THOMAS H. BARKER • *Department of Biomedical Engineering, University of Virginia, Charlottesville, VA, USA*

WILLIAM L. BERRY • *Department of Surgery, University of Oklahoma Health Sciences Center, Oklahoma City, OK, USA*

GRACE C. BINGHAM • *Department of Biomedical Engineering, University of Virginia, Charlottesville, VA, USA*

GERALD BURGSTALLER • *Institute of Lung Biology and Disease (ILBD) and Comprehensive Pneumology Center (CPC), Helmholtz Zentrum München, Member of the German Center for Lung Research (DZL), Munich, Germany; CPC-M bioArchive, Helmholtz Zentrum München, Comprehensive Pneumology Center Munich DZL/CPC-M, München, Germany*

JYOTSNA CHANDRA • *Department of Inflammation and Immunity, Lerner Research Institute, Cleveland Clinic Foundation, Cleveland, OH, USA*

TAMARA CRUZ • *Division of Pulmonary, Allergy, and Critical Care Medicine, University of Pittsburgh School of Medicine, Pittsburgh, PA, USA*

DINA DEJANOVIC • *Department of Inflammation and Immunity, Lerner Research Institute, Cleveland Clinic Foundation, Cleveland, OH, USA*

JÖRG H. W. DISTLER • *Department of Internal Medicine 3, Rheumatology and Immunology, Friedrich-Alexander-University Erlangen-Nürnberg and University Hospital Erlangen, Erlangen, Germany*

IAN M. C. DIXON • *Department of Physiology and Pathophysiology, Institute of Cardiovascular Sciences, University of Manitoba, St. Boniface Hospital Albrechtsen Research Centre, Winnipeg, MB, Canada*

OLIVER EICKELBERG • *Division of Pulmonary, Allergy, and Critical Care Medicine, Department of Medicine, University of Pittsburgh, Pittsburgh, PA, USA*

ADAM J. ENGLER • *Department of Bioengineering, University of California, San Diego, La Jolla, CA, USA; Engler Lab, Sanford Consortium for Regenerative Medicine, La Jolla, CA, USA*

JOAO FIRMINO • *Laboratory of Tissue Repair and Regeneration, Faculty of Dentistry, University of Toronto, Toronto, ON, Canada; Collaborative Advanced Microscopy Laboratories of Dentistry (CAMiLoD), Faculty of Dentistry, University of Toronto, Toronto, ON, Canada*

GIULIO GABBIANI • *Faculty of Medicine, Department of Pathology and Immunology, University of Geneva, Geneva, Switzerland*

MICHAEL GERCKENS • *Institute of Lung Biology and Disease (ILBD) and Comprehensive Pneumology Center (CPC), Helmholtz Zentrum München, Member of the German Center for Lung Research (DZL), München, Germany; CPC-M bioArchive, Helmholtz Zentrum München, Comprehensive Pneumology Center Munich DZL/CPC-M, Munich, Germany*

JAMES GRIFFITH • *Department of Surgery, University of Oklahoma Health Sciences Center, Oklahoma City, OK, USA*

AARON I. HAYAT • *Department of Medicine, McMaster University and The Research Institute of St. Joe's Hamilton, Firestone Institute for Respiratory Health, Hamilton, ON, Canada*

BORIS HINZ • *Laboratory of Tissue Repair and Regeneration, Faculty of Dentistry, University of Toronto, Toronto, ON, Canada*

PING HU • *Department of Biomedical Engineering, University of Virginia, Charlottesville, VA, USA*

MARK D. INMAN • *Department of Medicine, McMaster University and The Research Institute of St. Joe's Hamilton, Firestone Institute for Respiratory Health, Hamilton, ON, Canada*

GISLI JENKINS • *Division of Respiratory Medicine, NIHR Nottingham Respiratory Biomedical Research Centre, School of Medicine, University of Nottingham, Nottingham, UK*

MELANIE KÖNIGSHOFF • *CPC-M bioArchive, Helmholtz Zentrum München, , Comprehensive Pneumology Center Munich DZL/CPC-M, München, Germany; Division of Pulmonary, Allergy, and Critical Care Medicine, Department of Medicine, University of Pittsburgh, Pittsburgh, PA, USA; Comprehensive Pneumology Center (CPC), Research Unit Lung Repair and Regeneration, Helmholtz Zentrum München, Member of the German Center for Lung Research (DZL), Munich, Germany*

HOSSAM KADRY • *Department of Pharmaceutical Sciences, The University of Texas at El Paso, El Paso, TX, USA*

ANDRÁS KAPUS • *Keenan Research Centre for Biomedical Science of the St. Michael's Hospital; Department of Surgery, University of Toronto, Toronto, ON, Canada; Department of Biochemistry, University of Toronto, Toronto, ON, Canada*

NATALIE J. KIRKLAND • *Department of Bioengineering, University of California, San Diego, La Jolla, CA, USA; Engler Lab, Sanford Consortium for Regenerative Medicine, La Jolla, CA, USA*

MICHAEL KOFLER • *Keenan Research Centre for Biomedical Science of the St. Michael's Hospital, University of Toronto, Toronto, ON, Canada*

MARTIN R. J. KOLB • *Department of Medicine, Firestone Institute for Respiratory Health, McMaster University and The Research Institute of St. Joe's Hamilton, Hamilton, ON, Canada*

RAFAEL KRAMANN • *Institute of Experimental Medicine and Systems Biology, RWTH Aachen University Hospital, Aachen, Germany; Department of Internal Medicine, Nephrology and Transplantation, Erasmus Medical Center, Rotterdam, The Netherlands*

DAVID LAGARES • *Center for Immunology and Inflammatory Diseases, Division of Pulmonary and Critical Care Medicine, and Fibrosis Research Center, Massachusetts General Hospital, Harvard Medical School, Boston, MA, USA*

NATALIE M. LANDRY • *Department of Physiology and Pathophysiology, Rady Faculty of Health Sciences, Max Rady College of Medicine, Institute of Cardiovascular Sciences, University of Manitoba, Winnipeg, MB, Canada*

DARIO R. LEMOS • *Renal Division, Brigham and Women's Hospital, Boston, MA, USA; Harvard Medical School, Boston, MA, USA*

Jiannan Li • *Department of Inflammation and Immunity, Lerner Research Institute, Cleveland Clinic Foundation, Cleveland, OH, USA*
Markus Luber • *Department of Internal Medicine 3—Rheumatology and Immunology, Friedrich-Alexander-University Erlangen-Nürnberg and University Hospital Erlangen, Erlangen, Germany*
Christopher A. McCulloch • *Faculty of Dentistry, University of Toronto, Toronto, ON, Canada*
Christine McDonald • *Department of Inflammation and Immunity, Lerner Research Institute, Cleveland Clinic Foundation, Cleveland, OH, USA*
Olivia Mekhael • *Department of Medicine, Firestone Institute for Respiratory Health, McMaster University and The Research Institute of St. Joe's Hamilton, Hamilton, ON, Canada*
Sylvia Menzel • *Institute of Experimental Medicine and Systems Biology, RWTH Aachen University Hospital, Aachen, Germany*
Ana L. Mora • *Division of Pulmonary, Allergy, and Critical Care Medicine, University of Pittsburgh School of Medicine, Pittsburgh, PA, USA; Aging Institute, University of Pittsburgh School of Medicine, Pittsburgh, PA, USA*
Alejandro Moran-Horowich • *Renal Division, Brigham and Women's Hospital, Boston, MA, USA*
Veronique J. Moulin • *Centre LOEX de l'Université Laval, Research Center CHU de Qué bec—Université Laval and Faculty of Medicine, Surgery Department, Québec, QC, Canada*
Pura Muñoz-Cánoves • *Cell Biology Group, Department of Experimental and Health Sciences, Pompeu Fabra University (UPF), CIBER on Neurodegenerative Diseases (CIBERNED), Barcelona, Spain; Institució Catalana de Recerca i Estudis Avançats (ICREA), Barcelona, Spain; Centro Nacional de Investigaciones Cardiovasculares, Madrid, Spain*
Safaa Naiel • *Department of Medicine, Firestone Institute for Respiratory Health, McMaster University and The Research Institute of St. Joe's Hamilton, Hamilton, ON, Canada*
Sergey V. Plotnikov • *Department of Cell and Systems Biology, University of Toronto, Toronto, ON, Canada*
Joanne Porte • *Division of Respiratory Medicine, NIHR Nottingham Respiratory Biomedical Research Centre, School of Medicine, University of Nottingham, Nottingham, UK*
Michael Raghunath • *Center for Cell Biology and Tissue Engineering, Institute of Chemistry and Biotechnology, Zurich University of Applied Sciences (ZHAW), Wädenswil, Switzerland*
Sunil G. Rattan • *Department of Physiology and Pathophysiology, Rady Faculty of Health Sciences, Max Rady College of Medicine, Institute of Cardiovascular Sciences, University of Manitoba, Winnipeg, MB, Canada*
Aashreya Ravichandra • *Department of Medicine, Columbia University, New York, NY, USA*
Spencer D. Revill • *Department of Medicine, Firestone Institute for Respiratory Health, McMaster University and The Research Institute of St. Joe's Hamilton, Hamilton, ON, Canada*
Florian Rieder • *Department of Inflammation and Immunity, Lerner Research Institute, Cleveland Clinic Foundation, Cleveland, OH, USA; Department of Gastroenterology,*

Hepatology and Nutrition, Digestive Diseases and Surgery Institute, Cleveland Clinic Foundation, Cleveland, OH, USA

ALEIX RIUS RIGAU • *Department of Internal Medicine 3, Friedrich-Alexander-University Erlangen-Nürnberg and Universitätsklinikum Erlangen, Erlangen, Germany*

FERNANDO RODRIGUEZ-PASCUAL • *Centro de Biología Molecular "Severo Ochoa" Consejo Superior de Investigaciones Científicas (C.S.I.C.), Universidad Autónoma de Madrid, Madrid, Spain*

MAURICIO ROJAS • *Division of Pulmonary, Allergy, and Critical Care Medicine, University of Pittsburgh School of Medicine, Pittsburgh, PA, USA; Aging Institute, University of Pittsburgh School of Medicine, Pittsburgh, PA, USA*

TAMARA ROSELL-GARCÍA • *Centro de Biología Molecular "Severo Ochoa", Consejo Superior de Investigaciones Científicas (CSIC)/Universidad Autónoma de Madrid, Madrid, Spain*

FABIO M. V. ROSSI • *School of Biomedical Engineering and Biomedical Research Centre, University of British Columbia, Vancouver, BC, Canada*

BENEDIKT SABASS • *Theoretical Physics of Living Matter, Institute of Biological Information Processing and Institute of Advanced Simulation, Forschungszentrum Juelich, Juelich, Germany; Department of Veterinary Sciences, Institute for Infectious Diseases and Zoonoses, Ludwig-Maximilians-Universität München, Munich, Germany*

ROBERT F. SCHWABE • *Department of Medicine, Columbia University, New York, NY, USA; Institute of Human Nutrition, Columbia University, New York, NY, USA*

ANTONIO L. SERRANO • *Cell Biology Group, Department of Experimental and Health Sciences, Pompeu Fabra University (UPF), CIBER on Neurodegenerative Diseases (CIBERNED), Barcelona, Spain*

JAGATHPALA SHETTY • *Department of Biomedical Engineering, University of Virginia, Charlottesville, VA, USA*

HESHAM SOLIMAN • *School of Biomedical Engineering and Biomedical Research Centre, University of British Columbia, Vancouver, BC, Canada; Faculty of Pharmaceutical Sciences, Minia University, Minia, Egypt*

DONG OK SON • *Laboratory of Tissue Repair and Regeneration, Faculty of Dentistry, University of Toronto, Toronto, ON, Canada*

SIMON STEBLER • *Center for Cell Biology and Tissue Engineering, Institute of Chemistry and Biotechnology, Zurich University of Applied Sciences (ZHAW), Wädenswil, Switzerland*

AMANDA L. TATLER • *Division of Respiratory Medicine, NIHR Nottingham Respiratory Biomedical Research Centre, School of Medicine, University of Nottingham, Nottingham, UK*

VICTOR J. THANNICKAL • *Division of Pulmonary, Allergy and Critical Care Medicine, Department of Medicine, University of Alabama at Birmingham, Birmingham, AL, USA*

DANIEL J. TSCHUMPERLIN • *Department of Physiology and Biomedical Engineering, Mayo Clinic College of Medicine and Science, Rochester, MN, USA*

LIN WEI TUNG • *School of Biomedical Engineering and Biomedical Research Centre, University of British Columbia, Vancouver, BC, Canada*

FERNANDO R. VALENCIA • *Department of Cell and Systems Biology, University of Toronto, Toronto, ON, Canada*

MEGAN VIERHOUT • *Department of Medicine, Firestone Institute for Respiratory Health, McMaster University and The Research Institute of St. Joe's Hamilton, Hamilton, ON, Canada*

ALEXANDER J. WHITEHEAD • *Department of Bioengineering, University of California, San Diego, La Jolla, CA, USA; Engler Lab, Sanford Consortium for Regenerative Medicine, La Jolla, CA, USA*
SHUYING YANG • *Department of Cell and Systems Biology, University of Toronto, Toronto, ON, Canada*
CHIUAN-REN YEH • *Department of Biomedical Engineering, University of Virginia, Charlottesville, VA, USA*
FERESHTEH S. YOUNESI • *Laboratory of Tissue Repair and Regeneration, Faculty of Dentistry, University of Toronto, Toronto, ON, Canada*
MEGAN T. ZANGARA • *Department of Inflammation and Immunity, Lerner Research Institute, Cleveland Clinic Foundation, Cleveland, OH, USA*
RUOGANG ZHAO • *Department of Biomedical Engineering, University at Buffalo, The State University of New York, Buffalo, NY, USA*
YAN ZHOU • *Division of Pulmonary and Critical Care Medicine, Department of Medicine, Massachusetts General Hospital, Harvard Medical School, Boston, MA, USA*

Chapter 1

50 Years of Myofibroblasts: How the Myofibroblast Concept Evolved

Giulio Gabbiani

Abstract

Myofibroblasts are key cells in mediating normal wound contraction and promoting connective tissue deformations characteristic of fibrosis and scarring. Five decades ago, myofibroblasts were discovered in electron micrographs of wound granulation tissue as fibroblastic cells containing microfilaments that are organized in bundles like those present in smooth muscle. The contractile function of myofibroblasts was demonstrated by measuring the contraction of strips of granulation tissue in response to smooth muscle agonists and in cell culture. Although formation of contractile bundles already defines the myofibroblast, neo-expression of α-smooth muscle actin (α-SMA) in fibroblastic cells has become the most widely used myofibroblast marker. Because α-SMA incorporation into stress fibers mediates enhanced fibroblast contraction, it has been proposed and successfully tested as a drug target in therapeutic approaches to reduce tissue contractures. Other anti-fibrosis strategies target growth factor-, extracellular matrix-, and mechanical stress-induced pathways of myofibroblast activation from various precursors or aim to induce myofibroblast apoptosis. To understand the involved mechanisms of myofibroblast formation and function, critical experimental tools and animal models have been developed, which are made available in this collection of protocols by experts in the field.

Key words TGF-β1, α-Smooth muscle actin, Stress fiber, Myofibroblast origin, Actin, Myosin, Extracellular matrix, Wound healing, Fibrosis

The discovery of the myofibroblast originated from the unexpected electron microscopic observation that, contrary to fibroblasts of normal tissues, granulation tissue fibroblasts contain important amounts of microfilaments, organized in bundles similar to those present in smooth muscle, suggesting that they have acquired contractile features [1]. Later it was shown that strips of granulation tissue can contract or relax upon stimulation with agents capable to contract or relax smooth muscle tissue [2, 3]. It was also shown by means of a specific monoclonal antibody that granulation tissue fibroblasts, for which the name myofibroblasts was proposed, express α-smooth muscle actin, the actin isoform expressed in smooth muscle cells and very abundant in vascular smooth muscle cells [4, 5]. Later, it has been shown that α-smooth

Boris Hinz and David Lagares (eds.), *Myofibroblasts: Methods and Protocols*, Methods in Molecular Biology, vol. 2299, https://doi.org/10.1007/978-1-0716-1382-5_1,

muscle actin not only represents a good marker of the myofibroblast but is also instrumental in force generation by these cells [6] (Fig. 1).

The acceptance of the myofibroblast as a new connective tissue cell ended a long-lasting dispute about whether granulation tissue contraction is due to active cellular contraction or rearrangement of extracellular matrix, obviously in favor of the first hypothesis. These observations, together with the work of many laboratories around the world, contributed to the definition and description of the cellular apparatus responsible for the mechanical activity of non-muscle cells called cytoskeleton that was not known before the second half of the twentieth century [7]; now the biology of the cytoskeleton is a well-established cell biology topic and a chapter on cytoskeleton is present in every cell biology textbook. Myofibroblast contraction depends on classical contractile proteins, but is regulated differently from smooth muscle cell contraction and is mediated through Rho/Rho kinase; cytoskeletal structures, such as vinculin and tensin containing focal adhesions, allow force transmission to the extracellular matrix, thus resulting in creation of the tension that regulates tissue remodeling [8].

The regulation of the fibroblast-myofibroblast transition depends on many factors such as the presence of transforming growth factor β1 (TGF-β1) [9, 10], specific extracellular matrix proteins, such as the splice variant ED-A of cellular fibronectin [11] and local mechanical forces [12, 13]. Myofibroblasts may exhibit different phenotypes: in particular, one can distinguish proto-myofibroblasts which are equipped with many filaments but do not express α-smooth muscle actin and well-developed myofibroblasts whose filaments are rich in α-smooth muscle actin. Proto-myofibroblasts are present in certain normal structures, such as the wall of pulmonary alveoli and appear at the beginning of wound healing and fibrotic lesions; well-developed myofibroblasts are typical of granulation tissue and fibrotic lesions [3, 7].

The presence of myofibroblasts has been established in developing tissues, in normal tissues, and in all pathological situation characterized by tissue remodeling, in particular fibrotic diseases, such as fibromatoses, liver cirrhosis, and pulmonary and renal fibrosis [14], thus suggesting that the myofibroblast represents the main responsible cell for normal and pathological tissue remodeling. The origin of myofibroblasts has been the subject of intensive studies that surprisingly have shown a large spectrum of sources, from the local fibroblast population to the local epithelial and endothelial cells (epithelial- and endothelial-mesenchymal transition) and to bone marrow-derived circulating precursors, called fibrocytes [3, 14]; the origin of myofibroblasts from non-muscle cells has been recently confirmed by cell lineage tracing [15]. Myofibroblast origin varies according to the organ and the biological or pathological process. When myofibroblasts are no

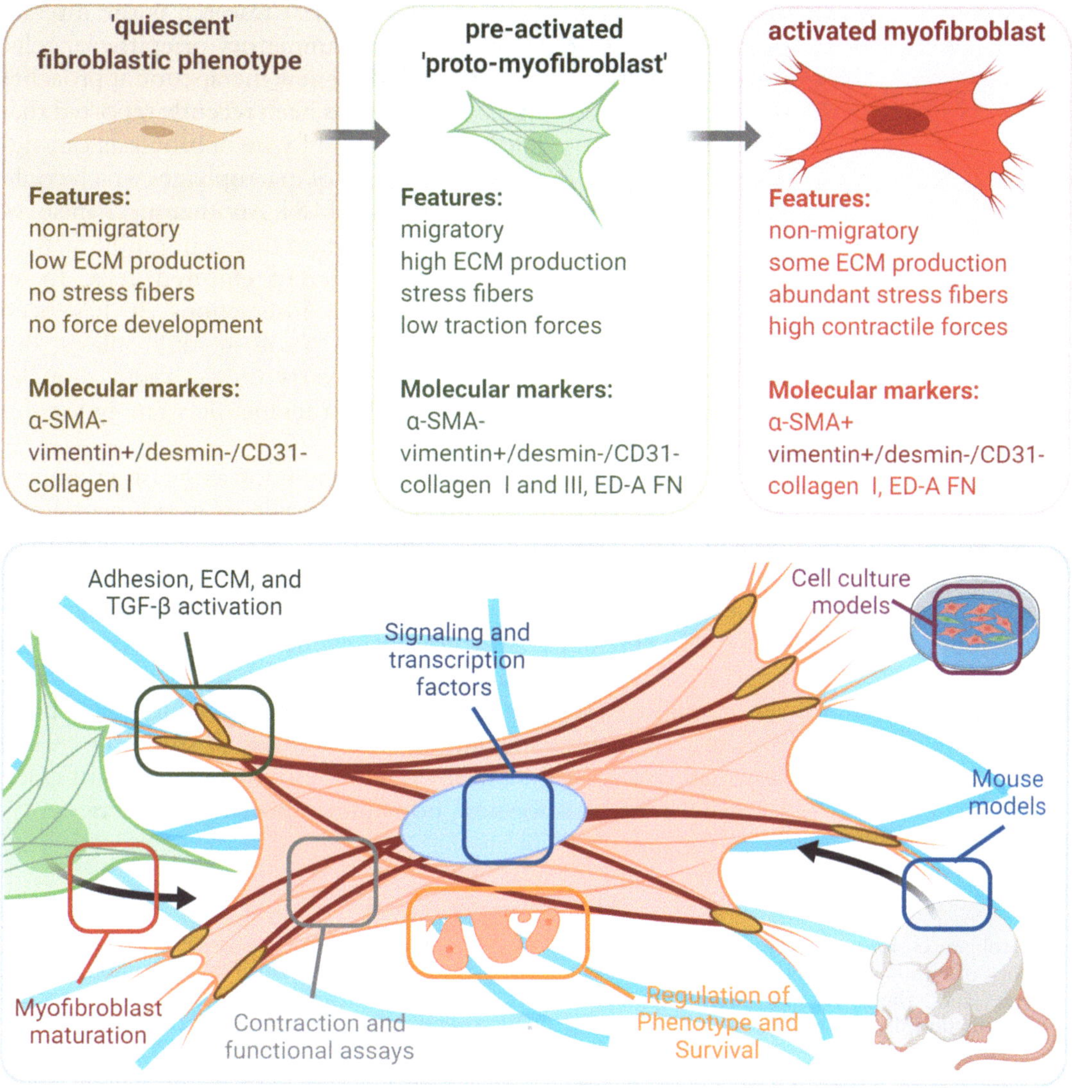

Fig. 1 Myofibroblast features and molecular targets. The minimal definition for a tissue myofibroblast is a fibroblastic (collagen-producing) cell with contractile features (stress fibers) and the ability to transmit force to the extracellular matrix. Because fibroblasts in standard cell culture always develop stress fibers, the term "proto-myofibroblast" has been proposed. Proto-myofibroblasts are a pre-stage of the fully activated myofibroblast, characterized by expression of α-smooth muscle actin (α-SMA) in stress fibers. Discriminating features of vimentin-positive and α-SMA-positive myofibroblasts to smooth muscle (express α-SMA and vimentin) and endothelial cells (express vimentin) are the lack of desmin and endothelial cell protein CD31. The scheme summarizes the topics of myofibroblast biology covered in this methods collection. Schemes were provided by B. Hinz (University of Toronto) using biorender (biorender.com)

more needed, they disappear mainly through apoptosis, such as in the transition between granulation tissue and the scar during the evolution of wound healing [16, 17].

An important aspect of myofibroblast research is the interactions between myofibroblasts and inflammatory or epithelial cells, thus opening the perspective of finding new therapeutic approaches to fibrotic diseases. In this respect, it has been recently reported that extracellular matrix deformation caused by myofibroblasts can significantly increase the migration speed of macrophages which could play a role in fibrotic lesions evolution [18]. Another mechanism of crosstalk is represented by communication through soluble factors, studied by using supernatant of different cultured cells. In an in vitro model of stroma-melanoma interactions, it has been shown that a supernatant of cultured proto-myofibroblasts, but not a supernatant of myofibroblasts, exerts a toxic action on melanoma cells, thus indicating a possible therapeutic strategy for this tumor [19].

As we have seen, the progress of the knowledge on myofibroblast biology and on the myofibroblast role in experimental and human pathology have been impressive during the half century elapsed from its discovery; however, a large amount of work will be needed in the next years in order to more deeply understand the processes at the basis of myofibroblast function; moreover there is a great need to find therapeutic strategies against diseases in which the myofibroblast plays a crucial role. This methodological book presenting numerous techniques and models to study the myofibroblast (Fig. 1) will be in my opinion instrumental in both cases, furnishing the basis for new, hopefully important discoveries.

References

1. Gabbiani G, Ryan GB, Majno G (1971) Presence of modified fibroblasts in granulation tissue and their possible role in wound contraction. Experientia 27:549–550
2. Gabbiani G, Hirschel BJ, Ryan GB, Statkov PR, Majno G (1972) Granulation tissue as a contractile organ. A study of structure and function. J Exp Med 135:719–734
3. Tomasek JJ, Gabbiani G, Hinz B, Chaponnier C, Brown RA (2002) Myofibroblasts and mechano-regulation of connective tissue remodelling. Nat Rev Mol Cell Biol 3:349–363
4. Darby I, Skalli O, Gabbiani G (1990) Alpha-smooth muscle actin is transiently expressed by myofibroblasts during experimental wound healing. Lab Investig 63:21–29
5. Skalli O, Ropraz P, Trzeciak A, Benzonana G, Gillessen D, Gabbiani G (1986) A monoclonal antibody against alpha-smooth muscle actin: a new probe for smooth muscle differentiation. J Cell Biol 103:2787–2796
6. Hinz B, Celetta G, Tomasek JJ, Gabbiani G, Chaponnier C (2001) Alpha-smooth muscle actin expression upregulates fibroblast contractile activity. Mol Biol Cell 12:2730–2741
7. Zampieri F, Coen M, Gabbiani G (2014) The prehistory of the cytoskeleton concept. Cytoskeleton (Hoboken) 71:464–471. https://doi.org/10.1002/cm.21177
8. Hinz B, McCulloch CA, Coelho NM (2019) Mechanical regulation of myofibroblast phenoconversion and collagen contraction. Exp Cell Res 379:119–128. https://doi.org/10.1016/j.yexcr.2019.03.027
9. Desmoulière A, Geinoz A, Gabbiani F, Gabbiani G (1993) Transforming growth factor-beta 1 induces alpha-smooth muscle actin expression in granulation tissue myofibroblasts and in quiescent and growing cultured fibroblasts. J Cell Biol 122:103–111
10. Rønnov-Jessen L, Petersen OW (1993) Induction of alpha-smooth muscle actin by transforming growth factor-beta 1 in quiescent

human breast gland fibroblasts. Implications for myofibroblast generation in breast neoplasia. Lab Investig 68:696–707

11. Serini G, Bochaton-Piallat ML, Ropraz P, Geinoz A, Borsi L, Zardi L, Gabbiani G (1998) The fibronectin domain ED-A is crucial for myofibroblastic phenotype induction by transforming growth factor-beta1. J Cell Biol 142:873–881
12. Arora PD, Narani N, McCulloch CA (1999) The compliance of collagen gels regulates transforming growth factor-beta induction of alpha-smooth muscle actin in fibroblasts. Am J Pathol 154:871–882
13. Goffin JM, Pittet P, Csucs G, Lussi JW, Meister JJ, Hinz B (2006) Focal adhesion size controls tension-dependent recruitment of alpha-smooth muscle actin to stress fibers. J Cell Biol 172:259–268
14. Hinz B, Phan SH, Thannickal VJ, Prunotto M, Desmouliere A, Varga J, De Wever O, Mareel M, Gabbiani G (2012) Recent developments in myofibroblast biology: paradigms for connective tissue remodeling. Am J Pathol 180:1340–1355. https://doi.org/10.1016/j.ajpath.2012.02.004
15. Eschenhagen T (2018) A new concept of fibroblast dynamics in post-myocardial infarction remodeling. J Clin Invest 128:1731–1733. https://doi.org/10.1172/JCI121079
16. Desmoulière A, Redard M, Darby I, Gabbiani G (1995) Apoptosis mediates the decrease in cellularity during the transition between granulation tissue and scar. Am J Pathol 146:56–66
17. Hinz B, Lagares D (2019) Evasion of apoptosis by myofibroblasts: a hallmark of fibrotic diseases. Nat Rev Rheumatol 16:11–31. https://doi.org/10.1038/s41584-019-0324-5
18. Pakshir P, Alizadehgiashi M, Wong B, Coelho NM, Chen X, Gong Z, Shenoy VB, McCulloch CA, Hinz B (2019) Dynamic fibroblast contractions attract remote macrophages in fibrillar collagen matrix. Nat Commun 10:1850. https://doi.org/10.1038/s41467-019-09709-6
19. Avagliano A, Ruocco MR, Nasso R, Aliotta F, Sanita G, Iaccarino A, Bellevicine C, Cali G, Fiume G, Masone S, Masullo M, Montagnani S, Arcucci A (2019) Development of a stromal microenvironment experimental model containing proto-myofibroblast like cells and analysis of its crosstalk with melanoma cells: a new tool to potentiate and stabilize tumor suppressor phenotype of dermal myofibroblasts. Cell 8:1435. https://doi.org/10.3390/cells8111435

Part I

Fundamental Methods to Study Myofibroblast Biology

Chapter 2

Myofibroblast Functions in Tissue Repair and Fibrosis: An Introduction

Victor J. Thannickal

Abstract

Our understanding of myofibroblast biology has advanced over the past two decades, and the seemingly antagonistic roles of these cells in both normal tissue repair and fibrotic diseases are better reconciled. An age-related loss of cellular plasticity that results in impaired capacity for de-differentiation and apoptosis susceptibility may predispose individuals to non-resolving and progressive fibrotic disorders involving diverse organ systems.

Key words Fibroblast, Injury, Regeneration, Apoptosis, Aging, Extracellular matrix

1 Myofibroblasts in Repair and Fibrosis

To grasp a more complete understanding of the biology of myofibroblasts, it may be useful to ask the "why" question before delving into the more experimentally tenable "how" questions. *Why* do myofibroblasts exist? *How* do myofibroblasts participate in tissue repair and fibrosis? While readers of this book are likely to be more interested in the latter question for pragmatic reasons, I posit that a teleological construct of the "why" can be useful to investigators in this field. Dismissing the question of why myofibroblasts exist in nature as philosophical rumination, untenable to the scientific method, may limit the formulation, and ultimately the impact, of experimental studies that test hypotheses of the "how." I have previously attempted to discuss the ultimate and proximate causation of tissue/organ fibrosis as an evolutionary trade-off that is subject to the loss of selective pressure in post-reproductive years [1].

There is now extensive experimental evidence supporting a fundamental role of myofibroblasts in *both* tissue repair and fibrosis. Fundamentally, tissue repair generally connotes a beneficial and desirable process, likely emerging from an adaptive response to

Boris Hinz and David Lagares (eds.), *Myofibroblasts: Methods and Protocols*, Methods in Molecular Biology, vol. 2299, https://doi.org/10.1007/978-1-0716-1382-5_2,

injury that results in restoration of cellular homeostasis, tissue architecture, and organ function. Fibrosis, on the other hand, typically denotes a pathologic process and a potentially crippling disorder that results from a maladaptive response to tissue injury. How do we reconcile these different roles for myofibroblasts in physiological repair versus pathological fibrosis? Alternatively, how does an adaptive response become maladaptive in the context of a pathological disease process? The answer to these questions may lie in understanding myofibroblast plasticity, the loss of which may convert an adaptive, physiological response to a maladaptive, pathological process.

2 Myofibroblast Plasticity

Normal tissue-repair responses are complex, multicellular processes that involve activation of resident stem cells to replace damaged cells and the recruitment of both resident mesenchymal progenitors and non-resident bone marrow-derived immune cells. Immune cell trafficking may serve several functions, including the surveillance of tissues for microbes that may have gained entry through disrupted epithelium, and the eventual phagocytic clearance of these microbes as well as remaining cellular debris/corpses. Normal wound healing, by definition, requires the regulated and orchestrated downregulation and "resolution" of these immune/inflammatory responses [2]. Similarly, the orchestrated and regulated resolution of repair responses involving tissue-resident mesenchymal progenitor cells are essential to normal tissue repair [3].

Despite a great deal of interest in the varied origins/sources of myofibroblasts [4], most recent studies support the concept that the majority of wound-associated myofibroblasts are derived from local, tissue-resident mesenchymal progenitors [5–7]. The question of whether the origin of myofibroblasts may determine its plasticity, defined here as the capacity for differentiation and de-differentiation, has not received much attention and remains unclear. The heterogeneity of fibroblasts and their plasticity has been long recognized from early ultrastructural studies [8]. While the capacity of fibroblasts to differentiate into multiple mesenchymal cell types, including adipocytes, chondrocytes, osteocytes, and myofibroblasts, has been appreciated for many years [9, 10], the concept that myofibroblasts may de-differentiate back to a less differentiated progenitor or to another mesenchymal lineage altogether is more recent [11–13].

3 Myofibroblast Apoptosis

Myofibroblast susceptibility to apoptosis could also be viewed within an expanded definition of "plasticity." After all, the property of cells to undergo spatially and temporally regulated apoptosis within (re)modeling tissues/organs is essential for organ development and normal tissue repair [14, 15]. The observation that apoptosis of myofibroblasts heralds the resolution of granulation tissue during wound healing was described over 25 years ago [16], and an understanding of the cellular and molecular events that regulate this critical process has advanced in recent years [17]. In contrast, the persistence and accumulation of an apoptosis-resistant myofibroblast phenotype is a consistent finding in almost all fibrotic diseases [17]. It is not surprising, then, that targeting this recalcitrant apoptosis-resistant myofibroblast phenotype has emerged as an attractive therapeutic target in organ fibrosis.

The development of effective therapeutics that target pathological myofibroblasts requires an understanding of: (a) the mechanisms that enable these cells to evade apoptotic signals (a cell autonomous property) and (b) the microenvironmental factors that favor survival over apoptosis in responsive cells (a nonautonomous mechanism). There is evidence for both cell autonomous and nonautonomous mechanisms could contribute to the persistence of myofibroblasts in fibrotic disorders [3, 17]. Here, I will briefly introduce one example of each, as these concepts are discussed in more detail in the chapters that follow.

Cellular senescence is an intrinsic property of cells that also acquire an apoptosis-resistant phenotype. Targeting myofibroblast senescence has been shown to be effective in preclinical animal models of lung fibrosis [18, 19]; a senolytic cocktail of dasatinib and quercetin appears to be well tolerated and potentially beneficial in a preliminary study of human subjects with idiopathic pulmonary fibrosis (IPF) [20]. Our group was the first to report on the feasibility of targeting this senescent, apoptosis-resistant phenotype of myofibroblasts by modulating redox imbalance involving both the reactive oxygen species (ROS)-generating enzyme, NADPH oxidase 4 (Nox4), and the Nrf2-dependent antioxidant response pathway [18]. Interestingly, Nox4 has been implicated both in suppressing Nrf2 activation [21, 22] and in inducing (myo)-fibroblast senescence [23–25].

The cellular microenvironment, in a cell nonautonomous manner, may influence the apoptosis susceptibility of myofibroblasts. Although this involves soluble pro-survival growth factors such as TGF-beta [26, 27], the biomechanical properties of the insoluble extracellular matrix (ECM) may also signal apoptosis-resistant and pro-survival pathways in myofibroblasts [28–30]. Several chapters in this series will address molecular methods on how to study the contractile and adhesive properties of myofibroblasts, as well as the biochemical and biomechanical properties of the surrounding ECM.

4 Reversibility of Fibrosis

The concept of myofibroblast plasticity and apoptosis susceptibility inform, in many ways, a paradigm shift in how we think about fibrosis. While fibrosis has traditionally been thought to represent an irreversible disease process, emerging data suggest that many forms of fibrosis are reversible or, at the very least, regress [3, 31]. Examples of the capacity of myofibroblasts to revert to a non-fibrogenic precursor has been demonstrated in the liver [32] and the lung [11, 33]. Therapeutic strategies that lower the threshold for myofibroblast apoptosis are effective in reversing established lung fibrosis in experimental animal models [18, 28, 34, 35]. Interestingly, the drug metformin may induce fibrosis regression by mechanisms that involve myofibroblast de-differentiation [36] and by recovery of apoptosis susceptibility [34]. Thus, strategies that target the cell intrinsic properties of myofibroblasts or the microenvironmental cues that regulate cellular plasticity and/or apoptosis may be effective in mediating fibrosis regression/reversal, resulting in the potential to not only arrest disease progression but also recover physiological organ function.

5 Teleology Informs Biology and Therapeutic Opportunity

Ignoring the teleological "why" in deference to the mechanistic "how" questions as they relate to myofibroblast biology may deprive us of exciting novel opportunities to target the pathological role of myofibroblasts in fibrotic diseases. Like many genes and signaling pathways, cellular phenotypes emerged through evolution to serve adaptive functions that were permissive for survival and reproductive success of the organism. This beneficial role early in life may become corrupted to serve disease-promoting roles in post-reproductive life when there are changes in the cellular microenvironment and accumulated intrinsic cell damage, both of which reduce cellular plasticity. These seemingly antagonistic pleiotropic functions of myofibroblasts may best explain their roles in both tissue repair and fibrosis.

Acknowledgments

This work was supported by NIH grants, P01 HL114470, R01 HL139617, and R01 AG046210; and the U.S. Department of Veterans Affairs Merit Award grant, I01BX003056.

References

1. Thannickal VJ, Zhou Y, Gaggar A, Duncan SR (2014) Fibrosis: ultimate and proximate causes. J Clin Invest 124:4673–4677. https://doi.org/10.1172/JCI74368
2. Nathan C, Ding A (2010) Nonresolving inflammation. Cell 140:871–882. https://doi.org/10.1016/j.cell.2010.02.029
3. Horowitz JC, Thannickal VJ (2019) Mechanisms for the resolution of organ fibrosis. Physiology (Bethesda) 34:43–55. https://doi.org/10.1152/physiol.00033.2018
4. Hinz B, Phan SH, Thannickal VJ, Galli A, Bochaton-Piallat ML, Gabbiani G (2007) The myofibroblast: one function, multiple origins. Am J Pathol 170:1807–1816. https://doi.org/10.2353/ajpath.2007.070112
5. Kanisicak O, Khalil H, Ivey MJ, Karch J, Maliken BD, Correll RN, Brody MJ, SC JL, Aronow BJ, Tallquist MD, Molkentin JD (2016) Genetic lineage tracing defines myofibroblast origin and function in the injured heart. Nat Commun 7:12260. https://doi.org/10.1038/ncomms12260
6. Kramann R, Schneider RK, DiRocco DP, Machado F, Fleig S, Bondzie PA, Henderson JM, Ebert BL, Humphreys BD (2015) Perivascular Gli1+ progenitors are key contributors to injury-induced organ fibrosis. Cell Stem Cell 16:51–66. https://doi.org/10.1016/j.stem.2014.11.004
7. Lama VN, Smith L, Badri L, Flint A, Andrei AC, Murray S, Wang Z, Liao H, Toews GB, Krebsbach PH, Peters-Golden M, Pinsky DJ, Martinez FJ, Thannickal VJ (2007) Evidence for tissue-resident mesenchymal stem cells in human adult lung from studies of transplanted allografts. J Clin Invest 117:989–996
8. Schmitt-Gräff A, Desmoulière A, Gabbiani G (1994) Heterogeneity of myofibroblast phenotypic features: an example of fibroblastic cell plasticity. Virchows Arch 425:3–24
9. Soundararajan M, Kannan S (2018) Fibroblasts and mesenchymal stem cells: two sides of the same coin? J Cell Physiol 233:9099–9109. https://doi.org/10.1002/jcp.26860
10. Pittenger MF, Mackay AM, Beck SC, Jaiswal RK, Douglas R, Mosca JD, Moorman MA, Simonetti DW, Craig S, Marshak DR (1999) Multilineage potential of adult human mesenchymal stem cells. Science 284:143–147
11. El Agha E, Moiseenko A, Kheirollahi V, De Langhe S, Crnkovic S, Kwapiszewska G, Szibor M, Kosanovic D, Schwind F, Schermuly RT, Henneke I, MacKenzie B, Quantius J, Herold S, Ntokou A, Ahlbrecht K, Braun T, Morty RE, Gunther A, Seeger W, Bellusci S (2017) Two-way conversion between lipogenic and myogenic fibroblastic phenotypes marks the progression and resolution of lung fibrosis. Cell Stem Cell 20:571. https://doi.org/10.1016/j.stem.2017.03.011
12. Plikus MV, Guerrero-Juarez CF, Ito M, Li YR, Dedhia PH, Zheng Y, Shao M, Gay DL, Ramos R, Hsi TC, Oh JW, Wang X, Ramirez A, Konopelski SE, Elzein A, Wang A, Supapannachart RJ, Lee HL, Lim CH, Nace A, Guo A, Treffeisen E, Andl T, Ramirez RN, Murad R, Offermanns S, Metzger D, Chambon P, Widgerow AD, Tuan TL, Mortazavi A, Gupta RK, Hamilton BA, Millar SE, Seale P, Pear WS, Lazar MA, Cotsarelis G (2017) Regeneration of fat cells from myofibroblasts during wound healing. Science 355:748–752. https://doi.org/10.1126/science.aai8792
13. Hecker L, Jagirdar R, Jin T, Thannickal VJ (2011) Reversible differentiation of myofibroblasts by MyoD. Exp Cell Res 317:1914–1921. https://doi.org/10.1016/j.yexcr.2011.03.016
14. Thannickal VJ, Horowitz JC (2006) Evolving concepts of apoptosis in idiopathic pulmonary fibrosis. Proc Am Thorac Soc 3:350–356
15. Chanda D, Otoupalova E, Smith SR, Volckaert T, De Langhe SP, Thannickal VJ (2019) Developmental pathways in the pathogenesis of lung fibrosis. Mol Asp Med 65:56–69. https://doi.org/10.1016/j.mam.2018.08.004
16. Desmouliere A, Redard M, Darby I, Gabbiani G (1995) Apoptosis mediates the decrease in cellularity during the transition between granulation tissue and scar. Am J Pathol 146:56–66
17. Hinz B, Lagares D (2020) Evasion of apoptosis by myofibroblasts: a hallmark of fibrotic

diseases. Nat Rev Rheumatol 16:11–31. https://doi.org/10.1038/s41584-019-0324-5

18. Hecker L, Logsdon NJ, Kurundkar D, Kurundkar A, Bernard K, Hock T, Meldrum E, Sanders YY, Thannickal VJ (2014) Reversal of persistent fibrosis in aging by targeting Nox4-Nrf2 redox imbalance. Sci Transl Med 6:231ra247. https://doi.org/10.1126/scitranslmed.3008182
19. Schafer MJ, White TA, Iijima K, Haak AJ, Ligresti G, Atkinson EJ, Oberg AL, Birch J, Salmonowicz H, Zhu Y, Mazula DL, Brooks RW, Fuhrmann-Stroissnigg H, Pirtskhalava T, Prakash YS, Tchkonia T, Robbins PD, Aubry MC, Passos JF, Kirkland JL, Tschumperlin DJ, Kita H, LeBrasseur NK (2017) Cellular senescence mediates fibrotic pulmonary disease. Nat Commun 8:14532. https://doi.org/10.1038/ncomms14532
20. Justice JN, Nambiar AM, Tchkonia T, Le Brasseur NK, Pascual R, Hashmi SK, Prata L, Masternak MM, Kritchevsky SB, Musi N, Kirkland JL (2019) Senolytics in idiopathic pulmonary fibrosis: results from a first-in-human, open-label, pilot study. EBioMedicine 40:554–563. https://doi.org/10.1016/j.ebiom.2018.12.052
21. Bernard K, Logsdon NJ, Miguel V, Benavides GA, Zhang J, Carter AB, Darley-Usmar VM, Thannickal VJ (2017) NADPH oxidase 4 (Nox4) suppresses mitochondrial biogenesis and bioenergetics in lung fibroblasts via a nuclear factor erythroid-derived 2-like 2 (Nrf2)-dependent pathway. J Biol Chem 292:3029–3038. https://doi.org/10.1074/jbc.M116.752261
22. Swamy SM, Rajasekaran NS, Thannickal VJ (2016) Nuclear factor-erythroid-2-related factor 2 in aging and lung fibrosis. Am J Pathol 186:1712–1723. https://doi.org/10.1016/j.ajpath.2016.02.022
23. Sanders YY, Liu H, Liu G, Thannickal VJ (2015) Epigenetic mechanisms regulate NADPH oxidase-4 expression in cellular senescence. Free Radic Biol Med 79:197–205. https://doi.org/10.1016/j.freeradbiomed.2014.12.008
24. Otoupalova E, Smith S, Cheng G, Thannickal VJ (2020) Oxidative stress in pulmonary fibrosis. Compr Physiol 10:509–547. https://doi.org/10.1002/cphy.c190017
25. Bernard K, Thannickal VJ (2020) NADPH oxidase inhibition in fibrotic pathologies. Antioxid Redox Signal 2020:455–479. https://doi.org/10.1089/ars.2020.8032
26. Horowitz JC, Lee DY, Waghray M, Keshamouni VG, Thomas PE, Zhang H, Cui Z, Thannickal VJ (2004) Activation of the pro-survival phosphatidylinositol 3-kinase/AKT pathway by transforming growth factor-beta1 in mesenchymal cells is mediated by p38 MAPK-dependent induction of an autocrine growth factor. J Biol Chem 279:1359–1367
27. Horowitz JC, Rogers DS, Sharma V, Vittal R, White ES, Cui Z, Thannickal VJ (2007) Combinatorial activation of FAK and AKT by transforming growth factor-beta1 confers an anoikis-resistant phenotype to myofibroblasts. Cell Signal 19:761–771
28. Zhou Y, Huang X, Hecker L, Kurundkar D, Kurundkar A, Liu H, Jin TH, Desai L, Bernard K, Thannickal VJ (2013) Inhibition of mechanosensitive signaling in myofibroblasts ameliorates experimental pulmonary fibrosis. J Clin Invest 123:1096–1108. https://doi.org/10.1172/JCI66700
29. Parker MW, Rossi D, Peterson M, Smith K, Sikstrom K, White ES, Connett JE, Henke CA, Larsson O, Bitterman PB (2014) Fibrotic extracellular matrix activates a profibrotic positive feedback loop. J Clin Invest 124:1622–1635. https://doi.org/10.1172/JCI71386
30. Zhou Y, Horowitz JC, Naba A, Ambalavanan N, Atabai K, Balestrini J, Bitterman PB, Corley RA, Ding BS, Engler AJ, Hansen KC, Hagood JS, Kheradmand F, Lin QS, Neptune E, Niklason L, Ortiz LA, Parks WC, Tschumperlin DJ, White ES, Chapman HA, Thannickal VJ (2018) Extracellular matrix in lung development, homeostasis and disease. Matrix Biol 73:77–104. https://doi.org/10.1016/j.matbio.2018.03.005
31. Campana L, Iredale JP (2017) Regression of liver fibrosis. Semin Liver Dis 37:1–10. https://doi.org/10.1055/s-0036-1597816
32. Kisseleva T, Cong M, Paik Y, Scholten D, Jiang C, Benner C, Iwaisako K, Moore-Morris-T, Scott B, Tsukamoto H, Evans SM, Dillmann W, Glass CK, Brenner DA (2012) Myofibroblasts revert to an inactive phenotype during regression of liver fibrosis. Proc Natl Acad Sci U S A 109:9448–9453. https://doi.org/10.1073/pnas.1201840109
33. Kato K, Logsdon NJ, Shin YJ, Palumbo S, Knox A, Irish JD, Rounseville SP, Rummel SR, Mohamed M, Ahmad K, Trinh JM, Kurundkar D, Knox KS, Thannickal VJ, Hecker L (2020) Impaired myofibroblast dedifferentiation contributes to non-resolving fibrosis in aging. Am J Respir Cell Mol Biol 62(5):633–644. https://doi.org/10.1165/rcmb.2019-0092OC
34. Rangarajan S, Bone NB, Zmijewska AA, Jiang S, Park DW, Bernard K, Locy ML,

Ravi S, Deshane J, Mannon RB, Abraham E, Darley-Usmar V, Thannickal VJ, Zmijewski JW (2018) Metformin reverses established lung fibrosis in a bleomycin model. Nat Med 24:1121–1127. https://doi.org/10.1038/s41591-018-0087-6

35. Lagares D, Santos A, Grasberger PE, Liu F, Probst CK, Rahimi RA, Sakai N, Kuehl T, Ryan J, Bhola P, Montero J, Kapoor M, Baron M, Varelas X, Tschumperlin DJ, Letai A, Tager AM (2017) Targeted apoptosis of myofibroblasts with the BH3 mimetic ABT-263 reverses established fibrosis. Sci Transl Med 9:eaal3765. https://doi.org/10.1126/scitranslmed.aal3765

36. Kheirollahi V, Wasnick RM, Biasin V, Vazquez-Armendariz AI, Chu X, Moiseenko A, Weiss A, Wilhelm J, Zhang JS, Kwapiszewska G, Herold S, Schermuly RT, Mari B, Li X, Seeger W, Gunther A, Bellusci S, El Agha E (2019) Metformin induces lipogenic differentiation in myofibroblasts to reverse lung fibrosis. Nat Commun 10:2987. https://doi.org/10.1038/s41467-019-10839-0

Chapter 3

Myofibroblast Markers and Microscopy Detection Methods in Cell Culture and Histology

Fereshteh S. Younesi, Dong Ok Son, Joao Firmino, and Boris Hinz

Abstract

The identification of myofibroblasts is essential for mechanistic in vitro studies, cell-based drug tests, and to assess the level of fibrosis in experimental animal or human fibrosis. The name myo-fibroblast was chosen in 1971 to express that the formation of contractile features-stress fibers is the essential criterion to define these cells. Additional neo-expression of α-smooth muscle actin (α-SMA) in stress fibers has become the most widely used molecular marker. Here, we briefly introduce the concept of different myofibroblast activation states, of which the highly contractile α-SMA-positive phenotype represents a most advanced functional stage. We provide targeted immunofluorescence protocols to assess this phenotype, and publicly accessible image analysis tools to quantify the level of myofibroblast activation in culture and in tissues.

Key words α-Smooth muscle actin, Stress fibers, ED-A fibronectin, Image analysis, Myofibroblast phenotype, Molecular markers, Immunofluorescence, Microscopy

1 Introduction

"Fibroblasts" comprise heterogeneous populations of connective tissue cells that reside in all organs as true multitaskers. In homeostasis, the major role of fibroblasts is to maintain the integrity of the tissue architecture and cell niches by secreting collagen-rich extracellular matrix (ECM) [1, 2]. In conditions of injury and inflammation, fibroblastic cells upregulate their ECM secretion and acquire novel remodeling activities to repair tissue defects and re-establish homeostasis. This change in function and phenotype is called myofibroblast activation [3]. When activation persists beyond the normal phase of tissue repair, a fibrotic state is reached, which has often severe consequences for organ function [4, 5].

Myofibroblasts were discovered five decades ago in rat wound granulation tissues and their name was chosen based on morphological features shared between ECM-producing fibroblasts and

Fereshteh Younesi and Dong Ok Son authors contributed equally to this work.

Boris Hinz and David Lagares (eds.), *Myofibroblasts: Methods and Protocols*, Methods in Molecular Biology, vol. 2299, https://doi.org/10.1007/978-1-0716-1382-5_3, © Springer Science+Business Media, LLC, part of Springer Nature 2021

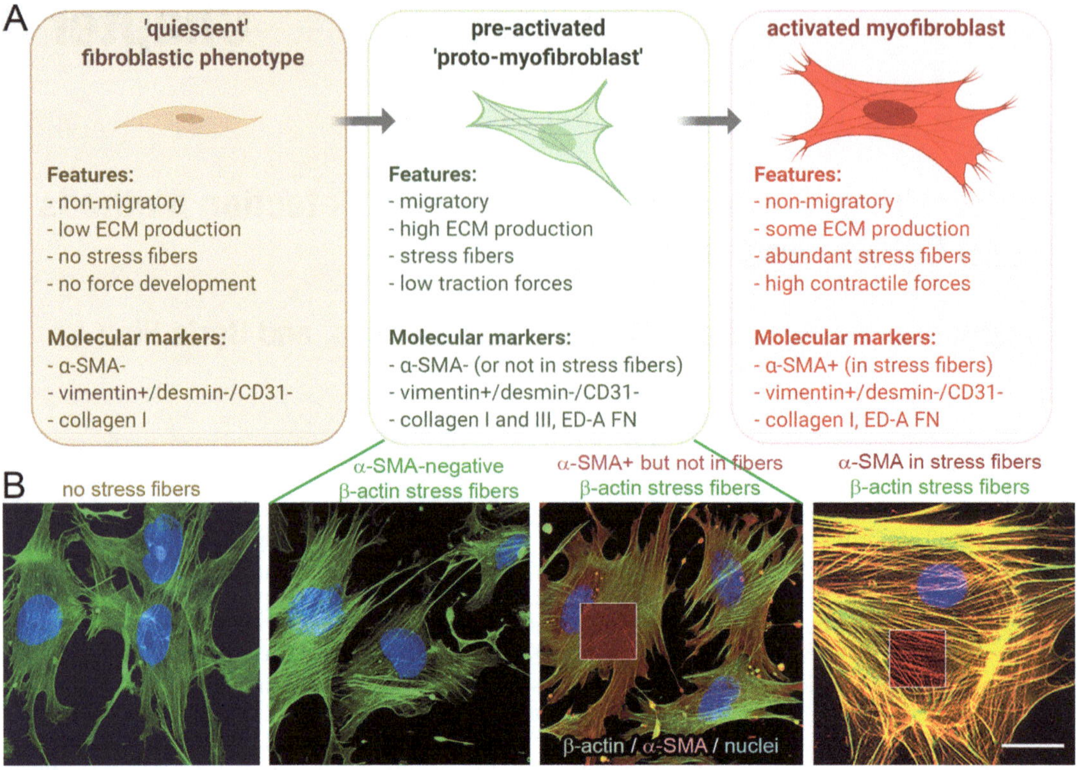

Fig. 1 Myofibroblast activation stages. (**a**) Schematic summary of myofibroblast activation stages and features. Scheme was produced with Biorender (biorender.com). (**b**) Immunofluorescence assessment of myofibroblast activation stages. Mouse embryonic fibroblasts were cultured on fibronectin-coated tissue culture plastic, except the first picture where fibronectin-coated soft polyacrylamide gels (elastic modulus of 1.0 kPa) were used as culture substrate. Cells were fixed and stained using antibodies against β-cytoplasmic actin (green) and α-SMA (red), and counterstained for nuclear DNA (blue). Insets blend out the green channel to highlight the organization level of α-SMA in the indicated region. Scale bar: 15μm. Abbreviations: *α-SMA* α-smooth muscle actin, *CD* cluster of differentiation, *ED-A FN* extradomain A fibronectin

contractile smooth muscle cells [6]. Myofibroblast activation from traditionally called "quiescent" fibroblasts and other mesenchymal cell sources is nowadays understood as a continuous process with intermediate phenotypes [7] (Fig. 1a). As part of the tissue repair program, activated fibroblasts initially migrate to sites of injury where they proliferate and start producing ECM rich in collagen types I and III [8]. This migratory and consequently already force-producing phenotype has been named "proto-myofibroblast"—based on the development of contractile bundles of filamentous actin and myosin, called stress fibers [7]. The stress fiber actin content of proto-myofibroblasts is largely composed of β-cytoplasmic actin filaments (Fig. 1b). The next myofibroblast activation stage is characterized by de novo expression of α-smooth muscle actin (α-SMA). Gradual incorporation of α-SMA into stress fibers grants them higher contractile activity than comparable levels of other actin isoforms [9]. The molecular and structural basis for higher contraction of α-SMA-positive versus α-SMA-negative stress fibers is not entirely understood. However, interfering with the α-SMA-specific N-terminus (which is the epitope for all α-SMA-specific antibodies) selectively removes α-SMA from stress fibers and reduces (but not completely abolishes) myofibroblast contraction [10–12]. From a metabolic and energetic perspective, it is unlikely that a highly contractile α-SMA-positive myofibroblast continues to produce and secrete high amounts of ECM. It is also reasonable to assume that the contractile phenotype pursues the ECM-producing stage since only ECM that is present can be contracted. In a close mechanical feedback loop, the degree of myofibroblast activation depends on the remodeling state of the ECM, which is "sensed" as intracellular stress caused by pulling on an increasingly "stiff" ECM [3, 13]. Consistently, recent single cell RNA sequencing data suggest that fibroblasts sorted for collagen αI promoter activity from fibrotic mouse lungs exhibit low α-SMA expression levels [14, 15] and that myofibroblasts exist in a spectrum of activation states [16].

If myofibroblast activation is a gradual process, how can we detect them in cultured cells and/or tissues and for which purposes? Identification of myofibroblasts is important from both a clinical and a research perspective. Localization and quantification of myofibroblast abundance is a widely used diagnostic tool to detect and rate the progress of fibrosis in various organs, including the stromal reaction to epithelial tumors [17–19]. In experimental animal models of fibrosis, myofibroblast accumulation is a direct measure for fibrosis severity, and identification of myofibroblasts helps to trace their cell origins [16, 20]. In vitro systems to study the mechanisms of myofibroblast activation or to screen antifibrotic compounds in industrial settings all depend on reliable and quantitative myofibroblast phenotyping. Here, we present detailed protocols to identify myofibroblasts in cultured cells and histological tissue sections using antibody staining protocols for immunofluorescence (IF) light microscopy imaging. More elaborate protocols and comparisons between different staining procedures have been published before [21]. We further provide image analysis routines to quantify myofibroblast accumulation using the open-source software Fiji [22].

Since there is no known unique myofibroblast marker, due to the heterogeneity of myofibroblast precursors and of myofibroblast activation states, which protein target should we aim for? One must revert to the original definition of the phenotype, where "myo" refers to the newly acquired contractile function during the activation process. The (ultra-)structural hallmark of fibroblast contraction is the formation of F-actin stress fibers as discussed above [6]. Stress fibers are containing high amounts of filamentous (F)-actin, which is discriminated from globular G-actin by using the F-

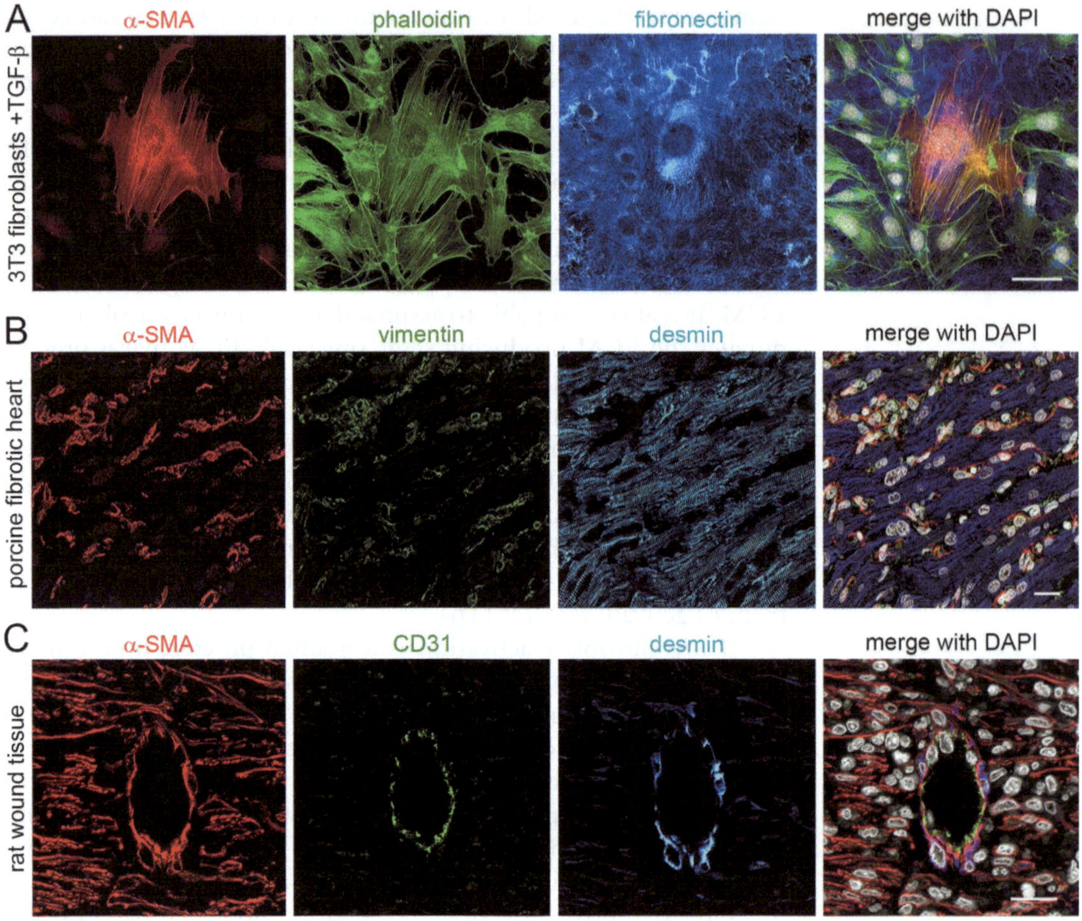

Fig. 2 Identification of myofibroblasts using α-SMA and exclusion markers. (**a**) 3T3 fibroblasts cultured on tissue culture plastic and treated for 4 days with TGF-β1 were stained for α-SMA (red), filamentous actin (phalloidin-green), and extracellular matrix protein fibronectin. (**b**) Fibrotic porcine right ventricular heart tissue was obtained after pulmonary vein banding, sectioned, and stained for the myofibroblast marker α-SMA (red), mesenchymal marker vimentin (green), and muscle (here: cardiac and smooth muscle) marker desmin (blue). (**c**) Granulation tissue from 9-day-old full-thickness rat wounds was sectioned and stained for α-SMA (red), endothelial cell marker CD31 (green), and muscle (here: smooth muscle) marker desmin (blue). All scale bars: 50μm. Abbreviations: *α-SMA* α-smooth muscle actin, *TGF-β1* transforming growth factor-β1 *CD* cluster of differentiation

actin-specific probe phalloidin, conjugated with a fluorochrome (Fig. 2a). In strict terms, phalloidin-positive fibroblastic cells are myofibroblasts and the intensity of phalloidin staining provides one measure for myofibroblast activation in vitro and in sections of frozen tissue [23]. However, all fibroblastic cells in standard cell culture will form F-actin stress fibers upon contact with stiff tissue culture surfaces which typically makes phalloidin staining alone a blunt tool. Thus, in these culture conditions and in tissues, expression of α-SMA in stress fibers has become the most commonly used molecular marker for myofibroblasts, and excellent monoclonal antibodies have been developed to discriminate between the specific N-termini of the different actin isoforms [24–26].

Two important considerations must be made when using α-SMA as molecular myofibroblast marker: (1) Only α-SMA localized to stress fiber qualifies as a myofibroblast marker. α-SMA can be expressed in noncontractile, stress-fiber negative cell states and remain as soluble α-SMA monomers (G-actin) without being incorporated into stress fibers (Fig. 1b). For instance, perivascular pericytes and bone-marrow-resident mesenchymal stromal cells have long been described to express α-SMA but are noncontractile in their normal (i.e., nonactivated) state [27–30]. During myofibroblast activation, α-SMA is expressed first and then gradually recruited to stress fibers, which may depend on the expression of additional stress fiber components, such as specific tropomyosin isoforms [31]. Therefore, we recommend to always co-stain with phalloidin or use a morphological assessment of stress fiber α-SMA (*see* Subheadings 3.2 and 3.5).

(2) Expression of α-SMA is not unique to myofibroblasts in tissues and in some primary tissue cell cultures at low passages. For instance, vascular smooth muscle cells express similar or higher levels of α-SMA compared to myofibroblasts, making discrimination based on α-SMA alone difficult. Hence, exclusion markers are required to distinguish between myofibroblasts, smooth muscle cells, and pericytes—all of which are α-SMA-positive [32]. Smooth muscle markers that are not expressed by myofibroblasts include desmin, smooth muscle myosin heavy chain, h-caldesmon, and smoothelin [32]. In culture, smooth muscle cells rapidly lose these muscle markers and become morphologically indistinguishable from myofibroblasts [33]. Vimentin usually fails as a discriminator since endothelial cells, smooth muscle, and some inflammatory cell populations all express this "fibroblast-specific" intermediate filament protein [32]. To identify myofibroblasts in fibrotic or neoplastic tissues, we recommend co-staining for desmin (an ubiquitous marker for muscle cells but not myofibroblasts), CD31 or VE-cadherin (specifically labels endothelial cells which associate with vessels), and α-SMA (myofibroblasts and smooth muscle) (*see* Subheading 3.4) [32, 34].

Of course, there are numerous other "myofibroblast markers" with different levels of usefulness depending on whether assessing a cell culture condition or tissue histology, including proteins of the myofibroblast ECM. In addition to collagen type I, the most frequently used proteins associating with the myofibroblast ECM include the extradomain-A (ED-A) splice variant of fibronectin, tenascin-C, CCN2 (formerly called connective tissue growth factor, CTGF), and periostin [8, 35–38]. Because they are typically stable for days, weeks, and even months, ECM proteins are an excellent indicator of past myofibroblast maturation stages. However, the minimal message you should take home from this introduction is that cells without contractile bundles may have been or will in future become myofibroblasts, but they are not myofibroblasts at the time of assessment.

2 Materials

2.1 Cell Culture and Preparation

1. Adherent fibroblastic cells of your choice (*see* **Note 1**).
2. Multi-well chamber slides, 22 × 22 mm coverslips or 25 mm diameter round coverslips placed in 35 mm diameter tissue culture plastic ware (*see* **Note 2**).
3. Dulbecco's Modified Eagle Medium (DMEM) or other recommended fibroblast culture medium, supplemented with penicillin/streptomycin (100 U/mL for each), and 10% fetal bovine serum (FBS) (*see* **Note 3**).
4. Trypsin-ethylenediaminetetraacetic acid (EDTA) (0.25%).
5. Sterile phosphate-buffered saline (PBS): 150 mM NaCl, 8.06 mM Na_2HPO_4, 1.74 mM NaH_2PO_4, adjust pH with 1 M NaOH to pH 7.4.
6. Gelatin (e.g., 1 mg/mL stock solution), fibronectin (e.g., 1 mg/mL stock solution), or collagen type I (e.g., 2.3 mg/mL stock solution) for surface coating (*see* **Note 2**).
7. Human transforming growth factor-β1 (TGF-β1) (*see* **Note 4**).

2.2 IF Staining of Cultured Cells

1. PBS buffer, pH 7.4.
2. Cell fixation buffer: 3% paraformaldehyde (PFA) in PBS (w/v) (*see* **Note 5**).
3. Cell permeabilization buffer: 0.2% Triton X-100 in PBS (*see* **Note 6**).
4. Blocking buffer: 3% bovine serum albumin (BSA) in PBS. BSA can be substituted with skim milk powder for most antibodies except those directed against phosphoproteins.
5. Antibody dilution buffer: 0.02% Triton X-100 with 0.2% BSA in PBS.
6. Washing buffer: 0.02% Triton X-100 in PBS.
7. Primary antibodies, e.g., anti-α-SMA (clone 1A4, mouse IgG2a), anti-fibronectin antibody (rabbit polyclonal), anti-ED-A fibronectin (clone IST-9, mouse IgG1, not reactive to mouse tissues or cells) (*see* **Note 7**).
8. Fluorochrome-conjugated secondary antibodies, e.g., Alexa Fluor 568 goat anti-mouse IgG2a to detect α-SMA, Alexa Fluor 488 goat anti-mouse IgG1 to detect ED-A fibronectin, or Alexa Fluor 488 goat anti-rabbit to detect total fibronectin.

9. Alexa Fluor 647-conjugated Phalloidin to detect F-actin and 4′,6-diamidino-2-phenylindole dihydrochloride (DAPI) or Hoechst to stain DNA in cell nuclei (*see* **Note 8**).
10. Mounting media suitable to preserve fluorochromes (e.g., Fluoromount-G).
11. 76 × 26 mm microscope slides (to mount cells grown on 22 × 22 or 24 mm diameter coverslips) or 22 × 50 mm glass coverslips (to mount cells grown on multi-chamber slides).

2.3 Tissue Preparation

2.3.1 Tissue Fixation

1. PBS buffer, pH 7.4.
2. 3% PFA in PBS.
3. 70% Ethanol.
4. 50 mL Reaction tubes.

2.3.2 Tissue Embedding

1. Paraffin embedding molds.
2. 100% Isopentane (2-methylbutane) in a glass beaker.
3. Liquid nitrogen.
4. Microtome.
5. 76 × 26 mm Microscope slides.

2.4 IF Staining of Fixed Tissues

1. Xylene.
2. 100% Ethanol.
3. Distilled water.
4. Citrate buffer: 10 mM sodium citrate in distilled water; adjust with HCl to pH 6.0.
5. Tris-buffered saline (TBS) buffer: 20 mM Tris, 150 mM NaCl; adjust with HCl to pH 7.4.
6. BSA.
7. Triton X-100.
8. Primary antibodies, e.g., anti-α-SMA (clone 1A4, mouse IgG2a), anti-CD31 antibody (rabbit polyclonal) or anti-vimentin antibody (rabbit polyclonal), and anti-desmin antibody (clone D33, mouse IgG1).
9. Fluorochrome-conjugated secondary antibodies, e.g., Alexa Fluor 488 goat anti-mouse IgG2a to detect the anti-α-SMA primary antibody, Alexa Fluor 568 goat anti-rabbit IgG to detect anti-CD31, and Alexa Fluor 647 goat anti-mouse IgG1 to detect anti-desmin.
10. Staining tray with lids, lined with wet tissue papers to maintain humidity.
11. 22 × 50 mm Glass coverslips.
12. Mounting media suitable to preserve fluorochromes (e.g., Fluoromount-G).

2.5 Image Analysis

1. Laptop or desktop computer, ideally with ≥8 GB RAM.
2. Fiji open-source software (https://imagej.net/Fiji/Downloads).
3. Macros for α-SMA expression analysis (available for download from https://github.com/CAMiLoD-UofT/alphaSMA-measurements).

3 Methods

3.1 Cell Culture

TGF-β1 is the most potent inducer of myofibroblast activation in vivo and in vitro and widely applied to induce α-SMA expression in various cultured cells in a dose- and time-dependent manner [39, 40] (Fig. 2a). Depending on FBS batches, TGF-β1 concentrations may need to be adapted to achieve the desired effect (*see* **Note 3**) but in our hands, incubation with 2 ng/mL TGF-β1 for 72–96 h usually suffices to enhance α-SMA protein expression in human, rat, and mouse fibroblasts. To reduce interference of serum components with myofibroblast activation (*see* **Note 3**), FBS content can be reduced to 2% during the time of TGF-β1 treatment if cells tolerate low serum levels. Some fibroblast types can spontaneously (i.e., without addition of TGF-β1) activate into myofibroblasts in low serum conditions as part of cellular stress responses [41].

1. Grow fibroblasts in standard culture medium containing 10% FBS for weekly passaging. Primary fibroblasts should be used no later than passage 5. Primary mouse fibroblasts can enter growth crisis and senescence early during passaging (passage 2–3), which can be delayed by increasing FBS content to 20% or growth under hypoxic (2–5% O_2) growth conditions (*see* **Note 9**). Always carefully monitor fibroblast morphologies and growth rates over passages—deviation from the norm always indicates shifts in fibroblast programs.
2. Either use uncoated tissue culture plastic surfaces or coat glass surfaces with adhesive proteins (e.g., fibronectin, gelatin, or collagen type I) diluted to 2μg/cm^2 in cell culture medium without FBS or PBS for at least 1 h; overnight coating is possible (*see* **Note 2**). Before seeding cells, aspirate coating solution gently—do not wash.
3. Trypsinize cells, count, and resuspend in medium 10% FBS. Plate cells to achieve ideally 30–50% confluence at the time of fixation and individually growing cells for better automatic quantification (*see* Subheading 3.5).
4. After cell attachment (at least 6 h, best overnight), replace media with fresh growth medium supplemented with FBS at the chosen percentage (*see* **Note 3** and above) with 2 ng/mL

TGF-β1 to induce myofibroblast activation and without TGF-β 1 for untreated controls (*see* **Note 10**).

5. Grow cells for at least another 72 h or up to one week in 21% O_2, 5% CO_2, and 37 °C incubators. If the cell culture time is no longer than 5 days, intermittent medium change with freshly added TGF-β1 may not be required; the TGF-β1 effect will last for up to 5 days [42]. α-SMA protein expression peaks around day 4 following TGF-β1 treatment [38, 39].

3.2 IF Staining of Fixed Cells

Fixation and permeabilization are essential steps for IF staining. However, the fixation and permeabilization buffer, and the order of the fixation and permeabilization steps can vary based on the target proteins and their location in order to optimize the resolution and clarity of distinct structures [43]. For IF staining of the standard myofibroblasts markers α-SMA, F-actin, and fibronectin in fixed cells (Fig. 2a), use a simple protocol:

1. Aspirate medium and wash the cells gently with pre-warmed PBS twice at room temperature (RT) (*see* **Note 11**). Do not let samples dry at any of the following steps until samples are mounted.
2. Fix cells in 3% PFA for 10 min at RT.
3. Carefully rinse with PBS for 10 min three times (*see* **Note 12**).
4. Permeabilize cells in permeabilization buffer for 5 min at RT.
5. Rinse cells with PBS for 10 min three times.
6. Incubate cells with blocking buffer for 1 h at RT to reduce nonspecific antibody binding. This step is optional and depends on the specificity of antibodies. Incubation with blocking buffer and adding BSA to the antibody dilution buffer (final concentration 0.2%) can reduce nonspecific antibody bindings. The antibody combination suggested in this protocol usually does not require BSA at any step.
7. Rinse twice with PBS for 5 min.
8. If your cells were grown in a multi-well chamber slide, take off the removable well cast now, which will allow you to work with smaller volumes of antibody (~25μL/well instead of ~100μL/well when keeping the cast).
9. Mix all primary antibodies in one micro reaction tube in dilution buffer on ice: anti-α-SMA (5μg/mL), anti-fibronectin (2.5μg/mL), or anti-ED-A fibronectin (10μg/mL) (*see* **Note 13**).
10. Incubate cells with diluted primary antibody solution (25μL per well of an 8-well chamber slide or 50μL per 22 × 22 or 25 mm diameter coverslip) for 60 min at RT (*see* **Note 14**).

Omit this step in one control sample to test nonspecific secondary antibody binding.

11. Wash samples with washing buffer (PBS buffer containing 0.02% Triton X-100) for 10 min three times at RT.
12. Prepare all fluorochrome-conjugated secondary antibodies (10μg/mL) along with fluorochrome-conjugated phalloidin (3 U/mL) and DAPI (5–10μg/mL) in dilution buffer in one vial and mix well (*see* **Note 15**). Continue all the following steps protected from direct strong light to reduce fluorochrome bleaching.
13. Incubate samples with antibodies in dilution buffer prepared at **step 12** for 60 min at RT with volumes defined in **step 8**.
14. Gently wash samples with washing buffer for 10 min three times at RT.
15. Carefully aspirate excess liquid off, without letting the sample dry.
16. Pipette one drop of mounting media (e.g., 20μL per previous of an 8-well chamber slide; 50μL for 22 × 22 or 25 mm diameter coverslips) into the center of the sample.
17. If cells were grown on the slide, e.g., on multi-well chamber slides: Place a clean 50 × 26 mm coverslip at the slide edge at 60° angle and slowly lower the coverslip onto the mounting medium using tweezers.
18. If cells were grown on coverslips: Using the same procedure, lower the coverslip with cells facing down onto a clean 76 × 26 mm microscopy slide (*see* **Note 16**).
19. Polymerize the mounting media at RT overnight protected from light. Store samples afterward for several months or even years at 4 °C, protected from light.

3.3 Tissue Preparation

3.3.1 Fixing and Paraffin-Embedding

1. Fix freshly dissected tissue samples with 3% PFA (or 10% formalin) (*see* **Notes 5 and 17**) for 24–48 h at RT to preserve tissue morphology and retain the antigenicity of the target molecules (*see* **Note 18**).
2. Wash tissues with PBS for 10 min three times and keep the samples in 70% ethanol at RT until moving to the next step.
3. Trim fixed tissues into the appropriate size and shape and place in embedding cassettes.
4. Dehydrate tissues gradually through graded ethanol: incubate with 70% ethanol twice for 1 h, then 95% ethanol twice for 1 h, followed by 100% ethanol twice for 1 h at RT.
5. Immerse tissues in xylene overnight at RT, followed by additional incubation with fresh xylene for 1 h at RT.

6. Infiltrate tissues with wax at 58 °C for 1 h three times with the last infiltration step performed under vacuum. If residues of xylene are remaining, add another wax infiltration step for 1 h under vacuum.
7. Open the cassette to view the tissue sample and choose a paraffin mold that best corresponds to the size of the tissue. A margin of at least 2 mm of paraffin surrounding all sides of the tissue gives best support during sample cutting.
8. Place a small amount of molten paraffin in the mold. Using warm forceps, transfer the tissue into a mold and orient as desired.
9. Transfer the mold to a cold plate, and gently press tissue flat. Add the labeled tissue cassette on top of the mold.
10. Dispense enough hot paraffin to the mold until the face of the plastic cassette is covered.
11. Cool the paraffin for 30 min and remove the paraffin block from the mold (*see* **Note 19**). It will be stable for several years, even decades once the tissue is embedded.

3.3.2 Sectioning Tissues

1. Turn on the water bath containing distilled water and set to 42 °C.
2. Trim paraffin blocks as necessary and place the paraffin blocks to be sectioned on ice.
3. Place a fresh blade on the microtome. Replace blade if cutting is not smooth.
4. Section the paraffin-embedded tissues in 2–10μm thickness (5μm is commonly used) using the microtome.
5. Pick up ribbons of consecutive sections with forceps and float on water in the water bath at 42 °C (*see* **Note 20**).
6. Mount the sections onto positively charged microscope slides (*see* **Note 21**).
7. Allow sections to air dry for 30 min at RT. Then bake slides at 37 °C overnight or 30 min at 60 °C to bond the tissue to the glass. Slides can be stored at RT.

3.4 Immunostaining of Fixed Tissues

1. Deparaffinize tissue sections in xylene for 5 min twice at RT, followed by rehydrating tissue sections with graded ethanol diluted in distilled water to 100% (twice), 95%, and 70% for 5 min each at RT (*see* **Note 22**).
2. Keep samples in distilled water at RT until citrate buffer is ready.
3. Perform antigen retrieval to unmask the antigenic sites using 95–100 °C citrate buffer at pH 6.0 for 15 min.

4. Cool the slides at RT in the same citrate buffer down to 40 °C by leaving the samples for 20–40 min at RT, depending on the volume of buffer. Then rinse in TBS buffer containing 0.025% Triton X-100 (TBS-T) twice for 5 min.
5. Draw a hydrophobic barrier using a barrier pen around the tissue.
6. Apply blocking buffer (1% BSA+10% normal goat serum in TBS) on tissue sections and incubate in a humidified chamber for 1 h at RT.
7. Mix all primary antibodies in one micro reaction tube in 1% BSA-TBS on ice: anti α-SMA (5μg/mL), anti-desmin (50μg/mL) and anti-CD31 (50μg/mL) or vimentin (10μg/mL) (*see* **Note 13**) (Fig. 2b, c).
8. Remove blocking buffer and incubate samples with diluted primary antibodies (prepared at **step 7**) overnight at 4 °C.
9. Wash slides for 10 min three times with TBS-T.
10. Mix all fluorochrome-conjugated secondary antibodies (2.5–5μg/mL) and DAPI (1μg/mL) in 1% BSA-TBS in one vial (*see* **Note 15**). Continue all following steps protected from direct strong light to reduce fluorochrome bleaching.
11. Incubate slides with the mixture of secondary antibodies and DAPI (prepared in **step 10**) for 1 h at RT protected from light.
12. Wash slides for 10 min three times with TBS-T at RT.
13. Mount stained tissues with an anti-fade mounting media. Polymerize the mounting media overnight at RT and store samples at 4 °C in dark.

3.5 Image Analysis

In this section, we are providing macro routines using the open-source image analysis platform Fiji [22] that consider both staining levels (average pixel intensity) and structure/alignment (coherency) of the fibrous cell morphology. To identify individual cells, cell boundary markers such as plasma membrane markers or phalloidin should be used. The advantage of assessing cells in the image field individually for their myofibroblast character is to obtain information on myofibroblast heterogeneity and different levels of activation. A widely used quantitative parameter for the assessment of myofibroblast activation in culture is the "percentage of myofibroblasts." To obtain this value, one must make a binary decision for every cell in the image field: is this a myofibroblast—yes or no? For some, including the authors, this decision can become an almost philosophical question that is much more difficult to answer than it may seem. As elaborated in the introduction, we consider formation of stress fibers as the single most important criterion to identify activated myofibroblasts. In culture, additional expression of α-SMA in stress fibers has been proven to be the more meaningful myofibroblast marker because all fibroblasts in standard culture form stress fibers (Fig. 1). For all practical aspects, we thus define in vitro myofibroblasts as fibroblastic cells expressing α-SMA in stress fibers. For image processing data post-analysis, a "cut-off" decision still needs to be made based on the minimum amount of structure that defines α-SMA-positive stress fibers in a cell. In the following section and the notes, we provide some guidance how to determine this threshold. However, eventually the user is left with the power and the dilemma to make the decision and properly report in publications how it was made.

In highly confluent cell monolayers, it will be difficult if not impossible to identify single cell boundaries, even by eye. For these conditions, we are providing a routine that disregards single cell features and segments images based on cell nuclei positions. Alternatively, the user can choose to analyze whole images for α-SMA intensity and structure rather than identifying individual cells. The outcome is an average myofibroblast behavior per image and information on myofibroblast heterogeneity is lost. As a rule, highly confluent cultures are not recommended for automated image analysis.

To quantify myofibroblasts in tissues, stress fiber structure is often not resolved from low magnification images (<40× objectives) or cannot be attributed to single cells due to high cell densities in fibrotic tissues. We provide two simple and robust procedures that rely on the users (1) to include regions of interest (ROI) within the tissue to then quantify α-SMA intensity per tissue area or (2) to identify ROIs in the image field that should be excluded from this analysis (e.g., vessels).

3.5.1 Quantification of α-SMA Expression in Cultured Cells Using Fiji

1. Take pictures using an epifluorescence or confocal microscope (*see* **Note 23**). Per fluorescence channel, use the same settings for exposure time (epifluorescence) or laser intensity (confocal) across all samples, e.g., control samples and TGF-β1-treated samples.
2. The described quantification procedures and automated routines (macros) are based on three-channel images with α-SMA staining in channel 1 (red); F-actin in channel 2 (green), and DAPI in channel 3 (blue). If you are using a different channel distribution, you will need to change the channel attribution in the macros (*see* **Note 24**).
3. Save images in the recommended commercial acquisition format of your microscope (e.g.,*.czi for images taken with Zeiss microscopes), which contains metadata such as image scaling and attributes one distinct channel per "color" (*see* **Note 25**). We do **NOT** recommend quantifying RGB TIF images, although the macros will be able to read and process this format

(*see* important **Note 26**). You cannot quantify image filetypes which may contain compressed data (e.g., JPG).

4. Place all images you want to process as one batch in one folder, which are typically all images from one experimental condition.
5. Install Fiji: Install the Fiji image processing package (https://fiji.sc/#download), and Java (https://www.java.com/en/download/). Make sure to work with the most recent versions of Fiji (V1.52p or higher) and Java (V1.8.0_172).
6. Update and install required plugins in Fiji: Start Fiji and from the main menu choose "*Help -› Update...". (Do NOT select "Update ImageJ...")* Confirm any user prompts that are asked during the process. Then click on the now available "*Manage update sites*" button. Activate the checkbox next to "BIG-EPFL", "ImageJ", "Fiji", "Java8", "IJPB-plugins", "PTBIOP", and "Bio-Formats". Then hit the "*Close*" button, followed by "*Apply changes.*" Restart Fiji.
7. Change or verify settings of the "*Bio-Formats*" *dialog box:* In the Fiji menu bar, select*: "Plugins → Bioformats → Bioformats Plugins Configuration*". Then select the "*Formats*" tab and highlight your file format (e.g., Zeiss CZI) in the list. Activate the "*Windowless*" checkbox and close the configuration window.
8. Download the following macros from https://github.com/CAMiLoD-UofT/alphaSMA-measurements and store in a folder of your choice: "CellAnalysis_ManualROI.ijm", "CellAnalysis_Individual.ijm", and "CellAnalysis_Voronoi.ijm".
9. To open a macro of your choice: *File -› Open*, browse the macro folder and select the macro file—or simply drag the macro file from the folder onto the Fiji menu.
10. Select the best macro for your condition (Fig. 3a): "CellAnalysis_ManualROI.ijm" will generate data with highest confidence but also requires time-consuming manual user input (*see* **Note 27**). The macro "CellAnalysis_Individual.ijm" will only be suitable for conditions where cells grow isolated and do not touch each other (*see* **Note 28**). If your biological question requires confluent cell cultures, "CellAnalysis_Voronoi.ijm" can be the macro of your choice to analyze image segments based on cell nuclei positions (*see* **Note 29**). All these macros deliver information per ROI, which is a cell or an image segment that includes a nucleus. Alternatively, you can use "CellAnalysis_Individual.ijm" for confluent cultures to obtain values that apply to the whole image (*see* **Note 30**). It is not advisable to directly compare data produced with different macros. It is good practice to test the macro on one or two images before processing a batch fully automated. Some

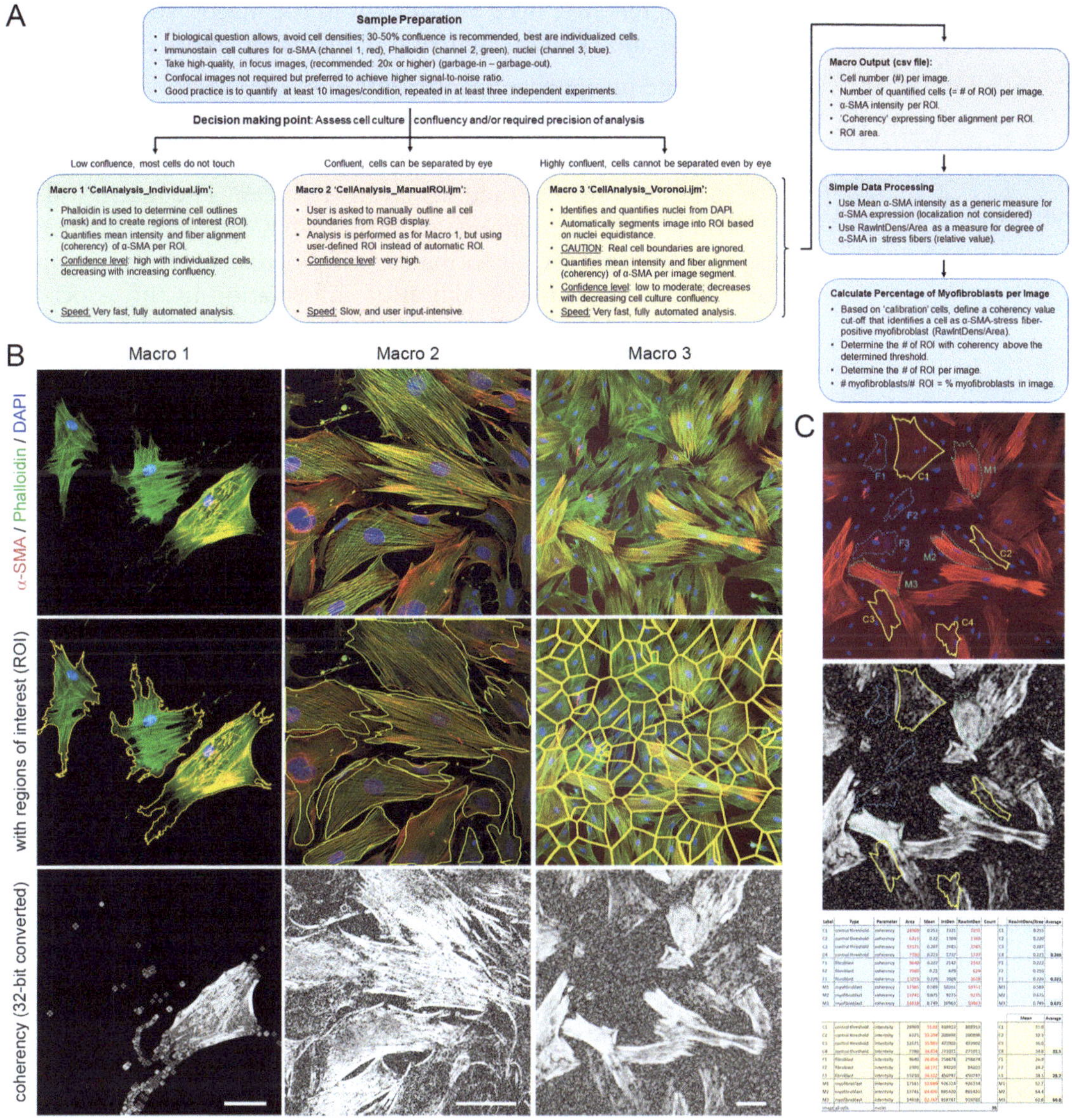

Fig. 3 Myofibroblast quantification in culture. (**a**) Workflow to quantify myofibroblast features from cultured cells, immunostained for α-SMA (red), filamentous actin (phalloidin-green), and DNA (DAPI, blue). (**b**) Shown are examples of fibroblasts in different culture densities and respective examples of the analysis process (top to bottom) using the three provided macros. Scale bar: 50μm. (**c**) The α-SMA channel (red) from the case presented for Macro 3 in (**b**) is shown as example how to determine the percentage of myofibroblasts from imaging data. Four cells from the image were selected to satisfy a minimum criterion for being "α-SMA stress fiber-positive" (C1–C4, yellow outlines). The respective coherency raw integrated density values were extracted from the coherency picture, divided by ROI area, and averages calculated in the data table (here: 0.246; *see* also Table 1). This value serves as the threshold to categorizing cells into "α-SMA stress fiber-positive" myofibroblast (M1–M3, green outlines, average: 0.671) or "α-SMA stress fiber-negative" fibroblast (F1–F3, blue outlines, average: 0.221) (Table 1 for better visibility)

parameters may need adjustment in the macro, such as thresholding and Gaussian blurring of nuclei (*see* **Note 31**).

11. Change image channel attribution in the macro if needed (*see* **Note 24**) and save the macro.

12. Click "*Run*" in the open macro window (bottom left button) and follow the instructions given on screen. The macro will ask you to calibrate your image (pixel-to-μm conversion factor) for every batch if the file format is *.tif. Microscopy software image formats will contain calibration information in the associated metadata. Measurements are dependent on the real dimensions of the sample (e.g., to identify nuclei and cells by their size).

13. For every processed image, all macros will produce: one zip file containing all ROI data, one PNG image with "burned-in" ROI overlay, and one PNG representation of the coherency data where high structure is shown as high intensity (brightness) (Fig. 3b). In addition, you will find one csv file per batch with data representing α-SMA intensity, α-SMA fibrous organization and alignment (coherency) and nuclei count per image. The csv file can be opened and organized using Excel (*see* **Note 32**). "Mean" (the sum of the gray values of all the pixels in the selection divided by the number of pixels; column D in the csv file) is the meaningful parameter for α-SMA intensity and "RawIntDen" (the sum of the coherency values of the pixels in the image or selection; column F in the csv file) should be used to represent the coherency value and divided by area (column C in the csv file). IntDens is the product of Area and Mean Gray/Coherency Value and not used in our analysis.

14. From the data table, calculate the percentages of myofibroblasts per image by dividing the number of ROIs including a myofibroblast by the total number of ROIs selected on the image (*see* **Note 33**). To determine the cut-off for a cell/ROI being considered as myofibroblast, we suggest the following calibration step: Select 5–10 cells from your image batch that satisfy your minimum criterion for being "α-SMA stress fiber-positive" and extract the respective coherency raw integrated density values from the data table. Use the average of these values as your threshold to categorizing cells into "α-SMA stress fiber-positive" (myofibroblast) or "α-SMA stress fiber-negative" (not a myofibroblast) (Fig. 3c). Quantification examples using fibroblasts from different species and organs with and without TGF-β1 treatment are provided in Fig. 4.

3.5.2 Quantification of Myofibroblasts in Tissue

The most widely used marker of myofibroblast α-SMA is also expressed in smooth muscle cells and pericytes. Therefore, myofibroblasts in tissue can be discriminated by being positive for α-SMA and negative for the muscle marker desmin and/or CD31 which is

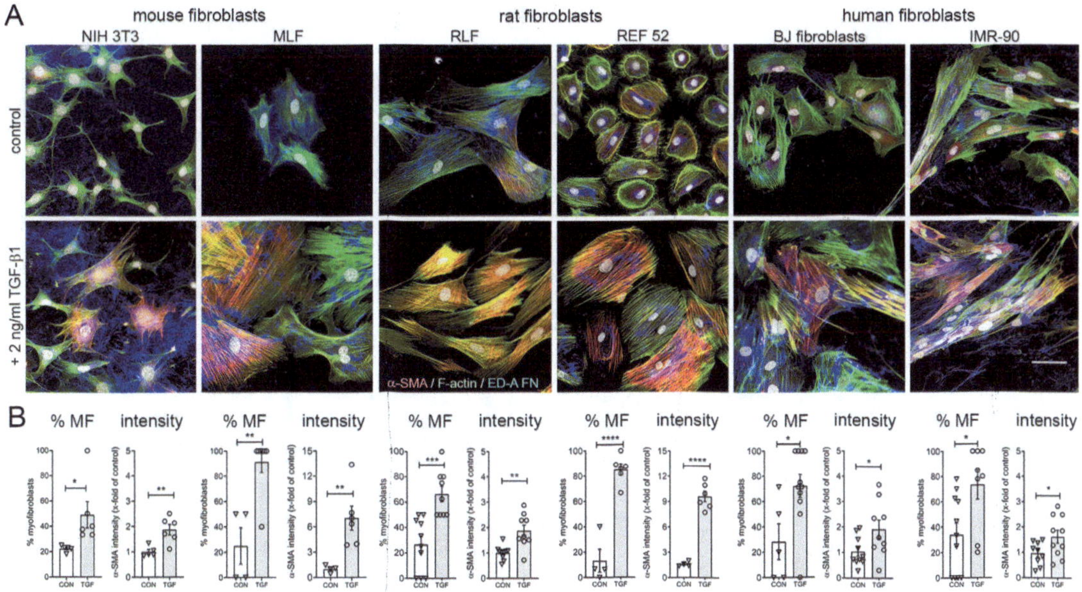

Fig. 4 Myofibroblast heterogeneity in cell cultures. (**a**) Primary and lineage fibroblasts from different species were cultured on tissue culture plastic with and without TGF-β1 for 4 days. Cells were then stained for α-SMA (red), filamentous actin (phalloidin-green), the extracellular matrix protein ED-A FN, and DNA (DAPI, white). Scale bar: 50μm. (**b**) Example pictures were quantified for average α-SMA intensity per cell and the percentage of α-SMA stress fiber-positive myofibroblasts using the "CellAnalysis_ManualROI.ijm" macro as described in Subheading 3.5.1. Every data point in the graphs represents values from one cell in the example images; standard deviation error bars are shown wit asterisks indicating statistical significance (*$p \leq 0.05$, **$p \leq 0.01$, unpaired Student *t*-test). Abbreviations: *α-SMA* α-smooth muscle actin, *ED-A FN* extradomain A fibronectin, *NIH* National Institutes of Health, *MLF* mouse lung fibroblast, *REF* rat embryonic fibroblast. BJ and IMR are cell batch/type identifiers

a marker of vascular endothelial cells. Myofibroblast content can be quantified from respectively stained tissue sections by subtracting the desmin and/or CD31 positive area from the total α-SMA area. This process is difficult to perform in a fully automated fashion if not using machine-learning software. We here provide two simple macro routines that provide maximum user control and high data confidence (Fig. 5).

1. Take pictures using an epifluorescence or confocal microscope (*see* **Notes 23 and 34**). Per fluorescence channel, use the same settings for exposure time (epifluorescence) or laser intensity (confocal) across all samples.

2. The described quantification procedures and automated routines (macros) are based on three-channel images with α-SMA staining in channel 1 (red); CD 31 in channel 2 (green), desmin in channel 3 (blue), and DAPI in channel 4 (white). If you are using a different channel distribution, you will need to change the channel attribution in the macro (*see* **Note 35**).

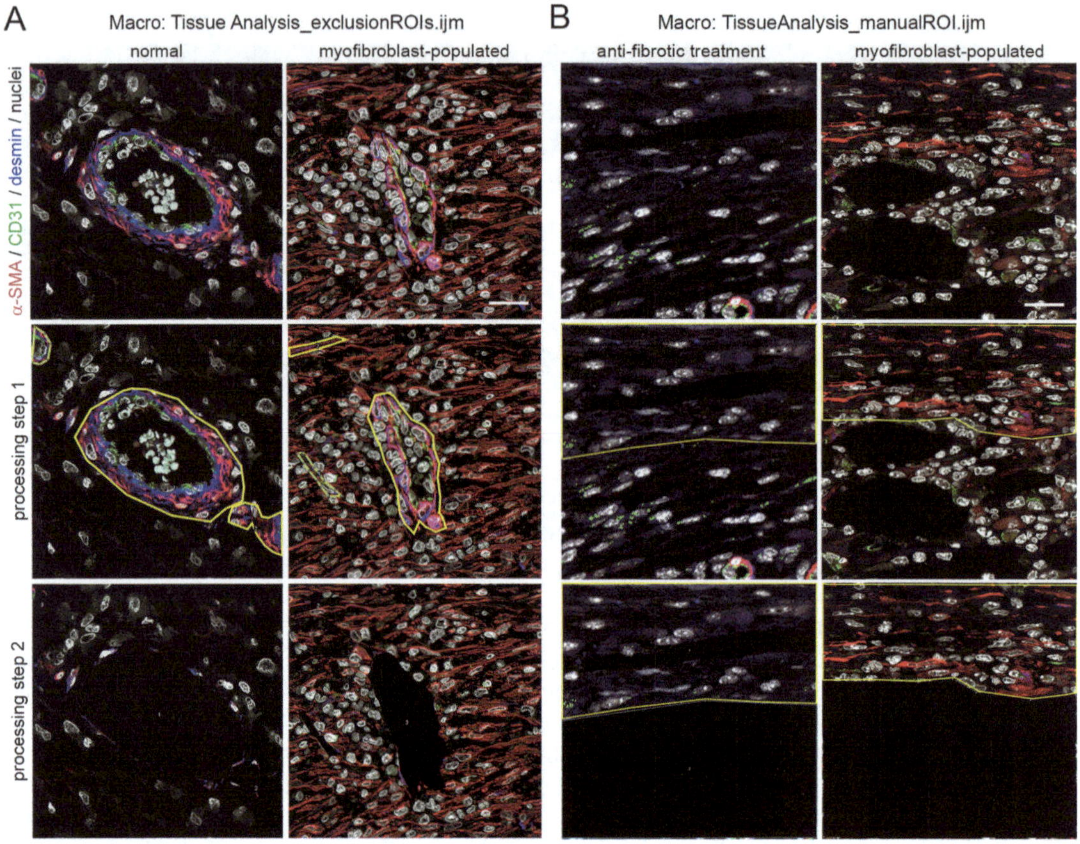

Fig. 5 Myofibroblast quantification from tissue sections. (**a**) Paraffin-embedded rodent normal skin and myofibroblast-populated 9-day-old wound granulation tissues was sectioned and stained for the myofibroblast marker α-SMA (red), endothelial cell marker CD31 (green), muscle marker desmin (blue), and nuclei (DAPI, white). Desmin stains vascular smooth muscle cells but not myofibroblasts in these examples. Macro analysis shown in panel A works on the principle of selecting regions of interest (ROI) (processing **step 1**) that are then excluded from the analysis (processing **step 2**, symbolized by blacking the excluded regions). (**b**) Shown are sections from connective tissue forming at the surface of silicone spheres implanted under the skin of rodents, either untreated (myofibroblast-populated) or treated with a myofibroblast-suppressing coating (anti-fibrosis treatment). The silicone surface is not shown and marked by the top of the image. Macro analysis shown in panel B works on the principle of selecting ROIs that are included for analysis; here the dense fibrous capsule forming around the implant (processing **step 1**). All non-selected image regions (here: loose subcutaneous connective tissue) are then excluded from the analysis (processing **step 2**, symbolized by blacking the excluded regions). Scale bar: 50μm. Abbreviations: *α-SMA* α-smooth muscle actin, *CD* cluster of differentiation

3. Save images in the recommended commercial acquisition format (e.g.,*.czi for images taken with Zeiss microscopes), which contains metadata and attributes one distinct channel per "color." Note that four channels cannot be unmixed from TIF images and this format is thus not supported by the macro (*see* **Notes 25** and **26**).

Table 1
Quantification of example images shown in Fig. 3c

Cell Label	Cell #Type	Parameter	ROI Area	Mean	IntDen	RawIntDen	RawIntDens/ Area	Average
C1	Control threshold	Coherency	28,969	0.253	7331	7331	0.253	
C2	Control threshold	Coherency	6221	0.22	1369	1369	0.220	
C3	Control threshold	Coherency	13,171	0.287	3783	3783	0.287	
C4	Control threshold	Coherency	7780	0.223	1737	1737	0.223	*0.246*
F1	Fibroblast	Coherency	9640	0.222	2142	2142	0.222	
F2	Fibroblast	Coherency	2989	0.21	629	629	0.210	
F3	Fibroblast	Coherency	13,210	0.229	3028	3028	0.229	*0.221*
M1	Myofibroblast	Coherency	17,581	0.589	10,351	10,351	0.589	
M2	Myofibroblast	Coherency	13,741	0.675	9275	9275	0.675	
M3	Myofibroblast	Coherency	14,638	0.749	10,963	10,963	0.749	*0.671*
							Mean	Average
C1	Control threshold	Intensity	28,969	31.03	898,913	898,913	31.0	
C2	Control threshold	Intensity	6221	32.294	200,898	200,898	32.3	
C3	Control threshold	Intensity	13,171	35.981	473,902	473,902	36.0	
C4	Control threshold	Intensity	7780	34.834	271,011	271,011	34.8	*33.5*
F1	Fibroblast	Intensity	9640	26.854	258,874	258,874	26.9	
F2	Fibroblast	Intensity	2989	28.171	84,203	84,203	28.2	
F3	Fibroblast	Intensity	13,210	34.122	450,747	450,747	34.1	*29.7*
M1	Myofibroblast	Intensity	17,581	52.689	926,334	926,334	52.7	
M2	Myofibroblast	Intensity	13,741	64.436	885,420	885,420	64.4	
M3	Myofibroblast	Intensity	14,638	62.767	918,781	918,781	62.8	*60.0*

4. Place all images that you want to process as one batch (typically all images from one experimental condition) in one folder.
5. Follow **steps 5–7** as outlined in Subheading 3.5.1 to install Fiji and required plugins.

6. Download the macro "TissueAnalysis_manualROI.ijm" and "Tissue Analysis_exclusionROIs.ijm" from https://github.com/CAMiLoD-UofT/alphaSMA-measurements and store in a folder of your choice.
7. To start Fiji and open the macro: *File -> Open, browse the folder containing the macro and select* "Tissue Analysis_exclusionROIs.ijm (analyzes the image for α-SMA content, *except* for what is included in ROIs) (Fig. 5a) or TissueAnalysis_manualROI.ijm" (analyzes selected ROIs for α-SMA content) (Fig. 5b).
8. Change image channel attribution in the macro if required (*see* **Note 35**) and save the macro.
9. Click "*Run*" in the open macro window (bottom left button) and follow the instructions given on screen.
10. Select "Input directory" and "Output directory" from your folders. Type in your "File suffix" (e.g., .czi) and click "OK." You will see your one of images and message window for the action required.
11. When using "TissueAnalysis_manualROI.ijm," draw ROIs manually around image areas that you want to have analyzed for α-SMA intensity, using the freehand selection tool following the instruction given on the screen. You can draw multiple ROIs in one image but must press "t" on your keyboard after every ROI. Press "OK" once done with ROI drawing. Repeat this procedure following the instruction given on the screen until your batch is completed.
12. When using "Tissue Analysis_exclusionROIs.ijm," draw your ROIs to *exclude* desmin (smooth muscle) and CD31 (endothelial cells) positive areas which are vessels (*see* **Note 36**).
13. Find results in the folder designated as "Output directory" in **step 10**. For every processed image, both macros will produce: one zip file containing all ROI data, one PNG image with "burned-in" ROI overlay. You will also find a csv file per batch with data representing α-SMA intensity. Column B contains your original sample labels which help you define data/sample. "Mean" (column D in the csv file) is the meaningful parameter for α-SMA intensity. Calculate the average of these mean values when you draw multiple ROIs with TissueAnalysis_manualROI.ijm macro.

4 Notes

1. Fibroblastic cells derived from different organs and species have characteristic and reproducible baseline levels of α-SMA expression in any given culture condition [44]. From our experience, baseline α-SMA expression increases in primary across

species for the same tissue origin: human, rat, and mouse (Fig. 4). Immortalized cell lines are often not well suited to study myofibroblast biology—being immortal and highly proliferative is not conducive to the contractile myofibroblast phenotype [13]. We here used commercially available human foreskin fibroblasts (BJ, ATCC® CRL-2522™) (20–40% α-SMA-positive at baseline), human fetal lung fibroblasts (IMR-90, ATCC® CCL-186™) (30–50% α-SMA-positive at baseline in 10% FBS-containing standard medium), primary rat lung fibroblasts (RLFs) (~30–60% α-SMA-positive at baseline) [9], rat embryonic fibroblasts (REF-52) (10–30% α-SMA-positive at baseline) [45], primary mouse lung fibroblasts (MLF) (20–50% α-SMA-positive at baseline) [46], and lineage NIH 3T3 fibroblast (NIH/3T3 ATCC® CRL-1658™) (5–20% α-SMA-positive at baseline). Percentage values will depend on the specific culture conditions established in other laboratories (e.g., serum content, growth time, cell density).

2. Fibroblastic cells adhere and spread better on tissue culture plastic surfaces compared to glass. Adhesiveness of plastic and glass culture ware can vary considerably between batches and manufactures which will affect myofibroblast phenotype. When using glass surfaces, pre-coating with ECM proteins like fibronectin, gelatin, or collagen type I (typically 2μg/cm^2) is recommended. Do not coat with poly-L-lysin which may promote cell attachment but will prevent integrin engagement.

3. Different FBS batches have different effects on baseline myofibroblast activation (α-SMA expression). You must re-establish baselines when switching batches. One of the main factors in serum antagonizing myofibroblast activation and TGF-β1 action (*see* **Note 4**) is fibroblast growth factor 2 (FGF-2), which enhances fibroblast proliferation [47].

4. Different vendors sell differently prepared TGF-β1 for cell culture applications with batch-to-batch differences. All require initial testing for myofibroblast activation responses in concentrations series at first use. Typically, recombinant TGF-β 1 is produced and sold as the latent pre-form, desiccated with citric acid salts added. Solubilization of the powder will automatically acidify the solution and activate latent TGF-β1. Be aware that active TGF-β1 loves to absorb to surfaces, ECM, and serum proteins which is why it is used in nM concentrations in cell culture but acts on cells in the pM range. Our lab and others have alternatively used the active TGF-β isoforms TGF-β2 and TGF-β3 with similar effects in vitro [48].

5. Dissolve 30 g PFA at 90 °C in 800 mL distilled water, add NaOH (32%) to clear the solution, add 100 mL of 10-times concentrated PBS, and adjust solution to pH 7.4 using

indicator paper. Do not use pH electrodes. Adjust volume to 1 L with additional distilled water. Store aliquots of 10–40 mL at −20 °C. Commercial soluble PFA can be used for convenience. PFA is toxic, which requires working with face masks and under fume hood at all stages of usages. Use hazardous waste container for the disposal of PFA.

6. Triton-X 100 is difficult to handle in its pure viscous form and it is not stable in high dilutions (e.g., 0.2%). Prepare 20% stock solution and keep at RT.
7. The 1A4 clone is derived from the original hybridoma produced and published by the Gabbiani lab [49]. Epitope for this antibody is the acetylated N-terminal sequence of α-SMA, which is different and discriminating among all actin isoforms [25]. Different vendors use different procedures to affinity-purify their product (or not); we can recommend this clone distributed by Sigma and Dako from personal testing. However, other vendors' products may work as well. Polyclonal rabbit antibodies are also commercially available. The IST-9 antibody is specific against the ED-A splice variant of fibronectin [50]; like for α-SMA multiple vendors distribute this clone and we have successfully worked with the antibody from Santa Cruz. Be aware that this antibody does not react with mouse ED-A fibronectin, which is why we have added total fibronectin as alternative to the mix.
8. Hoechst and DAPI are both popular for nuclear-specific staining of living or fixed cells and tissues. The dyes have minimal fluorescence in solution but become brightly fluorescent upon binding to DNA. Hoechst dyes are highly cell membrane permeant and they are typically used for staining at 1μg/mL. DAPI is less cell membrane permeant than Hoechst dyes and must be used at a higher concentration (usually 10μg/mL) for live cell staining. However, 1μg/mL of DAPI is enough for permeabilized cell staining. Hoechst and DAPI are extremely stable in water at 10 mg/mL and can be stored at 4 °C for years if they are protected from light.
9. Growth conditions that enhance fibroblast growth (e.g., hypoxia, high FBS concentrations, addition of growth factors such as FGF-2) most often reduce baseline myofibroblast activation levels [45, 47, 51]. High proliferation is not conducive to the contractile myofibroblast phenotype. Furthermore, cell density can dramatically—and often unpredictably—affect myofibroblast activation.
10. It is not required but recommended to change medium once fibroblasts have attached for two reasons: (a) to remove non-adherent cells and (b) to have one last control of successful cell seeding before adding expensive TGF-β1.

11. Washing is required. Keep living cells as undisturbed as possible by gently washing in serum-free medium or PBS that have been pre-warmed to at least RT. Do NOT use PBS buffers pre-made for cell culture, which typically are Ca^{2+} and Mg^{2+} free to weaken bivalent cation-dependent integrin adhesions and detach cells. Never pipette directly onto the cells as this may dislodge them even after fixation; apply solutions to the culture vessel walls.
12. Possible steps to suspend the protocol: After fixation, samples can be kept in PBS for approximately 1 week in sterile conditions at 4 °C. Because of possible epitope degradation, it is not recommended to store samples for longer periods or to store already-permeabilized cells before the staining procedure.
13. All primary antibodies can be combined in dilution buffer in one vial as long as they are from different species (e.g., mouse, rabbit, rat, and goat) or different isotypes (e.g., combinations of mouse IgG2a, mouse IgG1, and mouse IgG2b). If you are working with a new antibody for the first time, prepare primary antibody staining solution according to the manufacturer's instruction or at 10μg/mL. Most antibodies are sold at 1 mg/mL stock concentration and usually used at 1:100 dilution. Testing a broad range of antibody concentrations is also a good idea; test your primary antibody with a series of dilutions around the recommended concentration while keeping the secondary antibody concentration constant.
14. Incubation of primary antibodies overnight in 4 °C can sometimes help to improve the quality of IF staining. To decrease the risk of sample drying, put samples into small petri dishes along with wet tissue. However, good antibodies like the ones suggested here yield efficient staining already after 30 min incubation at RT.
15. Secondary antibodies must be specific against primary antibodies and the spectra of the fluorochromes must be separable by the microscopy setup. For observing samples through the microscope eyepieces (instead of cameras), fluorochromes emitting in the far-red are difficult to see with the naked eye.
16. For the selection of proper mounting media, different criteria should be considered, such as refractive index, which should be close to that of glass, polymerization degree, and fluorochrome preservation which are all important factors for preserving samples without changing staining properties [52]. Anti-fade mounting media, such as Fluoromount-G, do not fully polymerize but preserve fluorochrome intensity. With manual dexterity, you can always gently squish the coverslip against the glass to remove excess medium. Without manual dexterity, you will squish your cells.

17. Formalin, like PFA (*see* **Note 5**), should be handled with caution under a chemical hood. Formalin is a suspected carcinogen that can also irritate the eye, skin, and respiratory tracts. While PFA has not been identified as a carcinogen, it can affect lung and kidneys when inhaled and may be absorbed through the skin. Alcohol-based fixatives generally do not well preserve morphology but may be useful to preserve specific epitopes or for specific stains such as BrdU.

18. Make sure you have enough fixative to cover tissues (10–20 times volume). For example, use 1–2 mL of fixative per 100 mg of tissue. Fixation for 24–48 h can be applied for most tissues but the ideal fixation time will depend on tissue size. Over-fixation can mask the epitope. Antigen retrieval processes can overcome epitope masking, but there may be no signal even after antigen retrieval if the tissue has been fixed for too long. Under-fixation can cause uneven staining; strong signal on the edges and no/weak signal in the center of tissue sections.

19. When the paraffin block is completely cooled and hardened, it can be easily popped out of the mold and should not stick. If tissues are not well aligned or cracks appear in the paraffin wax, melt again and re-embed.

20. The water bath should be kept at 40–50 °C with distilled water. A warm water bath is used to float paraffin ribbons, stretch the sections, and remove wrinkles prior to being mounted onto a glass slide. If the water bath is too cold, wrinkles will not be removed in tissue sections. If the water bath is too hot, paraffin sections may melt and change tissue morphology. For fatty tissue, such as breast or brain, water temperature may need to be reduced.

21. Most animal tissues have a negative charge, owing to a small excess of acidic over basic amino acids in the structural proteins. Therefore, tissue samples are adhering well to the positively charged slide surface. Glass surfaces can be treated for a positive charge by coating the slide with a basic polymer or by a chemical reaction that leaves amino groups linked by covalent bonds to the silicon atoms of the glass.

22. Before moving to ethanol grades step, make sure to completely deparaffinize the sections. Additional 5 min incubation with Xylene will be required if the sections still have traces of paraffin. Uncompleted deparaffinization will give you high autofluorescence background. Xylene should be handled with caution under a chemical hood. Inhaling of xylene can cause headache, dizziness, nausea, and drowsiness.

23. There are different types of fluorescence microscopy which can be selected according to their availability and biological questions. Confocal microscopy is appropriate for generating

images with higher spatial and axial resolution and quality. Use a microscope with filters optimized for your choice of fluorochromes. Band-pass filters which collect emissions within a specific range may be more useful than long-pass filter sets which collect all emissions past a certain wavelength. The narrower the range of the band-pass filter, the better it can separate fluorophores with close emission spectra. Make sure to take only high-quality images with all segments of the image in focus. Always work with the optimal "Nyquist" resolution that most confocal microscope will suggest by default. Image analysis can turn image information into quantitative data, but it cannot do miracles: garbage-in-garbage-out.

24. To reattribute channel numbers to nuclei, α-SMA and phalloidin in the macro, open the respective macro in Fiji. Search for all comment text containing "//Change channel here" and edit preceding text. For instance, "run("Duplicate...", "duplicate channels = 3-3 title = Nuclei")" will become "run ("Duplicate...", "duplicate channels = 1-1 title = Nuclei")," if your nuclear staining is in channel 1 instead of channel 3.

25. Some commercial software uses pyramidal file formats rather than storing individual image files (e.g., LIFF formats from Leica microscopes). These formats cannot be directly read by the Fiji Bioformats plugin. One solution is to batch export 3-channel images from the commercial software as TIF files, which will be recognized and processed by the macros. *See* **Note 26** for important limitations and danger of this approach.

26. ATTENTION!!! From three-channel images with a red, green, and blue channel, commercial image acquisition software will be able to export RGB TIF images, where every color channel contains one "clean" 8-bit (grayscale) image of your respective staining. However, any other RGB TIF images may have information of one structure appearing in multiple channels (e.g., nuclei may appear in the green *and* blue channel). You will not be able to quantify these images. You also *cannot* use TIFF images that have been produced from more than three original channels. One quick test for the suitability of RGB images for quantification is as follows: Open RGB TIF image with Fiji. Then: "*Image → Color → Split channels.*" If the resulting three 8-bit grayscale images do not contain perfectly separated α-SMA, F-actin, and nuclei signals, DO NOT PROCEED with quantification.

27. The macro "CellAnalysis_ManualROI.ijm" requires the user to manually outline cells based on a blurred image of the phalloidin channel (Fig. 3b, "Macro 1"). Do not forget to press the key "t" on your keyboard after drawing every outline (ROI). Intensity and alignment (coherency) of α-SMA-positive

structures will be measured only over the defined ROIs. Consequently, including too much cell-free area can skew the results—rather go smaller than too large with your ROIs. This macro provides highest control over the analysis but is also the most time-consuming.

28. The macro "CellAnalysis_Individual.ijm" will process all images in the defined folder without any user input (Fig. 3b, "Macro 2"). It is critical that cells in the image do not touch to be recognized as individual cells. Touching cells will be recognized as one ROI (i.e., cell) and treated as such. This may be acceptable in conditions where there is little heterogeneity in α-SMA expression within one picture.

29. The macro "CellAnalysis_Voronoi.ijm" is to be used with care and only with 100% confluent cultures if your biological question does not allow working with lower cell densities (Fig. 3b, "Macro 3"). At high cell densities, it may not be possible to discriminate cell boundaries even by eye and manually drawing ROIs will become arbitrary in addition to be very time-consuming. This macro is based on the principle that individual nuclei will be automatically detected and used to segment the image into ROIs where nuclei are at equal distances. ROIs do NOT correspond to single cells in this approach but to image segments and respective data will have to be reported accordingly. This macro is highly dependent on proper detection of nuclei and you may need to adjust the respective parameters (*see* **Note 31**).

30. Another option to analyze highly confluent cultures is "ignoring" heterogeneity of α-SMA between different cells within one image field. Using the macro "CellAnalysis_Individual.ijm" on dense cultures will generate one ROI that outlines all cells within the image. This is the only condition, where you should divide α-SMA intensity and coherency by the total nuclear count in the image if you chose to normalize for cell numbers (*see* **Note 33**). The outcome of this analysis will be the average α-SMA intensity and coherency over all cells in the image.

31. The macros are designed to allow different conditions and image resolutions. However, they may not serve all resolutions/magnifications, illuminations, signal-to-noise ratios equally well. You may be required to adjust parameters within the macro code. Critical procedures are "setAutoThreshold ("*thresholding_type dark*")"; "Analyze Particles...", "*size=0-Infinity exclude add*"); ("Gaussian Blur...", "*sigma=10*") and parameters in italics may need to be changed. The reader is referred to the numerous Fiji forums and the Fiji help function for more details.

32. All macros generate data for the following parameters (identifier in column B): α-SMA intensity (suffix "-alphaSMA" in column B) which ignores whether α-SMA signal is diffuse or structured; coherency (suffix "-Coherency" in column B), which is a measure of fibrous structure, alignment, and intensity; and nuclei count per image (suffix "-Nuclei" in column B). Numbers following the colon in column B identify the individual ROIs. The continuous logging of data into rows may not be conducive to subsequent data analysis but restructuring is simple in Excel by using the identifiers in column B: (1) Insert empty column right of column B and then select column B. (2) In the "*Data*" tab, select "*text to columns*"→*Delimited*→*next*→*other:* ':'→Finish. (3) Insert empty column right of column B and then select column B. (4) In the "*Data*" tab, select "*text to columns*"→*Delimited*→*next*→*other:* '-'→Finish. (5) Select full table and in the "*Data*" tab, select "*Sort*"→*sort by column C*→*OK*. Take it from there.
33. IMPORTANT: Divide the number of α-SMA stress-fiber-positive or -negative ROIs by the number of ROIs in the image, NOT by the nuclear count per image which will be different if not all cells in the image are outlined by ROIs. "Coherency" is a relative parameter and absolute values will vary according to your imaging conditions. Be aware that any calibration step to threshold for α-SMA stress fiber-positive content in ROIs will introduce some user bias. You can reduce bias by asking a "blinded" (but not blind) colleague to identify α-SMA stress fiber-positive cells in the calibration image. Proper reporting of your chosen criteria and consistent application of these criteria across compared conditions will be essential. For an unbiased approach, you may report histograms with cells falling into classes of coherency value ranges. Alternatively, calculate averages of the raw coherency values per condition and express your results as percentage (x-fold) of your biological control.
34. Multiple tissue components such as red blood cells, elastin, and collagen can generate strong fluorescent signals making it difficult to discern between specific signal and background. Without proper controls, complex background autofluorescence can lead to misinterpretation of image data and generation of false positive results. Always include a negative control which incubated only with secondary antibody.
35. Search for comment text containing "//Change channel here" and change to the channel you want. For instance, "run ("Duplicate...", "title=alphaSMA duplicate channels=2")" will become "run("Duplicate...", "title=alphaSMA duplicate channels=1")," if your α-SMA staining is in channel 1 instead of channel 2.

36. The macro also offers an automated threshold function to exclude areas from the analysis that are not covered by tissue (negative in all channels). To activate this function, remove "/*" and "*/" as described in the line 42 in the macro "Tissue Analysis_exclusionROIs.ijm." Change the threshold value depending on your image sets. Try with the average threshold value of your images.

Acknowledgments

We would like to thank to Dr. Giulio Gabbiani for kindly providing α-SMA antibody and the staff of the Collaborative Advanced Microscopy Labs of Dentistry (CAMiLoD) (Faculty of Dentistry, University of Toronto) for technical support. Schemes were produced with Biorender (biorender.com). The research of BH is supported by a foundation grant from the Canadian Institutes of Health Research (#375597) and support from the John Evans Leadership funds (#36050 and #38861) and innovation funds ("Fibrosis Network, #36349") from the Canada Foundation for Innovation (CFI) and the Ontario Research Fund (ORF). FY is recipient of a Mary H. Beatty Fellowship from the School of Graduate Studies, University of Toronto. Fereshteh Younesi and Dong Ok Son contributed equally to this work.

References

1. Walraven M, Hinz B (2018) Therapeutic approaches to control tissue repair and fibrosis: extracellular matrix as a game changer. Matrix Biol 71–72:205–224. https://doi.org/10.1016/j.matbio.2018.02.020
2. Frangogiannis NG (2016) Fibroblast-extracellular matrix interactions in tissue fibrosis. Curr Pathobiol Rep 4:11–18. https://doi.org/10.1007/s40139-016-0099-1
3. Hinz B, McCulloch CA, Coelho NM (2019) Mechanical regulation of myofibroblast phenoconversion and collagen contraction. Exp Cell Res 379:119–128. https://doi.org/10.1016/j.yexcr.2019.03.027
4. Horowitz JC, Thannickal VJ (2019) Mechanisms for the resolution of organ fibrosis. Physiology (Bethesda) 34:43–55. https://doi.org/10.1152/physiol.00033.2018
5. Pakshir P, Hinz B (2018) The big five in fibrosis: macrophages, myofibroblasts, matrix, mechanics, and miscommunication. Matrix Biol 68–69:81–93. https://doi.org/10.1016/j.matbio.2018.01.019
6. Gabbiani G, Ryan GB, Majno G (1971) Presence of modified fibroblasts in granulation tissue and their possible role in wound contraction. Experientia 27:549–550
7. Tomasek JJ, Gabbiani G, Hinz B, Chaponnier C, Brown RA (2002) Myofibroblasts and mechano-regulation of connective tissue remodelling. Nat Rev Mol Cell Biol 3:349–363
8. Klingberg F, Hinz B, White ES (2013) The myofibroblast matrix: implications for tissue repair and fibrosis. J Pathol 229:298–309. https://doi.org/10.1002/path.4104
9. Hinz B, Celetta G, Tomasek JJ, Gabbiani G, Chaponnier C (2001) Alpha-smooth muscle actin expression upregulates fibroblast contractile activity. Mol Biol Cell 12:2730–2741
10. Hinz B, Gabbiani G, Chaponnier C (2002) The NH2-terminal peptide of alpha-smooth muscle actin inhibits force generation by the myofibroblast in vitro and in vivo. J Cell Biol 157:657–663

11. Chaponnier C, Goethals M, Janmey PA, Gabbiani F, Gabbiani G, Vandekerckhove J (1995) The specific NH2-terminal sequence Ac-EEED of alpha-smooth muscle actin plays a role in polymerization in vitro and in vivo. J Cell Biol 130:887–895
12. Clement S, Hinz B, Dugina V, Gabbiani G, Chaponnier C (2005) The N-terminal Ac-EEED sequence plays a role in {alpha}-smooth-muscle actin incorporation into stress fibers. J Cell Sci 118:1395–1404
13. Hinz B, Lagares D (2020) Evasion of apoptosis by myofibroblasts: a hallmark of fibrotic diseases. Nat Rev Rheumatol 16:11–31. https://doi.org/10.1038/s41584-019-0324-5
14. Tsukui T, Sun KH, Wetter JB, Wilson-Kanamori JR, Hazelwood LA, Henderson NC, Adams TS, Schupp JC, Poli SD, Rosas IO, Kaminski N, Matthay MA, Wolters PJ, Sheppard D (2020) Collagen-producing lung cell atlas identifies multiple subsets with distinct localization and relevance to fibrosis. Nat Commun 11:1920. https://doi.org/10.1038/s41467-020-15647-5
15. Valenzi E, Bulik M, Tabib T, Morse C, Sembrat J, Trejo Bittar H, Rojas M, Lafyatis R (2019) Single-cell analysis reveals fibroblast heterogeneity and myofibroblasts in systemic sclerosis-associated interstitial lung disease. Ann Rheum Dis 78:1379–1387. https://doi.org/10.1136/annrheumdis-2018-214865
16. Layton TB, Williams L, McCann F, Zhang M, Fritzsche M, Colin-York H, Cabrita M, Ng MTH, Feldmann M, Samson S, Furniss D, Xie W, Nanchahal J (2020) Cellular census of human fibrosis defines functionally distinct stromal cell types and states. Nat Commun 11:2768. https://doi.org/10.1038/s41467-020-16264-y
17. Sahai E, Astsaturov I, Cukierman E, DeNardo DG, Egeblad M, Evans RM, Fearon D, Greten FR, Hingorani SR, Hunter T, Hynes RO, Jain RK, Janowitz T, Jorgensen C, Kimmelman AC, Kolonin MG, Maki RG, Powers RS, Pure E, Ramirez DC, Scherz-Shouval R, Sherman MH, Stewart S, Tlsty TD, Tuveson DA, Watt FM, Weaver V, Weeraratna AT, Werb Z (2020) A framework for advancing our understanding of cancer-associated fibroblasts. Nat Rev Cancer 20:174–186. https://doi.org/10.1038/s41568-019-0238-1
18. Affo S, Yu LX, Schwabe RF (2017) The role of cancer-associated fibroblasts and fibrosis in liver cancer. Annu Rev Pathol 12:153–186. https://doi.org/10.1146/annurev-pathol-052016-100322
19. Liao Z, Tan ZW, Zhu P, Tan NS (2019) Cancer-associated fibroblasts in tumor microenvironment—accomplices in tumor malignancy. Cell Immunol 343:103729. https://doi.org/10.1016/j.cellimm.2017.12.003
20. Kisseleva T (2017) The origin of fibrogenic myofibroblasts in fibrotic liver. Hepatology 65:1039–1043. https://doi.org/10.1002/hep.28948
21. Smith-Clerc J, Hinz B (2010) Immunofluorescence detection of the cytoskeleton and extracellular matrix in tissue and cultured cells. Methods Mol Biol 611:43–57. https://doi.org/10.1007/978-1-60327-345-9_4
22. Schindelin J, Arganda-Carreras I, Frise E, Kaynig V, Longair M, Pietzsch T, Preibisch S, Rueden C, Saalfeld S, Schmid B, Tinevez JY, White DJ, Hartenstein V, Eliceiri K, Tomancak P, Cardona A (2012) Fiji: an open-source platform for biological-image analysis. Nat Methods 9:676–682. https://doi.org/10.1038/nmeth.2019
23. Hinz B, Mastrangelo D, Iselin CE, Chaponnier C, Gabbiani G (2001) Mechanical tension controls granulation tissue contractile activity and myofibroblast differentiation. Am J Pathol 159:1009–1020
24. Darby I, Skalli O, Gabbiani G (1990) Alpha-smooth muscle actin is transiently expressed by myofibroblasts during experimental wound healing. Lab. Invest 63:21–29
25. Chaponnier C, Gabbiani G (2016) Monoclonal antibodies against muscle actin isoforms: epitope identification and analysis of isoform expression by immunoblot and immunostaining in normal and regenerating skeletal muscle. F1000Res 5:416. https://doi.org/10.12688/f1000research.8154.2
26. Dugina V, Zwaenepoel I, Gabbiani G, Clement S, Chaponnier C (2009) Beta and gamma-cytoplasmic actins display distinct distribution and functional diversity. J Cell Sci 122:2980–2988. https://doi.org/10.1242/jcs.041970
27. El Agha E, Kramann R, Schneider RK, Li X, Seeger W, Humphreys BD, Bellusci S (2017) Mesenchymal stem cells in fibrotic disease. Cell Stem Cell 21:166–177. https://doi.org/10.1016/j.stem.2017.07.011
28. Lemos DR, Duffield JS (2018) Tissue-resident mesenchymal stromal cells: implications for tissue-specific antifibrotic therapies. Sci Transl Med 10:eaan5174. https://doi.org/10.1126/scitranslmed.aan5174
29. Soliman H, Paylor B, Scott RW, Lemos DR, Chang C, Arostegui M, Low M, Lee C, Fiore D, Braghetta P, Pospichalova V, Barkauskas CE, Korinek V, Rampazzo A, MacLeod K,

Underhill TM, Rossi FMV (2020) Pathogenic potential of Hic1-expressing cardiac stromal progenitors. Cell Stem Cell 26:205–220.e208. https://doi.org/10.1016/j.stem.2019.12.008

30. Di Carlo SE, Peduto L (2018) The perivascular origin of pathological fibroblasts. J Clin Invest 128:54–63. https://doi.org/10.1172/JCI93558
31. Prunotto M, Bruschi M, Gunning P, Gabbiani G, Weibel F, Ghiggeri GM, Petretto A, Scaloni A, Bonello T, Schevzov G, Alieva I, Bochaton-Piallat ML, Candiano G, Dugina V, Chaponnier C (2015) Stable incorporation of alpha-smooth muscle actin into stress fibers is dependent on specific tropomyosin isoforms. Cytoskeleton (Hoboken) 72:257–267. https://doi.org/10.1002/cm.21230
32. Arnoldi R, Chaponnier C, Gabbiani G, Hinz B (2012) In: Hill J (ed) Muscle: fundamental biology and mechanisms of disease. Elsevier Inc, Amsterdam, pp 1183–1195
33. Christen T, Bochaton-Piallat ML, Neuville P, Rensen S, Redard M, van Eys G, Gabbiani G (1999) Cultured porcine coronary artery smooth muscle cells. A new model with advanced differentiation. Circ Res 85:99–107
34. Hinz B (2016) In: Ågren MS (ed) Wound healing biomaterials. Vol. I: Therapies and regeneration. Woodhead Publishing Limited, Springer Science and Business Media Publishing, Cambridge, UK, pp 69–100
35. Taha IN, Naba A (2019) Exploring the extracellular matrix in health and disease using proteomics. Essays Biochem 63:417–432. https://doi.org/10.1042/EBC20190001
36. Herrera J, Henke CA, Bitterman PB (2018) Extracellular matrix as a driver of progressive fibrosis. J Clin Invest 128:45–53. https://doi.org/10.1172/JCI93557
37. Schiller HB, Fernandez IE, Burgstaller G, Schaab C, Scheltema RA, Schwarzmayr T, Strom TM, Eickelberg O, Mann M (2015) Time- and compartment-resolved proteome profiling of the extracellular niche in lung injury and repair. Mol Syst Biol 11:819. https://doi.org/10.15252/msb.20156123
38. Serini G, Bochaton-Piallat ML, Ropraz P, Geinoz A, Borsi L, Zardi L, Gabbiani G (1998) The fibronectin domain ED-A is crucial for myofibroblastic phenotype induction by transforming growth factor-beta1. J Cell Biol 142:873–881
39. Desmouliere A, Geinoz A, Gabbiani F, Gabbiani G (1993) Transforming growth factor-beta 1 induces alpha-smooth muscle actin expression in granulation tissue myofibroblasts and in quiescent and growing cultured fibroblasts. J Cell Biol 122:103–111. https://doi.org/10.1083/jcb.122.1.103
40. Ronnov-Jessen L, Petersen OW (1993) Induction of alpha-smooth muscle actin by transforming growth factor-beta 1 in quiescent human breast gland fibroblasts. Implications for myofibroblast generation in breast neoplasia. Lab Investig 68:696–707
41. Hinz B, Gabbiani G (2003) Cell-matrix and cell-cell contacts of myofibroblasts: role in connective tissue remodeling. Thromb Haemost 90:993–1002
42. Klingberg F, Chow ML, Koehler A, Boo S, Buscemi L, Quinn TM, Costell M, Alman BA, Genot E, Hinz B (2014) Prestress in the extracellular matrix sensitizes latent TGF-beta1 for activation. J Cell Biol 207:283–297. https://doi.org/10.1083/jcb.201402006
43. Smith-Clerc J, Hinz B (2010) Histology protocols. Springer, New York, pp 43–57
44. Desmoulière A, Rubbia-Brandt L, Abdiu A, Walz T, Macieira-Coelho A, Gabbiani G (1992) Alpha-smooth muscle actin is expressed in a subpopulation of cultured and cloned fibroblasts and is modulated by gamma-interferon. Exp Cell Res 201:64–73
45. Hinz B, Dugina V, Ballestrem C, Wehrle-Haller B, Chaponnier C (2003) Alpha-smooth muscle actin Is crucial for focal adhesion maturation in myofibroblasts. Mol Biol Cell 14:2508–2519
46. Lodyga M, Cambridge E, Karvonen HM, Pakshir P, Wu B, Boo S, Kiebalo M, Kaarteenaho R, Glogauer M, Kapoor M, Ask K, Hinz B (2019) Cadherin-11-mediated adhesion of macrophages to myofibroblasts establishes a profibrotic niche of active TGF-beta. Sci Signal 12:eaao3469. https://doi.org/10.1126/scisignal.aao3469
47. Papetti M, Shujath J, Riley KN, Herman IM (2003) FGF-2 antagonizes the TGF-beta1-mediated induction of pericyte alpha-smooth muscle actin expression: a role for myf-5 and Smad-mediated signaling pathways. Invest Ophthalmol Vis Sci 44:4994–5005
48. Serini G, Gabbiani G (1996) Modulation of alpha-smooth muscle actin expression in fibroblasts by transforming growth factor-beta isoforms: an in vivo and in vitro study. Wound Repair Regen 4:278–287
49. Skalli O, Ropraz P, Trzeciak A, Benzonana G, Gillessen D, Gabbiani G (1986) A monoclonal antibody against alpha-smooth muscle actin: a new probe for smooth muscle differentiation. J Cell Biol 103:2787–2796

50. Carnemolla B, Borsi L, Zardi L, Owens RJ, Baralle FE (1987) Localization of the cellular-fibronectin-specific epitope recognized by the monoclonal antibody IST-9 using fusion proteins expressed in E. coli. FEBS Lett 215:269–273
51. Modarressi A, Pietramaggiori G, Godbout C, Vigato E, Pittet B, Hinz B (2010) Hypoxia impairs skin myofibroblast differentiation and function. J Invest Dermatol 130:2818–2827. https://doi.org/10.1038/jid.2010.224
52. Ravikumar S, Surekha R, Thavarajah R (2014) Mounting media: an overview. J Dr NTR Univ Health Sci 3:1

Chapter 4

Fibroblast and Myofibroblast Subtypes: Single Cell Sequencing

Hesham Soliman, Lin Wei Tung, and Fabio M. V. Rossi

Abstract

The stroma constitutes the structural framework of an organ and plays crucial roles in health and following organ damage. The major player of the stroma with respect to extracellular matrix deposition, maintenance, and remodeling is the fibroblast and its activated derivative, the myofibroblast. It has long been recognized that there is considerable variability to the fibroblast phenotype. The recent advent of new single cell "omics" technologies has revolutionized our understanding and appreciation of cellular heterogeneity of fibroblasts been revolutionized. With these tools, the nature and defining characteristics of the cells comprising the stroma is finally being defined not just through a few markers, but by taking a wholistic look at transcriptional programs. It is now apparent that stromal cells are not only transcriptionally diverse, but also functionally, epigenetically, and spatially heterogeneous. Studying populations at single cell resolution has enabled identification of new clusters of cells with unique transcriptional signatures. Whether these clusters truly represent distinct subpopulations or different states of the same population remains to be clarified. In this chapter, we first describe a procedure for purification and preparation of a single cell suspension from tissue samples (in this case the heart) for single cell RNA sequencing. We also describe preparation of high-quality tissue sections for spatial transcriptomics. Secondly, we outline a workflow for computational analysis of single cell RNA sequencing and spatial transcriptomics data, as well as integrating them together, to explore the heterogeneity within fibroblasts/myofibroblasts and identify different subtypes and their locations in the heart.

Key words Single cell RNA sequencing, Spatial transcriptomics, Fibroblasts and myofibroblast subsets, Visium, 10× Genomics, Computational analysis, Heart

1 Introduction

Fibroblasts and myofibroblasts play a key role in health and disease as the main extracellular matrix (ECM)-producing cells. The ECM not only provides structural support to maintain the integrity of the organ, but also plays essential roles in inter- and intracellular signaling either by transducing biochemical or mechanical cues [1–5]. Myofibroblasts were first identified by Giulio Gabbiani and coworkers in 1971 [6] as ECM-producing cells that also possess a contractile phenotype and are usually characterized by the

Boris Hinz and David Lagares (eds.), *Myofibroblasts: Methods and Protocols*, Methods in Molecular Biology, vol. 2299,
https://doi.org/10.1007/978-1-0716-1382-5_4,

expression of alpha smooth muscle actin (α-SMA) [7]. These cells are induced in granulation tissue following injury and are believed to be generally absent in healthy undamaged organs [8].

The diversity of fibroblasts has been described since long ago, especially those derived from different organs. Although it was quite evident that fibroblasts possessed discriminating organ-specific properties [9], an appreciation of fibroblast heterogeneity within organs and tissues developed only recently with the advent of newer technologies. Results generated with these new technologies support that fibroblasts are functionally, transcriptionally, epigenetically, and spatially heterogeneous within the same organ. This heterogeneity increases even more following damage [2, 9].

Single cell RNA sequencing (scRNAseq) has the ability to transcriptionally profile each cell separately using different approaches, unlike bulk RNA sequencing where the readout of any gene is an average of all cell populations, including populations that may not even express the gene [10–12]. Additionally, scRNA-seq enables unbiased clustering of cell populations into transcriptionally distinct clusters, which facilitates dissecting heterogeneity and identification of rare populations or multiple activation, metabolic, and functional cell states within the same population [10–13]. Moreover, cell lineage relationships between different clusters may be inferred by performing trajectory analysis [12–16].

There are several different platforms for performing scRNAseq, each with its own advantages and disadvantages [11, 12]. One of the most important differences among the various techniques is the extent of transcript coverage. While there are methods that produce (nearly) full-length transcript coverage, such as SMART-seq [2], SUPeR-seq, and MATQ-seq, other techniques cover only the 3′, such as Drop-seq, InDrop, chromium sequencing, Cel-seq [2], and SPLiT-seq, or only the 5′ end such as STRT-seq [17–24]. Full-length scRNAseq methods are more appropriate for studying alternative splice variants, transcriptome annotation, and analysis of allele-specific expression using single nucleotide polymorphisms owing to the superiority of their transcript coverage [11, 12]. These methods, especially SMART-seq2 and MATQ-seq, also seem to possess greater sensitivity as they can detect a greater number of genes per cell, as well as lower dropout probability, the probability of failing to detect a gene that is biologically expressed in a cell due to technical limitations [11]. Conversely, 3′ and 5′ end RNA sequencing protocols sacrifice full-length coverage in order to substantially increase the throughput of library generation by orders of magnitude using multiplexing cDNA amplification after tagging cells with cell-specific barcodes [17–19, 21, 25]. This approach also permits the use of unique molecular identifiers (UMI), which are unique random oligonucleotide tags that label individual mRNA molecules, enabling their detection and quantification. Use of UMI is very important to identify amplified fragments and therefore correct for amplification bias [26]. Similar to UMIs, cell hashing, a recently established method of "multiplet" detection by Stoeckius et al., further reduces errors arising from pooling samples in droplet-based systems [27]. Prior to sample pooling, cells are labeled with sample-specific hashtag antibodies against surface proteins. Sequencing of these hashtag oligoes (HTOs) will distinguish cross-sample multiplets from singlets. In addition, 3′ and 5′ end sequencing methods reduce sequencing costs per cell compared to full-length protocols [11]. This advantage, in addition to their higher throughput, makes them more suitable for identification of rare subpopulations in complex tissues in health and disease. In this chapter, we focus on the 10× Genomics Chromium™ system as it is the most commonly used platform optimized for high throughput and cost effectiveness [11].

2 Materials

Prepare all solutions using ultrapure water and analytical grade reagents.

2.1 Digestion of Heart to Purify Myo(Fibroblasts)

1. Platelet-derived growth factor receptor alpha (Pdgfra) H2BeGFP mice, B6.129S4-Pdgfratm11(EGFP)Sor/J (Jax stock number 007669), or C57Bl/6J mice (Jax stock number 000664).
2. Surgical tools.
3. Avertin (tribromoethanol): Prepare a 1.25%, v/v working solution by mixing 2.5 g of 2,2,2-tribromoethanol with 5 mL of tert-amyl alcohol and dissolving it in 200 mL of water using a magnetic stirrer. Make sure the tribromoethanol is fully dissolved. This probably may take overnight stirring. The pH of final working solution should be >5; pH <5 is toxic.
4. 10 mL syringes.
5. 1 mL syringes.
6. Syringe needles 23 gauge (G) or 26G.
7. Centrifuge.
8. Tube rotator.
9. Phosphate-buffered saline (PBS), pH 7.4: To prepare 1 L of 1× PBS, dissolve 8 g NaCl, 0.2 g KCl, 1.44 g Na_2HPO_4, and 0.24 g KH_2PO_4 in 800 mL of ultrapure water. Adjust pH to 7.4 using HCl and complete to 1 L with ultrapure water.
10. *N*-[Tris(hydroxymethyl)methyl]-2-aminoethanesulfonic acid (TES) buffer: 50 mM TES, titrated with NaOH to reach pH 7.4 and then sterile filtered.

11. Hank's balanced salt solution (HBSS): dissolve 8.0 g NaCl, 0.4 g KCl, 0.14 g $CaCl_2$, 0.1 g $MgSO_4 \bullet 7H_2O$, 0.1 g $MgCl_2 \cdot 6H_2O$, 0.06 g $Na_2HPO_4.2H_2O$, 0.06 g KH_2PO_4, 1.0 g glucose, and 0.35 g $NaHCO_3$ in 1000 mL of ultrapure water and sterile filter.
12. HEPES-buffered saline: 50 mM HEPES/KOH, 150 mM NaCl, pH 7.4.
13. Ice.
14. Collagenase from Clostridium histolyticum type II solution: prepare solution in phosphate-buffered saline (PBS; pH 7.4) or N-[Tris(hydroxymethyl)methyl]-2-aminoethanesulfonic acid (TES) buffer containing 50 mM TES (pH 7.4). Dissolve enough powder to make a solution with activity of 2.5 collagen digestion units (CDU)/mL. Sterile filter with 0.22μm membrane filter (*see* **Note 1**).
15. Collagenase D from Clostridium histolyticum solution: prepare solution in phosphate-buffered saline (PBS; pH = 7.4) or Hank's balanced salt solution (HBSS). Dissolve enough powder to make a solution with activity of three collagen digestion units/mL. Sterile filter with 0.22μm membrane filter (*see* **Note 2**).
16. Dispase II solution: dissolve the lyophilized dispase II enzyme in HEPES-buffered saline (50 mM HEPES/KOH pH 7.4, 150 mM NaCl) to make a working solution with activity of 2.4 U/mL (*see* **Note 2**). Sterile filter with 0.22μm membrane filter (*see* **Note 3**).
17. $CaCl_2$ solution: prepare a 10 mM stock (100×) solution in water. Sterile filter with 0.22μm membrane filter and aliquot and store at room temperature. Solutions are stable for up to 6 months at room temperature.
18. 40μm cell strainers.
19. Fluorescence-activated cell sorting (FACS buffer): prepare in PBS by adding 2 mM EDTA and 2% fetal bovine serum (FBS). Sterile filter with 0.22μm membrane filter. Solutions are stable for up to 1–2 weeks at 2–8 °C. Store FBS at −20 °C.
20. 15 mL and 50 mL cylindrical tubes.
21. Siliconized centrifuge tubes.
22. Disposable bulb pipettes.
23. Eppendorf tubes (1.8 mL and 0.2 mL).
24. Hemocytometer.
25. Trypan blue solution: 0.4% solution prepared in PBS and sterile filtered.

2.2 Fluorescence-Activated Cell Sorting (FACS) of (Myo) Fibroblasts for scRNAseq

1. FACS machine.
2. Polypropylene 5-mL round-bottom FACS tubes.
3. Polystyrene 5-mL round-bottom FACS tubes with cell strainer snap caps.
4. Hoechst 33342 solution: prepare in ultrapure water as 20 mM stock solution and dilute to a 4μM working solution. Solutions could be stored for up to 6 months at 2–8 °C. Store in −20 °C for longer durations. Protect solutions from light (*see* **Note 4**).
5. Propidium iodide (PI) solution: prepare in ultrapure water as 1 mg/mL stock solution and dilute to 1–2μg/mL working solution. Solutions could be stored for at least 6 months at 2–8 °C. Store in −20 °C for longer durations. Protect solutions from light (*see* **Note 5**).
6. Sorting media: prepare with Dulbecco's minimum essential medium (DMEM) supplemented with 20% FBS, 2 mM L-glutamine, 1% (v/v) penicillin/streptomycin.
7. Fluorochrome-conjugated anti-CD45-APC or anti-CD45-FITC antibody (clone 30-F11, 1:500 and 1:1000, respectively).
8. Fluorochrome-conjugated anti-CD31-APC or anti-CD31-FITC antibody (clone MEC13.3, 1:300).
9. Barcoded hashing antibodies against mouse major histocompatibility I (MHC I) and CD45 proteins (Totalseq®, A301-A304, Biolegened; 1μg/10^6 cells in 100μL volume).

2.3 Preparation of Tissue Sections for Spatial Transcriptomic Profiling

1. Platelet-derived growth factor receptor alpha (Pdgfra) H2BeGFP mice, B6.129S4-Pdgfratm11(EGFP)Sor/J (Jax stock number 007669), or C57Bl/6J mice (Jax stock number 000664).
2. PBS.
3. Ice.
4. Disposable plastic cryomold.
5. Optimal cutting temperature (OCT) compound, which is a medium for frozen tissue specimens to ensure optimal cutting temperature.
6. Metal canister.
7. Isopentane.
8. Liquid nitrogen.
9. Sealable plastic bag.
10. Cylindrical slide holder with a tight cap.
11. Paintbrush.
12. Anti-roll plate.
13. Visium® spatial gene expression slide.

2.4 Computational Analysis of scRNAseq and Spatial Transcriptomics to Identify Fibroblast Subsets

All the computational analyses in the rest of this chapter are performed in R, with required packages and sample data.

1. R language and environment for statistical computing and graphics: available as free software to download from the official website of R project: https://www.r-project.org/.
2. Download R language version 3.4 or later.
3. RStudio: an integrated development environment (IDE) for R that can be downloaded from: https://rstudio.com/products/rstudio/.
4. Seurat: R toolkit for single cell transcriptomic analysis is designed for the quality control and analysis of scRNAseq data, as well as integrating scRNAseq data with other types of data such as spatial transcriptomics to explore cellular heterogeneity [16, 28]. Seurat package version 3.2 or above is required for spatial transcriptomic analysis and is available on Comprehensive R Archive Network (CRAN): https://cran.r-project.org/.

3 Methods

3.1 Digestion of Heart to Purify Myo(Fibroblasts)

1. Induce terminal anesthesia in the mouse by injecting Avertin (tribromoethanol; 400 mg/kg) intraperitoneally and wait until animal reaches surgical anesthesia. Test reflexes by toe pinching (*see* **Note 6**).
2. Make a small vertical incision inferior to the xiphoid process and cut through skin and muscle. Gently cut through diaphragm and ribs in the way to expose the heart.
3. Make a small incision in the right atrium and gently perfuse the heart with 5–10 mL of cold PBS by inserting a syringe needle (preferably 23–26 gauge) in the left ventricle through the apex (*see* **Note** 7).
4. Gently excise the heart, rinse in cold PBS on ice, remove atria if not needed, and clean from surrounding attached tissues.
5. Gently cut the ventricles into 2–3 mm pieces and place them in an Eppendorf tube (*see* **Note 8**).
6. Activate the collagenase type II solution by adding 10μL of 10 mM $CaCl_2$ for every 1 mL of enzyme solution. Mix well and add 500μL per heart of the activated enzyme solution to the cardiac tissue pieces.
7. Incubate at 37 °C for 30 min while gently rotating using a tube rotator.
8. After the 30 min incubation, gently mechanically dissociate/loosen any undigested tissue and mix the tissue (*see* **Note 9**).

9. Add 5 mL of ice-cold PBS and centrifuge at 850 rpm for 5 min at 4 °C to pellet the cells, then decant the supernatant.
10. Repeat **step 9** one more time.
11. During the spinning step, mix 500μL of collagenase D solution with 500μL of dispase II solution per heart and activate by adding 10μL of 10 mM $CaCl_2$ for every 1 mL of the collagenase D/dispase II solution.
12. Resuspend the cells in the collagenase D/dispase II solution (1 mL per heart) and incubate for 60 min at 37 °C while gently rotating using a tube rotator (*see* **Note 10**).
13. After 60 min, add 5 mL of FACS buffer and mix gently but thoroughly with a disposable bulb pipet (*see* **Note 11**).
14. Filter the cell suspension through a 40μm cell strainer placed on top of a 50 mL conical tube.
15. Wash the tube where the cell suspension was in and the cell strainer with another 5 mL of FACS buffer and add them to the cell suspension.
16. Centrifuge at 500 × *g* for 5 min at 4 °C to pellet the cells, then decant the supernatant (*see* **Note 12**).
17. Repeat **step 16** one more time.
18. Proceed immediately with the FACS protocol detailed below.

3.2 Fluorescence-Activated Cell Sorting (FACS) of (Myo) Fibroblasts for scRNAseq

1. Incubate with the conjugated CD45 and CD31 antibodies in 2 mL of FACS buffer for 20–30 min on ice (*see* **Note 13**).
2. Wash twice with FACS buffer, spinning at 500 × *g* for 5 min at 4 °C and decanting supernatant after each wash.
3. Reconstitute cells with 500μL FACS buffer containing Hoechst 33342 (4μM) and propidium iodide (1 mg/mL) and incubate for 15 min.
4. Filter the cell suspension into a 5-mL round-bottom FACS tube with 35μm cell strainer snap cap (*see* **Note 14**).
5. Insert the sample in a FACS machine and start acquiring events. Gate cells based on their forward and side scatter (*see* **Note 15**).
6. Then gate on the PI negative and Hoechst 33342 mid events (*see* **Note 16**).
7. Gate on CD31/CD45 negative and GFP positive (if using Pdgfra H2BeGFP mice) events and collect those cells. Alternatively, all cells from **step 6** may be collected if immune and endothelial cells are to be included in the scRNAseq. If C57Bl/6J mice are used instead, then either a fluorescently labeled anti-PDGFRa antibody could be used or more simply all cells that are CD31/CD45 negative may be collected. In the latter

case, identification of fibroblasts would be then performed computationally by analyzing the RNA sequencing results (*see* **Note 17**).

8. Collect cells in polypropylene or polystyrene 5-mL round-bottom FACS tubes containing 1 mL sorting buffer for a preliminary sort.
9. Perform a second sort with the collected cells. More stringent gating strategy and gates should be used this time to increase purity of the cell suspension and remove debris. If the target cell population is known to be fragile or cell numbers are low in case of rare populations, then we recommend omitting the preliminary sort.
10. Collect cells in 35μL of sterile filtered sorting media in a microcentrifuge tube.
11. Mix the sorted cell suspension well with trypan blue solution in a 1:1 ratio. Load 20μL to a hemocytometer and count viable cells per μL (*see* **Note 18**).
12. Calculate the volume required to reach the target number of cells intended to be recovered and sequenced (*see* **Note 19**).

3.3 Preparation of Tissue Sections for Spatial Transcriptomic Profiling

1. Perform **steps 1–4** as described in Subheading 3.1.
2. Once the heart is harvested and washed with cold PBS, make sure to remove any excess connective tissue and any attached unwanted tissue.
3. Embed the heart in a disposable plastic cryomold filled with Optimal cutting temperature (OCT) compound, which is a medium for frozen tissue specimens to ensure optimal cutting temperature. Note the heart orientation in OCT (*see* **Note 20**).
4. Freeze the embedded heart by slowly lowering the cryomold into a metal canister of isopentane chilled by liquid nitrogen (*see* **Note 21**).
5. Store the frozen blocks in a sealable bag at −80 °C overnight and seal the bag.
6. Prepare the cryostat and mount the tissue block in the correct orientation for cross-sections (*see* **Note 22**).
7. Remove the OCT until the tissue is visible. Resize the tissue block to no more than 6.5 mm × 6.5 mm by cutting away OCT around the tissue (*see* **Note 23**).
8. Cut sections of 10μm thickness using an anti-roll plate (*see* **Note 24**).
9. To adhere sections onto the capture area of a Visium slide, the slide is positioned very close to the section, within a few millimeters, while being held from the edges with the thumbs and middle fingers and the index finger is placed on the back of the

slide where the 6.5 × 6.5 mm capture area is, to melt the section as it adheres to the slide. Keep the slide on the cryobar in the cryostat until sectioning is completed. Use a paintbrush to carefully move sections on the stage if needed (*see* **Note 25**).

10. Quickly transfer the slides in a sealable conical slide holder to store at −80 °C (*see* **Note 26**).

3.4 Computational Analysis of scRNAseq to Identify Fibroblast Subsets

This section describes a general workflow to analyze scRNAseq data obtained using the 10× Genomics platform with Seurat. We will use a dataset that we published recently [14] and is publicly available (Gene Expression Omnibus [GEO: GSE141929]). The dataset was generated from scRNAseq of FACS-sorted Pdgfra H2BeGFP+ cells from healthy hearts and hearts that underwent left anterior descending coronary artery (LAD) ligation and collected 7 d later. LAD ligation is a model of acute myocardial infarction that leads to death of myocardium and subsequent replacement by scar tissue. This injury model also allows the studying of fibroblast and myofibroblast heterogeneity following damage. The workflow provides preprocessing steps, including data retrieval, HTO demultiplexing, and quality control (QC), followed by a series of user-dependent exploratory analyses. Principal component analysis (PCA) is a linear dimensionality reduction technique that summarizes data by finding fitted dimensions through a collection of points while preserving data variance. From the PCA space, cells are clustered using K-nearest neighbors, an algorithm that classifies groups of data points based on similarity. Cell clusters are projected into two dimensions with uniform manifold approximation and projection (UMAP), a non-linear dimensionality reduction algorithm, to illustrate global similarity among clusters [29]. The workflow also provides downstream analyses such as differential gene expression and various graphical representations of data.

1. To install the latest version of Seurat, enter in R or RStudio and perform the command:

```
install.packages("Seurat")
```

2. Likewise, install all other required software packages:

```
install.packages("dplyr")
install.packages("RColorBrewer")
install.packages("ggplot2")
install.packages("cowplot")
```

3. Load all required packages:

```
library(Seurat)
library(dplyr)
```

```
library(ggplot2)
library(RColorBrewer)
library(cowplot)
```

4. Use Read10× to load the feature-barcode matrix outputted by the 10× Genomics pipeline, Cell Ranger, into R by setting the directory to the appropriate file location. This will create two matrices, one for gene expression and the other for HTOs. Users may skip demultiplexing with HTOs if not applicable (**steps 7–12**).

```
sampleData <- Read10X(data.dir = "E:/Springer review/Sample
data")
```

5. Optional: If merging or integration of multiple datasets is required, prefix cell barcodes with sample-specific names as some may be shared across datasets.

```
colnames(x = exprData) <- paste("sameplData", colnames(x =
exprData), sep = "_")
colnames(x = HTOData) <- paste("sameplData", colnames(x =
HTOData), sep = "_")
```

6. Convert the gene expression matrix to a Seurat object with CreateSeuratObject (*see* **Note 27**).

```
exprData <- CreateSeuratObject(counts = sampleData$'Gene
Expression', min.cells = 3, min.features = 200, project =
"sampleData")
```

7. Subset the HTO matrix using cell barcodes shared with the gene expression matrix.

```
sharedCells <- intersect(colnames(exprData),
colnames(sampleData$'Antibody Capture'))
HTOData <- sampleData$'Antibody Capture'[, sharedCells]
```

8. To process HTO data, append the HTO matrix to the Seurat object using CreateAssayobject and normalize HTOs with NormalizeData via center log-ratio transformation (CLR). Compute for HTO thresholds and annotate cells as doublets, negatives, and singlets with HTODemux (*see* **Note 28**).

```
exprData[["HTO"]] <- CreateAssayObject(counts = HTOData)
exprData <- NormalizeData(exprData, assay = "HTO",
normalization.method = "CLR")
exprData <- HTODemux(exprData, assay = "HTO", positive.quan-
tile = 0.99)
table(exprData$HTO_classification.global)
```

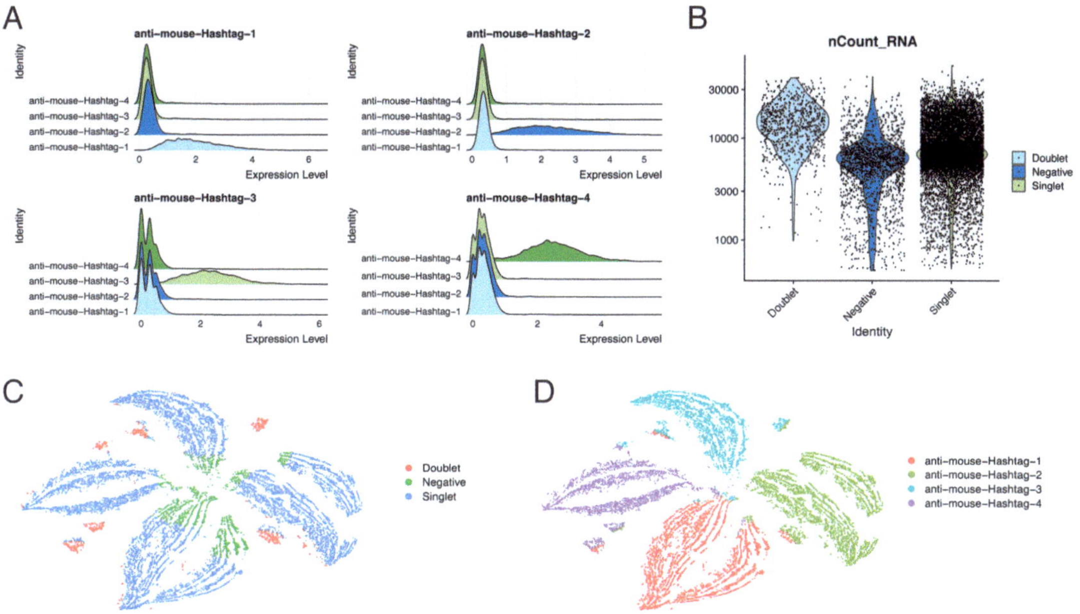

Fig. 1 Demultiplexing with HTOs. (**a**) Ridge plot depicting the level of HTOs detected across groups. (**b**) Violin plot depicting the distribution of HTO annotations. (**c**) tSNE plot colored by annotations. (**d**) tSNE plot colored by HTO groups

9. To visualize HTO levels and annotations, set the active identification to the highest HTO of each cell and active assay to the HTO matrix. Cells are identified and allotted to groups based on the highest HTO detected. HTO levels of each group and their exclusivity can be visualized with a ridge plot (Fig. 1a).

```
Idents(exprData) <- "HTO_maxID"
exprData@active.assay <- "HTO"
RidgePlot(exprData, assay = "HTO", features =
rownames(exprData[["HTO"]])[1:4], ncol = 2, group.by =
"HTO_maxID")
```

10. The distribution of annotations (doublets, negatives, and singlets) can be shown using a violin plot (Fig. 1b).

```
VlnPlot(exprData, features = "nCount_RNA", pt.size = 0.1, log
= TRUE)
```

11. To show the relationship between annotations and HTO levels, HTO-based distance matrix is computed to generate t-distributed stochastic neighbor embedding (tSNE) plots colored by the HTO identification of each cell together with the annotation (Fig. 1a, d).

```
hto.dist.mtx <- as.matrix(dist(t(GetAssayData(object = ex-
prData,
```

```
assay = "HTO"))))
exprData_tSNE <- RunTSNE(exprData, distance.matrix = hto.dist.
mtx,
perplexity = 100)
DimPlot(exprData_tSNE, group.by = "HTO_classification.glo-
bal")
DimPlot(exprData_tSNE)
```

12. To subset out negatives and doublets, set the active identification to HTO annotations and filter for singlets with subset (*see* **Note 28**). This marks the end of demultiplexing with HTOs. Be sure to set the active assay back to gene expression and return cell identifications to default before moving onto the next step.

```
 Idents(exprData) <- "HTO_classification.global"
exprData <- subset(exprData, idents = "Singlet")
exprData@active.assay <- "RNA"
Idents(exprData) <- "orig.ident"
```

13. In addition to HTOs, low-quality cells are filtered out based on three metrics: number of expressed genes/cell (nFeature_RNA), number of total counts/cell (nCount_RNA), and % mitochondrial gene counts/cell (percent.mt). PercentageFeatureSet is used to compute and store percent.mt in the Seurat object (*see* **Note 29**). Distribution of each metric can be visualized with a violin plot (Fig. 2a). Similarly, the relationship between two metrics can be summarized with a scatter plot (Fig. 2b). With subset, users may choose thresholds to remove potential noise/outliers (*see* **Note 30**).

```
    exprData[["percent.mt"]] <- PercentageFeatureSet(exprDa-
ta, pattern
= "^mt-")
VlnPlot(exprData, features = c("nFeature_RNA", "nCount_RNA",
```

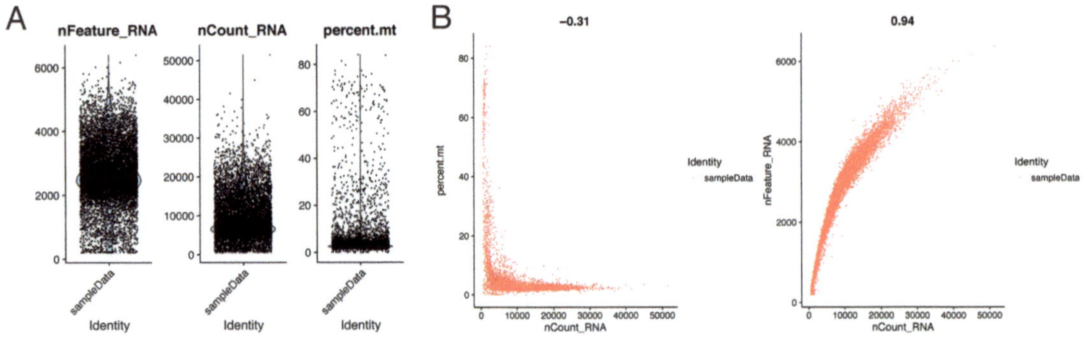

Fig. 2 QC of scRNAseq gene expression data. (**a**) Violin plot summarizing the distribution of QC metrics. (**b**) Scatter plot showing the relationship between QC metrics

```
"percent.mt"), ncol = 3)
plot1 <- FeatureScatter(exprData, feature1 = "nCount_RNA",
feature2 = "percent.mt")
plot2 <- FeatureScatter(exprData, feature1 = "nCount_RNA",
feature2 = "nFeature_RNA")
CombinePlots(plots = list(plot1, plot2))
exprData <- subset(exprData, subset = nFeature_RNA > 200 &
percent.mt < 10 & nCount_RNA < 25000)
```

14. Optional: Adaptively-thresholded low rank approximation (ALRA) is a method based on low-rank approximation for imputing technical dropouts (transcripts that are expressed, usually at low levels, but that are not detected by the platform, 10× Genomics platform in this case, and hence assigned a zero expression value) in scRNAseq data [30], while preserving genes not biologically expressed (biological zeros) at zero levels. Here we show the imputed expression for fibroblast activation protein (*Fap*), a proposed marker of activated fibroblasts [31], is increased by ALRA (Fig. 3a).

```
    exprData.ALRA <- RunALRA(object = exprData)
 a&lt;-RidgePlot(exprData, features = "Fap")) + NoLegend()
 b&lt;-RidgePlot(exprData.ALRA, features = "Fap")) + NoLegend
()
 a + b
```

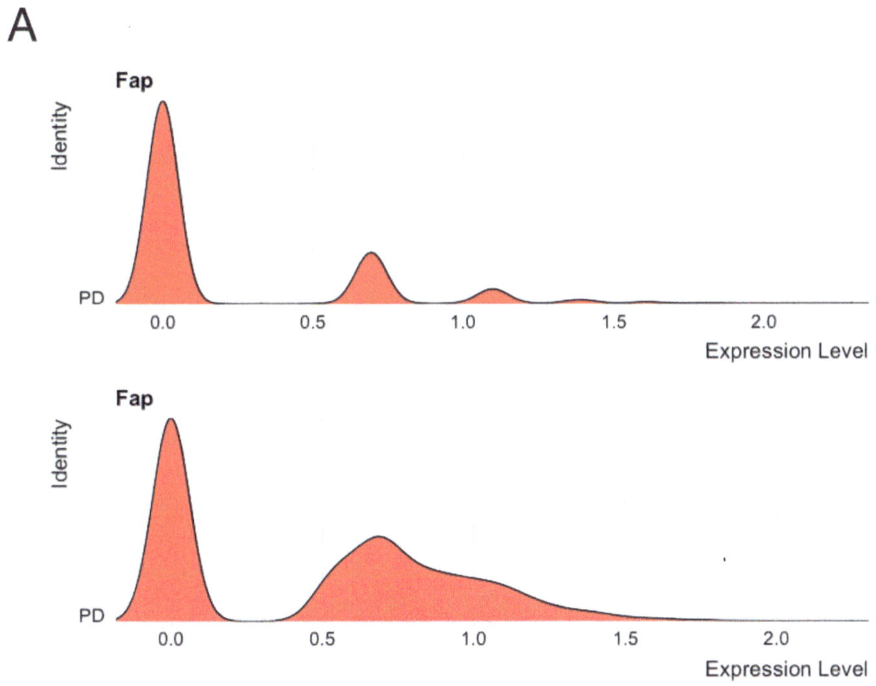

Fig. 3 Imputed expression by ALRA. (**a**) Ridge plot showing the expression of *Fap* before and after imputation

15. Count normalization is applied to individual cells using NormalizeData. The default method of normalization involves standardizing counts per gene to total counts per cell, followed by scaling to a factor of 10,000 and natural-log transformation.

```
exprData <- NormalizeData(exprData, normalization.method =
"LogNormalize", scale.factor = 10000)
```

16. To identify the top variable genes in the dataset, FindVariableFeatures is used. The default method involves variance-stabilization transformation to compute standardized variance of gene expression and selects the top 2000 variable genes, which will be used for PCA in **step 19**.

```
exprData <- FindVariableFeatures(exprData, selection.method =
"vst", nfeatures = 2000)
```

17. Optional: The Seurat workflow also includes several methods to analyze multiple scRNAseq datasets. In cases where variables such as experimental conditions are different, integration aims to identify cells of similar type and state based on mutual nearest neighbors [16]. This approach establishes pairing of cells, or "anchors," across datasets in a shared space (*see* **Note 31**).

```
anchors <- FindIntegrationAnchors(object.list = list(control,
experimental), dims = 1:20)
exprData <- IntegrateData(anchorset = anchors, dims = 1:20)
```

18. As a prerequisite for PCA, genes are scaled using ScaleData. Effects associated with technical features such as "nCount_RNA" and "percent.mt" can be regressed out if supplied to the argument "vars.to.regress" (*see* **Note 32**).

```
exprData <- ScaleData(exprData, vars.to.regress = c("percent.
mt",
"nCount_RNA"), features = rownames(exprData), block.size =
2000)
```

19. To compute PCA, perform RunPCA and plot the percentage of variance in data explained by each component using ElbowPlot (Fig. 4a). The inflection point, the "elbow," indicates the number of dimensions where most of the data variance is captured. This value will be supplied to clustering (**step 20**) and dimensionality reduction via UMAP (**step 21**). In this workflow, 20 PCs are selected.

```
    exprData <- RunPCA(exprData, features = VariableFeatures
(object =
exprData), npcs = 30, verbose = F)
ElbowPlot(exprData, ndims = 30)
```

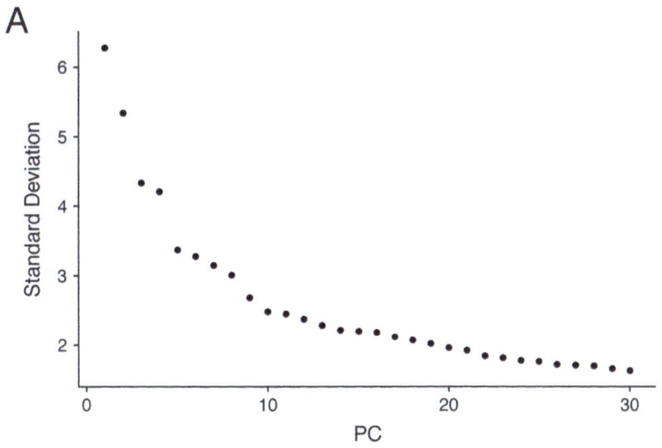

Fig. 4 PCA. (**a**) Elbow plot depicting the % of data variability accounted by each PC

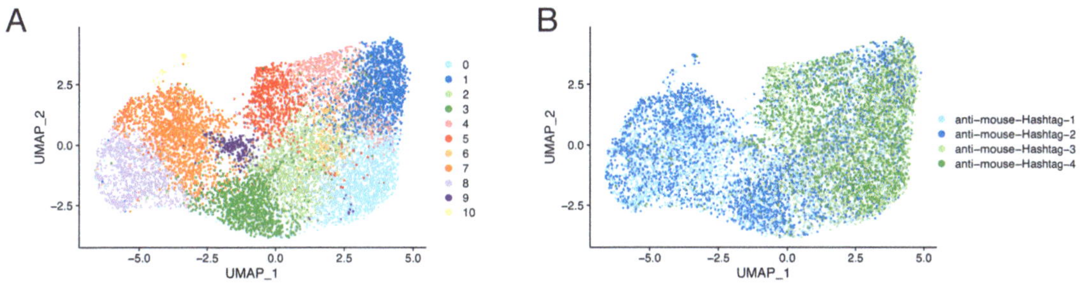

Fig. 5 Clustering and dimensionality reduction with UMAP. (**a**) UMAP displaying clustering of cells. (**b**) UMAP colored by HTO identification

20. Cluster cells using FindNeighbors and FindClusters (*see* **Note 33**). Specify the number of components determined in the previous step for "dims." Increasing "resolution" in FindClusters will result in more communities during the optimization of modularity, and thus increase the number of clusters.

```
exprData <- FindNeighbors(exprData, dims = 1:20, verbose =
F)
exprData <- FindClusters(exprData, resolution = 0.8, verbose =
F)
```

21. To portray clusters in two dimensions, UMAP is performed with RunUMAP, which is subsequently plotted with DimPlot (*see* Fig. 5a). Likewise, supply "dims" with the appropriate number of components. UMAP can also be colored by HTO identification (Fig. 5b). The separation between HTOs 1&2 versus 3&4 reflects the hashing procedure where cells from damaged hearts are hashed with 1&2 and those from healthy hearts are hashed with 3&4.

```
    exprData <- RunUMAP(exprData, dims = 1:20, verbose = F)
DimPlot(exprData, reduction = "umap", label = TRUE, pt.size =
1,
repel = T, label.size = 6)
DimPlot(exprData, reduction = "umap", label = TRUE, pt.size =
1,
repel = T, label.size = 6, group.by = "HTO_maxID")
```

22. Optional: The processed Seurat object can be saved and loaded with saveRDS and readRDS, respectively.

```
saveRDS(exprData, file = "E:/Springer review/Sample
analysis/computed_Seurat_object.rds")
exprData <- readRDS(file = "E:/Springer review/Sample
analysis/computed_Seurat_object.rds")
```

23. Normalized expression of genes across clusters is best presented using FeaturePlot to identify cell types and states (Fig. 6a). Here we show the expression of markers associated with fibro/adipogenic progenitors (FAPs) and/or their more fibrogenic progeny, the fibroblasts/myofibroblasts (Fib) [4, 9] fibroblast.

```
genesList = c("Pdgfra", "Col1a1", "Postn", "Acta2")
FeaturePlot(exprData, features = genesList, pt.size = 0.25,
reduction = "umap", min.cutoff = 0, max.cutoff = "q99", cols =
c("lightgrey", "blue"), ncol = 2)
```

24. Names are assigned to the clusters in the order from 0th to nth cluster and plotted on the UMAP (Fig. 6b). Users may wish to relevel the annotations to order them on the UMAP (code not shown).

```
    new.cluster.ids <- c("FAP-1", "Fib-1", "FAP-2", "FAP-3",
"FAP-4",
"Fib-2", "FAP-5", "FAP-6", "FAP-7", "Fib-3", "Fib-4")
names(new.cluster.ids) <- levels(exprData)
exprData <- RenameIdents(exprData, new.cluster.ids)
DimPlot(exprData, reduction = "umap", label = TRUE, pt.size =
1,
label.size = 6)
```

25. Other plots for visualizing gene expression can be generated as such (Fig. 7a–c):

```
    # Violin plot
VlnPlot(object = exprData, features = genesList, same.y.lims =
TRUE, sort = F, ncol = 4)
# Ridgeplot
RidgePlot(exprData, features = genesList, same.y.lims = T,
ncol = 4)
```

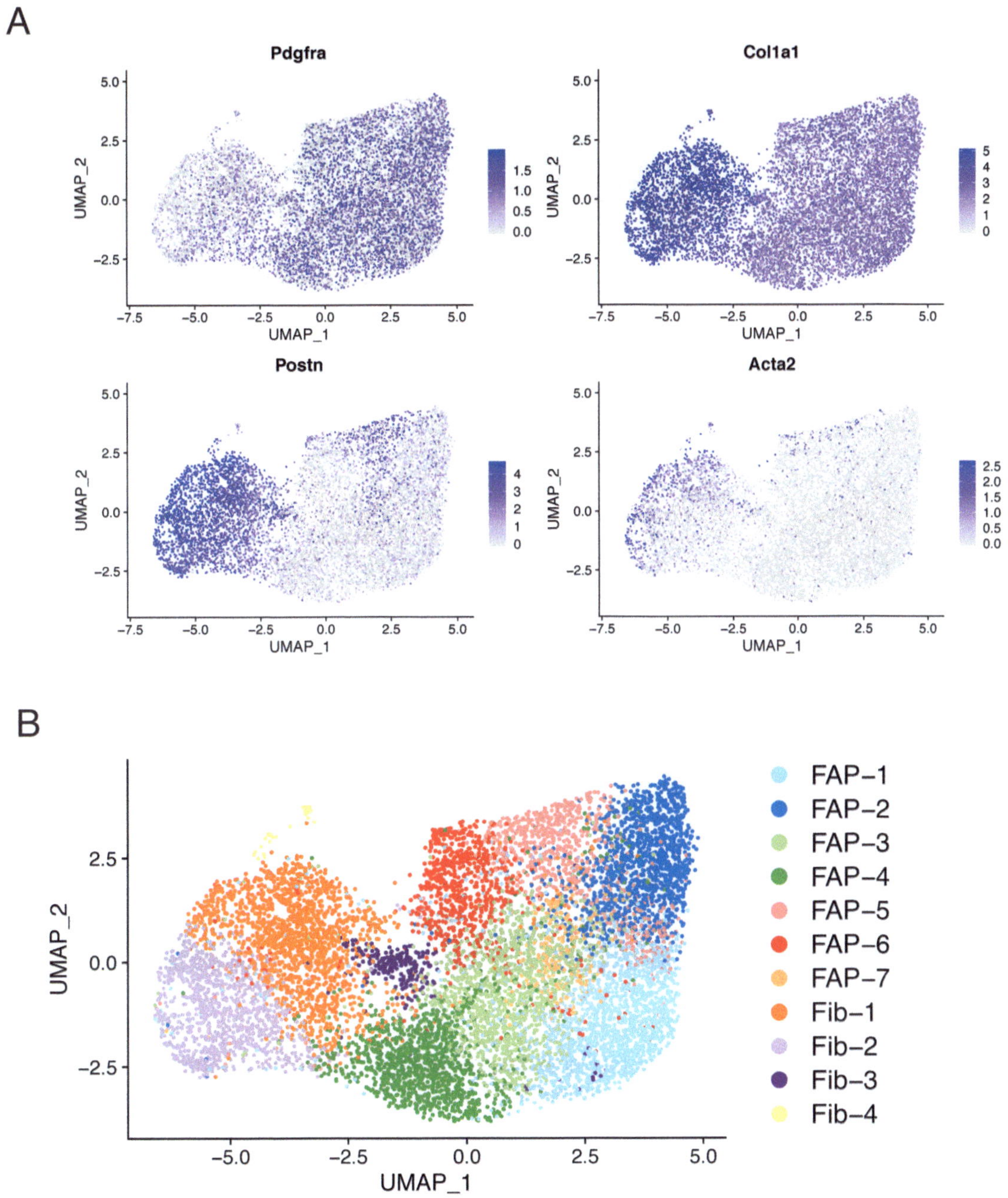

Fig. 6 Assignment of cell type and/or state based on marker expression. (**a**) Feature plots depicting markers associated with FAPs and/or fibroblast/myofibroblasts (Fib). (**b**) UMAP of the annotated clusters

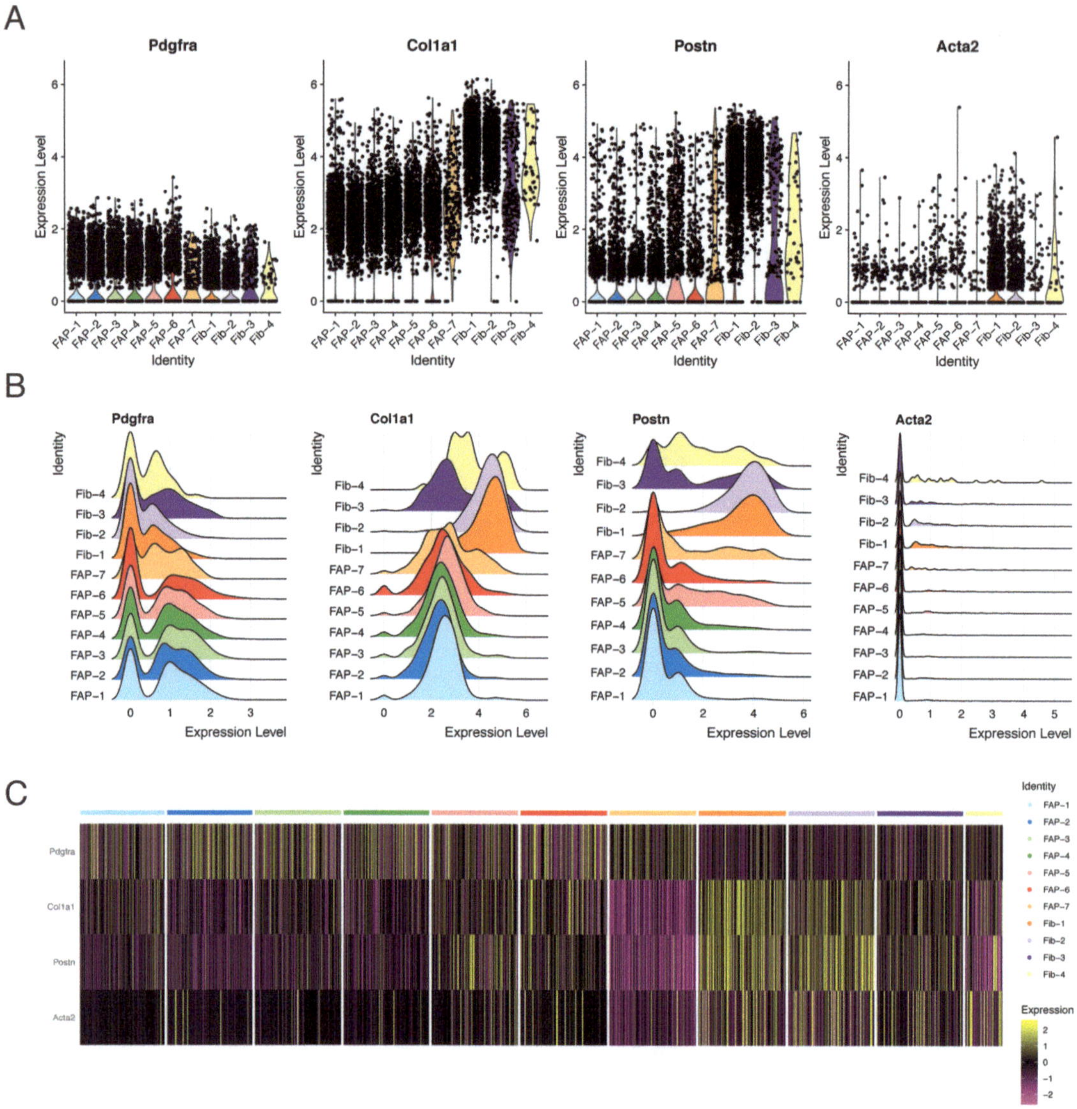

Fig. 7 Different types of visualizations on marker expression. (**a**) Violin plot displaying normalized expression of selected markers. (**b**) Ridge plot displaying normalized expression of markers. (**c**) Heatmap displaying scaled expression of markers

```
# Heatmap of scaled genes
DoHeatmap(subset(exprData, downsample = 100), features =
genesList, size = 3)
```

26. Differential gene expression between selected individual or groups of clusters is computed with FindMarkers. Arguments "ident.1" and "ident.2" are used to specify groups of comparison. In the example below, FAP-1 is compared with FAP-2. On the other hand, FindAllMarkers computes for differentially expressed genes (DEGs) for each cluster vs. all other clusters. "min.pct" can be adjusted to test for genes expressed by at least

a certain % of cells per cluster. Setting "only.pos" to true will return only positive DEGs. Users may check the expression of DEGs on feature plots.

```
clusterMarkers <- FindMarkers(exprData, ident.1 = c("FAP-
1"),
 ident.2 = c("FAP-2"),
 min.pct = 0.25, only.pos = T)
 allClusterMarkers <- FindAllMarkers(exprData, only.pos =
TRUE,
min.pct = 0.25, logfc.threshold = 0.25, verbose = F)
```

27. Specific clusters can be subsetted with SubsetData and must be reanalyzed starting from **step 16**.

```
subsettedData <- SubsetData(object = exprData, ident.use = c
("FAP-1", "FAP-2", "FAP-3"))
```

3.5 Spatial Transcriptomic Analysis with Visium

This section describes a workflow for analysis of spatially resolved RNA sequencing data obtained using the 10× Genomics spatial transcriptomics platform, Visium. Similar to the workflow on the analysis of scRNAseq data described in Subheading 3.4, the workflow provides steps in processing spatial transcriptomic data, followed by basic exploratory analyses. However, instead of log normalization, a recent normalization method by Hafemeister and Satija [32], SCTransform, is adopted here. SCTransform models gene expression with regularized negative binomial regression and outperforms the standard approach of log normalization as the latter does not take biological heterogeneity across the tissue into account. In addition to differential gene expression, the identification of spatially variable genes (SVGs) across regions of the tissue is also described. It is important to note that each spot on the Visium slide does not represent a single cell. In fact, each spot is 55μm in diameter and may include a variable number of cells depending on the size of cells in the tissue. We will use a dataset generated from a heart 7 days after being subjected to ischemic injury by LAD ligation.

1. If all packages are installed as described in Subheading 3.4, load the required packages:

```
library(Seurat)
library(dplyr)
library(ggplot2)
library(RColorBrewer)
library(cowplot)
```

2. First, load the image and filtered gene-spot matrix from the appropriate directory with Read10×_Image and Load10×_-Spatial, respectively (*see* **Note 34**).

```
mage_dir <- 'E:/Springer review/Visium sample data/spa-
tial'
```

```
Read10X_Image(image.dir=image_dir, filter.matrix = TRUE)
data_dir <- 'E:/Springer review/Visium sample
data/spatial/gene_spot_matrix'
Spatial_Heart<-Load10X_Spatial(data.dir = data_dir,
assay="Spatial", filename = "filtered_feature_bc_matrix.h5",
slice
= "Spatial_Heart", filter.matrix = T)
```

3. To observe the variation in the number of expressed genes/spot (nFeature_Spatial) and number of total counts/spot (nCount_Spatial) across the dataset, we can look at their distributions using a violin plot or a spatial feature plot (Fig. 8a, b).

```
p1<-VlnPlot(object = Spatial_Heart, features =
c("nFeature_Spatial"), ncol = 2)+FontSize(0)+NoLegend()
p2<-VlnPlot(object = Spatial_Heart, features =
c("nCount_Spatial"))+FontSize(0)+NoLegend()
```

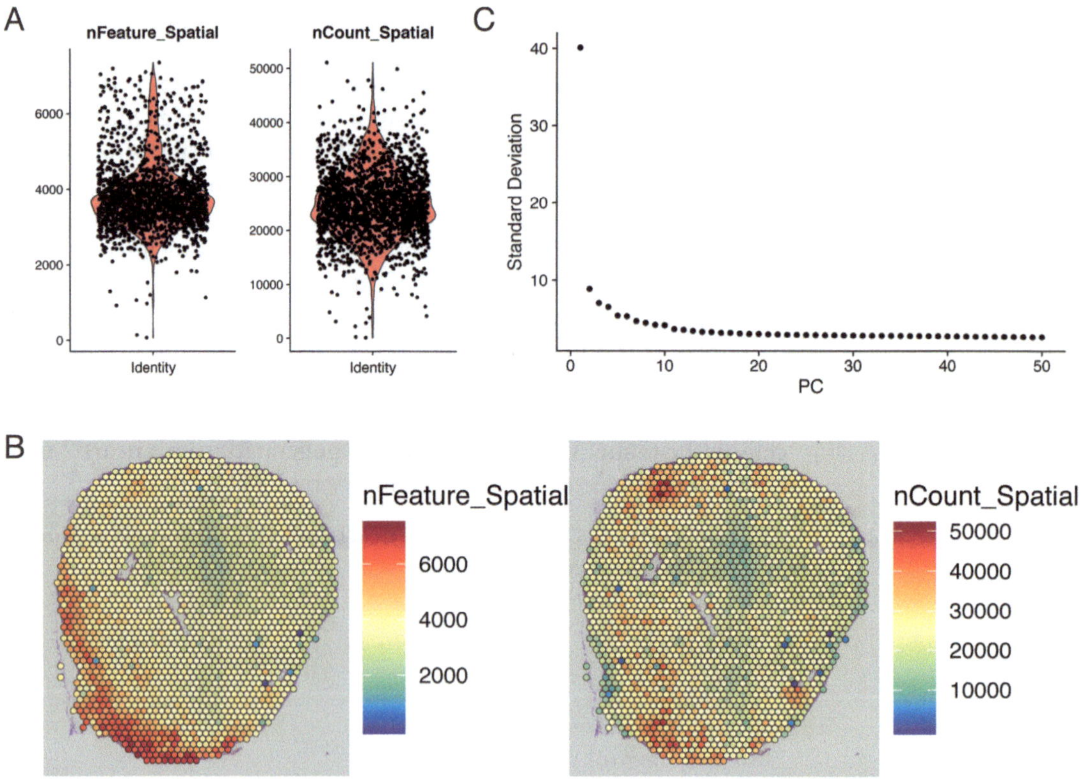

Fig. 8 QC of Visium gene expression data and PCA. (**a**) Violin plot summarizing the distribution of QC metrics. (**b**) Spatial feature plot displaying QC metrics on the tissue section. (**c**) PCA elbow plot depicting % data variability accounted PCs

```
    plot(p1+p2)
p3<-SpatialFeaturePlot(Spatial_Heart, features =
"nFeature_Spatial", pt.size.factor = 2) + theme(legend.posi-
tion =
"right")
p4<-SpatialFeaturePlot(Spatial_Heart, features = "nCount_Spa-
tial",
pt.size.factor = 2) + theme(legend.position = "right")
p3+p4
```

4. To normalize data, SCTransform is performed. This function also substitutes the ScaleData and FindVariableFeatures functions (*see* **Note 35**).

```
Spatial_Heart <- SCTransform(Spatial_Heart, assay = "Spatial",
verbose = FALSE)
```

5. Next, PCA is computed on the SCT assay, which stores normalized counts after SCTransform. To optimize the number of PCs for downstream steps, plot the % data variance accounted by PCs with ElbowPlot (Fig. 8c).

```
 Spatial_Heart <- RunPCA(Spatial_Heart, assay = "SCT",
verbose = FALSE)
p5<-ElbowPlot(object = Spatial_Heart, ndims = 50)
p5
```

6. Visium spots are clustered using the FindNeighbors and FindClusters functions.

```
 Spatial_Heart <- FindNeighbors(Spatial_Heart, reduction =
"pca", dims = 1:20, nn.method = "rann", annoy.metric =
"euclidean")
 Spatial_Heart <- FindClusters(Spatial_Heart, verbose = FALSE)
```

7. To project clusters into two dimensions, RunUMAP is performed, followed by DimPlot and SpatialDimPlot to plot clusters on the UMAP and tissue section, respectively (Fig. 9a) (*see* **Note 36**).

```
    Spatial_Heart <- RunUMAP(Spatial_Heart, reduction = "pca",
dims =
1:20, verbose = FALSE)
 p6 <- DimPlot(Spatial_Heart, reduction = "umap", label =
TRUE,
pt.size = 1, label.size=6) + NoLegend()
 p7 <- SpatialDimPlot(Spatial_Heart, label = T, label.size =
4,
crop = T, alpha = c(0.1, 1.5), pt.size.factor = 2, stroke =
0.1) +
NoLegend()
p6+p7
```

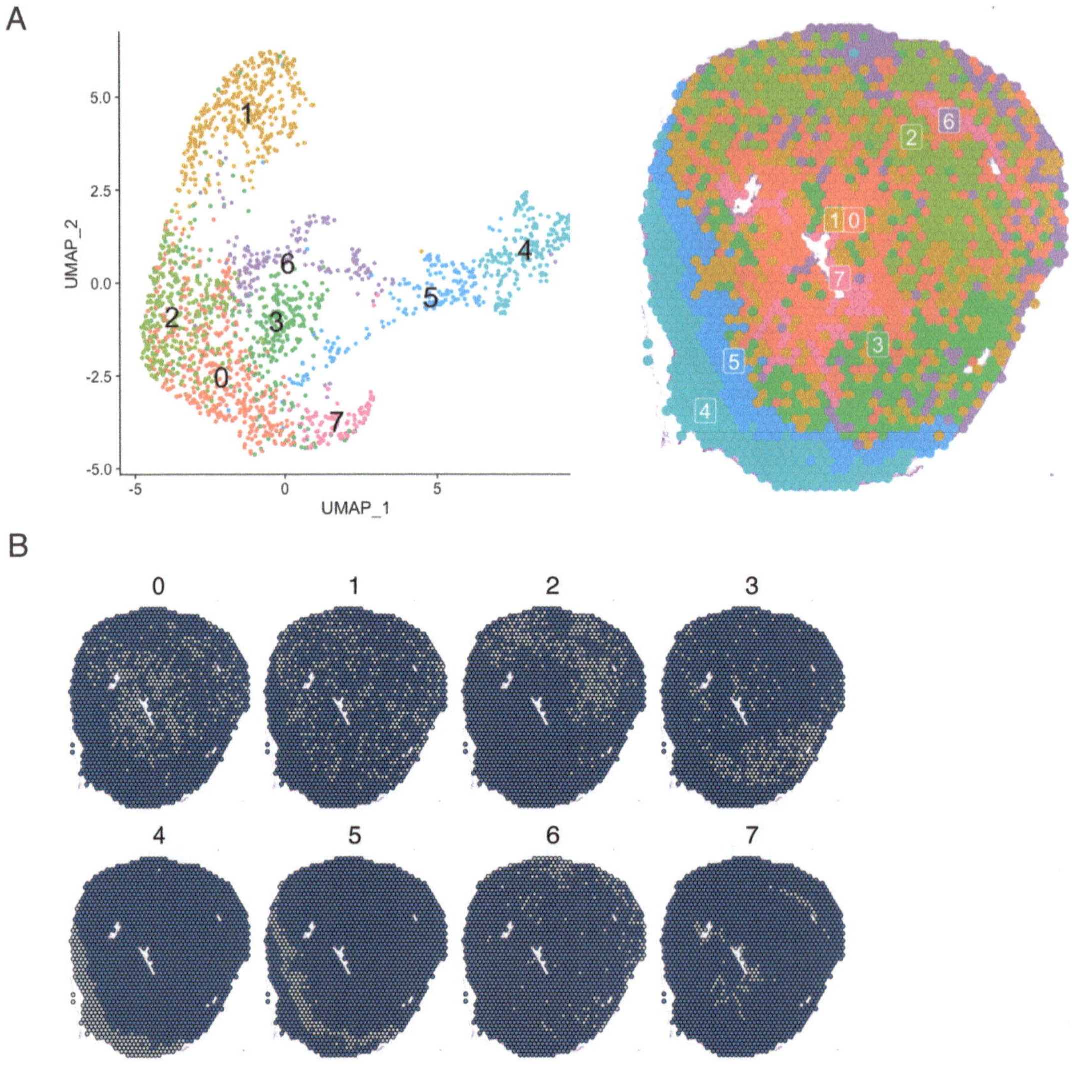

Fig. 9 Clustering, UMAP, and annotated tissue section. (**a**) UMAP displaying clustering of spots (left) and tissue section overlaid with clustered spots (right). (**b**) Highlighting spots of each cluster on the tissue section

8. To visualize each cluster separately, we can supply the argument "cell.highlight" in SpatialDimPlot with cluster annotations (Fig. 9b).

```
SpatialDimPlot(Spatial_Heart, cells.highlight =
cellsByIdentities(object = Spatial_Heart), facet.highlight =
TRUE,
crop = T, ncol=3,alpha = c(0.1, 1.5), pt.size.factor = 2, stroke
=
0.5)
```

9. Normalized expression of marker genes described in Subheading 3.4 can be explored with FeaturePlot and SpatialFeaturePlot (Fig. 10a, b).

```
genesList = c("Pdgfra", "Col1a1", "Postn", "Acta2")
p8 <- SpatialFeaturePlot(object = Spatial_Heart, features =
genesList, alpha = c(0.1, 1.5), crop = T, pt.size.factor = 2,
stroke = 0.1, max.cutoff = "q99", combine = F)
for(i in 1:length(p8)) {
 p8[[i]] &lt;- p8[[i]] + scale_fill_gradient(low= "yellow",
high =
"red3")
}
cowplot::plot_grid(plotlist = p8)
 FeaturePlot(Spatial_Heart, features = genesList, pt.size =
0.25,
reduction = "umap", min.cutoff = 0, max.cutoff = "q99", cols =
c("lightgrey", "blue"), ncol = 2)
```

10. To determine DEGs among clusters, FindMarkers is applied. In the example below, DEGs in cluster 4, the infarction zone, are obtained from the comparison against all other clusters. However, groups can be specified if supplied to both "ident.1" and "ident.2." The expression of DEGs is checked with SpatialFeaturePlot and FeaturePlot. Only the top 6 DEGs are plotted here (Fig. 11a, b).

```
Spatial_Heart_c4 <- FindMarkers(Spatial_Heart, ident.1 =
4,
verbose = F)
p9<-SpatialFeaturePlot(object = Spatial_Heart, features =
rownames(Spatial_Heart_c4)[1:6], alpha = c(0.1, 1.5), max.
cutoff =
"q99", combine = F)
for(i in 1:length(p9)) {
 p9[[i]] &lt;- p9[[i]]+scale_fill_gradient(low = "yellow",
high =
"red3")+theme(legend.position = "right")
}
cowplot::plot_grid(plotlist = p9)
 FeaturePlot(Spatial_Heart, features =
rownames(Spatial_Heart_c4[1:6,]), pt.size = 0.25, reduction =
"umap", min.cutoff = 0, max.cutoff = "q99", cols = c("light-
grey",
"blue"), ncol = 3)
```

11. To determine SVGs, FindSpatiallyVariableFeatures is applied. Likewise, their expression is checked with SpatialFeaturePlot and FeaturePlot. Only the top 8 SVGs are plotted here (Fig. 12a, b).

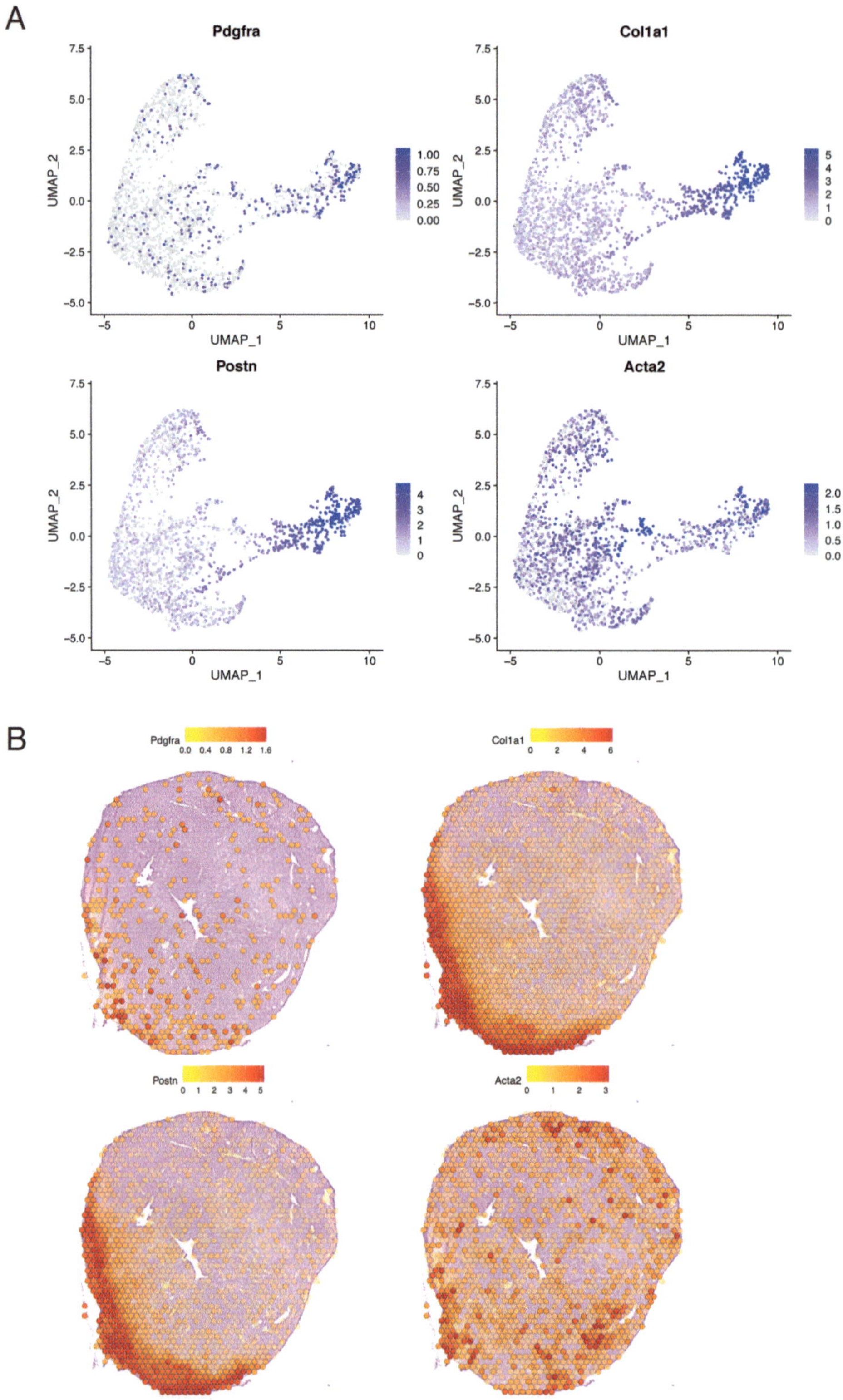

Fig. 10 Visualizing the expression of markers. (**a**) Feature plots depicting the expression of markers on UMAP. (**b**) Expression of markers depicted on the tissue section

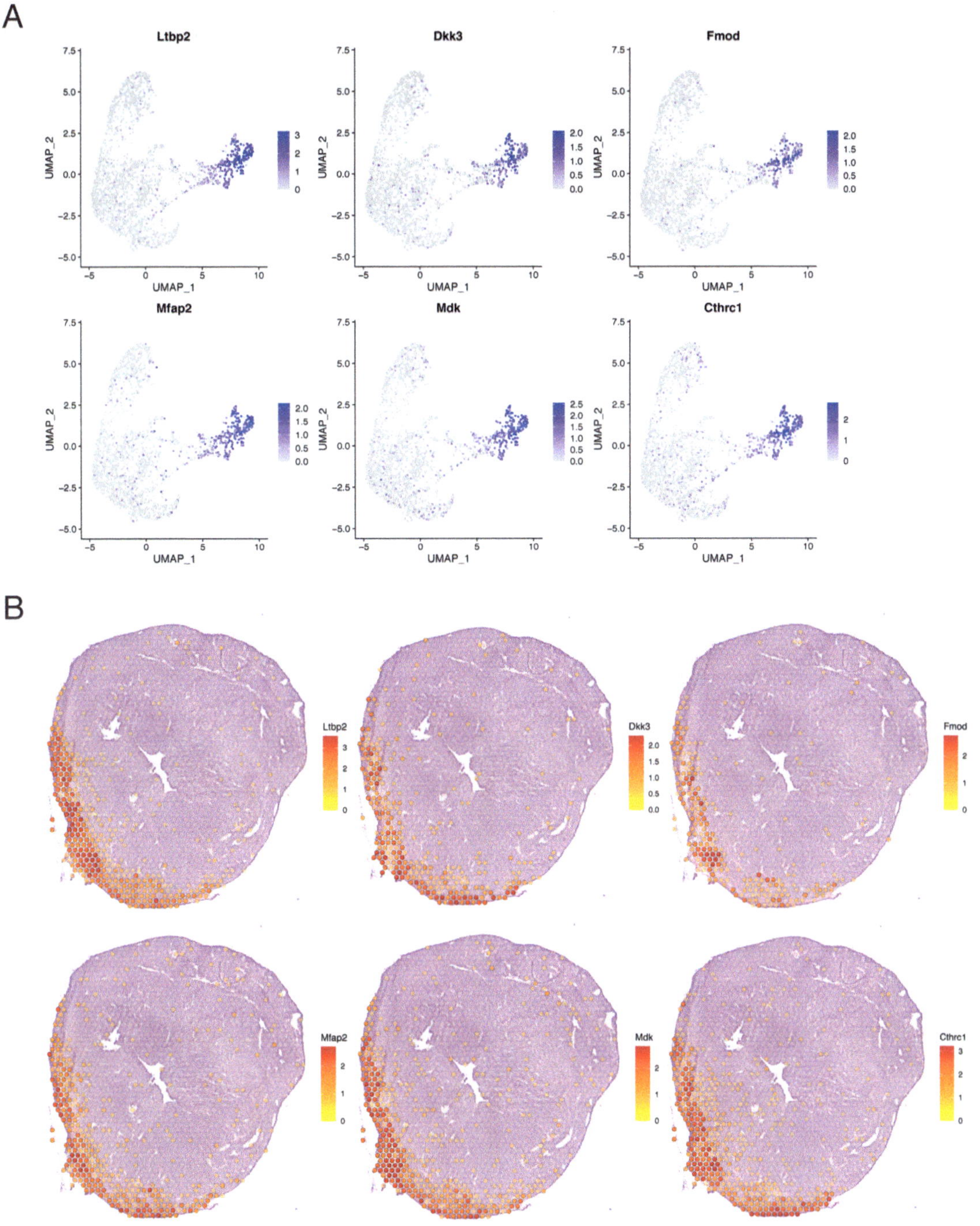

Fig. 11 Visualizing the expression of DEGs. (**a**) Feature plots depicting the expression of the top six DEGs in cluster 4 on the UMAP. (**b**) Expression of DEGs on the tissue section

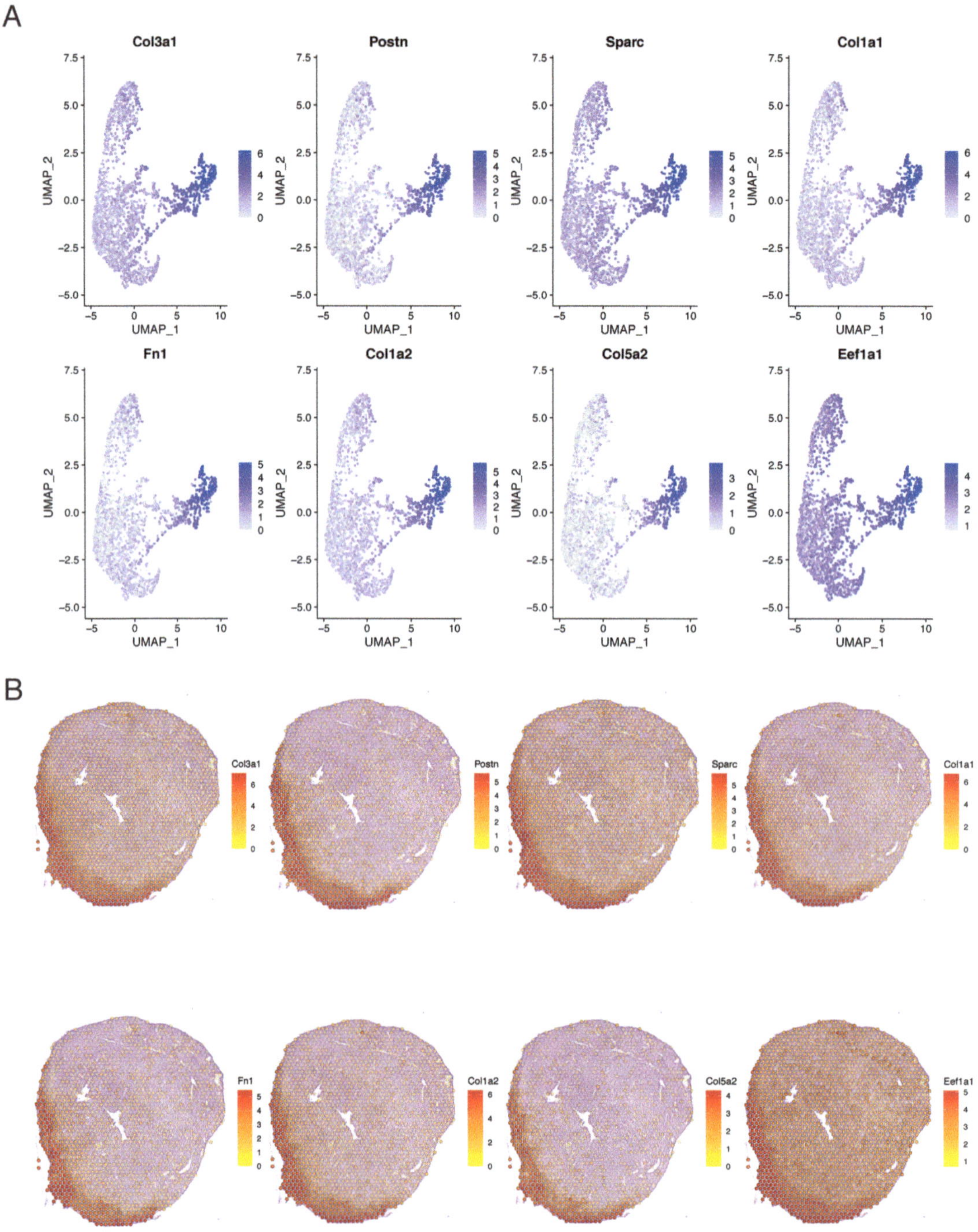

Fig. 12 Visualizing the expression of SVGs. (**a**) Feature plots depicting the expression of the top eight SVGs in cluster 4 on the UMAP. (**b**) Expression of SVGs on the tissue section

```
    Spatial_Heart <- FindSpatiallyVariableFeatures(Spatial_-
Heart,
assay = "SCT", features = VariableFeatures(Spatial_Heart)
[1:1000],
 selection.method = "markvariogram")
 Heart_var <- head(SpatiallyVariableFeatures(Spatial_Heart,
selection.method = "markvariogram"), 8)
p10<-SpatialFeaturePlot(Spatial_Heart, features = Heart_var,
alpha = c(0.1, 1.5), combine = F)
for(i in 1:length(p10)) {
 p10[[i]] &lt;- p10[[i]] + scale_fill_gradient(low= "yellow",
high
= "red3")+theme(legend.position = "right")
}
cowplot::plot_grid(plotlist = p10, ncol = 4)
FeaturePlot(Spatial_Heart, features = Heart_var, pt.size =
0.25,
reduction = "umap", min.cutoff = 0, max.cutoff = "q99", cols =
c("lightgrey", "blue"), ncol = 4)
```

12. Optional: If multiple Visium datasets are present and the user wishes to compare them, they can be merged with the merge function after SCTransform. Prior to merging, we add a new column called "group" to the metadata of each object to distinguish them after merging. Here we merge the section processed above with another section from the same damaged heart and compute for the top variable genes separately before appending to the merged object. This second section derived from an area closer to the apex of the heart (*see* **Note 37**). **Steps 5–9** are repeated with the merged object (code not shown) (Fig. 13).

```
Spatial_Heart@meta.data$group <- "Spatial_Heart"
Spatial_Heart2@meta.data$group <- "Spatial_Heart2"
Spatial_Heart_merge <- merge(Spatial_Heart,Spatial_Heart2)
VariableFeatures(Spatial_Heart_merge) <-
c(VariableFeatures(Spatial_Heart),
VariableFeatures(Spatial_Heart2))
```

13. Optional: Clusters of interest in a merged object can be subsetted and examined in the absence of other clusters. In the example below, clusters associated with the infarction and peri-infarction zones are subsetted (Fig. 14a) (*see* **Note 38**).

```
Infarction <- subset(Spatial_Heart_merge, idents = c(4,5,7))
SpatialDimPlot(Infarction, crop = T, label = TRUE, pt.size.
factor = 2, alpha = c(0.1,1.5), stroke = 0.1)
```

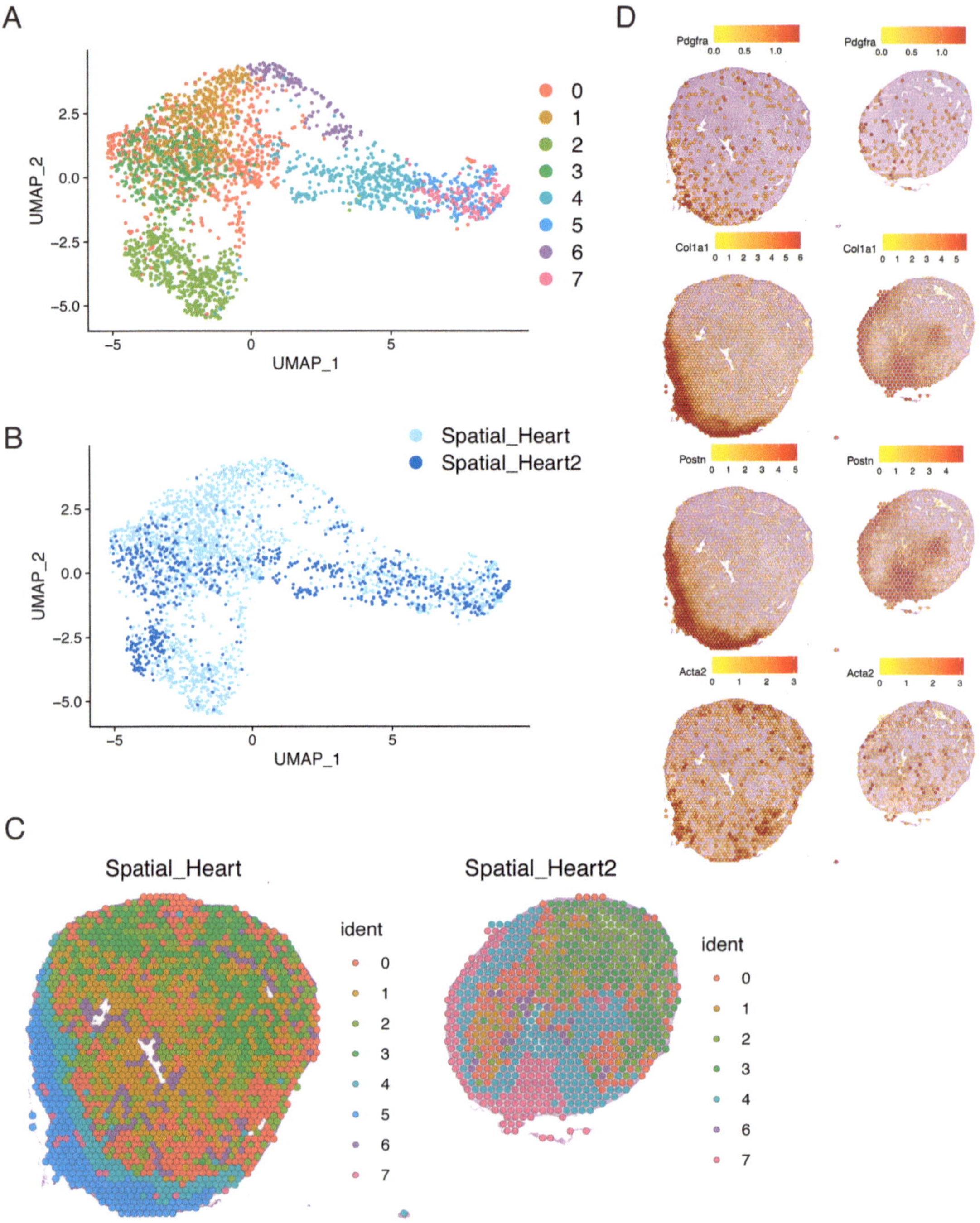

Fig. 13 Merging of Visium datasets. (**a**) UMAP of clusters on merged data. (**b**) UMAP of merged data colored by tissue section. (**c**) Clusters projected onto the tissue sections. (**d**) Expression of markers depicted on the tissue section

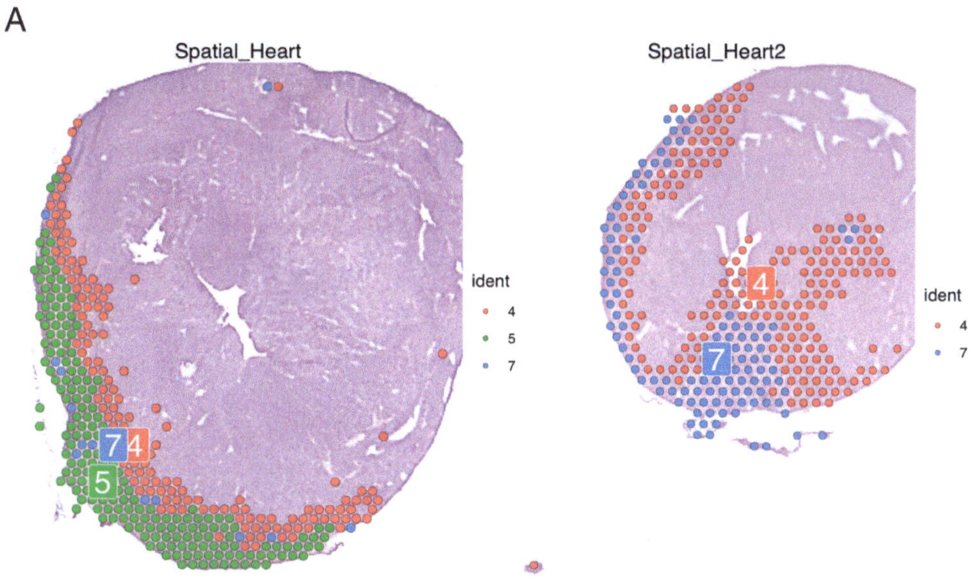

Fig. 14 Visualizing subsetted clusters of merged Visium data. (**a**) Subsetted clusters projected on tissue sections

3.6 Integration of scRNAseq and Spatial Transcriptomics Datasets

Although each data point consists of multiple cells that fit in individual Visium spot, Seurat offers "anchor-based" integration (described in **step 17** of Subheading 3.4) between Visium and scRNAseq datasets. This involves a probabilistic approach that transfers information such as cell annotations from a reference (scRNAseq) dataset to a query (Visium) dataset. In the example below, we loaded the processed Seurat objects from Subheadings 3.4 and 3.5 (code not shown) and applied integration. These are data collected from the same time point after LAD ligation.

1. To integrate scRNAseq and Visium datasets, perform label transfer with FindTransferAnchors and TransferData. This creates an assay containing the prediction scores that a class (cell annotation) is mapped to each Visium spot.

```
anchors <- FindTransferAnchors(reference = exprData, query
=
Spatial_Heart, normalization.method = "SCT")
predictions.assay <- TransferData(anchorset = anchors, refdata
=
exprData$seurat_clusters, prediction.assay = TRUE,
weight.reduction = Spatial_Heart[["pca"]])
```

2. Append the prediction assay to the integrated object and set it as the default assay for downstream analyses.

```
Spatial_Heart[["predictions"]] <- predictions.assay
DefaultAssay(Spatial_Heart) <- "predictions"
```

3. Plot the prediction scores for specific cell annotations using SpatialFeaturePlot. Here, we plot the prediction scores of selected FAP and fibroblast clusters (Fig. 15a, b).

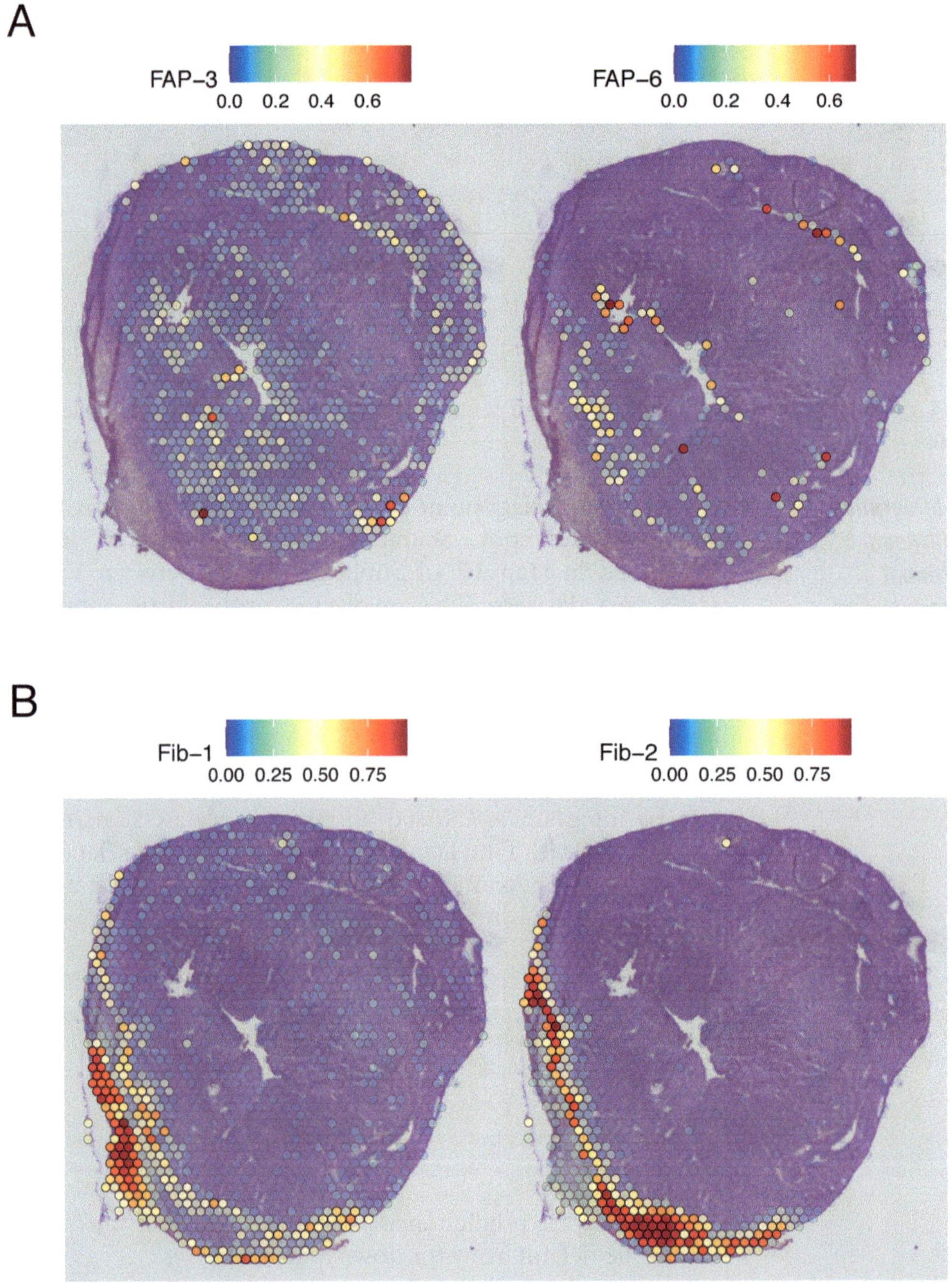

Fig. 15 Prediction scores of cell annotations. Prediction scores of selected FAP (**a**) and Fib (**b**) clusters on the tissue section

```
SpatialFeaturePlot(Spatial_Heart, features = c("FAP-3",
"FAP-6"),
pt.size.factor = 2, ncol = 2, crop = TRUE, alpha = c(0.1,1.5),
stroke=0.1, max.cutoff = "q99")
SpatialFeaturePlot(Spatial_Heart, features = c("Fib-1","
Fib-2"),
pt.size.factor = 2, ncol = 2, crop = TRUE, alpha = c(0.1,1.5),
stroke=0.1, max.cutoff = "q99")
```

4 Notes

1. One collagen digestion unit is defined as the amount that liberates peptides from collagen from bovine Achilles tendon equivalent in ninhydrin color to 1 μmol of leucine in 5 h at pH 7.4 at 37 °C in the presence of calcium ions. Calcium chloride is added fresh just before use to activate the enzyme. After preparing Collagenase II solution, aliquot and store at −20 °C.
2. It is recommended that only the amount of lyophilizate needed for immediate use be reconstituted. The reconstituted Collagenase D solution can be stored at −15 to −25 °C for up to 1 week. Avoid repeated freezing and thawing since activity decreases after reconstitution.
3. One unit of Dispase II is the enzyme activity which liberates Folin-positive amino acids and peptides (react with Folin-Ciocalteu reagent) corresponding to 1 μmol (181 μg) tyrosine in 1 min under assay conditions at pH 7.5 and 37 °C. Reconstituted solutions are stable at 2–8 °C for up to 2 weeks. For longer storage periods, aliquot and store at −20 °C.
4. Do not use PBS to reconstitute Hoechst 33342 to prepare stock solution as it may precipitate. However, PBS can be used to dilute working solutions of Hoechst 33342. Hoechst 33342 is a potential mutagen and should be handled with care. The dye must be disposed of safely and in accordance with applicable local regulations.
5. Propidium iodide is a potential mutagen and should be handled with care. The dye must be disposed of safely and in accordance with applicable local regulations.
6. Avertin is preferred over CO_2 and cervical dislocation as we find that the viability of dissociated cells is much better.
7. This step is best performed quickly to prevent the formation of intramural clots which reduce the viability of cells. Do not use forceps while trying to insert the needle in the apex of the heart as this will damage the outer surface of the heart.

8. The purpose of cutting into small pieces is to expose as much tissue as possible to the enzyme solutions. However, do not exaggerate cutting the heart as this will lead to excessive digestion of tissue when incubated with the enzyme solutions.
9. The purpose of this step is to loosen any undigested tissue to allow for the second enzyme solution to penetrate and dissociate further. This step is best done using the flat end of a 1 mL syringe plunger to gently mash the tissue after putting it in a small well/dish.
10. Ensure that at the beginning of the second digestion step with collagenase D/dispase II that you have loose tissue pieces instead of completely dissociated cells in suspension as this may lead to over digestion and poor viability.
11. After adding FACS buffer, you may observe that some cells clump together. If this happens, gently disperse the clump with a disposable bulb pipet.
12. We find that transferring the cell suspension back to a 15 mL tube following filtration and before spinning makes it much easier to observe the cell pellet.
13. 1 mL of cell suspension should contain approximately 1–2 million cells.
14. Pre-wetting the filter caps with two drops of FACS buffer is important to facilitate the filtration step.
15. Fibroblasts are situated between on 25,000–150,000 in the FSC axis. Gate on events in this range.
16. Viable cells are not permeable for PI. The Hoechst negative events are probably cell debris and the Hoechst high events are probably doublets.
17. Anti-PDGFRa antibodies do not always produce satisfactory staining. Make sure that you use a recently opened vial within a month that has been properly stored according to the manufacturer's instructions.
18. The recommended cell suspension concentration as specified by 10× Genomics is 700–1200 cells per μL of at least 70% viability. If the concentration is lower, spin the cells down at 500 × *g* for 5 min at 4 °C and adjust volume to reach desired concentration. If the suspension is too concentrated, add more sorting media to reach desired concentration.
19. The final volume that contains the desired number of cells depends on the cell viability, concentration of the cell suspension, and the efficiency of cell capture in droplets which is approximately 50–65%. For example, if the target number of cells for sequencing is 10,000 cells and the cell suspension concentration is 700 cell/μL with a 70% viability, the volume needed would be approximately 40μL of cell suspension

assuming a 50% cell capture. The maximum volume of cell suspension that could be loaded is 46.6μL but the volume may change with newer versions of the single cell kit.

20. Be sure to submerge the tissue completely in OCT and remove any air bubble with a needle to avoid cracks during sectioning.

21. Allow the OCT to freeze slowly from the bottom up. Do not submerge the cryomold completely until the whole block is frozen.

22. Begin with a chamber and objective temperature of −20 °C and −23 °C, respectively, and adjust according to the integrity of sections. Ensure all tools and slides are precooled in the cryostat chamber prior to sectioning.

23. Resizing the block is necessary to prevent sections from overlapping on the Visium slide, which may interfere with tissue permeabilization and mRNA capture. If the embedded tissue is larger than 6.5 × 6.5 mm, then make an incision in the block to split the tissue to the required size.

24. If the sections are cracked, increase the objective and chamber temperature. Likewise, reduce the temperature if the sections are crumpled. Adjust the temperature until sections appear flat and undamaged. A slice thickness of 10μm is recommended for the heart. For other tissues, optimization may be required to determine the optimum thickness.

25. Avoid any contact with the slide surface. Complete adherence to the slide may take 15–45 s. Practice with plain glass slides as needed.

26. Do not expose the slide to room temperature and try to quickly transfer the slide from the cryostat to the −80 °C freezer, making sure that the slide holder is closed tightly.

27. Thresholds can be supplied to "min.cells" and "min.features" to remove lowly expressed genes and cells expressing few genes, respectively.

28. The default threshold determining HTO detection is set at the 99% quantile for each HTO. Negatives are cells that do not pass the threshold on any of the HTO. Singlets are cells that pass the threshold for one of the HTOs. Doublets are cells positive for more than one HTO, indicating the presence of multiple cells in the same gel bead in emulsion (GEM) during library preparation.

29. percent.mt is an indicator of cell stress and viability. In apoptotic/lysed cells, cytoplasmic transcripts are released and degraded due to disrupted cellular membrane. However, mitochondrial transcripts may remain intact within the mitochondria. As a result, partitioning of lysed cells during GEM formation will likely lead to an increased detection of mitochondrial transcripts.

30. Removal of noise such as low-quality cells and lowly expressed genes may improve cell clustering. Multiplets, multiple cells partitioned into the same GEM, are undiscernible from cells in an activated state as both scenarios may manifest in high cellular counts. If cell hashing is applied, previous steps on HTO-based preprocessing will remove multiplets. We recommend users to apply soft thresholds to remove noticeable outliers.
31. One scenario of applying integration maybe the comparison between control and experimental datasets, as shown in the example. Users may also try SCTransform, another workflow that aims to remove technical differences while maintaining biological heterogeneity across datasets.
32. By default, ScaleData scales the variable genes identified in **step 15** if the "feature" argument is not supplied. Unscaled genes cannot be plotted on the heatmap, and thus scaling all genes may be of interest to users. Regression of technical features is demanding in terms of computational power. Users may choose to bypass "vars.to.regress" or remove specific technical feature and examine the differences.
33. Clustering of cells in Seurat applies the K-nearest neighbor approach on Euclidean distance in PCA space. Iterative grouping of cells via Louvain method of community detection is used to optimize the number of clusters.
34. Unlike scRNAseq data, each data point in Visium data is a spot rather than an individual cell, and thus the gene expression matrix manifests as a "gene-spot matrix" rather than a "gene-cell matrix." In addition to the gene-spot matrices, the output of the Space Ranger pipeline includes a spatial folder that contains the image of the tissue section that was sequenced.
35. The SCTransform function normalizes the data, detects variable features, and stores the data in the SCT assay.
36. The argument "pt.size.factor" determines the size of the spots on the tissue, the argument alpha determines the minimum and maximum opacity of the points and the argument stroke controls the width of the border around the spots.
37. By combining different sections from the same damaged heart, we can study how gene expression in fibroblasts changes with the severity of damage at different regions of the heart. Data for the second tissue section was loaded and processed as that of the first tissue section.
38. Notice that the cluster numbers have changed when the section was merged with another section and re-clustering was performed so that before combining the infarct zone was cluster 4, while after merging the infarct zone became cluster 5.

References

1. Travers JG, Kamal FA, Robbins J et al (2016) Cardiac fibrosis: the fibroblast awakens. Circ Res 118:1021–1040. https://doi.org/10.1161/CIRCRESAHA.115.306565
2. Tallquist MD (2020) Cardiac fibroblast diversity. Annu Rev Physiol 82:63–78. https://doi.org/10.1146/annurev-physiol-021119-034527
3. Hinz B (2007) Formation and function of the myofibroblast during tissue repair. J Invest Dermatol 127:526–537. https://doi.org/10.1038/sj.jid.5700613
4. Furtado MB, Nim HT, Boyd SE, Rosenthal NA (2016) View from the heart: cardiac fibroblasts in development, scarring and regeneration. Development 143:387–397. https://doi.org/10.1242/dev.120576
5. Saucerman JJ, Tan PM, Buchholz KS et al (2019) Mechanical regulation of gene expression in cardiac myocytes and fibroblasts. Nat. Rev Cardiol 16:361–378. https://doi.org/10.1038/s41569-019-0155-8
6. Gabbiani G, Ryan GB, Majno G (1971) Presence of modified fibroblasts in granulation tissue and their possible role in wound contraction. Experientia 27:549–550. https://doi.org/10.1007/BF02147594
7. Hinz B, Celetta G, Tomasek JJ et al (2001) Alpha-smooth muscle actin expression upregulates fibroblast contractile activity. Mol Biol Cell 12:2730–2741. https://doi.org/10.1091/mbc.12.9.2730
8. Willems IEMG, Havenith MG, De Mey JGR, Daemen MJAP (1994) The a-smooth muscle actin-positive cells in healing human myocardial scars. Am J Pathol 145:868–875
9. Soliman H, Rossi FMV (2020) Cardiac fibroblast diversity in health and disease. Matrix Biol 91–92:75–91. https://doi.org/10.1016/j.matbio.2020.05.003
10. Ackers-Johnson M, Tan WLW, Foo RSY (2018) Following hearts, one cell at a time: recent applications of single-cell RNA sequencing to the understanding of heart disease. Nat Commun 9:4434. https://doi.org/10.1038/s41467-018-06894-8
11. Ziegenhain C, Vieth B, Parekh S et al (2017) Comparative analysis of single-cell RNA sequencing methods. Mol Cell 65:631–643. e4. https://doi.org/10.1016/j.molcel.2017.01.023
12. Chen G, Ning B, Shi T (2019) Single-cell RNA-seq technologies and related computational data analysis. Front. Genetics 10:317. https://doi.org/10.3389/fgene.2019.00317
13. Satija R, Farrell JA, Gennert D et al (2015) Spatial reconstruction of single-cell gene expression data. Nat Biotechnol 33:495–502. https://doi.org/10.1038/nbt.3192
14. Soliman H, Paylor B, Scott RW et al (2020) Pathogenic potential of Hic1-expressing cardiac stromal progenitors. Cell Stem Cell 26:205–220. e8. https://doi.org/10.1016/j.stem.2019.12.008
15. Farbehi N, Patrick R, Dorison A et al (2019) Single-cell expression profiling reveals dynamic flux of cardiac stromal, vascular and immune cells in health and injury. elife 8:e43882. https://doi.org/10.7554/eLife.43882
16. Stuart T, Butler A, Hoffman P et al (2019) Comprehensive integration of single-cell data. Cell 177:1888–1902. e21. https://doi.org/10.1016/j.cell.2019.05.031
17. Hashimshony T, Senderovich N, Avital G et al (2016) CEL-Seq2: sensitive highly-multiplexed single-cell RNA-Seq. Genome Biol 17:77. https://doi.org/10.1186/s13059-016-0938-8
18. Rosenberg AB, Roco CM, Muscat RA et al (2018) Single-cell profiling of the developing mouse brain and spinal cord with split-pool barcoding. Science (80-) 360:176–182. https://doi.org/10.1126/science.aam8999
19. Zheng GXY, Terry JM, Belgrader P et al (2017) Massively parallel digital transcriptional profiling of single cells. Nat Commun 8:1–12. https://doi.org/10.1038/ncomms14049
20. Klein AM, Mazutis L, Akartuna I et al (2015) Droplet barcoding for single-cell transcriptomics applied to embryonic stem cells. Cell 161:1187–1201. https://doi.org/10.1016/j.cell.2015.04.044
21. Macosko EZ, Basu A, Satija R et al (2015) Highly parallel genome-wide expression profiling of individual cells using nanoliter droplets. Cell 161:1202–1214. https://doi.org/10.1016/j.cell.2015.05.002
22. Sheng K, Cao W, Niu Y et al (2017) Effective detection of variation in single-cell transcriptomes using MATQ-seq. Nat Methods 14:267–270. https://doi.org/10.1038/nmeth.4145
23. Fan X, Zhang X, Wu X et al (2015) Single-cell RNA-seq transcriptome analysis of linear and circular RNAs in mouse preimplantation embryos. Genome Biol 16:148. https://doi.org/10.1186/s13059-015-0706-1

24. Islam S, Kjällquist U, Moliner A et al (2012) Highly multiplexed and strand-specific single-cell RNA 5′ end sequencing. Nat Protoc 7:813–828. https://doi.org/10.1038/nprot.2012.022
25. Hashimshony T, Wagner F, Sher N, Yanai I (2012) CEL-Seq: single-cell RNA-Seq by multiplexed linear amplification. Cell Rep 2:666–673. https://doi.org/10.1016/j.celrep.2012.08.003
26. Sena JA, Galotto G, Devitt NP et al (2018) Unique molecular identifiers reveal a novel sequencing artefact with implications for RNA-Seq based gene expression analysis. Sci Rep 8:1–13. https://doi.org/10.1038/s41598-018-31064-7
27. Stoeckius M, Zheng S, Houck-Loomis B et al (2018) Cell Hashing with barcoded antibodies enables multiplexing and doublet detection for single cell genomics. Genome Biol 19:224. https://doi.org/10.1186/s13059-018-1603-1
28. Butler A, Hoffman P, Smibert P et al (2018) Integrating single-cell transcriptomic data across different conditions, technologies, and species. Nat Biotechnol 36:411–420. https://doi.org/10.1038/nbt.4096
29. Mcinnes L, Healy J, Saul N, Großberger L (2018) UMAP: uniform manifold approximation and projection. J Open Source Softw 3(29):861. https://doi.org/10.21105/joss.00861
30. Linderman GC, Zhao J, Kluger Y (2018) Zero-preserving imputation of scRNA-seq data using low-rank approximation. bioRxiv. https://doi.org/10.1101/397588
31. Tillmanns J, Hoffmann D, Habbaba Y et al (2015) Fibroblast activation protein alpha expression identifies activated fibroblasts after myocardial infarction. J Mol Cell Cardiol 87:194–203. https://doi.org/10.1016/j.yjmcc.2015.08.016
32. Hafemeister C, Satija R (2019) Normalization and variance stabilization of single-cell RNA--seq data using regularized negative binomial regression. Genome Biol 20:296. https://doi.org/10.1186/s13059-019-1874-1

Chapter 5

Myofibroblast Adhesome Analysis by Mass Spectrometry

Christopher A. McCulloch

Abstract

Myofibroblasts form adhesions to their underlying extracellular matrices, which is an essential step in their formation and differentiation. These adhesions comprise protein-rich aggregates of a wide variety of signaling, cytoskeletal, cell adhesion, and matrix proteins that interact with one another to enable bidirectional flow of information between the cell and the surrounding extracellular matrix. The concentrated repertoire of the proteins in matrix adhesions of myofibroblasts (i.e., over 450 different proteins) and their important role in regulating the metabolic activities of myofibroblasts, has motivated in-depth analysis of their protein complement and how this repertoire is influenced by experimental conditions.

In this protocol I describe in detail: (1) the method for isolating focal adhesion-associated proteins using matrix ligand-bound magnetite beads; (2) the method for eluting the proteins from the beads and their preparation for mass spectrometry (Fig. 1). I also briefly consider the mass spectrometry methods including the use of isobaric tags to enable multifactorial experiments and the analysis of the identified proteins. I consider the advantages of these approaches, and the challenges and pitfalls that are encountered with these methods.

Key words Focal adhesions, Fibroblasts, Mass spectrometry, Isobaric tags, Signaling, Extracellular matrix

1 Introduction

Anchorage-dependent cells rely on adhesion to extracellular matrix ligands for their growth, metabolism, and determination of fate, which include differentiation pathways. Myofibroblasts are anchorage-dependent cells that use chemical and mechanical signals derived from extracellular matrix ligands to inform their formation, function, and future. Many of the signals that flow through adhesions are dependent on the composition of the protein repertoire in the adhesions which these cells express. Therefore, a detailed understanding of the proteins in adhesions, the so-called "myofibroblast adhesome," could enable new insights into the signaling systems that regulate the metabolism and fate of these cells.

Boris Hinz and David Lagares (eds.), *Myofibroblasts: Methods and Protocols*, Methods in Molecular Biology, vol. 2299, https://doi.org/10.1007/978-1-0716-1382-5_5,

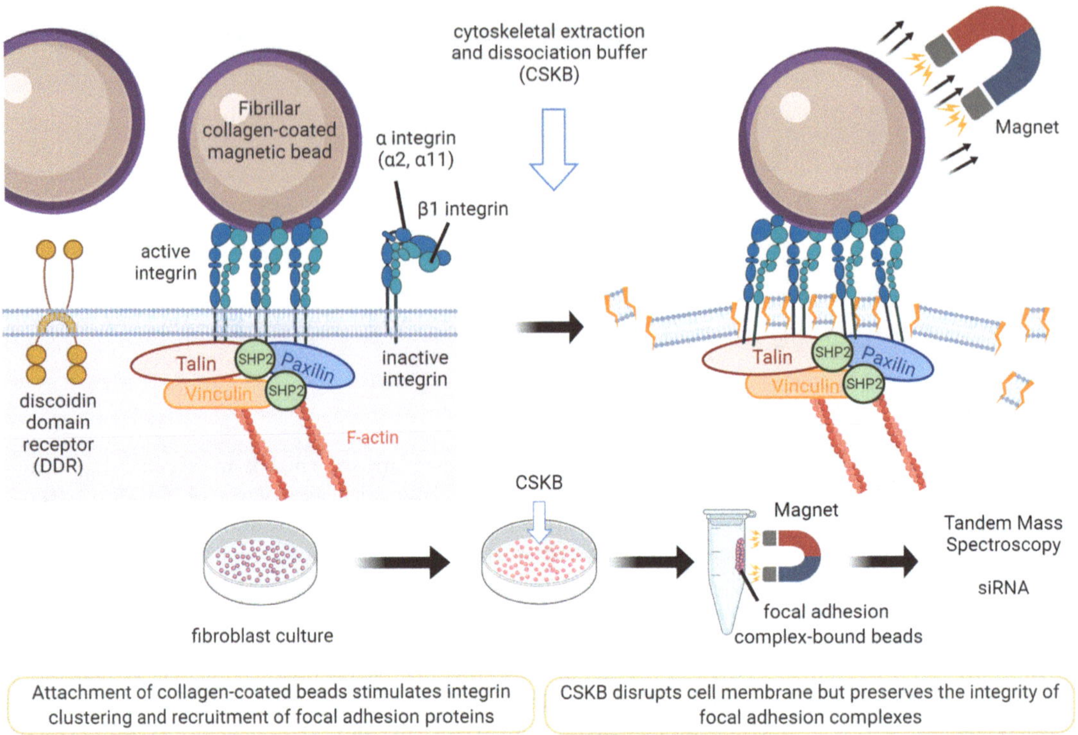

Fig. 1 Diagram to illustrate the steps involved in the preparation of focal adhesion proteins that can be analyzed by mass spectrometry. Collagen-coated magnetite beads are incubated with cultured fibroblasts. On their dorsal surface, cells form focal adhesion-like structures around the fibrillar collagen-coated beads that can either be based on integrin-mediated attachment or on discoidin domain receptors (DDR, not considered here). The cells with attached beads are lysed in cytoskeletal extraction and dissociation buffer (CKSB) and using a permanent magnet, the proteins and cell components that are not part of the adhesion complex are washed away. Subsequent steps for analysis of the adhesion proteins by mass spectrometry and functional analysis by siRNA are shown on the lower right. Figure edits by M. Lodyga and B. Hinz (Faculty of Dentistry, University of Toronto) using Biorender (biorender.com)

Cell adhesions exhibit considerable heterogeneity both within and between different cell types. Indeed, within a single cell after plating on a specific matrix ligand, there is an ordered repertoire of proteins that changes over time. The adhesions that form in response to substrate adherence in cultured cells are called “focal adhesions,” which are actin-enriched organelles exhibited by a wide variety of cell types including fibroblasts, endothelial cells, synoviocytes, and chondrocytes [1]. Focal adhesions is the collective name of highly dynamic organelles that mature through stages that include focal contacts, focal adhesions, and fibrillar adhesions, each with a distinctive appearance and molecular composition [2]. This dynamic nature of focal adhesions is readily observed in fibroblasts adhering to a substratum coated with collagen or fibronectin; the cells will adhere, spread, and form progressively more “mature” focal adhesions after plating [3, 4].

In the context of signaling and the control of the myofibroblastic phenotype, it is worth noting that spatial co-localization of interacting signaling molecules provides a pivotal regulatory locus for signal transduction. Once thought to be strictly structural adhesive complexes, focal adhesions are now recognized for their role in signal transduction [2, 5]. For example, a very large number of signaling and structural molecules have been detected in nascent focal adhesions [6, 7], which include the interleukin 1 receptor [8]. Engagement of interleukin-1R_I by interleukin-1β leads to tyrosine phosphorylation of the focal adhesion proteins talin [9] and focal adhesion kinase [10]. Interleukin-1-induced Ca^{2+} flux and activation of the canonical Ras-Raf-MEK- ERK signaling pathway also require focal adhesions [11]. These conditions correlate with maximal interleukin-1-induced activation of the extracellular signal-regulated kinase activation, secretion of matrix metalloproteinases, and promotion of the myofibroblast phenotype itself.

Cognizant of the complex collection of proteins in focal adhesions that contribute to adhesive, structural, and signaling processes, we developed specialized fractionation and proteomic methods to examine the proteins in these structures that regulate interleukin-1 signaling. We focused on interleukin-1 signaling because in fibroblasts, chondrocytes, and synoviocytes, the transduction of interleukin-1 signaling relies on the *a priori* formation of focal adhesions by the target cells [10]. We first devised a preparative method to isolate adhesion-associated proteins [12] and tested the efficacy of this system in the purification of known focal adhesion proteins [13]. We then used this preparative method to look for novel proteins that were enriched in focal adhesions of fibroblasts and discovered that the actin capping protein was indeed enriched in focal adhesions [7] and played fundamental roles in regulating focal adhesion-dependent cell migration and collagen remodeling.

Other methods for preparation of cell adhesions have been developed and the interested experimentalist should consult the interesting approach devised by Waterman and colleagues [14] for a discussion of how this method enables purification of these structures. We should emphasize here that any preparative method provides a somewhat biased glimpse into which subset of proteins may be expressed in the organelle of interest. The method described here almost certainly does not capture the whole repertoire of focal adhesion proteins and may under- or over-report certain proteins. As is indicated in the accompanying notes section of this chapter, the interested investigator is encouraged to optimize certain key steps of the protocol for her/his own special needs and choice of cell type. Notably, the structures that form in response to collagen or fibronectin-coated beads are not the same as the structures that form on planar solid supports, such as a glass surface coated with collagen. Thus, it is probably most accurate to describe these structures as focal adhesion-like structures.

2 Materials

We suggest that all solutions that are needed should be prepared using ultrapure water and analytical grade reagents. All reagents should be prepared and stored at room temperature unless indicated otherwise. From the standpoint of laboratory safety, it is essential that all locally applicable regulations for the disposal of wastes should be followed.

2.1 Coated Bead Preparations

1. Magnetite (ferric oxide) beads (<5 μm in diameter) (*see* **Note 1**).
2. Bovine type I collagen stock solution: 6 mg/mL (e.g., Advanced Biomatrix) in acetic acid, pH = 1. Store at 4 °C, for bead preparations the final concentration should be 1 mg/mL (*see* **Note 2**).
3. NaOH (1 N): used for neutralization of acidified collagen.
4. Phosphate-buffered saline (PBS): 150 mM NaCl, 8.06 mM Na_2HPO_4, 1.74 mM NaH_2PO_4, adjust pH with 1 M NaOH, 0.9 mM $CaCl_2$ to pH 7.4.
5. Electronic particle counting device (e.g., Coulter Channelyzer, Coulter Electronics, Hialeah, FL).
6. Porcine plasma fibronectin: 1 μg/mL in PBS, pH = 7.4.
7. Bovine serum albumin (BSA): 1 mg/mL in PBS, pH = 7.4.
8. Poly-L-lysine: 1 mg/mL in PBS, pH = 7.4 (*see* **Note 3**).
9. Antibodies to detect collagen coating: type I collagen (i.e., rabbit anti-bovine antiserum; Fisher Scientific) and a tetramethylrhodamine B isothiocyanate-conjugated donkey anti-rabbit second antibody (e.g., Jackson Laboratories).

2.2 Protein Preparation and Elution from Beads

1. Cytoskeleton extraction and dissociation buffer: 0.5% Triton X-100, 50 mM NaCl, 300 mM sucrose, 3 mM $MgCl_2$, 20 μg/mL aprotinin, 1 μg/mL leupeptin, 1 μg/mL pepstatin, 1 mM PMSF, and 10 mM PIPES, at pH 6.8.
2. Laemmli sample buffer (2×): 4% SDS, 20% glycerol, 0.004% bromphenol blue, 0.125 M Tris–Cl, pH 6.8, 10% 2-mercaptoethanol, add immediately before use.
3. Carbonate buffer: 25 mM NH_4HCO_3, pH 7.5.
4. Magnet: high flux density ceramic bar magnet, 6″ × 4″ × 1″ (here purchased form Jobmaster Magnets Canada Inc., 1420 Cornwall, Units 1&2, Oakville, Ontario) (*see* **Note 4**).
5. Dulbecco's minimal essential medium (DMEM) or α-minimal essential medium (α-MEM) for fibroblast culture.
6. Tabletop vortexer.

7. Cell culture scraper.
8. Laemmli buffer.

2.3 Preparation for Mass Spectrometry

1. 10 mM dithiothreitol (DTT).
2. Iodoacetamide solution: (55 mM), prepare fresh immediately prior to use; protect from light.
3. Trypsin solution: 10 mg/mL in carbonate buffer at pH = 8.0.

3 Methods

Carry out all procedures at room temperature unless otherwise specified.

3.1 Bead Preparations

1. Collagen solution (1 mL of 6 mg/mL) is neutralized to pH 7.4 with 100 μL of 1 N NaOH. Under these conditions, the collagen will polymerize around the beads (*see* **Note 5**).
2. Magnetite beads (0.4 g) are incubated for 1 h with 1 mL of bovine type I collagen solution (6 mg/mL) at a final concentration of 1 mg/mL in PBS (pH 7.4) at 37 °C to induce collagen fibril assembly on beads.
3. The beads are sonicated with a sonifier to eliminate clumps.
4. Analysis of bead size is performed with an electronic particle counting device (*see* **Note 6**).
5. Beads are rinsed in PBS, washed three times, and resuspended in calcium-free PBS.
6. Verification of collagen coating on beads is performed using an antibody to type I collagen and a tetramethyl rhodamine B isothiocyanate-conjugated donkey anti-rabbit second antibody. Collagen-coated beads are imaged by fluorescence microscopy to ensure that there is binding of collagen to the beads.
7. For some experiments in which investigators wish to examine fibronectin-induced focal adhesions, porcine plasma fibronectin can be used at 1 ug/mL in phosphate-buffered saline (pH = 7.4), which is used for bead incubations with 0.4 g of magnetite beads. The beads are incubated with the fibronectin coating solution for 30 min at 37 °C.
8. To facilitate distinction of proteins that are not involved in specific collagen adhesion, a separate sample of magnetite beads (0.4 g) are coated with BSA (0.15% in PBS at 37 °C for 30 min).
9. As an additional control, a third sample of magnetite beads is prepared by incubation with poly-L-lysine (1 mg/mL in PBS at 37 °C for 30 min). This molecule is highly adherent to cells but does not induce the formation of focal adhesions.

10. Evaporator or lyophilizer.
11. Reversed phase high-performance liquid chromatography (RP-HPLC) or ZIP tips (e.g., Millipore).

3.2 Protein Preparation and Elution from Beads

1. Cells of interest (e.g., human gingival fibroblasts or any cell type that is known to form focal adhesions in culture) are grown in either α-MEM or DMEM containing 10% fetal bovine serum and antibiotics. These culture conditions may need to be adjusted depending on the normal cell culture requirements of the cells of interest.
2. Prior to the bead incubation step (*see* below), cells are incubated at ~10^4 cells per cm^2 of culture ware plastic and preferably in 100 mm diameter dishes for 3 h prior to bead incubation (*see* **Note 7**).
3. Collagen-coated magnetite beads (0.4 g) are incubated for 1 h with fibroblast cultures (or other cell types of interest). Note that the beads are gently added to the cells, which means that they will attach to the dorsal surface of the cells. For comparisons with collagen, fibronectin-coated beads may be used here. For the determination of the specificity of the focal adhesion preparations, instead of collagen-coated beads, BSA or poly-L-lysine-coated beads may be used and the data obtained with these nonspecific controls is compared with the collagen or fibronectin-coated beads (*see* **Note 8**).
4. If investigators wish to examine the effect of a certain cell treatment (e.g., interleukin-1β, TGF-β) on the myofibroblast adhesome, then the treatment(s) can be added at this time directly to the cell cultures.
5. After any treatments are completed, cells are washed twice with cold (4 °C) PBS to remove unbound or loosely bound beads.
6. A conventional cell culture scraper is used to physically dislodge the cells with the attached beads from the bottom of the culture dish. The scraped cells are incubated with cytoskeleton extraction buffer (1 mL for a 100 mm dish) at 4 °C.
7. The scraped cells and the cytoskeleton extraction buffer are suctioned from the bottom of the dish, placed in a 1.5 mL Eppendorf type of tube and vortexed for 30 s with a tabletop vortexer at room temperature.
8. The beads are isolated from the lysate using a high flux density magnet in which the pole face of the magnet is placed in direct contact with the Eppendorf tube (*see* **Note 9**). The tube is maintained in contact with the magnet as the cytoskeleton extraction buffer is suctioned away.
9. Fresh, 4 °C cold cytoskeleton buffer (1 mL) is added to the tube, vortexed for 30 s, and the application of the magnet to the side of the tube is repeated, which is followed by the

suctioning away of the cytoskeleton extraction buffer. This step should be repeated so that the beads undergo three separate wash procedures (*see* **Note 10**).

10. The bead-associated proteins produced by this process are removed by boiling in Laemmli buffer (1 mL) for 10 min. Proteins can also be eluted from beads using 50 mM glycine buffer (pH 2.3–2.5) for 36 h. The choice of which procedure to use will depend upon the extent of protein linearization that is desired for mass spectrometry, which is greater with the Laemmli buffer.
11. The beads are separated from the buffer by the application of the magnet to the side of the tube as indicated in point # 7 above followed by the suctioning away of the Laemmli buffer.
12. The proteins that are eluted from the beads are dialyzed for 36 h in 1 mL carbonate buffer pH 7.5.
13. The concentration of the protein in the sample tubes is analyzed with standard, commercially available bicinchoninic acid (BCA) or Bradford assays.

3.3 Protein Analysis by Mass Spectrometry

1. It may be necessary, based on the abundance of disulfide bonds that are anticipated in the proteins of interest, to reduce disulfide bonds and to alkylate free cysteine residues (*see* **Note 11**).
2. The procedure for reduction of disulfide bonds can be done by adding DTT to the sample at a final concentration of 10 mM and incubate for 30 min at room temperature with agitation (e.g., end-over-end rotation, vortexing, sonication, or shaking).
3. Add iodoacetamide to a final concentration of 10 mM final and incubate at room temperature in the dark for 30 min.
4. To create protein fragments that can be analyzed by mass spectrometry, add trypsin at a final concentration of 1 μg/20 μg eluted target protein in 1 mL; pH = 8.0 in 1.5 mL Eppendorf tubes. The protein samples are digested by the trypsin solution and rotated overnight at 37 °C. Trypsin cleaves proteins into peptides with an average size of 700–1500 daltons, which is in the ideal range for mass spectrometry (*see* **Note 12**).
5. The digest is allowed to cool to room temperature and the digestion is stopped by acidification with acetic acid to 0.1% (vol/vol) followed by air drying with an evaporator or lyophilizer.
6. Lyophilized samples may be further processed by liquid chromatography as is routinely used prior to mass spectrometry to remove salt (which complicates the mass spectrometry analysis

of the charged peptides) and reduces the complexity of the sample. One commonly used method employs reversed-phase high performance liquid chromatography which concentrates and desalts intact proteins, either on its own or downstream from other types of chromatographic separations. Reversed phase (RP)-HPLC is a popular choice for proteomic workflows, as the mobile phases, which are volatile, are compatible with mass spectroscopy and may also be easily removed via vacuum centrifugation, if necessary. Alternatively, ZIP tips can be used to remove salts and Tris, which also complicates the mass spectrometry analysis.

7. The samples are analyzed by liquid chromatography-electrospray ionization tandem mass spectrometry system that is available through most institutional Mass Spectrometry Facilities. Details on the actual mass spectrometry analysis of these samples is not discussed in detail here as interested users should contact their own local facilities and discuss in detail their biological problem, their expected fractionation approach (discussed in detail above), and what components of the analysis will be undertaken by the mass spectrometry facility. Some facilities, for example, will conduct the preparation of the tryptic digests.
8. Scaffold 4.0 (Proteome Software, Portland, OR) is used for analyzing search results, calculating p values for each peptide match, matching peptide spectra, and enabling comparison between the collagen, fibronectin, BSA and poly-L-lysine-coated bead preparations and other experimental factors, including soluble cytokines that drive myofibroblast differentiation (e.g., TGF-β) (*see* **Note 13**).
9. Data from these runs are analyzed using the National Center for Biotechnology Information (NCBI) database and the MASCOT search engine. Scaffold 2.0 proteome software (Portland, OR, USA) is used to analyze the search results.
10. Only proteins that can be positively identified from the tryptic peptides with high certainty (>95%) and that show at least a twofold increase in the number of positively identified peptides are generally considered worthy of subsequent confirmation with conventional biochemical (i.e., immunoblot) and/or immunohistochemical assessments.

4 Notes

1. The use of ferric oxide (magnetite) beads for physical separation of the focal adhesion-associated proteins is central to this protocol, and accordingly the choice of the source of the ferric oxide beads is an important consideration. The ferric oxide

particles that can be obtained from Sigma provide the highest ferric oxide content of any manufacturer, but the particle size distribution is very broad. In contrast, Dynabeads (Fisher Scientific) have much lower ferric oxide content (and therefore lower magnetic moment) but their size distribution is narrow. We have prepared focal adhesion proteins using both manufacturers' particles and those produced by Bangs Laboratories (Fishers, IN) who produce beads of a wide range of different sizes and with different magnetite content. Because of the greatly reduced cost of the Sigma beads, it may be worthwhile starting with these beads, and then depending on the consistency of the focal adhesion protein preparations as determined by comparisons of the protein complements from replicate analysis, it may be worth trying more homogeneous-sized bead populations.

2. Collagen solutions derived from bovine skin or rat tail are widely available from several different manufacturers (e.g., Advanced Biomatrix, Roche). If a working solution of 1 mg/mL of collagen in aqueous solution can be produced and with appropriate neutralization, routinely polymerize on the bead surface, there is no pressing need to try different manufacturers.

3. Poly-L-lysine and BSA are used as coating solutions to assess nonspecific protein aggregation to the beads. With any preparative procedure, there will be considerable nonspecific signals that are generated which will be more apparent upon analysis of the bead-associated proteins. This point will be considered later in the analysis of the proteins that are detected. In our experience, we have seen classical focal adhesion proteins (e.g., vinculin) in beads coated with bovine serum albumin while the proteins seen with beads coated with poly-L-lysine were consistent with nonspecific adherence since we did not see recruitment of classical, focal adhesion proteins. In addition, if integrin receptors for the ligand of interest (e.g., collagen) are limited in the cells undergoing analysis, it is worthwhile plating the cells on poly-L-lysine prior to the addition of the collagen-coated beads on the dorsal surfaces of the cells. With this approach, when collagen receptors are limiting, they are not involved in cell attachment of the ventral surface of the cell and are therefore more available for binding to collagen-coated beads on the dorsal surface.

4. The magnet is used to physically restrict the magnetite beads (which are covered with focal adhesion proteins) to the side of the Eppendorf tube during the wash steps. As it is important that the beads are not inadvertently removed during washing, the magnet should be powerful enough to retain the beads firmly in place. The magnets from Jobmaster are ceramic

magnets and exhibit very high flux density. The Eppendorf tube can be held right against the side of the magnet during the washing stage, which will ensure no bead loss.

5. The excessive clumping of beads when bound to the surface of the cells will compromise the utility of the preparation of the proteins. Ideally, the beads should be monodisperse [12]. If the beads are clumping excessively, then the sonification step (Subheading 3.1) should be prolonged and the intensity of the sonification should be increased until clumping is eliminated.
6. Beads obtained from Sigma typically exhibit a heterogeneous size distribution with a pronounced modal peak at 5 μm, although there may be many particles with smaller diameters. For assessing the size distribution of the beads, if a particle analyzer is not available, then some estimates of the size distribution can be made with a microscope equipped with a 40× objective lens.
7. It is not advisable to use closed flasks for these experiments, which will restrict the use of cell scrapers. Instead, 100 mm dishes should be used as they have open access for complete scraping of the cells from the bottom of the dish.
8. Individuals who contemplate the use of this approach for analysis of the myofibroblast adhesome may wish to establish multifactorial experiments to improve the specificity of the mass spectrometry analysis. To improve specificity, it is recommended to include for each experimental run, at least one nonspecific control using poly-L-lysine and/or bovine serum albumin-coated beads. Comparisons of the peptide count data sets from these two controls and extracellular matrix ligand positive sample (e.g., collagen or fibronectin) will provide insights into which proteins are specific to the adhesions (i.e., present in the collagen or fibronectin bead data) but will not be seen in the nonspecific controls. This approach can be greatly extended through the use of tandem mass tag mass spectrometry in which isobaric tags [7, 15] are used to label the lysates from the different experimental groups. Subsequently, data analysis can be used to quantitatively assess the proteins that are specifically expressed in focal adhesions.
9. The ceramic magnet is used to generate tensile force that pulls the magnetite beads to the side of the tube in which the cell lysates are incubated. For all experiments the pole face of the magnet is parallel with and in direct contact with the side of the tube containing the cell lysate. At this distance, the field strength is ~400 Gauss (G) with a vertical gradient of ~100 G/cm as measured by a fluxgate magnetometer (Jobmaster, Oakville, ON).

10. During adhesion of cells to beads, adhesion-associated proteins organize around the beads. The nonspecific attachment of beads to cells is, in the first instance, reduced with the wash steps. The number of wash steps may need to be optimized depending on the cell type so that a balance is obtained in which specific bead attachment is maximized and excessive, indiscriminate bead detachment is minimized. Further, the use of the cytoskeleton buffer as described and the magnetic isolation that is accompanied by additional wash steps needs to be optimized to enable specific protein attachment to beads, but without wholesale loss of all bound beads.
11. Reduction and alkylation are not absolutely required prior to trypsin cleavage but this approach will enhance sequence coverage of the proteins. Further, these treatments will help to unfold the protein, which facilitates the cleavage. The concentrations and incubation times for DTT and iodoacetamide may vary depending on the amount of protein to be treated but the usual concentrations are 5–10 mM for DTT and 10–50 mM for iodoacetamide. Incubation times may range from 5 min to 1 h for the reduction step and 10 min to 1 h for the alkylation step. The reduction can be performed at room temperature, 37 °C or 56 °C. These parameters will need to be optimized depending on the nature of the protein sample.
12. The amount of trypsin needed for digestion will vary depending on the amount of protein that is in the sample and what is the expected speed of digestion. Many commercial vendors sell proteomics-grade trypsin in aliquots (e.g., 1–25 μg vials), which obviate the need to weigh out the enzyme. Second, it is helpful to reduce the tendency for the trypsin to undergo autolysis. If this occurs, extra peaks will be added to the mass spectrometry spectra. It is always a good idea to use the minimum amount of trypsin. In general, a ratio of 1:20 (w/w) will be enough for complete digestion of 10 μg protein within 4–8 h. If the concentration of protein in the sample is quite high, or if trypsin cannot easily obtain access to cleavage sites (e.g., poor solubility), it may help to add a second aliquot of trypsin. As trypsin activity is optimized at pH = 8.0, the pH of the solution should be examined prior to the addition of the enzyme.
13. Depending on the experimental conditions in the culture (e.g., addition of vehicle or stimulant of a specific cell function, such as interleukin-1) and as mentioned above, the bead-associated proteins can be compared by tandem mass spectrometry. However, this approach generates considerable uncertainty because, even with the use of negative control coatings (i.e., poly-L-lysine, bovine serum albumin), because there are so many peptides, the probability of correct identification (i.e., >95%)

may be compromised. However, as described earlier [7], this problem can largely be overcome in which cultures are differentially labeled with isobaric tags that are specific for each treatment. Up to 12 different tags (different molecular mass) are now widely available (e.g., Applied Biosystems) for the conduct of these types of experiments. Based on these types of experimental designs and comparative analyses, it is possible to detect a much smaller population of proteins that are differentially recruited to adhesion complexes under those specific conditions. The functional importance of the differentially recruited proteins can be examined by knockdown of specific proteins using RNA interference of cultured cells. Further, using specialized imaging methods, such as total internal reflection fluorescence microscopy, the localization and recruitment of newly identified proteins into focal adhesion structures can be examined at very high z-axis resolution in cultured cells.

References

1. Critchley DR (2000) Focal adhesions - the cytoskeletal connection. Curr Opin Cell Biol 12:133–139. https://doi.org/10.1016/s0955-0674(99)00067-8
2. Wehrle-Haller B, Imhof B (2002) The inner lives of focal adhesions. Trends Cell Biol 12:382–389
3. von Wichert G, Jiang G, Kostic A, De Vos K, Sap J, Sheetz MP (2003) RPTP-alpha acts as a transducer of mechanical force on alphav/beta3-integrin-cytoskeleton linkages. J Cell Biol 161:143–153
4. Partridge MA, Marcantonio EE (2006) Initiation of attachment and generation of mature focal adhesions by integrin-containing filopodia in cell spreading. Mol Biol Cell 17:4237–4248. https://doi.org/10.1091/mbc.e06-06-0496
5. Burridge K, Chrzanowska-Wodnicka M (1996) Focal adhesions, contractility, and signaling. Annu Rev Cell Dev Biol 12:463–518
6. Aplin AE, Juliano RL (1999) Integrin and cytoskeletal regulation of growth factor signaling to the MAP kinase pathway. J Cell Sci 112 (Pt 5):695–706
7. Wang Q, Delcorde J, Tang T, Downey GP, McCulloch CA (2018) Regulation of IL-1 signaling through control of focal adhesion assembly. FASEB J 32:3119–3132. https://doi.org/10.1096/fj.201700966R
8. Qwarnstrom EE, Page RC, Gillis S, Dower SK (1988) Binding, internalization, and intracellular localization of interleukin-1 beta in human diploid fibroblasts. J Biol Chem 263:8261–8269
9. Qwarnstrom EE, MacFarlane SA, Page RC, Dower SK (1991) Interleukin 1 beta induces rapid phosphorylation and redistribution of talin: a possible mechanism for modulation of fibroblast focal adhesion. Proc Natl Acad Sci U S A 88:1232–1236. https://doi.org/10.1073/pnas.88.4.1232
10. Arora PD, Ma J, Min W, Cruz T, McCulloch CA (1995) Interleukin-1-induced calcium flux in human fibroblasts is mediated through focal adhesions. J Biol Chem 270:6042–6049
11. Lo YY, Luo L, McCulloch CA, Cruz TF (1998) Requirements of focal adhesions and calcium fluxes for interleukin-1-induced ERK kinase activation and c-fos expression in fibroblasts. J Biol Chem 273:7059–7065
12. Glogauer M, Arora P, Yao G, Sokholov I, Ferrier J, McCulloch CA (1997) Calcium ions and tyrosine phosphorylation interact coordinately with actin to regulate cytoprotective responses to stretching. J Cell Sci 110 (Pt 1):11–21
13. Glogauer M, Arora P, Chou D, Janmey PA, Downey GP, McCulloch CA (1998) The role of actin-binding protein 280 in integrin-dependent mechanoprotection. J Biol Chem 273:1689–1698. https://doi.org/10.1074/jbc.273.3.1689
14. Kuo JC, Han X, Yates JR 3rd, Waterman CM (2012) Isolation of focal adhesion proteins for biochemical and proteomic analysis. Methods

Mol Biol 757:297–323. https://doi.org/10.1007/978-1-61779-166-6_19

15. Mei L, Montoya MR, Quanrud GM, Tran M, Villa-Sharma A, Huang M, Genereux JC (2020) Bait correlation improves interactor identification by tandem mass tag-affinity purification-mass spectrometry. J Proteome Res 19:1565–1573. https://doi.org/10.1021/acs.jproteome.9b00825

Chapter 6

Myofibroblast TGF-β Activation Measurement In Vitro

Joanne Porte, Gisli Jenkins, and Amanda L. Tatler

Abstract

Myofibroblasts are critical to processes involved in normal wound healing and during pathological fibrosis. They transdifferentiate from fibroblasts, and in doing so become contractile and capable of secreting large amounts of extracellular matrix proteins. Transforming growth factor-beta (TGFβ) is a key cytokine involved in wound healing and fibrogenesis. TGFβ signaling has long been the subject of experimental therapeutic approaches to inhibit fibrosis in a variety of organ systems. Inhibition of TGFβ can reduce myofibroblast transdifferentiation, contractility, and matrix production. Importantly, TGFβ is released from cells and sequestered in the extracellular matrix in a latent form that requires activation for biological function. There have been multiple mechanisms of TGFβ activation described in a variety of cell types and in cell free systems; however, myofibroblasts have previously been shown to activate TGFβ via cell surface integrins, particularly αvβ5 integrins. This chapter will provide detailed protocols for accurately measuring activation of TGFβ by myofibroblasts in vitro. Levels of active TGFβ usually represent a small proportion of the total amount of latent TGFβ present in the matrix. Methods to measure active TGFβ therefore need to be sensitive and specific to detect the active cytokine only.

Key words Integrins, Latent transforming growth factor-β, Latency-associated peptide, Fibrosis

1 Introduction

The ubiquitously expressed cytokine transforming growth factor-β (TGFβ) is critical to wound healing processes and fibrogenesis. It is a pluripotent cytokine that has a wide variety of effects ranging from promoting extracellular matrix production and angiogenesis, influencing cellular proliferation and apoptosis, and suppressing inflammation and the immune system. The crucial role of TGFβ signaling in fibrosis is highlighted by the many diverse ways in which the pathway is being targeted as a potential anti-fibrotic therapy [1].

All three mammalian isoforms of TGFβ undergo posttranslational processing during which the pro-peptide is cleaved from the active cytokine homodimer [2]. The propeptide, also known as the latency-associated peptide (LAP), binds to the active cytokine homodimer via a non-covalent interaction, rendering the homodimer inactive. The latent TGFβ complex, also known as the small

Boris Hinz and David Lagares (eds.), *Myofibroblasts: Methods and Protocols*, Methods in Molecular Biology, vol. 2299, https://doi.org/10.1007/978-1-0716-1382-5_6,

latent complex, then binds to latent TGFβ binding proteins (LTBP) to form the large latent complex prior to release from the cell [3]. LTBPs serve to tether the large latent complex to extracellular matrix proteins. It is the extracellular activation of latent TGFβ that limits the bioavailability of TGFβ [4].

There have been many mechanisms of TGFβ activation described in vitro including extremes of the physical environment such as temperature and pH, and proteolytic cleavage of the latent complex to liberate the active TGFβ homodimer [5]. Additionally, some proteins have been shown to activate TGFβ by inducing a conformational change in the latent complex to either liberate the active homodimer or allow binding of active TGFβ1 to its cell surface receptor to initiate signaling. Such proteins include thrombospondin-1 and certain αv containing integrins [6–9]. In vivo the primary mechanism of TGFβ activation is via integrins, as multiple complex in vivo studies have demonstrated that interrupting integrin-mediated TGFβ activation has profound effects on both development and fibrotic processes [10, 11]. Integrin-mediated TGFβ activation also represents a feed-forward pathway as it has been shown to regulate expression of at least one of the key TGFβ activating integrins [12].

TGFβ is critical for the function of myofibroblasts. It induces their transdifferentiation from fibroblasts, thereby enhancing their contractility, and increases the synthesis and secretion of extracellular matrix proteins leading to matrix deposition [13]. Myofibroblasts are capable of activating TGFβ, and activation of their own endogenous TGFβ is thought to be crucial in their transdifferentiation [14]. Increased mechanical tension and/or contraction results in integrin-mediated TGFβ activation by myofibroblasts, specifically via the αvβ3 and αvβ5 integrins [8]. Furthermore, myofibroblasts can activate TGFβ via αvβ5 integrins in response to exogenous stimuli [15].

Due to the relatively small proportion of total TGFβ that gets activated and the fact that once released TGFβ is rapidly bound to its high-affinity cell surface receptors, specific methods of measuring TGFβ activation have been developed. Such methods allow the accurate detection of only the active portion of TGFβ enabling the researcher to be confident that the TGFβ they are measuring is biologically active. This chapter describes the methods that can be employed to detect TGFβ accurately and sensitively that is activated by myofibroblasts but can also be applied to most in vitro cell cultures. Such methods include using a transformed mink lung cells (TMLC) reporter cell assay and measuring downstream TGFβ signaling pathways as a proxy for directly measuring active TGFβ levels.

The TMLC reporter cell assay has become a widely used, sensitive assay for measuring TGFβ activation in vitro [7, 9, 16–22]. The assay utilizes mink lung epithelial cells stably transfected with a plasminogen-activator inhibitor-1 (PAI-1) promoter driving a firefly luciferase gene, which allows the detection of active TGFβ by measuring luciferase activity [23]. The response of TMLCs to all three isoforms of TGFβ is highly specific and they show no response to other inducers of PAI-1 [23]. The assay can be performed using either conditioned media from other cells or tissue samples to stimulate the TMLCs and quantify the amount of active TGFβ present in the sample [16]. Alternatively, the TMLCs can be used in a coculture assay with myofibroblasts (or other cell types) to sensitively measure integrin-mediated TGFβ activation where any active TGFβ is so rapidly bound to the TGFβ receptor that it may not be detectable in conditioned media samples [9, 21]. In both applications the use of a TGFβ1 standard curve acts as a positive control for the assay and allows for the quantification of TGFβ concentrations activated. The TMLC assay is somewhat limited when using some inhibitors as they can have an effect on the PAI-1 promoter independently of the TGFβ receptor (personal observation) or can affect viability of the TMLCs [21]; therefore it is crucial to include an experimental control that directly tests inhibitors on the response of the TMLCs without the presence of an additional cell type in coculture.

Smad2 is a TGFβ signaling molecule that becomes phosphorylated following the binding of TGFβ to its cell surface receptor. Phosphorylation of Smad2 is specific to TGFβ signaling and does not occur in response to signaling cascades initiated by other members of the TGFβ superfamily of proteins. Levels of phosphorylated Smad2 (PSmad2) can be detected either semiquantitatively by western blotting or quantitatively by commercially available ELISA kits. The advantage of measuring PSmad2 levels over the TMLC assay is that the sensitivity is increased due to not relying on the proximity of the TMLC reporter cells to the myofibroblasts.

Each of the following methods has been included in a previous Methods in Molecular Biology book chapter [24].

2 Materials

2.1 Materials for Transformed Mink Lung Cell (TMLC) Assay

1. TMLC culture medium: Dulbecco's modified Eagle medium (DMEM) supplemented with 10% (v/v) fetal calf serum, 4 mM L-glutamine, and 250 μg/mL G418 antibiotic.
2. Cell culture equipment and supplies (incubator, bio-safety cabinet, cell culture dishes, pipettes, etc.).
3. Recombinant TGFβ.
4. Luciferase assay kit: most products contain cell lysis buffer, luciferase assay substrate, and luciferase assay buffer (we use the Luciferase Assay System from Promega).

5. White opaque plate.
6. Luminometer.

2.2 Materials for Phosphorylated Smad2 Western Blotting

1. Recombinant TGFβ.
2. Protein molecular weight marker (we use Rainbow molecular weight marker).
3. Polyacrylamide gel casting apparatus.
4. Electrophoresis apparatus.
5. Transfer apparatus.
6. Phosphate-buffered saline (PBS): 150 mM NaCl, 8.06 mM Na_2HPO_4, 1.74 mM NaH_2PO_4, adjust pH with 1 M NaOH to pH 7.4.
7. Lysis buffer: 20 mM Tris–HCl, pH 7.4, 137 mM NaCl, 1% Triton X-100, 2 mM ethylenediaminetetraacetic acid (EDTA), 10% (w/v) glycerol, 25 mM β-glycerophosphate, 1 mM $NaVO_4$. Just prior to use, add protease inhibitor cocktail (we use P1860 from Sigma Aldrich) and phosphatase inhibitor cocktail (we use #37492 from Active Motif).
8. Cell scraper.
9. SDS-PAGE gels can be purchased pre-casted, or casted in advance using the following reagents: 30% bis acrylamide, distilled H_2O, ammonium persulfate, tetramethyl ethylenediamine (TEMED).
10. Buffer 1: 1.5 M Tris–HCl, 0.4% SDS, pH 8.8.
11. Buffer 2: 0.5 M Tris–HCl, 0.4% SDS, pH 6.8.
12. Tris-buffered saline (TBS) supplemented with Tween-20 (TBS-T): 25 mM Tris pH 7.4–7.6, 150 mM NaCl, 0.1% (v/v) Tween-20.
13. Blocking solution: 5% (v/v) non-fat milk diluted in TBS-T.
14. Laemmli buffer: 125 mM Tris–HCl, pH 6.8, 20% (v/v) glycerol, 2% (w/v) sodium dodecyl sulfate (SDS), 2 mg/mL bromophenol blue. Just before use, add 60 μL/mL β-mercaptoethanol.
15. Running buffer: 25 mM Tris, 190 mM glycine, 0.1% (w/v) SDS.
16. Transfer buffer: 25 mM Tris, 190 mM glycine and 20% (v/v) methanol.
17. Polyvinylidene difluoride (PVDF) membrane.
18. Primary antibody anti-Phospho-Smad2 (we use an.d anti-Ser456/467, #3108 from Cell Signaling, diluted 1:1000 in blocking solution).

19. Secondary antibody conjugated to horseradish peroxidase (we use a goat anti-rabbit-HRP #P0448 from Dako, diluted to 333 μg/mL with blocking solution).
20. Saran wrap.
21. Dark room.
22. X-ray film cassette.
23. Tape.
24. Developing fluid.
25. Fixer solution.
26. Marker pen.
27. Electrogenerated chemiluminescence (ECL) western blotting detection liquid.
28. ECL Hyperfilm.

3 Methods

3.1 Method for Transformed Mink Lung Cell (TMLC) Assay

3.1.1 Measurement of Global TGFβ Activity in Conditioned Medium

1. Maintain TMLCs in TMLC culture medium at 37 °C, 5% CO_2.
2. Trypsinize TMLCs, resuspend them in growth supplement-free DMEM, and count cells using a hemocytometer.
3. Seed the TMLC in a 96-well plate at a density of 1×10^5 cells/well and allow them to adhere for 3 h at 37 °C, 5% CO_2.
4. Conditioned media from either in vitro cell experiments (*see* **Note 1**).
5. Prepare a little more than 300 μL of each standard: dilute recombinant TGFβ to 250, 500, and 1000 ng/mL in growth supplement-free DMEM.
6. Aspirate and discard the TMLC medium from **step 3** and add 100 μL of the test samples and TGFβ1 standards in triplicate to the plate.
7. Incubate the plate at 37 °C, 5% CO_2 for 16 h.
8. Remove the medium by aspiration.
9. Wash the wells with PBS and remove all liquid.
10. Add 50 μL of 1× reporter lysis buffer per well to disrupt the cells (*see* **Note 2**).
11. Transfer the cell lysates to a white opaque plate.
12. Analyze using a luminometer: inject 100 μL of luciferase substrate into each well prior to the luminescence measurement.
13. Calculate TGFβ concentration in test samples based on the standard curve obtained from reading of standards run in parallel.

3.1.2 Measurement of Integrin-Mediated TGFβ Activation by TMLC Coculture Assay

1. Culture the experimental cell of interest in a 96-well plate until fully confluent.
2. Maintain TMLC cells in TMLC culture medium at 37 °C, 5% CO_2.
3. Switch the experimental cells' medium to supplement-free medium for 24 h to induce a growth arrest.
4. Trypsinize the TMLCs, count them using a hemocytometer, and resuspend them in growth supplement-free medium at a cell density of 0.5×10^6 cells/mL.
5. Aspirate the medium from the experimental cells and seed TMLCs directly on top by adding 100 μL/well of the cell suspension prepared in **step 4**.
6. Prepare the test samples and standards: dilute test compounds in serum-free DMEM to the desired concentration. Prepare a TGFβ standard curve by diluting recombinant TGFβ1 in serum-free DMEM to 500, 1000, and 2000 ng/mL.
7. Add 100 μL of the test samples and TGFβ1 standards to the plate, in triplicate.
8. Incubate the plate at 37 °C, 5% CO_2 for 16 h.
9. Remove all media by aspiration and wash all wells with PBS, removing all liquids.
10. Add 50 μL of 1× reporter lysis buffer per well to disrupt the cells (*see* **Note 2**).
11. Transfer cell lysates to a white opaque plate.
12. Analyze using a luminometer. Inject 100 μL of luciferase substrate into each well prior to the luminescence measurement.
13. Calculate TGFβ concentration in test samples based on the standard curve obtained from reading of standards run in parallel.

3.2 Method for Measurement of pSMAD2 by Western Blotting

1. In vitro experiments on growth arrested (i.e., cultured 24 h in serum-free medium) test cells are performed in duplicate in 12-well plates.
2. Stimulate two wells with 2 ng/mL TGFβ1 as a positive control for the assay. Incubate the cells for 4 h at 37 °C, 5% CO_2.
3. Place the plates on ice, remove the culture medium, and wash the wells in ice-cold sterile PBS.
4. Add 50 μL ice-cold lysis buffer to each well, ensuring that all cells are covered, and crap the monolayer with a cell scraper.
5. Collect samples from each well pooling the samples from duplicate wells, thus giving a final volume of 100 μL per treatment.
6. Quantify the protein concentration of each sample using a protein assay kit. Calculate volume of samples required to yield 100 μg protein.

7. Set up a 10% polyacrylamide gel either by using purchased pre-casted gels (then proceed directly to **step 9**) or by preparing gels according to the following steps:
8. Prepare a 10% polyacrylamide resolving gel by mixing 6.66 mL of 30% Bis acrylamide, 5.2 mL Buffer 1 mL, and 7.92 mL distilled H_2O. Add 200 μL of 10% ammonium persulfate and 50 μL of TEMED just prior to casting the gel.
9. Pour the resolving gel between the glass plates of the gel casting apparatus, leaving a 1.5 cm gap between the top level of the resolving gel and the top of the plates. Gently pipet 1 mL of water on top of the resolving gel to prevent gel from crusting.
10. Incubate at room temperature (RT) until the polyacrylamide resolving gel is set (*see* **Note 3**).
11. Pour off the water.
12. Prepare a polyacrylamide stacking gel by mixing 1.3 mL 30% bis acrylamide, 2.5 mL buffer 2 mL, and 6.1 mL distilled H_2O. Add 50 μL of 10% ammonium persulfate and 10 μL of TEMED just prior to casting the gel.
13. Pour the stacking gel onto the solidified resolving gel up to the top of the glass plates; insert a plastic comb to produce wells and allow the gel to solidify at RT.
14. Gently remove the comb.
15. Insert the polyacrylamide gel into the electrophoresis tank and add 1× running buffer to the middle compartment and to the outside compartment of the tank.
16. Mix the corresponding volume of 100 μg of each protein lysate 1:3 (v/v) with Laemmli buffer in a microtube and heat samples to 95 °C for 5 min.
17. Briefly centrifuge the microtubes, and pipet the samples into the wells of the polyacrylamide gel. Add 8 μL of molecular weight markers to one well of the gel from control.
18. Apply 150 mV across the gel for 90 min.
19. Cut a PVDF membrane and filter paper to the polyacrylamide gel's dimensions.
20. Activate the PVDF membrane by soaking it in methanol for 30 s and then in tap water.
21. Soak the electrophoresed gel, PVDF membrane and all sponges and filter paper to be used in the transfer in transfer buffer for 2 min.
22. Lay the electrophoresed gel onto the PVDF membrane and insert the assembly into the transfer apparatus sandwiched between filter paper and a sponge layer. Prior to closing the

transfer cassette, gently run a roller over the "sandwich" to expel air bubbles, which can interfere with efficient transfer of the proteins to the PVDF membrane.

23. Place the transfer assembly into the transfer tank containing ice-cold transfer buffer, ensuring the terminals are loaded in the correct orientation (i.e., cathode to cathode) (*see* **Note 4**).
24. Insert an ice block into the tank (*see* **Note 5**) and apply 110 V through the transfer apparatus for 35 min.
25. Disassemble the transfer assembly and incubate the membrane in blocking solution on a rocker for 1 hour at RT.
26. Wash the membrane three times with TBS-T for 5 min each time.
27. Apply the primary antibody diluted in blocking solution to the membrane and incubate overnight at 4 °C on a rocker.
28. Wash the membrane three times with TBS-T for 5 min each time.
29. Apply the secondary antibody diluted in blocking solution to the membrane and incubate for 1 h on a rocker at RT.
30. Wash the membrane three times with TBS-T for 5 min each time.
31. Prepare the ECL detection reagent by mixing equal volumes of Reagent 1 and Reagent 2.
32. Blot excess liquid from the membrane and apply the ECL reagent to the protein side of the membrane. Incubate for 45 s.
33. Blot ECL detection reagent from the membrane and wrap the membrane in Saranwrap, ensuring that no air bubble is trapped.
34. In a dark room, tape the wrapped membrane securely into an X-Ray film cassette and lay a sheet of X-ray Hyperfilm on top of the membrane. Close the cassette.
35. Expose the film to the membrane for an appropriate time until a good signal is found (approximately 20–30 min) (*see* **Note 6**).
36. In the dark room, remove the film from the cassette and place it in a tank of developing fluid (1:5 dilution in water) until bands appear.
37. Transfer the film to a tank of water for 3 min to wash, then to a tank of fixer solution (1:5 dilution in water) until the cloudy background of the film appears clear.
38. Wash the film in water and allow it to dry.
39. Lay the dry film over the membrane and mark the presence of the molecular weight markers on the film with a marker pen.

4 Notes

1. Samples can be stored at −80 °C until the assay when they should be thawed on ice. These samples are used neat in the assay, or can be heated to 100 °C for 5 min, then cooled prior to using in the assay to determine levels of total TGFβ (i.e., latent plus active) present [16].
2. Lysed plates can be stored at −80 °C until analysis.
3. To check if polyacrylamide resolving gel is set, gently tilt the gel casting apparatus sideways: if the gel is set, the water will collect in the corner but the gel's upper level will not move and from a straight line.
4. The SDS-denatured proteins are charged negatively. Therefore, they will migrate toward the positive pole when current is applied. The PVDF membrane should then be placed between the gel and the positively charged electrode (i.e., the anode).
5. Where not available the transfer tank can be stood in ice to ensure the transfer stays cool.
6. The appropriate time will need to be determined empirically as membrane exposure can be repeated.

References

1. Walton KL, Johnson KE, Harrison CA (2017) Targeting TGF-beta mediated SMAD signaling for the prevention of fibrosis. Front Pharmacol 8:461. https://doi.org/10.3389/fphar.2017.00461
2. Dubois CM, Laprise MH, Blanchette F, Gentry LE, Leduc R (1995) Processing of transforming growth factor beta 1 precursor by human furin convertase. J Biol Chem 270:10618–10624
3. Rifkin DB (2005) Latent transforming growth factor-beta (TGF-beta) binding proteins: orchestrators of TGF-beta availability. J Biol Chem 280:7409–7412. https://doi.org/10.1074/jbc.R400029200
4. Annes JP, Munger JS, Rifkin DB (2003) Making sense of latent TGFbeta activation. J Cell Sci 116:217–224
5. Tatler AL, Jenkins G (2012) TGF-beta activation and lung fibrosis. Proc Am Thorac Soc 9:130–136. https://doi.org/10.1513/pats.201201-003AW
6. Schultz-Cherry S, Murphy-Ullrich JE (1993) Thrombospondin causes activation of latent transforming growth factor-beta secreted by endothelial cells by a novel mechanism. J Cell Biol 122:923–932
7. Jenkins RG, Su X, Su G, Scotton CJ, Camerer E, Laurent GJ, Davis GE, Chambers RC, Matthay MA, Sheppard D (2006) Ligation of protease-activated receptor 1 enhances alpha (v)beta6 integrin-dependent TGF-beta activation and promotes acute lung injury. J Clin Invest 116:1606–1614
8. Wipff PJ, Rifkin DB, Meister JJ, Hinz B (2007) Myofibroblast contraction activates latent TGF- 1 from the extracellular matrix. J Cell Biol 179:1311–1323. https://doi.org/10.1083/jcb.200704042
9. Tatler AL, John AE, Jolly L, Habgood A, Porte J, Brightling C, Knox AJ, Pang L, Sheppard D, Huang X, Jenkins G (2011) Integrin alphavbeta5-mediated TGF-beta activation by airway smooth muscle cells in asthma. J Immunol 187:6094–6107. https://doi.org/10.4049/jimmunol.1003507
10. Yang Z, Mu Z, Dabovic B, Jurukovski V, Yu D, Sung J, Xiong X, Munger JS (2007) Absence of integrin-mediated TGFbeta1 activation in vivo recapitulates the phenotype of TGFbeta1-null mice. J Cell Biol 176:787–793. https://doi.org/10.1083/jcb.200611044
11. Henderson NC, Arnold TD, Katamura Y, Giacomini MM, Rodriguez JD, McCarty JH,

Pellicoro A, Raschperger E, Betsholtz C, Ruminski PG, Griggs DW, Prinsen MJ, Maher JJ, Iredale JP, Lacy-Hulbert A, Adams RH, Sheppard D (2013) Targeting of alphav integrin identifies a core molecular pathway that regulates fibrosis in several organs. Nat Med 19:1617–1624. https://doi.org/10.1038/nm.3282

12. Tatler AL, Goodwin AT, Gbolahan O, Saini G, Porte J, John AE, Clifford RL, Violette SM, Weinreb PH, Parfrey H, Wolters PJ, Gauldie J, Kolb M, Jenkins G (2016) Amplification of TGFbeta induced ITGB6 gene transcription may promote pulmonary fibrosis. PLoS One 11:e0158047. https://doi.org/10.1371/journal.pone.0158047

13. Scotton CJ, Chambers RC (2007) Molecular targets in pulmonary fibrosis: the myofibroblast in focus. Chest 132:1311–1321

14. Hinz B, Phan SH, Thannickal VJ, Galli A, Bochaton-Piallat ML, Gabbiani G (2007) The myofibroblast: one function, multiple origins. Am J Pathol 170:1807–1816

15. Scotton CJ, Krupiczojc MA, Konigshoff M, Mercer PF, Lee YC, Kaminski N, Morser J, Post JM, Maher TM, Nicholson AG, Moffatt JD, Laurent GJ, Derian CK, Eickelberg O, Chambers RC (2009) Increased local expression of coagulation factor X contributes to the fibrotic response in human and murine lung injury. J Clin Invest 119:2550–2563. https://doi.org/10.1172/JCI33288

16. Tatler AL, Porte J, Knox A, Jenkins G, Pang L (2008) Tryptase activates TGFbeta in human airway smooth muscle cells via direct proteolysis. Biochem Biophys Res Commun 370:239–242. https://doi.org/10.1016/j.bbrc.2008.03.064

17. Jolly L, Stavrou A, Vanderstoken G, Meliopoulos VA, Habgood A, Tatler AL, Porte J, Knox A, Weinreb P, Violette S, Hussell T, Kolb M, Stampfli MR, Schultz-Cherry S, Jenkins G (2014) Influenza promotes collagen deposition via alphavbeta6-integrin mediated transforming growth factor beta activation. J Biol Chem 289(51):35246–35263. https://doi.org/10.1074/jbc.M114.582262

18. Bedke N, Sammut D, Green B, Kehagia V, Dennison P, Jenkins G, Tatler A, Howarth PH, Holgate ST, Davies DE (2012) Transforming growth factor-beta promotes rhinovirus replication in bronchial epithelial cells by suppressing the innate immune response. PLoS One 7:e44580. https://doi.org/10.1371/journal.pone.0044580

19. Xu MY, Porte J, Knox AJ, Weinreb PH, Maher TM, Violette SM, McAnulty RJ, Sheppard D, Jenkins RG (2009) Lysophosphatidic Acid Induces {alpha}v{beta}6 Integrin-Mediated TGF-{beta} Activation via the LPA2 Receptor and the Small G Protein G{alpha}q. Am J Pathol 174:1264–1279. https://doi.org/10.2353/ajpath.2009.080160

20. Munger JS, Huang X, Kawakatsu H, Griffiths MJ, Dalton SL, Wu J, Pittet JF, Kaminski N, Garat C, Matthay MA, Rifkin DB, Sheppard D (1999) The integrin alpha v beta 6 binds and activates latent TGF beta 1: a mechanism for regulating pulmonary inflammation and fibrosis. Cell 96:319–328

21. Porte J, Jenkins G (2014) Assessment of the effect of potential antifibrotic compounds on total and alphaVbeta6 integrin-mediated TGF-beta activation. Pharmacol Res Perspect 2:e00030. https://doi.org/10.1002/prp2.30

22. Tatler AL, Barnes J, Habgood A, Goodwin A, McAnulty RJ, Jenkins G (2016) Caffeine inhibits TGFbeta activation in epithelial cells, interrupts fibroblast responses to TGFbeta, and reduces established fibrosis in ex vivo precision-cut lung slices. Thorax 71:565–567. https://doi.org/10.1136/thoraxjnl-2015-208215

23. Abe M, Harpel JG, Metz CN, Nunes I, Loskutoff DJ, Rifkin DB (1994) An assay for transforming growth factor-beta using cells transfected with a plasminogen activator inhibitor-1 promoter-luciferase construct. Anal Biochem 216:276–284

24. John AE, Porte J, Jenkins G, Tatler AL (2017) Methods for the assessment of active transforming growth factor-beta in cells and tissues. Methods Mol Biol 1627:351–365. https://doi.org/10.1007/978-1-4939-7113-8_23

Chapter 7

Contraction Measurements Using Three-Dimensional Fibrillar Collagen Gel Lattices

James Griffith and William L. Berry

Abstract

The ability of cells to contract is an important feature of many cell types. Methods to quantitate changes in the degree of contraction are important to study how cells respond to stimuli or change due to various pathologies. Here we describe a method of embedding cells in three-dimensional collagen lattices to measure contractile properties of cells in vitro.

Key words Collagen, Lattice, Contraction, $NaHCO_3$, NaOH

1 Introduction

Myofibroblasts and vascular smooth muscle cells (VSMC) are examples of two widely studied contractile cell types found in nature. However, there are other cell types that can acquire a contractile phenotype. Myofibroblasts are likely the primary cells responsible for the deposition of collagen during the fibrotic process. Myofibroblasts can form in response to a multitude of stimuli and can be derived from many other differentiated cell types [1]. However, vascular smooth muscle cells are also capable of depositing collagen and have been implicated in fibrotic diseases [2]. Although both of these cell types can remodel the extracellular matrix (ECM) and are contractile [3], methods to assay changes in contractile properties of cells are of great importance to study biological processes and the development of specific diseases.

Here we describe methods that can be used to assess rapid (free-floating) and/or delayed (attached lattice) contractile properties of cells. Free-floating lattices are more useful for measuring the ability of cells to contract in the absence of mechanical tension, whereas attached lattices are better at analyzing contraction under the influence of mechanical tension [4]. These two methods require embedding cells in a collagen-rich solution that polymerizes while

Boris Hinz and David Lagares (eds.), *Myofibroblasts: Methods and Protocols*, Methods in Molecular Biology, vol. 2299,
https://doi.org/10.1007/978-1-0716-1382-5_7,

settling on a plastic tissue culture dish [5, 6]. It has been shown that the ability of cells to contract either free-floating and/or attached collagen lattices can be both cell type specific and cell signaling dependent [7, 8]. Changes in contraction can occur through loss-of-function and gain-of-function mechanisms. These changes can be analyzed by measuring the diameter of collagen lattices over time. In summary, these methods are critical for studying myofibroblast formation and function under various conditions.

2 Materials

2.1 Preparation of Reagents for Collagen Lattices

1. Prepare a filter-sterilized (0.1μm filter) 10× solution of your desired growth media from powdered growth medium.
2. Based on the chosen growth medium, you should determine the amount of sodium bicarbonate ($NaHCO_3$) to make an equivalent 10× solution to maintain physiological pH in a 5–10% CO_2 environment (*see* **Note 1**).
3. Prepare a 0.1 N Solution of NaOH in water and filter sterilize (0.1μm filter). This will be necessary to aid in neutralizing the acidic collagen.
4. Stock concentration of rat tail collagen I (preferred EMD Millipore 08-115) varies from lot to lot but is often between 3.5 and 4.5 mg/mL. The final desired concentration can be adjusted based on experimental parameters. For simplification, the sample calculations will assume a stock concentration of 4.0 mg/mL and a final working solution using 1 mg/mL collagen.
5. Store prepared solutions at 4 °C until ready to use. These solutions can be stored and used for many months if proper sterile technique is maintained throughout preparation.
6. Tissue Culture Dishes (*see* **Note 2**).

2.2 Cell Culture

1. Cell line of choice.
2. Desired Growth Medium (Prepared in ultrapure water).
3. Plate warmer set to 37 °C.

2.3 Imaging

1. Olympus SZX16 Stereo Microscope with 0.5× Objective and a color camera or comparable setup from another manufacturer.
2. CellSens for Olympus microscopes or other comparable image acquisition software.
3. Microsoft Excel or comparable software for data collection and analyses.

3 Methods

Perform all steps in a tissue culture hood under sterile conditions. Begin by warming up plates on a clean plate warmer set to 37 °C. When working with collagen and final cell suspension use wide-bore 1 mL tips. Standard 1 mL tips are too narrow to use efficiently. We advise performing experiments in triplicate and repeated three times with biological replicates.

3.1 Preparation of Collagen Lattice Solution

All steps for the collagen solution should be prepared on ice (4 °C) and kept cold until plating. The example calculations below will be sufficient for 12–14 collagen lattices using 0.250 mL of a 4 mL cell suspension. Adjust volumes to fit your desired experiments.

1. Use a wide bore pipet tip to add collagen I to a prechilled 15 mL tube. The appropriate volume should be based on the collagen concentration of the stock. If we were to use a final volume of 4 mL, from a stock containing 4 mg/mL collagen I, then we would use 1 mL to achieve 1 mg/mL in the final mixture.
2. Add 0.1 N NaOH. Using 0.1 N NaOH you would add an amount equal to the collagen multiplied by 0.2. In the presented example case, we would need 0.2 mL.
3. Add sterile-filtered 10× growth medium. The amount of 10× used is determined by adding the volume of collagen and NaOH, then multiplying by 0.1. (1 mL collagen + 0.2 mL NaOH) × 0.1 = 0.120 mL 10× medium.
4. Add equivalent volume of 10× $NaHCO_3$. We would add an amount equal to the 10× media –0.12 mL.
5. Agitate gently to mix solution. Do not vortex! Avoid bubble formation.
6. Leave mixture on ice while cells are prepared.

3.2 Preparation of Cells for Collagen Lattices

1. Obtain cells by trypsinization or per usual laboratory methods and collect in a 15 mL tube.
2. Centrifuge to pellet cell suspension and carefully aspirate media.
3. Resuspend in growth media using whatever volume is necessary to ease counting procedure.
4. Count cells using your preferred method. Adequate cell number for collagen lattices should be optimized for cell type, growth conditions, and duration of experiment. As an example: 125,000 cells per milliliter final solution, or 31,250 per final collagen lattice (0.25 mL).

5. Resuspend in appropriate volume of growth media to achieve desired cell concentration. This will be equal to the remainder (2.56 mL) of the final collagen lattice cell suspension (4 mL) minus the collagen mixture (1.44 mL).
6. Pipet thoroughly to obtain single cell suspension and then add to the previously prepared collagen mixtures and keep on ice until ready to plate. Mix very gently with a pipet to avoid introducing bubbles. Do not vortex!

3.3 Seeding the Lattices

1. Using a wide bore pipet tip seed 0.25 mL cell/collagen solution carefully to the center of a warm dish and gently cover with lid (*see* **Note 3**). If possible, keep the warming tray in the tissue culture hood to maintain the plates at 37 °C (*see* **Note 4**).
2. Let collagen lattices sit undisturbed on warming tray for 5–10 min.
3. Move the lattices to an incubator under standard (5–10% CO_2) conditions for 1 h. It can be helpful to place dishes on a clean pre-warmed tray to move all at once, rather than risk moving them one at a time.
4. After 1 h incubation, gently add 2 mL of warm growth medium to the edge of dishes. Cold or room temperature media has greater surface tension, potentially hindering complete immersion of the lattice.

3.4 Release and Measurement

1. Release collagen lattices immediately after 1 h incubation in medium if performing free-floating lattice contraction assay. Release later 4–5 days culture if performing fixed collagen lattice contraction assays (recommended to preserve myofibroblast contractile features).
2. Use a scalpel to make small cuts along the edge of lattice.
3. Use a 200μL pipet to gently force media under and around cuts until lattice is released.
4. Measure diameter and/or image immediately after release, and additionally at specified time points. Depending on cell type and experimental conditions, contraction can occur rapidly or over a long period of time. It may be necessary to measure every 5–10 min after release, or every few hours. In general, we measure 0.5, 1, 3, 5, and 24 h after release (Fig. 1a, b). Changes in the diameter of the lattice can be determined by using a stereomicroscope equipped with a camera and software capable of real-time measurements that are linked to the changes in magnification because of focusing. Alternatively, a ruler can be used. Values should be recorded for future analyses. We typically analyze our data as changes in the percent of the original diameter.

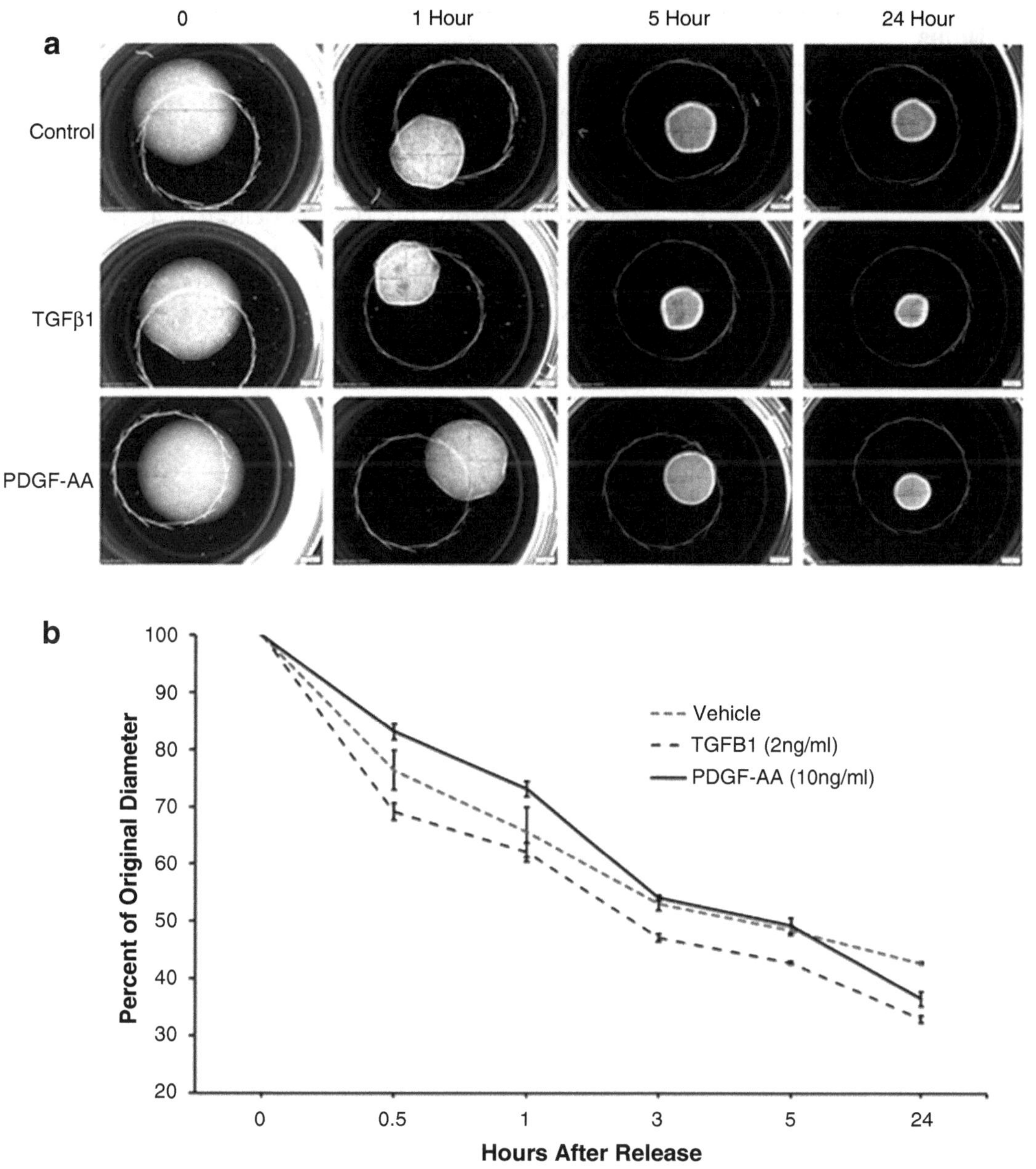

Fig. 1 Example Experiment Measuring Contraction of Treated Cells Grown in Fixed Collagen Lattices. Cells were cultured in an attached collagen lattice for 5 day in the presence of vehicle, TGFβ1, or PDGF-AA. Lattices were released from the plate on day 5 using a scalpel blade. Panels (**a**) and (**b**) photographs and measurements were taken at the indicated time points. Data is analyzed at a percent of the original diameter of the lattice. Error bars indicate the standard deviation

4 Notes

1. For DMEM, 3.7 g/L $NaHCO_3$ will provide a pH of 7.4 in a 10% CO_2 environment.
2. We recommend using plates from TPP.
3. Be extra careful to do this step as delicately as possible to produce perfectly circular lattices. An unfortunate bump can distort the lattice for the remainder of the experiment, impacting any measurements.
4. It may be worthwhile to make extra dishes to test collagen stability.

References

1. van Caam A, Vonk M, van den Hoogen F, van Lent P, van der Kraan P (2018) Unraveling SSc pathophysiology; the myofibroblast. Front Immunol 9:2452. https://doi.org/10.3389/fimmu.2018.02452
2. Doran AC, Meller N, McNamara CA (2008) Role of smooth muscle cells in the initiation and early progression of atherosclerosis. Arterioscler Thromb Vasc Biol 28:812–819. https://doi.org/10.1161/atvbaha.107.159327
3. Donovan J, Shiwen X, Norman J, Abraham D (2013) Platelet-derived growth factor alpha and beta receptors have overlapping functional activities towards fibroblasts. Fibrogenesis Tissue Repair 6:10. https://doi.org/10.1186/1755-1536-6-10
4. Montesano R, Orci L (1988) Transforming growth factor beta stimulates collagen-matrix contraction by fibroblasts: implications for wound healing. Proc Natl Acad Sci U S A 85:4894–4897. https://doi.org/10.1073/pnas.85.13.4894
5. Tomasek JJ, Haaksma CJ, Eddy RJ, Vaughan MB (1992) Fibroblast contraction occurs on release of tension in attached collagen lattices: dependency on an organized actin cytoskeleton and serum. Anat Rec 232:359–368. https://doi.org/10.1002/ar.1092320305
6. Ehrlich PH, Sun B, Kainth KS, Kromah F (2006) Elucidating the mechanism of wound contraction: rapid versus sustained myosin ATPase activity in attached-delayed-released compared with free-floating fibroblast-populated collagen lattices. Wound Repair Regen 14:625–632. https://doi.org/10.1111/j.1743-6109.2006.00170.x
7. Tingstrom A, Heldin CH, Rubin K (1992) Regulation of fibroblast-mediated collagen gel contraction by platelet-derived growth factor, interleukin-1 alpha and transforming growth factor-beta 1. J Cell Sci 102(Pt 2):315–322
8. Brown RA, Sethi KK, Gwanmesia I, Raemdonck D, Eastwood M, Mudera V (2002) Enhanced fibroblast contraction of 3D collagen lattices and integrin expression by TGF-beta1 and -beta3: mechanoregulatory growth factors? Exp Cell Res 274:310–322

Chapter 8

Techniques to Assess Collagen Synthesis, Deposition, and Cross-Linking In Vitro

Tamara Rosell-García and Fernando Rodriguez-Pascual

Abstract

Synthesis, deposition, and cross-linking of collagen are hallmarks of fibroblast to myofibroblast differentiation. Standard methods for determining collagen from tissue samples are not directly applicable to cell culture conditions, where the overall synthesis and deposition of collagen is clearly unfavorable, mainly due to quantity limitations and dilution of required extracellular remodeling factors. In this chapter, we describe the methods we have established to analyze collagen production and deposition into the extracellular matrix by cultured myo/fibroblasts, as well as to determine lysyl oxidase (LOX) activity in cell supernatants as an index of the capacity of the cell to cross-link collagen in vitro.

Key words Collagen deposition, Collagen cross-linking, Lysyl oxidase, Extracellular matrix, Fibroblast

1 Introduction

Collagens are the most abundant extracellular matrix proteins and fundamental to providing the connective tissues with their biomechanical properties. In vertebrates, 28 types of collagens have been described (I–XXVIII) and classified into several families, including the fibrillary collagens (I–III, V, XI XXIV, and XXVII) and the basement membrane collagen IV (reviewed in [1]). An important part of the knowledge on the biochemical and structural features of collagen has been obtained from studies with animal or human tissue samples, where collagen constitutes more than 80% of the organic matter. Using this material, investigations dating back half a century and more have delineated that fibrillary collagens form associations of three α-chains conforming the triple helical structure, its most representative feature, flanked by short non-helical segments, the telopeptides [2]. The abundance of collagen in these samples facilitated also the analysis of the complex repertoire of posttranslational modifications of collagen, including prolyl- and lysyl-hydroxylation, glycosylation, lysine oxidation, and

Boris Hinz and David Lagares (eds.), *Myofibroblasts: Methods and Protocols*, Methods in Molecular Biology, vol. 2299, https://doi.org/10.1007/978-1-0716-1382-5_8, © Springer Science+Business Media, LLC, part of Springer Nature 2021

cross-linking [3–5]. Despite the substantial progress, a more precise information about the collagen biosynthetic process has required studies with cultured cells, mainly fibroblasts, which have provided (and still does) remarkable details of the intracellular synthesis and chain association, the transit through the Golgi network and extracellular secretion, and the proteolysis and cross-linking that ultimately facilitate fibril assembly [6]. Giving the quantity limitations intrinsic to cell culture methodologies, these studies demand alternative approaches in order to analyze collagen metabolism. This type of experiments features also an unfavorable picture when assessing collagen deposition onto the insoluble extracellular matrix, giving the fact that the diluted microenvironment and the transitory nature of cell culture experimentation does not favor the action of extracellular collagen remodeling enzymes [7]. This is well illustrated when considering the formation of collagen cross-links by the action of lysyl oxidases (LOX). These covalent associations, important contributors to the stiffness of the collagen fibril, are generated by LOX-mediated oxidation of lysine (or hydroxylysine) residues within the telopeptide domains to the corresponding aldehydes. In a second step, these telopeptide aldehydes spontaneously condense with helical lysines (or hydroxylysines) to form immature cross-links, which further react over months/years with other aldehydes or additional residues to form a variety of permanent cross-links [8]. When assaying collagen cross-links from tissues, where the extracellular matrix is fully formed, the analysis of cross-linking metabolites by HPLC allows qualitative and quantitative characterization of the cross-linking profile [9]. However, under cell culture conditions, this process hardly progresses to advanced maturation products and their minute amounts make difficult, if not impossible, to apply HPLC methodology for the analysis. In this case, the determination of LOX activity in the cell supernatants provides a reliable index of the capacity of the cells to cross-link collagen.

In this chapter, we describe the methods we have established to analyze collagen synthesis and deposition by cultured myo/fibroblasts, as well as the analysis of LOX activity in cell supernatants.

2 Materials

We have established our methods with the human fibroblast cell line CCD-19 Lu and the murine fibroblast cell line NIH 3T3, where the former synthesizes and deposits increased amounts of collagen under our experimental conditions. The method can nevertheless be applied to any other cell type with the capacity to express and secrete collagen to the extracellular medium, including other primary or immortalized myo/fibroblasts, as well as keratinocytes, tenocytes, or chondrocytes from diverse sources.

2.1 Assessment of Collagen Synthesis and Deposition in Cultured Fibroblasts

1. Human fibroblast cell line CCD-19Lu.
2. Mouse fibroblast cell line NIH 3T3.
3. Tissue Culture Dishes (100 mm).
4. Screw cap Eppendorf tubes (2 mL).
5. Cell culture media: Dulbecco's Modified Eagle Medium (DMEM) without phenol red, calf serum and supplements, 500 KDa dextran sulfate and L-ascorbic acid 2-phosphate.
6. Phosphate-buffered saline (PBS): 0.138 M NaCl, 0.0027 M KCl, pH 7.4.
7. Transforming growth factor-β1 (TGF-β1).
8. Sircol™ Soluble Collagen Assay kit (Biocolor).
9. Acetic acid.
10. Pepsin.
11. Hydrochloric acid (HCl).
12. Sodium hydroxide (NaOH).
13. Acetate-citrate buffer (0.140 M citric acid, 0.453 M sodium acetate).
14. Chloramin-T.
15. 4-(Dimethylamino) benzaldehyde (DMAB).
16. Hydroxyproline.
17. Absorbance multiplate reader (570 nm wavelength).
18. Centrifuge.
19. Block heater.

2.2 Analysis of Lysyl Oxidase (LOX) Activity in Cell Supernatants

1. Amicon-Ultra 4 centrifugal filters (Ultracel-10K).
2. Lysyl oxidase activity assay kit (Fluorometric, Abcam).
3. β-Aminopropionitrile (BAPN) monofumarate.
4. Fluorescence multiplate reader (excitation and emission wavelengths of 525 and 580–640 nm)

3 Methods

3.1 Assessment of Collagen Synthesis and Deposition in Cultured Fibroblasts

Consistent with the amounts of collagen expected to be produced by cultured fibroblasts, we have set up colorimetric-based detection methods for the analysis of the synthesis and deposition of collagen. Sircol™, a dye binding to the Gly-X-Y tripeptides within the helical structure, is the only commercially available reagent reported to show enough sensitivity to accurately quantify the levels of soluble collagen from cell preparations. Extraction procedures, as described here, are established to convert the various fractions representing the sequential steps in the collagen biosynthetic process in a form

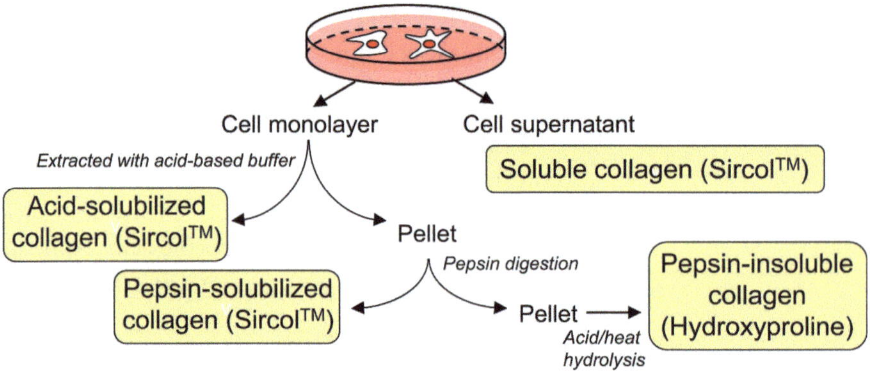

Fig. 1 Schematic flowchart summarizing the protocol established for the analysis of collagen synthesis and deposition in culture fibroblasts. The method, as depicted here, allows for a quantitative determination of the various fractions of collagen representing the sequential steps in the biosynthetic process. Cell supernatants and acid- or pepsin-solubilized collagens are assayed using Sircol™, as the dye have access to the helical portion of the molecule. More mature (cross-linked) pepsin-insoluble collagen must be broken down by heat/acid hydrolysis and further assayed by the content of hydroxyproline

suitable for the analysis with Sircol™ (Fig. 1) [10]. Finally, insoluble collagen, not accessible to Sircol™ dye, is quantified by the colorimetric measurement of hydroxyproline upon heat/acid hydrolysis.

1. Seed fibroblasts on 100-mm dishes in culture until confluency. This would require some days (almost a week) depending on the starting passage ratio and the proliferative capacity of the cell type used.
2. Prepare standard Dulbecco's Modified Eagle Medium (DMEM) media without serum and phenol red. Add 100 μg/mL 500 KDa dextran sulfate (D×S) and 29 μg/mL L-ascorbic acid 2-phosphate (*see* **Note 1**). Wash cell plates with phosphate-buffered saline (PBS) and incubate with 5 mL of this medium for 2–4 days. Longer times can be assayed to further increase collagen deposition but note that NIH 3T3 fibroblasts will die when cultured in low serum or serum-free medium for extended periods of time. Consider the addition of TGF-β1 at 5 ng/mL to increase the synthesis and deposition of collagen.
3. Take cell supernatants and assay soluble collagen with Sircol™ Soluble Collagen Assay kit following manufacturer's instructions.
4. Scrap cell layers in 2 mL of acid-base buffer (0.5 M acetic acid) in screw-cap Eppendorf Tubes and extract overnight at 4 °C with agitation.
5. Centrifuge for 15 min at 9000 × *g* and assay supernatants for acid-solubilized collagen with Sircol™ Soluble Collagen Assay kit.

6. Resuspend pellets in 0.5 mg/mL pepsin in 10 mM HCl for overnight digestion at 4 °C with agitation.
7. Centrifuge again and save supernatants for assessment of pepsin-solubilized collagen with Sircol™ Soluble Collagen Assay kit. Resuspend pellets in 12 M HCl for overnight hydrolysis at 100 °C with agitation.
8. Neutralize samples with NaOH and analyze by hydroxyproline assay [11, 12]. Hydroxyproline contained within samples is oxidized by addition of 0.05 M chloramine T in acetate-citrate buffer (0.140 M citric acid, 0.453 M sodium acetate, pH 6.5) followed by incubation at room temperature for 20 min. Then, add DMAB 1 M in 2-propanol/perchloric acid (2:1, v/v) and mix extensively. Incubate at 65 °C for 20 min and then read plate in a spectrophotometer at an absorbance wavelength between 550 and 565 nm [2] (*see* **Note 2**).

3.2 Analysis of Lysyl Oxidase (LOX) Activity in Cell Supernatants

Enzyme reaction catalyzed by LOX results in the oxidation of peptidyl lysine in collagens (and elastin) to the corresponding aldehydes, the initiation products for the cross-linking formation. Traditional methods to assess LOX-mediated reactions used acid/heat hydrolysis of collagen samples coupled to detection and quantification of cross-linking metabolites by HPLC [13]. This has been the method of choice for tissue samples where there is almost no limitation in the levels of collagen available for the analysis. Under culture conditions, with cells expressing and depositing limited amounts of collagen, HPLC-based technologies become problematic. To overcome this constraint, Palamakumbura and Trackman established a method that exploits the concomitant production of hydrogen peroxide in the reaction catalyzed by LOX by using Amplex red, a fluorometric probe specific for peroxide [14]. The method provided a sensitive assay for monitoring LOX reactions in samples from cultures. However, cellular sources for peroxide production are countless, including mitochondrial respiratory chain, the endoplasmic reticulum, the peroxisomes, and diverse oxidases. Therefore, adequate controls must be considered to accurately measure that peroxide coming specifically from LOX. What follows is the protocol we have developed to analyze the LOX enzymatic activity from supernatants of cultured fibroblasts (*see* **Note 3**).

1. Seed fibroblasts on 100-mm dishes in culture until confluency. Wash cell plates with PBS and incubate with 5 mL of DMEM media without serum and phenol red for 2–4 days.
2. Take cell supernatants and concentrate down to about 200 μL using Amicon Ultracentrifugal filters.
3. Transfer 50 μL of concentrated supernatant to wells of a 96 black well plate. Prepare controls with DMEM media only or with preparations of recombinant LOX protein (for example, supernatants from LOX-overexpressing cells) in the same way.

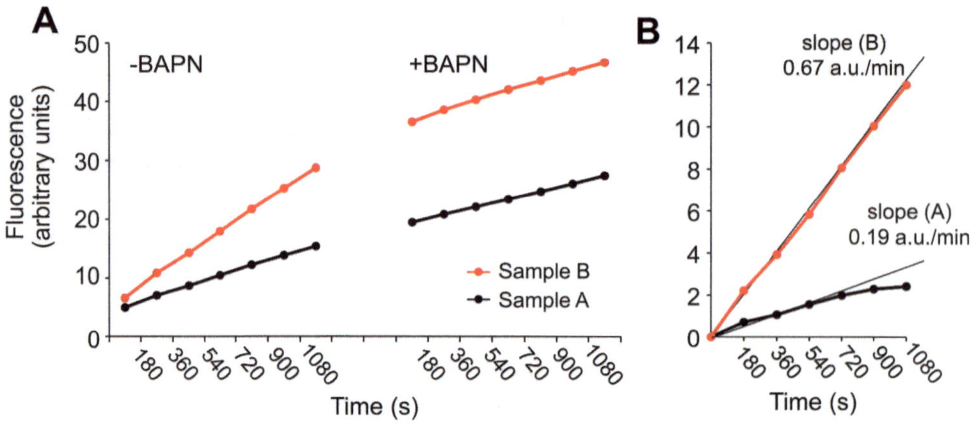

Fig. 2 Determination of lysyl oxidase (LOX) activity in cell supernatants. The figure shows an example of the protocol established to determine LOX activity from samples A and B by fluorometry. The method is based on the detection of peroxide, a product of the reaction catalyzed by LOX, using Amplex red. Specificity of the assay is gained by performing a sequential determination of the peroxide production in the absence and presence of the LOX inhibitor, β-aminopropionitrile (BAPN), and further subtraction of the peroxide coming from non-LOX sources

4. Add into each well an equivalent volume of reaction mixture containing a proprietary LOX substrate as well as horseradish peroxidase and buffer (lysyl oxidase activity assay kit) (*see* **Note 4**).
5. Insert plate in the fluorescence multiplate reader and take measurements of fluorescence (excitation and emission wavelengths of 525 and 580–640 nm) every 3 min for a total period of 18 min.
6. Remove plate from reader and add to each well a solution of the LOX inhibitor, BAPN for a final concentration of 0.3 mM (*see* **Note 5**).
7. Record fluorescence for an additional 18 min with measurements every 3 min.
8. LOX enzymatic activity is expressed as fluorescence arbitrary units/min from the slope of the traces of LOX-specific activity obtained upon subtraction of the values measured in the presence of BAPN (Fig. 2) (*see* **Note 6**).

4 Notes

1. Addition of L-ascorbic acid 2-phosphate and dextran sulfate is fundamental to provide conditions for cells to synthesize and deposit collagen into the extracellular matrix. Ascorbic acid is a cofactor required for several collagen-modifying enzymes, including prolyl and lysyl hydroxylases [4]. Instead of

L-ascorbic acid, our protocol uses the oxidation resistant analog, L-ascorbic acid 2-phosphate, thereby avoiding its replenishment in the culture medium. On the other hand, dextran sulfate has been described to promote collagen deposition into the matrix based on the principle of macromolecular crowding [15, 16]. In our hands, the addition of dextran sulfate increases pepsin-soluble and insoluble fractions, both representing maturation products of collagen, at the expense of the accumulation of soluble collagen in the extracellular medium. In general, as mentioned above, collagen synthesis and deposition is unfavorable under culture conditions. In addition to macromolecular "crowders," overall production can be increased with TGF-β1, with the exogenous addition of modifying enzymes or under hypoxic conditions [10, 17–19].

2. Hydroxyproline can also be analyzed using an amino acid analyzer, but, as mentioned above for Sircol™, the colorimetric method is much more sensitive, and therefore more applicable for cell culture samples.
3. Our protocol, as described here, performs well with cell supernatants, which are supposed to express the mature form of the LOX isoforms. Methods have also been reported to determine LOX activity from cell extracts [20]. However, unprocessed forms or lacking specific posttranslational modifications may account for the intracellular expression of these enzymes, therefore raising concerns on the validity of this type of measurements. In these circumstances, in addition to LOX activity assays, it is advisable to determine the expression levels of the LOX isoforms by western blotting using specific antibodies that recognize, for instance, the precursor and mature forms.
4. Instead of a commercial kit, a LOX enzymatic assay can be set up by separately purchasing a LOX substrate, cadaverine or benzylamine, horseradish peroxidase and the fluorometric Amplex red probe, mounting the reaction in a 50 mM sodium borate, pH 8.2 buffer. Nevertheless, the commercial LOX activity assay kit used here provides a superior sensitivity compared to homemade systems.
5. The addition of the LOX inhibitor, BAPN, serves as an internal control for every sample, taking into account other sources for hydrogen peroxide and the spontaneous oxidation of the fluorescent probe.
6. Arbitrary units can be easily converted to enzymatic activity (pmol or nmol of converted substrate/generated product per minute) by performing a calibration curve with known amounts of peroxide.

References

1. Kadler KE, Baldock C, Bella J, Boot-Handford RP (2007) Collagens at a glance. J Cell Sci 120 (Pt 12):1955–1958. https://doi.org/10.1242/jcs.03453
2. Fratzl P (2008) Collagen: structure and mechanics, an introduction. In: Fratzl P (ed) Collagen. Springer US, New York, pp 1–13. https://doi.org/10.1007/978-0-387-73906-9_1
3. Trackman PC (2005) Diverse biological functions of extracellular collagen processing enzymes. J Cell Biochem 96(5):927–937. https://doi.org/10.1002/jcb.20605
4. Gjaltema RAF, Bank RA (2017) Molecular insights into prolyl and lysyl hydroxylation of fibrillar collagens in health and disease. Crit Rev Biochem Mol Biol 52(1):74–95. https://doi.org/10.1080/10409238.2016.1269716
5. Yamauchi M, Sricholpech M (2012) Lysine post-translational modifications of collagen. Essays Biochem 52:113–133
6. Hulmes DJS (2008) Collagen diversity, synthesis and assembly. In: Fratzl P (ed) Collagen. Springer US, New York, pp 15–47. https://doi.org/10.1007/978-0-387-73906-9_2
7. Chen CZC, Peng YX, Wang ZB, Fish PV, Kaar JL, Koepsel RR, Russell AJ, Lareu RR, Raghunath M (2009) The Scar-in-a-Jar: studying potential antifibrotic compounds from the epigenetic to extracellular level in a single well. Br J Pharmacol 158(5):1196–1209. https://doi.org/10.1111/j.1476-5381.2009.00387.x
8. Rodriguez-Pascual F, Slatter DA (2016) Collagen cross-linking: insights on the evolution of metazoan extracellular matrix. Sci Rep 6:37374. https://doi.org/10.1038/srep37374. https://www.nature.com/articles/srep37374#supplementary-information
9. Yamauchi M, Shiiba M (2008) Lysine hydroxylation and cross-linking of collagen. Methods Mol Biol 446:95–108
10. Rosell-Garcia T, Rodriguez-Pascual F (2018) Enhancement of collagen deposition and cross-linking by coupling lysyl oxidase with bone morphogenetic protein-1 and its application in tissue engineering. Sci Rep 8(1):10780. https://doi.org/10.1038/s41598-018-29236-6
11. Kesava Reddy G, Enwemeka CS (1996) A simplified method for the analysis of hydroxyproline in biological tissues. Clin Biochem 29 (3):225–229. https://doi.org/10.1016/0009-9120(96)00003-6
12. Cissell DD, Link JM, Hu JC, Athanasiou KA (2017) A modified hydroxyproline assay based on hydrochloric acid in Ehrlich's solution accurately measures tissue collagen content. Tiss Eng Part C Methods 23(4):243–250. https://doi.org/10.1089/ten.tec.2017.0018
13. Eyre DR, Koob TJ, Van Ness KP (1984) Quantitation of hydroxypyridinium crosslinks in collagen by high-performance liquid chromatography. Anal Biochem 137 (2):380–388. https://doi.org/10.1016/0003-2697(84)90101-5
14. Palamakumbura AH, Trackman PC (2002) A fluorometric assay for detection of lysyl oxidase enzyme activity in biological samples. Anal Biochem 300(2):245–251. https://doi.org/10.1006/abio.2001.5464
15. Lareu RR, Arsianti I, Subramhanya HK, Yanxian P, Raghunath M (2007) In vitro enhancement of collagen matrix formation and crosslinking for applications in tissue engineering: a preliminary study. Tissue Eng 13 (2):385–391. https://doi.org/10.1089/ten.2006.0224
16. Lareu RR, Subramhanya KH, Peng Y, Benny P, Chen C, Wang Z, Rajagopalan R, Raghunath M (2007) Collagen matrix deposition is dramatically enhanced in vitro when crowded with charged macromolecules: the biological relevance of the excluded volume effect. FEBS Lett 581(14):2709–2714. https://doi.org/10.1016/j.febslet.2007.05.020
17. Gilkes DM, Bajpai S, Chaturvedi P, Wirtz D, Semenza GL (2013) Hypoxia-inducible factor 1 (HIF-1) promotes extracellular matrix remodeling under hypoxic conditions by inducing P4HA1, P4HA2, and PLOD2 expression in fibroblasts. J Biol Chem 288 (15):10819–10829. https://doi.org/10.1074/jbc.M112.442939
18. Chen CP, Yang YC, Su TH, Chen CY, Aplin JD (2005) Hypoxia and transforming growth factor-beta 1 act independently to increase extracellular matrix production by placental fibroblasts. J Clin Endocrinol Metab 90 (2):1083–1090. https://doi.org/10.1210/jc.2004-0803
19. Hanna C, Hubchak SC, Liang X, Rozen-Zvi B, Schumacker PT, Hayashida T, Schnaper HW (2013) Hypoxia-inducible factor-2alpha and TGF-beta signaling interact to promote normoxic glomerular fibrogenesis. Am J Physiol Renal Physiol 305(9):F1323–F1331. https://doi.org/10.1152/ajprenal.00155.2013
20. Trackman PC, Bais MV (2018) Chapter 8: Measurement of lysyl oxidase activity from small tissue samples and cell cultures. In: Mecham RP (ed) Methods in cell biology, vol 143. Academic Press, New York, pp 147–156. https://doi.org/10.1016/bs.mcb.2017.08.009

Chapter 9

Methods for Studying Myofibroblast Apoptotic Pathways

Yan Zhou and David Lagares

Abstract

Evasion of apoptosis by myofibroblasts is a hallmark of fibrotic diseases, ultimately leading to persistent myofibroblast activation, extracellular matrix (ECM) deposition, and remodeling. Targeting myofibroblast apoptosis is emerging as a novel therapeutic strategy to reverse established fibrosis. We have recently discovered that in the process of fibroblast-to-myofibroblast transdifferentiation driven by matrix stiffness, the "mitochondrial priming" (readiness to undergo apoptosis) is dramatically increased in stiffness-activated myofibroblasts. Thus, myofibroblasts, traditionally viewed as apoptosis-resistant cells, appear poised to die when survival pathways are blocked, a cellular state we call "primed for death." This apoptosis-prone phenotype is driven by high levels of pro-apoptotic proteins loaded in myofibroblast's mitochondria, which require concomitant upregulation of pro-survival BCL-2 proteins to suppress mitochondrial apoptosis and ensure survival. Here, we describe a method called BH3 profiling which measures myo/fibroblast apoptotic priming as well as their antiapoptotic dependencies for survival. In addition, we describe how BH3 profiling can be used to predict myofibroblast responses to therapeutic agents targeting pro-survival BCL-2 proteins, also known as BH3 mimetic drugs. Finally, we describe methods to assess myofibroblast sensitivity to extrinsic apoptosis via Annexin V staining.

Key words Myofibroblasts, Apoptosis, Fibrosis, BH3 profiling, BH3 mimetic drugs, Annexin V

1 Introduction

Apoptosis, a form of programmed cell death, is a complex cellular process tightly regulated by two interconnected molecular pathways: the extrinsic and intrinsic (mitochondrial) pathways. Deregulation in both extrinsic and intrinsic pathways of apoptosis are associated with myofibroblast persistence and the development of fibrosis. Here, we describe methods to assess myofibroblast apoptosis pathways using Annexin V staining and the so-called BH3 profiling assay, a functional tool that predicts myofibroblast apoptotic responses to peptide and drug treatments.

Boris Hinz and David Lagares (eds.), *Myofibroblasts: Methods and Protocols*, Methods in Molecular Biology, vol. 2299, https://doi.org/10.1007/978-1-0716-1382-5_9,

1.1 Control of Mitochondrial Apoptosis by the BCL-2 Family of Proteins

The intrinsic or mitochondrial pathway of apoptosis is controlled by more than fifteen members of the BCL-2 family of proteins, categorized by their structure and function into pro-apoptotics, antiapoptotics, BH3-only activators, and BH3-only sensitizers [1, 2](Fig. 1). Functionally, the pro-apoptotic BCL-associated X protein (BAX) and BCL-2 homologous antagonist/killer (BAK) proteins, also known as "effectors," are responsible for generating pores in the outer mitochondrial membrane and inducing mitochondrial outer membrane permeabilization (MOMP). BAX/BAK-mediated MOMP results in the release of cytochrome c and subsequent activation of caspases, which execute the apoptotic program [3]. Worth noting, the pro-apoptotic activities of BAX/-BAK are tightly regulated by the so-called BH3-only "activator" proteins including BCL-2-like protein 11 (BCL2L11; also known as BIM) and BH3-interacting domain death agonist (BID), which directly bind to and activate BAX/BAK to induce MOMP (Fig. 1). Even in the presence of high levels of pro-apoptotic "effectors" and "activators," cells can still block mitochondrial apoptosis by upregulating levels of "antiapoptotic" proteins such as BCL-2, BCL-W, BCL-XL, BFL-1, and MCL-1 (also known as "pro-survivals"). Antiapoptotic proteins directly bind to and sequester pro-apoptotic activators BIM/BID (Fig. 1), thus preventing their

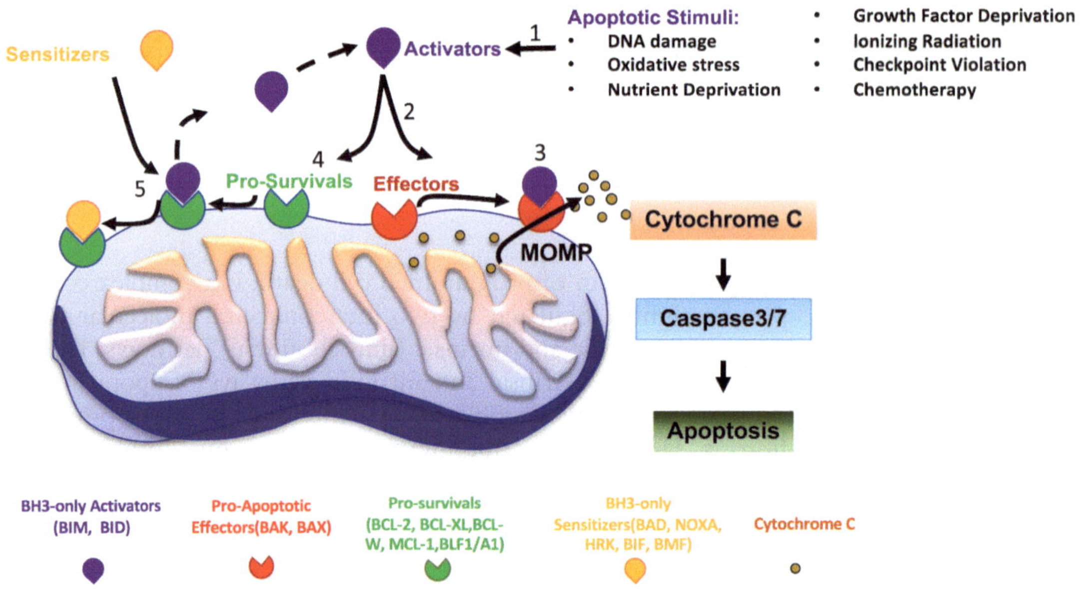

Fig. 1 The mitochondrial apoptosis pathway. In this simplified schematic, (*1*) cellular stress or damage signals increase levels of proapoptotic proteins (BH3-only "activators" of apoptosis), (*2*) which can either activate BAX and/or BAK ("effectors") (*3*). Activation of BAX or BAK causes mitochondrial outer membrane permeabilization (MOMP), resulting in the release of cytochrome c from mitochondria and consequent activation of caspases for dismantling of the cell (*4*). Antiapoptotics (pro-survivals) can bind and sequester activators (antiapoptotics such as BCL-2, BCL-XL or MCL-1) (*5*). BH3-only "Sensitizers" of apoptosis can be bound by pro-survival proteins, releasing activators to activate BAX and/or BAK

binding to BAX/BAK and initiation of mitochondrial apoptosis. More recently, antiapoptotic proteins have been shown to bind and inhibit effectors BAX/BAK as well [4–6]. The mitochondrial apoptotic pathway is further regulated by another group of proteins called "sensitizers," a class of BH3-only proteins that bind and sequester antiapoptotic proteins, thus indirectly promoting apoptosis by freeing activators (Fig. 1) [7]. Together, the mitochondrial apoptotic pathway is tightly regulated by expression levels and specific molecular interactions among effectors, activators, antiapoptotic, and sensitizers, which control the so-called "mitochondrial priming" of a cell [7, 8].We have recently found that fibroblast-to-myofibroblast differentiation induced by matrix stiffness associates with increased mitochondrial priming, augmenting the sensitivity of these cells to undergo apoptosis [9, 10].

1.2 Priming States

Based on relative levels and interactions of effectors, activators, pro-survivals, and sensitizers, mitochondrial priming (closeness to the apoptosis threshold) can be divided into three major classes (Fig. 2). Cells that lack functional BAX/BAK due to mutations, insufficient expression, or homozygous deletion of both proteins are unable to undergo MOMP (may still be able to die via the extrinsic apoptosis pathway). These cells are considered "apoptosis refractory." A cell that has functional BAX/BAK but expresses enough pro-survival proteins that can efficiently sequester activators BIM/BID is less likely to undergo apoptosis, or "unprimed." Finally, a cell that has functional BAX/BAK and is loaded with high levels of activators BIM/BIM requires upregulation of specific

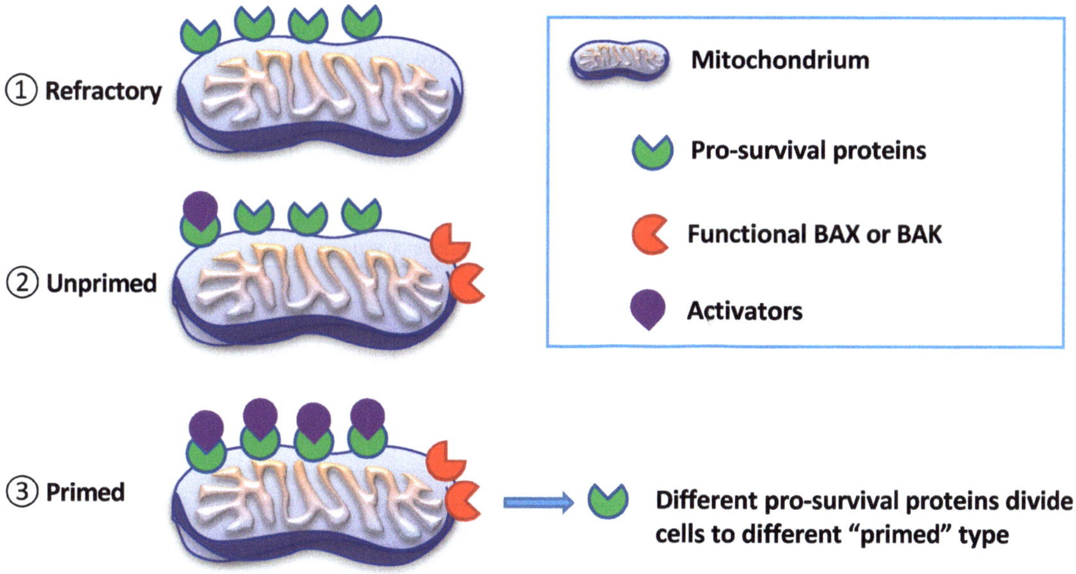

Fig. 2 Mitochondrial priming states. Priming states can be divided into three major classes based on levels and interactions among BCL-2 family of proteins: ① Apoptosis refractory; ② Unprimed; ③ Primed

pro-survival factors that can sequester BIM/BID to ensure survival. These cells are "primed" for apoptosis and die when the major pro-survival factor is inhibited. In the context of fibrosis, we have identified that quiescent fibroblasts appear to be unprimed while stiffness-activated myofibroblasts display increased mitochondrial priming [9]. Myofibroblasts are primed for death due to upregulation of activator pro-apoptotic protein BIM, which is sequestered by antiapoptotic protein BCL-X_L, ensuring their survival. For these studies, we applied BH3 profiling to determine myofibroblasts' apoptotic priming and survival dependencies.

1.3 Overview of BH3 Profiling Assay

Expression levels of each individual BCL-2 family protein alone is not enough to predict whether a cell will survive or undergo apoptosis in response to cellular perturbations. In this regard, BH3 profiling assay has been applied as a functional tool to determine the extent of MOMP that occurs in response to synthetic pro-apoptotic peptides that bind to specific BCL-2 proteins (Fig. 3). Peptides are derived from pro-apoptotic BH3-only proteins, mimicking pro-apoptotic activities of full-length BH3-only proteins. Digitonin is used to gently permeabilize the plasma membrane while keeping the integrity of mitochondria intact. Permeabilized cells are exposed to BH3 peptides, which diffuse and bind to mitochondrial BCL-2 proteins, modulating their activity. Upon peptide exposure, paraformaldehyde is used to fix cells and terminate peptide incubation. MOMP is now measured as a function of cytochrome c retained within the mitochondria in response to pro-apoptotic BH3 peptides, which is detected by immunostaining and quantified by flow cytometry (Fig. 4). The more cytochrome c released in response to BH3 peptide treatment, the more primed the cell was for apoptosis. BH3 profiling assay uses an array of BH3 peptides that can determine not only overall apoptotic priming but also antiapoptotic dependencies for survival.

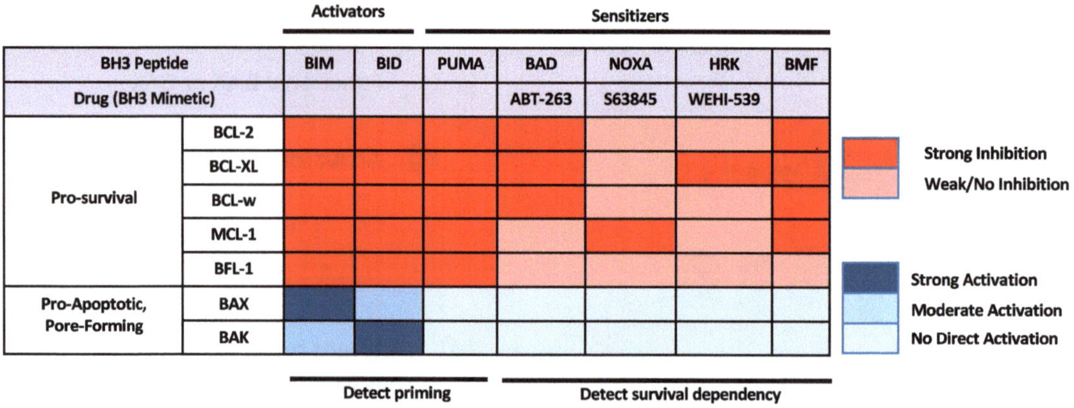

Fig. 3 Binding affinities of BH3 peptides and BH3 mimetic drugs to BCL-2 family of proteins. Binding affinities of BH3 peptides derived from activator or sensitizer BH3-only proteins to their pro-survival and pro-apoptotic partners from the BCL-2 family of proteins. BH3 mimetic drugs interaction with BCL-2 proteins

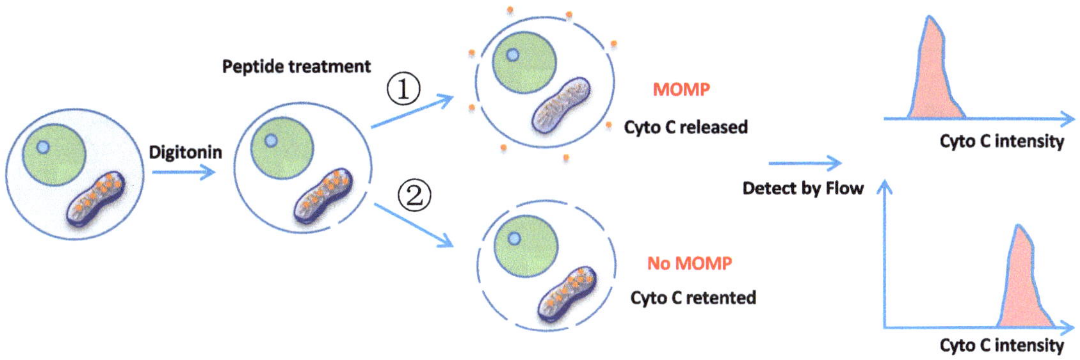

Fig. 4 BH3 profiling assay. Schematic of cell level events during BH3 profiling. In this assay, cells are permeabilized and exposed to different doses of pro-apoptotic BH3 peptides. Cytochrome c staining is used to monitor mitochondrial permeabilization (MOMP) as a read-out. Initially, digitonin permeabilizes the plasma membrane of cells while leaving mitochondria intact, thus permitting diffusion of BH3 peptides into cells and their interaction with BCL-2 family of proteins. If peptide treatment triggers MOMP, cytochrome c will be released (*1*). On the other hand, if mitochondria do not response to BH3 peptide treatment, cytochrome c will be retained within mitochondria. Healthy mitochondria retain cytochrome c and are therefore positive for cytochrome c staining (*2*). After peptide exposure and fixation, MOMP is measured as a function of cytochrome c retained within the mitochondria, which is detected by immunostaining and quantified by flow cytometry.

1.4 Measuring Overall Mitochondrial Apoptotic Priming and Antiapoptotic Dependencies

Overall mitochondrial priming is measured with BH3 peptides derived from BIM, BID, or PUMA proteins. BIM BH3 peptide not only binds to and activates pro-apoptotic BAX/BAK, but also binds to and inhibits any antiapoptotic proteins expressed in the cell (Fig. 3). BID BH3 peptide is functionally similar to BIM peptide and can be used to assess overall mitochondrial priming. Worth noting, BID preferentially activates BAK while BIM preferentially activates BAX [11]. Contrarily, PUMA BH3 peptide binds to and inhibits all antiapoptotic proteins but cannot directly activate BAX/BAK (Fig. 3) [12]. Overall mitochondrial priming can be assessed by titrating these peptides and determining the dose required to induce MOMP, which inversely correlates with the level of priming. For instance, low doses of BIM peptide required to induce MOMP indicate increased mitochondrial priming, whereas high doses of BIM required to induce MOMP indicates low level of priming. "Refractory" cells cannot undergo MOMP no matter how high concentration of these peptides are used (Fig. 5).

BH3 profiling can also determine which antiapoptotic BCL-2 protein(s) maintains cell survival by using peptides derived from BH3-only sensitizer proteins that selectively inhibit specific antiapoptotic members. Functionally, sensitizer BH3 peptides bind to pro-survival protein(s), releasing sequestered BIM/BID which can now bind to BAX/BAK to induce apoptosis. Given their selectivity at binding one or multiple pro-survival factors, sensitizer BH3 peptides can determine whether a cell uses one or multiple pro-survival proteins to prevent apoptosis. *See* interaction panel of

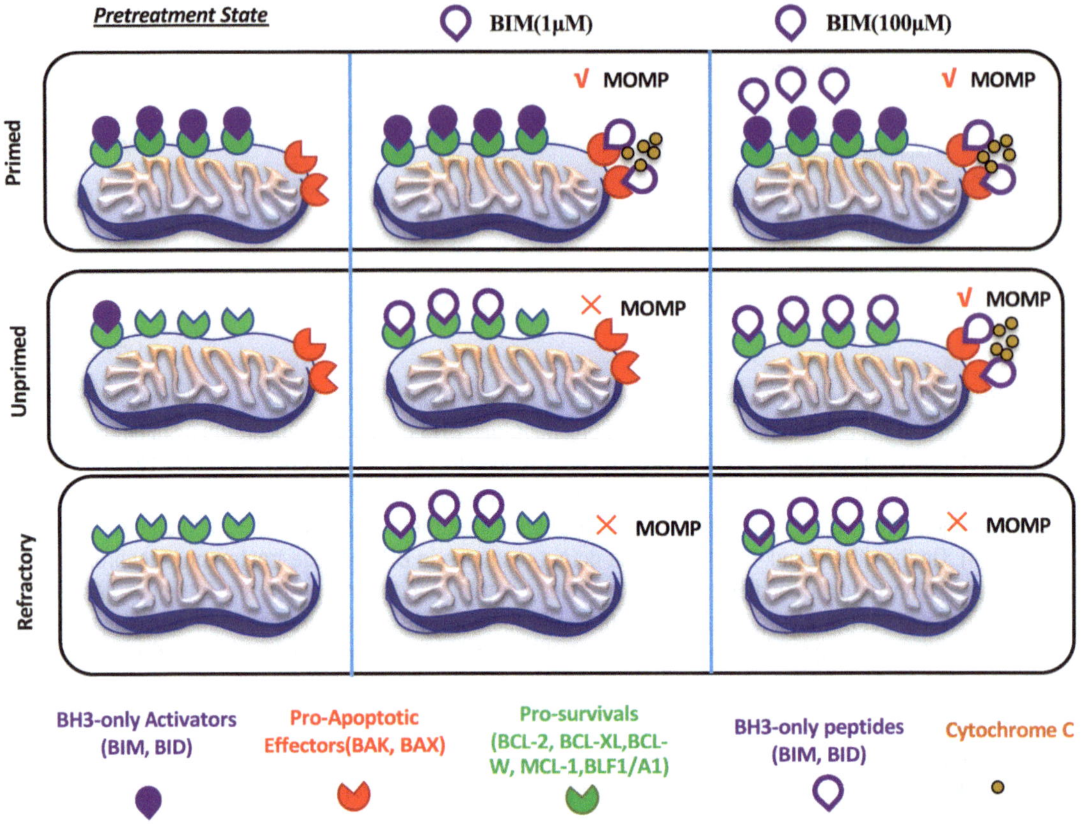

Fig. 5 Mitochondrial apoptotic priming state determined by BH3 profiling. Using BH3 proapoptotic peptides such as BIM or BID peptides, BH3 profiling assay can assess overall mitochondrial priming of cells. Mitochondria that have less unbound pro-survival proteins are quickly and fully depolarized (loss of cytochrome c) by even low doses of BIM (1μM) and are considered "primed" for apoptosis. Mitochondria that have high level of unbound pro-survival proteins and only respond to high concentration of peptides (100μM) are considered "unprimed" for apoptosis. Mitochondria that lack BAX or BAK are apoptosis incompetent and are considered "Refractory" for apoptosis

sensitizer BH3 peptides in Fig. 3. For example, since HRK peptide only binds to BCL-X_L protein, MOMP induced by HRK peptide indicates a dependency on BCL-X_L for survival. Similarly, NOXA BH3 peptide specifically assesses MCL-1 dependency (Fig. 6). In addition, survival dependencies can be assessed by BH3 profiling using the so-called BH3 mimetic drugs, which are small molecules that bind to and inhibit specific BCL-2 survival proteins, functionally mimicking BH3 peptides (Fig. 3). BH3 mimetic drugs including ABT-263 has been shown to reverses established in mouse models of fibrosis by inducing apoptosis of myofibroblasts [9, 10]. Therefore, BH3 profiling acts a functional biomarker assay that not only determines myofibroblast mitochondrial priming and survival dependencies but also predicts whether cells will undergo apoptosis in response to specific BH3 mimetic drugs or other therapeutic interventions.

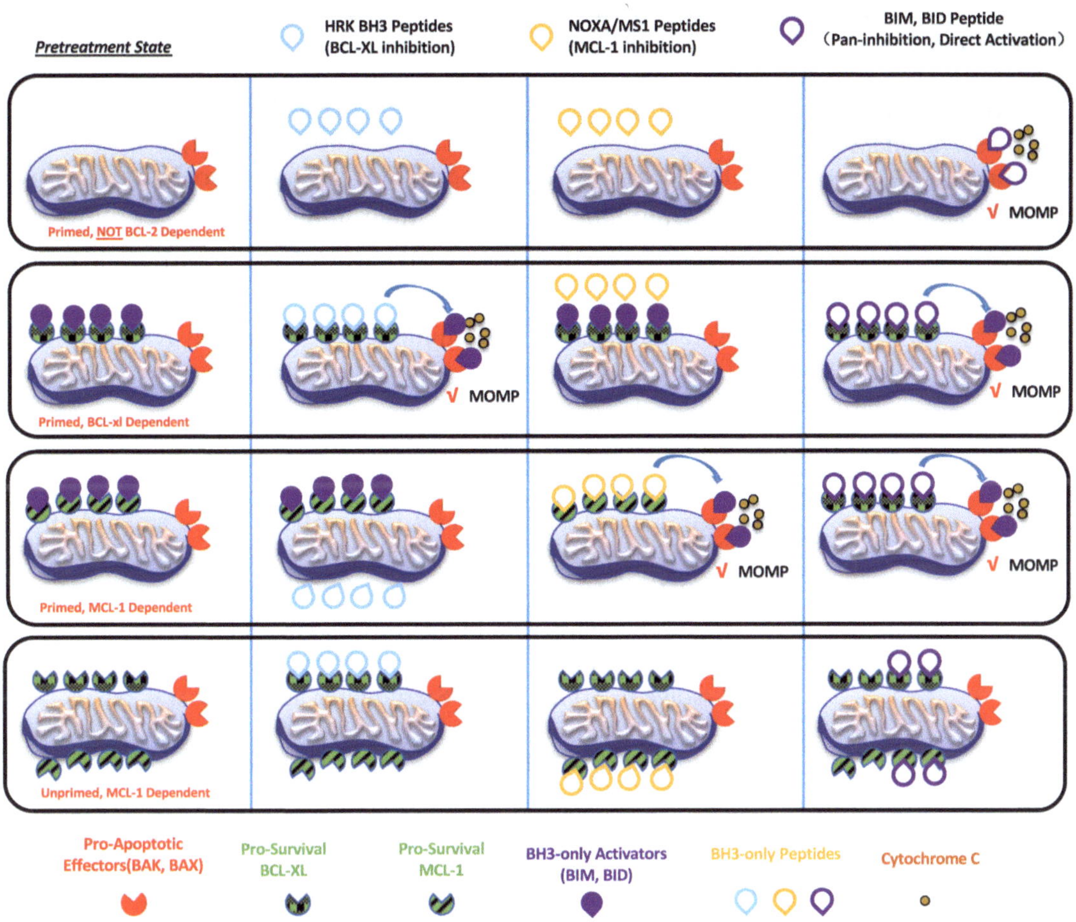

Fig. 6 Diagnosing dependencies on pro-survival BCL-2 family proteins by BH3 profiling. Using the BH3-only sensitizer peptides, the BH3 profiling assay can detect if a cell has a particular dependence on one or multiple antiapoptotic BCL-2 family proteins for survival. The sensitizer peptides can selectively inhibit particular antiapoptotic proteins. If a cell is primed, but does not express a particular antiapoptotic protein at high levels, the sensitizer peptides will have little effect. If a cell is primed and expresses BCL-2 or MCL-1 at a high level, treatment with the HRK or NOXA peptides, respectively, will inhibit the antiapoptotic and allow for the activation of BAX/BAK

1.5 Extrinsic Pathway of Apoptosis

The extrinsic pathway of apoptosis is activated by the binding of extracellular death ligands such as FAS ligand (FASL), Tumor Necrosis Factor (TNF), and TNF ligand superfamily member 10 (TRAIL) to death receptors. In this chapter, we also describe classic methods to assess myofibroblast apoptosis via Annexin V staining in response to treatment with Fas ligand or activating anti-Fas antibodies.

2 Materials

2.1 Assessment of Intrinsic Apoptotic Pathways via BH3 Profiling Assay

2.1.1 Buffers and Reagents

1. Mannitol Experimental Buffer (MEB) buffer: 150 mM mannitol, 10 mM HEPES pH 7.5, 150 mM KCl, 1 mM EGTA, 1 mM EDTA, 0.1% BSA, 5 mM succinate (*see* **Note 1**).
2. 5% Digitonin: 50 mg/mL digitonin in DMSO (*see* **Note 2**).
3. Alamethicin dissolved in DMSO at 5 mg/mL (100× final concentration).
4. BH3 Peptides or BH3 mimetic drugs, typically resuspended in DMSO (100× final concentration) (*see* **Note 3**, Fig. 3 and Table 1).
5. 4% Paraformaldehyde in PBS.
6. "N2" Buffer [1.7 M Tris, 1.25 M Glycine pH 9.1]: 1.7 M Tris base, 1.25 M glycine, pH 9.1. For 50 mL of buffer, add 10.3 g Tris base (M.W. 121.11) and 4.69 g glycine (M.W. 75.07). Adjust pH to 9.1.
7. 10× Intracellular Staining Buffer (10% BSA (define BSA), 2% Tween20): Per 50 mL, combine 1 mL Tween20, 5 g BSA, and dissolve in 50 mL PBS. Sterile filter and store at 4 °C.

2.1.2 Staining

1. Zombie Aqua™ Fixable Viability Kit (BioLegend) (or any live cell/dead cell discrimination dye).
2. Cytochrome c Antibody: We recommend the 6H2.B4 Clone from BioLegend. Both Alexa Fluor® 647 anti-cytochrome c

Table 1
Frequently used peptides and their amino acid sequences

Peptides	Sequence	Extinction coefficient at 280 nm
hBIM	Acetyl-MRPEIWIAQELRRIGDEFNA-Amide	5500 cm^{-1} M^{-1}
hBID-Y	Acetyl-EDIIRNIARHLAQVGDSMDRY-Amide	1490 cm^{-1} M^{-1}
mBAD	Acetyl-LWAAQRYGRELRRMSDEFEGSFKGL-Amide	6990 cm^{-1} M^{-1}
mNoxaA	Acetyl-AELPPEFAAQLRKIGDKVYC-Amide	1490 cm^{-1} M^{-1}
Puma	Acetyl-EQWAREIGAQLRRMADDLNA-Amide	5500 cm^{-1} M^{-1}
Bmf-Y	Acetyl-HQAEVQIARKLQUADQFHRY-Amide	1490 cm^{-1} M^{-1}
W-Hrk	Acetyl-SSAAQLTAARLKALGDELHQY-Amide	5500 cm^{-1} M^{-1}
Puma2A	Acetyl-EQWAREIGAQARRMAADLNA-Amide	5500 cm^{-1} M^{-1}

Ac Acetyl, *NH_2* Amide, -Y and W- designate added residues for UV absorbance measurements at C or N term, respectively

Antibody and FITC anti-cytochrome c Antibody have been validated for this assay in our laboratory.

2.1.3 Equipment

1. Flat-bottom 96-well clear NBS (Corning).
2. Flow cytometer.

2.2 Measuring Extrinsic Apoptotic Pathway via Annexin V Staining

2.2.1 Buffers and Reagents

1. FasL (Calbiochem, CA) or activating anti-Fas antibody CH11 (Millipore, Billerica, MA).
2. Annexin V Binding Buffer (Biolegend).

2.2.2 Staining

FITC-Annexin V antibody (Biolegend).

2.2.3 Equipment

1. Flow cytometer.

2.3 Human Primary Dermal Fibroblasts

3 Methods

3.1 Measuring Myofibroblast Intrinsic Apoptotic Pathways via BH3 Profiling Assay

We have recently discovered that stiffness-activated myofibroblasts increased their mitochondrial priming and became susceptible to apoptosis upon treatment with BH3 mimetic drugs [9]. Here, we describe how to apply BH3 profiling to study mitochondrial priming in fibroblasts and myofibroblasts, which can be directly isolated from tissue samples obtained from experimental models of fibrosis or human biopsies obtained from patients with fibrotic disease, or from any experimental condition that modulates the biology of fibroblasts in vitro including matrix stiffness or growth factor stimulation. Here, we will simply use human primary dermal fibroblasts cultured in petri dishes in order to show the steps required to establish BH3 profiling in the laboratory (for overview of the workflow, *see* Fig. 7).

3.1.1 Viability Staining

Fibroblasts are initially trypsinized from petri dish. Single-cell suspensions are then labeled with Zombie Aqua dye following this protocol:

1. Dilute Zombie Aqua™ dye at 1:200 in PBS. Resuspend a maximum of 10×10^6 cells in 100µL of diluted Zombie Aqua™ solution.
2. Incubate cells at room temperature, in the dark, for 30 min.

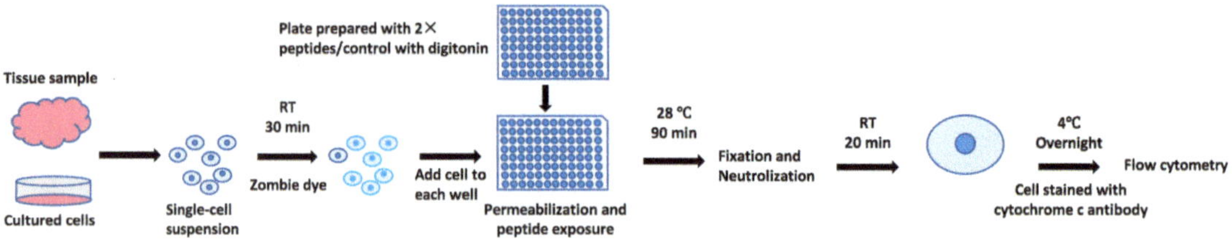

Fig. 7 Flow cytometry BH3 profiling workflow. Tissue or cells are collected and prepared into single-cell suspension. Cells are labeled with Zombie Aqua dye and washed with PBS. Cells are then permeabilized with digitonin and loaded into a 96-well plate that contains BH3 peptides or controls and incubated for 90 min, 28 °C. Cells are then fixed and stained for cytochrome c overnight, 4 °C. Cells are then analyzed by flow cytometry

3. Fibroblasts are spun down (1000 rpm or 180 × *g*, 5 min) and washed once with 2% FBS, which quenches the reaction.
4. Resuspend cells at 2 × 10^6/mL in MEB buffer (50,000 cells in 25μL are needed per well of 96-well plate).

3.1.2 Preparation of 96-Well Plate with Permeabilization Buffer and BH3 Peptides and Drugs

The basis of BH3 profiling has been described in Subheading 1.3. As discussed above, digitonin initially permeabilizes the plasma membrane of cells while leaving mitochondria intact, thus permitting diffusion of BH3 peptides or drugs into cells and their interaction with BCL-2 family in the outer mitochondria membrane. If any of the treatments with BH3 peptides or drugs triggers MOMP, cytochrome c will be released. On the other hand, if mitochondria do not response to BH3 peptide or drug treatment, cytochrome c will be retained within mitochondria. After peptide exposure and fixation, MOMP is measured as a function of cytochrome c retained within the mitochondria, which is detected by immunostaining and quantified by flow cytometry. Prepare a 96-well plate for BH3 profiling assay as follows:

1. Preparation of treatments at 2× final concentration (BH3 peptides and/or BH3 mimetic drugs) and controls on permeabilization MEB buffer: For reference, 25μL are required per well of 96-well plate.
   (a) Prepare 0.005% digitonin in MEB buffer by adding 1μL of 5% digitonin per 1 mL of MEB buffer.
   (b) Prepare BH3 peptides/drugs in 0.005% digitonin MEB buffer at 2× final concentration. For instance, if the desired final concentration is 100μM and the peptide stock concentration is 10 mM, simply add 2μL of 10 mM peptide to 98μL MEB buffer containing 0.005% digitonin, leaving peptide at 200μM (2× desired final concentration, since we will add the same volume of cell suspension solution culture with peptides/drugs). We typically test the following peptides:

- BH3 peptides use to assess priming state (desired final concentration): BIM (0.3μM, 1μM, 3μM); BID (0.3μM, 1μM, 3μM); PUMA (10μM, 25μM, 50μM).
- BH3 peptides to assess survival mechanisms (desired final concentration): BAD (100μM), NOXA (100μM), HRK (100μM), BMF (10μM).
- BH3 mimetic drugs (desired final concentration): ABT-263 (1μM), S63845 (1μM), WEHI-539 (1μM).

(c) Prepare Negative control: 1% DMSO or the inert PUMA2A peptide (dilute in the 0.005% digitonin MEB buffer) at the highest used peptide concentration as controls for complete cytochrome c retention.

(d) Prepare Positive control: Alamethicin at 50μg/mL (dilute in the 0.005% digitonin MEB buffer) as a complete cytochrome c release positive control (*see* **Note 3**).

2. Add 25μL of treatment/control solutions to each well of 96-well plate (*see* **Note 5**).
3. Keep 96-well plate on ice while cell suspensions are prepared.

3.1.3 BH3 Profiling Assay

1. Add 25μL of resuspend cells in MEB buffer to each well (already containing peptides) while keeping the 96-well plate on ice (*see* **Note 6**).
2. The assay is controlled by temperature. In order to initiate BH3 profiling assay, incubate plate for 90 min at 28 °C. Be aware that peptide responses increase as a function of time and temperature.
3. Terminate treatment exposure by adding 15μL of 4% formaldehyde to each well (final concentration about to 1%), allow plate to rest 10 min at room temperature.
4. Neutralize formaldehyde: Add 30μL of N_2 buffer to each well, allow plate to rest 10 min at room temperature.

3.1.4 Cytochrome c Staining

1. Prepare cytochrome c antibody (Alexa Fluor® 647 anti-cytochrome c) at 1:400 dilution on 10× intracellular staining solution, add 10μL to each well (final concentration about 1:4000).
2. Stain with cytochrome c antibody overnight at 4 °C: Cover plate with adherent plate cover and protect from light.
3. Use a fluorescence minus one (FMO) control without the addition of an isotype or cytochrome c antibody as a negative control. Because a matched isotype control for the cytochrome c antibody occasionally stain brighter than some treatments resulting in cytochrome c release values greater than 100% (*see* **Note 7**).

4. Proceed to flow cytometry, standard compensation controls are those typically used in standard Flow cytometry experiments.

3.1.5 Flow Cytometry Analysis

1. Identify cells of interest by excluding doublets and dead cells.
2. Draw gates in the cytochrome c Florescence channel (Fig. 8).
   - **Identify cytochrome c negative population:** Alamethicin and unstained FMO should be negative for cytochrome c staining. Use these wells to set a gate for cytochrome c negative cells.
   - **Identify cytochrome c positive population:** DMSO and Puma-2A will not cause cytochrome c loss. Use this population to set a gate for cytochrome c positive. The cytochrome c negative and positive gates should not overlap.
   - **Run treatment samples.** Once the gate is set, run all other treatment samples using the same gates and look at the percentage of cytochrome c negative cells caused by the treatments. This can be plot as percentage of cytochrome c loss after treatment (Fig. 8).

3.2 Assessment of Myofibroblast Extrinsic Pathway of Apoptosis via Annexin V Staining

Activation of the extrinsic pathway involves death receptors such as Fas and can be investigated by stimulating fibroblasts and myofibroblasts with recombinant Fas ligand or activating anti-Fas antibodies. Apoptosis in response to these treatments is measured by Annexin V-staining by flow cytometry, which involves binding to extracellular-facing phosphatidylserine (PS) on the surface of apoptotic cells.

3.2.1 Fas Ligand Treatment

Primary human dermal fibroblasts obtained from healthy controls or patients with fibrotic diseases cultured in DMEM with 1% FBS are treated with FasL (50 ng/mL) or activating anti-Fas antibody (250 ng/mL) for 24 h.

3.2.2 Sample Collection and Staining Procedure

1. Collect cell supernatants and save them on a 15 mL falcon tube. Keep them on ice.
2. Wash cells with 1 mL of cold PBS and add it to the same 15 mL tube.
   (a) Trypsinize fibroblasts and transfer cell suspension into the same tube of supernatants and PBS. Centrifuge for 5 min at 1000 rpm (4 °C).
   (b) Wash pellet twice with cold PBS, and then resuspend cells in Annexin V Binding Buffer at a concentration of $0.25–1.0 \times 10^7$ cells/mL.
   (c) Transfer 100μL of cell suspension into a 1.5 mL tube and stain cells with 2μL of FITC-Annexin V (1:50 dilution).

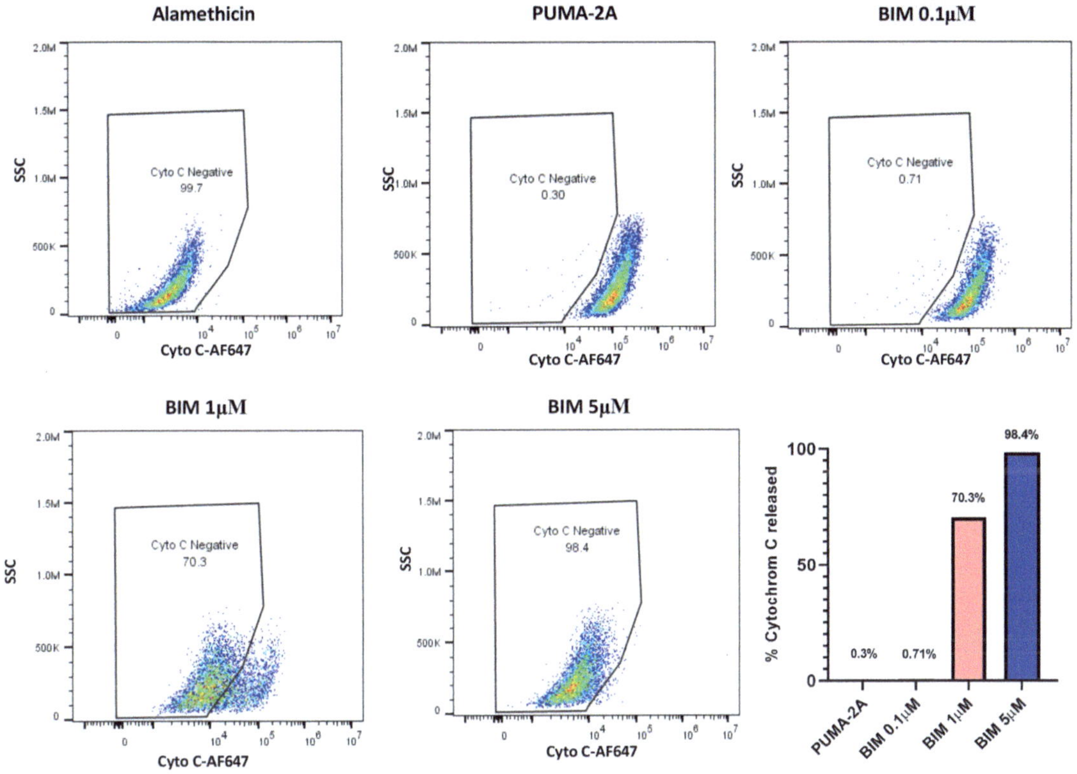

Fig. 8 Flow cytometry BH3 profiling analysis. After selecting the population of interest, cytochrome c staining is analyzed by flow cytometry. PUMA-2A treated cells should be cytochrome c positive. Alamethicin causes total cytochrome c loss and therefore cells should be cytochrome c negative. Increasing dosage of the pro-apoptotic BIM BH3 peptide results in increased cytochrome c loss, indicating MOMP. Results are typically plot as % cytochrome c released in response to peptide treatment

(d) Gently vortex cells and incubate for 15 min at room temperature (25 °C) in the dark.

(e) Add 400μL of Annexin V Binding Buffer to each tube.

(f) Proceed to flow cytometry within 4 h after the initial incubation period.

3.2.3 Flow Cytometry Analysis

1. Exclude doublets.
2. Draw a gate in the Annexin V fluorescence channel: Under normal physiological conditions, PS is mainly located on the inner side of cell membrane. At the beginning of apoptosis, PS loses its asymmetric distribution in phospholipid bilayer and it is transferred to the outside of cell membrane. Once PS reaches the outer surface of the membrane, it can be detected in a calcium-dependent manner by fluorescent-labeled annexin V.
3. Record the percentage of Annexin V^+ cells.

4 Notes

1. Mannitol Experimental Buffer (MEB) is a solution that contains non-ionic solutes to maintain osmolarity because high salt concentrations tend toward spontaneous release of cytochrome c. They also contain a carbon source, such as succinate, and often low concentrations of EGTA and/or EDTA to sequester calcium in particular.
2. Digitonin is a mild detergent made of a steroid head and a polysaccharide tail. It makes pore large enough to bleed the cytoplasm out of the cell within minutes but leaves the organelles intact. This is necessary to ensure all cells and their mitochondria get even exposure to the peptides so that they can be compared to each other.
3. Alamethicin is a pore forming peptide that can make holes in both the plasma and mitochondrial membranes. Alamethicin results in BAX/BAK-independent MOMP and full release of cytochrome c followed by depolarization. It is highly recommended as the only means to discriminate failure to release cytochrome c and operator error.
4. Peptides should be 95% pure or greater and should be made as TFA salts. Dissolving lyophilized peptide powder at 50 mg/mL in DMSO will generally produce peptide stocks of 10–15 mM and then verify peptide concentration by UV absorbance at 280 nm. Master stocks store at −80 °C. Avoid frequent freeze-thaws.
5. During 90 min incubation, fibroblasts can easily attach to the plate. The NBS coatings are used for iBH3 because they yield 2–3 times as many cells over standard polystyrene or polypropylene. Polystyrene and polypropylene generally lead to cell loss due to cell attachment to the plate walls via hydrophobic interactions. Moreover, MEB buffer contains low concentrations of EGTA and EDTA to sequester calcium. Also, to avoid cell loss and clumping, mix cells properly before fixation with formaldehyde.
6. Keep in mind that increasing cell number may decrease peptide efficacy (for 96-well plate, we typically use 10,000–50,000 cells per well). Also, be aware that peptide responses increase with incubation time and temperature. Make sure cell numbers, incubation time, and temperature are standardized throughout your experiments.
7. It is very important to remember that iBH3 profiling measures the amount of cytochrome c retained in the mitochondria. The pores made by digitonin in the plasma membrane are so large that any cytochrome c released from the mitochondria into the cytoplasmic space will immediately leave the cell completely.

References

1. Hotchkiss RS, Strasser A, McDunn JE, Swanson PE (2009) Cell death. N Engl J Med 361:1570–1583. https://doi.org/10.1056/NEJMra0901217
2. Moldoveanu T, Follis AV, Kriwacki RW, Green DR (2014) Many players in BCL-2 family affairs. Trends Biochem Sci 39:101–111. https://doi.org/10.1016/j.tibs.2013.12.006
3. Czabotar PE, Lessene G, Strasser A, Adams JM (2014) Control of apoptosis by the BCL-2 protein family: implications for physiology and therapy. Nat Rev Mol Cell Biol 15:49–63. https://doi.org/10.1038/nrm3722
4. Borras C et al (2020) BCL-xL, a mitochondrial protein involved in successful aging: from C. elegans to human centenarians. Int J Mol Sci 21:418. https://doi.org/10.3390/ijms21020418
5. Leber B, Lin J, Andrews DW (2010) Still embedded together binding to membranes regulates Bcl-2 protein interactions. Oncogene 29:5221–5230. https://doi.org/10.1038/onc.2010.283
6. Llambi F et al (2011) A unified model of mammalian BCL-2 protein family interactions at the mitochondria. Mol Cell 44:517–531. https://doi.org/10.1016/j.molcel.2011.10.001
7. Singh R, Letai A, Sarosiek K (2019) Regulation of apoptosis in health and disease: the balancing act of BCL-2 family proteins. Nat Rev Mol Cell Biol 20:175–193. https://doi.org/10.1038/s41580-018-0089-8
8. Kuehl T, Lagares D (2018) BH3 mimetics as anti-fibrotic therapy: Unleashing the mitochondrial pathway of apoptosis in myofibroblasts. Matrix Biol 68–69:94–105. https://doi.org/10.1016/j.matbio.2018.01.020
9. Lagares D et al (2017) Targeted apoptosis of myofibroblasts with the BH3 mimetic ABT-263 reverses established fibrosis. Sci Transl Med 9:eaal3765. https://doi.org/10.1126/scitranslmed.aal3765
10. Hinz B, Lagares D (2020) Evasion of apoptosis by myofibroblasts: a hallmark of fibrotic diseases. Nat Rev Rheumatol 16:11–31. https://doi.org/10.1038/s41584-019-0324-5
11. Sarosiek KA et al (2013) BID preferentially activates BAK while BIM preferentially activates BAX, affecting chemotherapy response. Mol Cell 51:751–765. https://doi.org/10.1016/j.molcel.2013.08.048
12. Ryan J, Montero J, Rocco J, Letai A (2016) iBH3: simple, fixable BH3 profiling to determine apoptotic priming in primary tissue by flow cytometry. Biol Chem 397:671–678. https://doi.org/10.1515/hsz-2016-0107

Chapter 10

Determination of Senescent Myofibroblasts in Precision-Cut Lung Slices

Tamara Cruz, Ana L. Mora, and Mauricio Rojas

Abstract

Fibroblast-to-myofibroblast transdifferentiation and the acquisition of a senescent phenotype are hallmarks of fibrotic diseases. The study of the localization of senescent myofibroblasts as well as their interactions with other cell types in the fibrotic tissue has been hindered by the lack of methods to detect these cells *in vivo*. Here, we describe methods to detect tissue localization of senescent myofibroblasts in precision-cut lung slices (PCLS) by combining β-galactosidase staining with immunofluorescence techniques.

Key words Fibrosis, Myofibroblasts, Senescence, β-Galactosidase, PCLS

1 Introduction

Cellular senescence is defined by increased expression of cell cycle inhibitors p21 and p16^{INK4a}, which leads to cell cycle arrest. Senescent cells are also resistant to apoptosis, display increased β-galactosidase activity, and produce a wide array of secreted factors collectively known as the senescence-associated secretory phenotype (SASP) [1, 2]. The accumulation of senescent cells has been recently described in healthy aging and lung age-related fibrotic diseases including idiopathic pulmonary fibrosis (IPF) [3, 4]. In this context, the SASP produced by senescent cells is a source of pro-fibrotic mediators that can promote fibroblast-to-myofibroblast transdifferentiation, thus contributing to the development and progression of lung fibrosis [5, 6].

Precision-cut lung slices (PCLS) are being increasingly used as an ex vivo approach to visualize the position of live cells in fibrotic lung tissues and to investigate the biological interactions between different cell types [7, 8]. PCLS can also be cultured, offering the possibility of testing different interventions, including evaluating therapeutic candidates [8–10]. Here, we provide methods to detect senescent cells including senescent myofibroblasts in PCLS by

Boris Hinz and David Lagares (eds.), *Myofibroblasts: Methods and Protocols*, Methods in Molecular Biology, vol. 2299, https://doi.org/10.1007/978-1-0716-1382-5_10,

performing β-galactosidase staining. β-Galactosidase is a hydrolase enzyme that catalyzes the hydrolysis of β-galactosides into monosaccharides only in senescent cells. Thus, this method requires intact activity of β-galactosidase in order to catalyze the conversion of a chromogenic substrate into a blue-colored final product. The use of PCLS also offers the possibility to combine β-galactosidase staining with immunofluorescence in order to determine cellular senescence in a cell-specific manner.

2 Materials

2.1 Preparation of PCLS

(For details, *see* Chapters 19 and 30). Our protocol was optimized for 300μm thick mouse PCLS (*see* **Note 1**).

2.2 β-Galactosidase Activity: X-Gal Staining

1. 4% Paraformaldehyde (PFA) in phosphate-buffered saline (PBS): 0.138 M NaCl, 0.0027 M KCl, pH 7.4.
2. Detergent for tissue permeabilization: 0.1 M Na_2HPO_4 pH 7.3, 0.1 M Na_2HPO_4 pH 7.3, 2 mM $MgCl_2$, 0.1% sodium deoxycholate, 0.02% Nonidet P-40, 0.05% bovine serum albumin (BSA), in ddH_2O (*see* **Note 2**).
3. Senescence β-Galactosidase Staining kit (Cell Signalling, cat.9860) (*see* **Note 2**).
4. Parafilm.
5. Shaker.
6. Standard light microscope.

2.3 OCT Blocks and Sectioning

1. 30% Sucrose.
2. 2-Methylbutane.
3. Plastic beaker.
4. Styrofoam box with liquid nitrogen.
5. Optimal cutting temperature (OCT) media.
6. Embedding cassettes.
7. Freezing box or bag.
8. Cryostat.

2.4 Immunofluorescence Staining

1. 0.5% BSA in PBS (PBB) and serum.
2. Primary antibodies (α-Smooth Muscle Actin (α-SMA, clone 1A4), PDGFRα (Clones clone D1E1E), CD90 (clone JF10-09), TE-7 (clone TE-7).
3. Secondary antibody.
4. Antifade DAPI prolong gold mounting media.
5. Cover glass.

2.5 Imaging and Quantification

1. Fluorescence microscope.
2. Image J software.

3 Methods

Our protocol was optimized for 300μm mouse PCLS (*see* **Note 1**).

3.1 β-Galactosidase Staining

1. Wash PCLS slices in PBS.
2. Fix PCLS slices in fresh cold 4% PFA for 1 h at 4 °C.
3. Permeabilized PCLS slices with detergent two times for 20 min at room temperature on a shaker (*see* **Notes 2** and **3**).
4. Wash for 10 min in PBS to remove any rest of detergent.
5. During the wash in PBS prepare the β-galactosidase staining solution following the senescence β-Galactosidase Staining kit instructions (*see* **Note 4**).

   Add β-galactosidase staining solution at room temperature. Ensure tissues are entirely covered, seal with paraffin, and protect from light. Place at 37 °C (no CO_2) overnight on a shaker.
6. The following morning examine the samples using a standard light microscope. At this point the tridimensional structure of the PCLS makes difficult to focus, but it must be easy to visualize dark spots in the tissue. If the staining pattern is satisfactory, proceed to the next step. If no stain is detectable, then fresh β-galactosidase staining solution can be added, and the reaction restarted at 37 °C for the rest of the day.
7. Wash PCLS in PBS for 20 min in the dark, three times, at room temperature on a shaker.
8. Fix PCLS with 4% PFA overnight at 4 °C, on a shaker, in the dark. The next day, wash the samples three times for 1 min with PBS.

3.2 OCT Block Generation and Cryostat Sectioning

1. Place PCLS in 30% sucrose for 24 h. Replace sucrose 2 times× within 24 h period. Provide enough 30% sucrose to submerge PCLS.
2. Place a plastic beaker containing 2-Methylbutane in a Styrofoam box, or equivalent container, containing a few inches of liquid nitrogen. Allow the beaker with 2-methyl butane to cool for 5 min and maintain it submerged in the liquid nitrogen during the entire freezing protocol.
3. Samples can be embedded in OCT compound in a small embedding cassette.

4. Pick up the cassette with the sample in OCT and completely immerse in the liquid nitrogen cooled 2-Methylbutane for 30 s.
5. Remove samples from the beaker and immerse them in the liquid nitrogen for an additional 10 s.
6. Immerse an adequately labeled freezing box or bag in the liquid nitrogen for a few seconds to cool container to the same temperature as the tissue.
7. Place samples into the freezing box or bag on dry ice until placing the samples into a –80 °C storage. Samples to be transferred between labs or waiting sectioning should be stored on dry ice to avoid thawing. Store until ready for sectioning in the cryostat.
8. Cryostat section PCLS samples to generate 10μm slides.

3.3 Immuno-fluorescence Staining

1. Rehydrate frozen tissue sections with two washes of PBS at room temperature, keep samples in solution at all time, do not let them dry.
2. Washes samples in PBB (0.5% BSA in PBS) three times for 1 min.
3. Block with 5% serum diluted in PBB for 1 h at room temperature. Serum must be from the species in which your secondary antibodies are generated. *See* **Note 5**.
4. Dilute primary antibodies in PBB at concentrations described below and gently drop 70–100μL of antibody solution over section, so the section is fully covered (*see* **Note 6**). Incubate overnight at 4 °C (*see* **Note 7**). The following molecular markers are recommended for labeling lung fibroblasts and myofibroblasts.
   (a) Myofibroblasts:
      - α-Smooth Muscle Actin (α-SMA). Mouse monoclonal antibody, clone 1A4, use at 1:1000.

   (b) Fibroblasts:
      - PDGFRα: Rabbit monoclonal. Cones D1E1E or D13C6, used at 1:500.
      - CD90: Rabbit monoclonal antibody, clone JF10-09, used at 1:50–1:200 (*see* **Note 8**).
      - Anti-fibroblast clone TE-7: Mouse monoclonal, used at 1:100.
5. Wash sections five times with PBB.
6. Secondary antibody: Add secondary antibody at the recommended dilution in PBB to the section for 60 min. Secondaries can be added together if using more than one, but they must be against two separate species (*see* **Note 6**).

7. Wash sections with five times with PBB.
8. Wash sections with five times with PBS (*see* **Note 9**).
9. Add DAPI prolong to the slices place slides horizontal in a slide box, and allow the cover glass to adhere to slide overnight at 4 °C in the dark.

3.4 Imaging and Quantification

1. Take bright-field and fluorescence images using a fluorescence microscope (Fig. 1).
2. Use imageJ to quantify the total number of cells, the number of fibroblasts and myofibroblasts. The image of each channel (DAPI and all the antibodies added) must be analyzed independently, but the analysis strategy must be the same for all the different samples.
3. Generate the mask that fits the cells in each stack (channel). Apply the same mask to all the samples.
   (a) Every stack must be in an 8-bit format.
   (b) Apply a threshold that fits better with the samples.

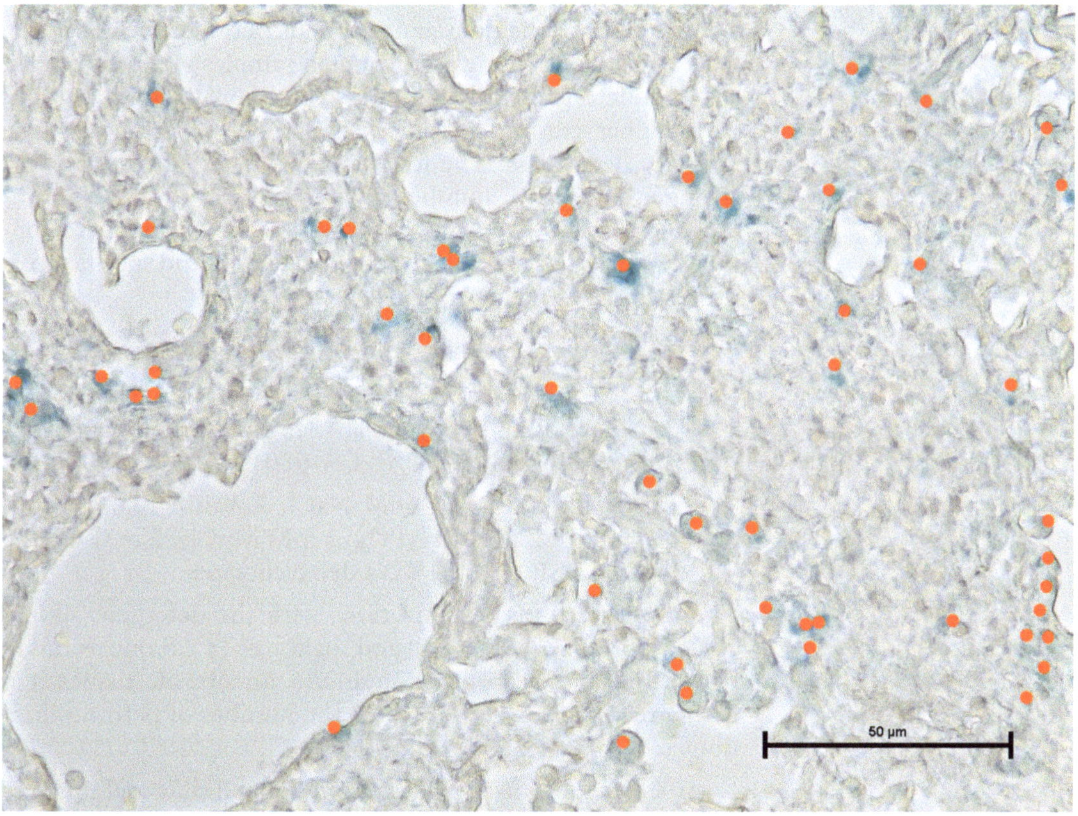

Fig. 1 Representative image of a section of fibrotic PCLS stained with β-Gal for the detection of the senescent cells (blue) and quantified by imageJ cell counter (red dots)

   (c) We have optimized the fitting using the ImageJ options noise, filter, and binary.
4. Automatic cell counter using analyzed particles option.
   (a) Limit the size to 2-infinity.
   (b) Select the show mask option that can superpose it with the original image to double-check the fitness of the mask.
5. Count the βb-galactosidase positive cells using the ImageJ cell counter, *see* **Note 10** (Fig. 1).
6. Generate the proportion of senescence cells in the total number of cells in the lung, myofibroblast or fibroblast.

4 Notes

1. PCLS of 300–700μm depth are recommended. The depth of the PCLS must be adjusted depending on the integrity of the tissue. Healthy lungs can usually be cut at shallower depths, while fibrotic and other unhealthy tissue may need to be cut at deeper depths. Deeper depths will allow you to get a higher number of sections for different co-stainings.
2. The strength of the detergent depends between species and in healthy vs. disease lungs. For human samples it is recommend stronger detergents than for mouse. For fibrotic lungs or lungs with higher content of myofibroblast we recommend even stronger detergents. The following two alternatives represent stronger detergents, number one can be used for human samples and number two for fibrotic tissues, but we highly recommend optimizing it according to your sample and specific post-processing.
   (a) Detergent wash: Sodium phosphate dibasic (0.5 M; pH 7.0–7.5) 0.1 M, sodium phosphate monobasic (0.5 M; pH 7.0–7.5) 5 mM, $MgCl_2 \cdot 6H_2O$ (1 M) 3 mM, sodium deoxycholate 1.5 mM, octylphenoxypolyethoxyethanol (IGEPAL CA-630) 3.0%.
   (b) Detergent wash: 137 mM NaCl, 2.7 mM KCl, 10 mM Na_2HPO_4, 2 mM KH_2PO_4, 2 mM $MgCl_2$, 5 mM EGTA, 0.02% IGEPAL, 0.01% Na deoxycholate.
3. The number of permeabilizations with the detergent should also be optimized. For small and shallower PCLS one round of permeabilization can be enough and for fibrotic tissues it is highly recommended to increase the number of permeabilizations to 3.
4. Keep X-Gal solution protected from light all time and it is very important to adjust the pH at 6.0 (a pH 5.9–6.1 is acceptable).

pH differences can affect staining, a low pH can result in false positives and high pH can result in false negatives.

5. If the protein of interest is nuclear, a permeabilization step should be added before blocking. Permeabilization can be performed with 0.3% Triton-X for 15 min to 1 h. Optimization must be performed for each sample and antibody.
6. Spin down primary and secondary antibodies for 5 min to pellet any aggregates.
7. Put the samples in a humid environment (wet paper towel within slide box).
8. Add an antibody against CD45. Fibroblast should be negative for CD45, and this marker will allow the exclusion of lymphocytes and NK cells also positive for CD90.
9. Remaining BSA would bind DAPI indiscriminately.
10. The automatic counting does not work great with bright-field images.

References

1. Coppé JP et al (2011) Tumor suppressor and aging biomarker p16(INK4a) induces cellular senescence without the associated inflammatory secretory phenotype. J Biol Chem 286:36396–36403. https://doi.org/10.1074/jbc.M111.257071
2. Mosteiro L, Pantoja C, de Martino A, Serrano M (2018) Senescence promotes in vivo reprogramming through p16. Aging Cell 17: e12711. https://doi.org/10.1111/acel.12711
3. Campisi J (2016) Cellular senescence and lung function during aging. Yin and Yang. Ann Am Thorac Soc 13(Suppl 5):S402–S406. https://doi.org/10.1513/AnnalsATS.201609-703AW
4. Merkt W, Bueno M, Mora AL, Lagares D (2020) Senotherapeutics: targeting senescence in idiopathic pulmonary fibrosis. Semin Cell Dev Biol 101:104–110. https://doi.org/10.1016/j.semcdb.2019.12.008
5. Mellone M et al (2016) Induction of fibroblast senescence generates a non-fibrogenic myofibroblast phenotype that differentially impacts on cancer prognosis. Aging (Albany NY) 9:114–132. https://doi.org/10.18632/aging.101127
6. Untergasser G et al (2005) Profiling molecular targets of TGF-beta1 in prostate fibroblast-to-myofibroblast transdifferentiation. Mech Ageing Dev 126:59–69. https://doi.org/10.1016/j.mad.2004.09.023
7. Lyons-Cohen MR, Nakano H, Thomas SY, Cook DN (2018) Imaging precision-cut lung slices to visualize leukocyte localization and trafficking. Methods Mol Biol 1799:237–246. https://doi.org/10.1007/978-1-4939-7896-0_18
8. Sanderson MJ (2011) Exploring lung physiology in health and disease with lung slices. Pulm Pharmacol Ther 24:452–465. https://doi.org/10.1016/j.pupt.2011.05.001
9. Alsafadi HN et al (2020) Applications and approaches for 3D precision-cut lung slices: disease modeling and drug discovery. Am J Respir Cell Mol Biol 62(6):681–691. https://doi.org/10.1165/rcmb.2019-0276TR
10. Liberati TA, Randle MR, Toth LA (2010) In vitro lung slices: a powerful approach for assessment of lung pathophysiology. Expert Rev Mol Diagn 10:501–508. https://doi.org/10.1586/erm.10.21

Chapter 11

The Scar-in-a-Jar: In Vitro Fibrosis Model for Anti-Fibrotic Drug Testing

Simon Stebler and Michael Raghunath

Abstract

Excessive deposition of type I collagen follows in the wake of chronic inflammation processes in dysregulated tissue healing and causes fibrosis that can ultimately lead to organ failure. While the *development* of antifibrotic drugs is targeting various upstream events in collagen matrix formation (synthesis, secretion, deposition, stabilization, remodeling), the *evaluation* of drug effects would use as net read-out of the above effects the presence of a deposited collagen matrix by activated cells, mainly myofibroblasts. Conventional methods comprise lengthy and labor-intensive protocols for the quantification of deposited collagen, some with sensitivity and/or specificity issues. Here we describe the Scar-in-a-Jar assay, an in vitro fibrosis model for anti-fibrotic drug testing that benefits from a substantially accelerated extracellular matrix deposition employing macromolecular crowding and a collagen-producing cell type of choice (e.g., lung fibroblasts like WI-38). The system can be aided by activating compounds such as transforming growth factor-β1, a classical inducer of the myofibroblast phenotype in fibroblasts. Direct image analysis of the well plate not only eliminates the need for matrix extraction or solubilization methods, but also allows for direct imaging and monitoring of phenotypical markers and offers the option for high-content screening applications when adapted to well formats compatible with a screening format.

Key words Macromolecular crowding, Ultra-flat 3D, Collagen quantitation, Extracellular matrix, Drug discovery, High-content screening, Bioimaging, Immunocytochemistry

1 Introduction

During scarring, an ill-regulated process of wound healing, excessive amounts of extracellular matrix (ECM) (especially type I collagen) are deposited at the site of tissue injury. This process, when locally contained, will form a defined scar, but when perpetuated and spreading through an organ can destroy its microarchitecture, and ultimately its function. Anti-fibrotic drugs therefore aim to disrupt local or organ-wide collagen build-up at a transcriptional, translational, or posttranslational level. In the last decades various methods have been described to quantify the amount of deposited

Boris Hinz and David Lagares (eds.), *Myofibroblasts: Methods and Protocols*, Methods in Molecular Biology, vol. 2299,
https://doi.org/10.1007/978-1-0716-1382-5_11,

collagen and used in combination with in vitro and ex vivo fibrosis models that allowed researchers to assay for anti-fibrotic properties of drug candidates [1].

A classical method for the colorimetric quantification of collagen is the determination of 4-hydroxyproline, the abundant and unique signature amino acid of collagen (about 14%) [2]. Following total acid hydrolysis of dried and weighed tissue material 4-hydroxyproline is reacted with dimethylaminobenzaldehyde (Ehrlich's reagent) which yields a product that can be determined photometrically at an absorption maximum of $\lambda_{max} = 558 \pm 2$ nm [3]. Several refinements of this hydrolysis method have been introduced aiming at shorter processing times, increased sensitivity, and alternative detection methods such as HPLC or LC-MS [4–6]. Another popular method to quantify soluble and solubilized collagen is the Sircol Collagen Assay. The assay's central reagent is the anionic dye Sirius red [7]. The original method suffers from specificity issues as the dye binds to basic amino acids regardless of the source and must have led to many overestimations in the literature when performed in serum-containing material [8]. This can be remedied by an additional pepsin digestion and ultrafiltration step followed by the normal Sircol Collagen Assay protocol to yield accurate collagen concentrations [8]. Other quantitative analyses include MS [9], ELISA [10], SDS-PAGE in combination with densitometry [11] or image-based analyses by confocal microscopy [12–14], two-photon excitation fluorescence and second and third harmonic generation microscopy [15, 16] that make use of the nonlinear optical property of fibrillar collagen.

The above described collagen quantification methods have been used in combination with various in vitro and ex vivo fibrosis models some of which are highlighted below. Emulating the formation of fibrous-like tissue with excessive ECM build-up potential was achieved in vitro using long-term fibroblast cultures, and allowing them to form multiple cell layers under stimulation with transforming growth factor-β1 (TGFβ1) [17], a known main driver of fibrosis [18]. Another approach for anti-fibrotic drug testing using kidney-derived cells in fact consists of two separate fibrosis models, as under stimulus of TGFβ1, human mesangial cells either formed monolayers (2D) or nodules (3D) solely depending on the growth substrate [19]. More recently, sophisticated examples for in vitro fibrosis models featured lung organoids derived from human embryonic stem cells with engineered gene mutations [20], or from induced pluripotent stem cells (iPSC)-derived human mesenchymal cells cultured in the interstitial spaces between collagen-functionalized hydrogel beads in the presence of TGFβ1 [21]. Biomechanical properties such as stiffening and contraction of progressive fibrotic tissue have been emulated using micropillars or rods that are deflected by contractile forces exerted by contiguous fibroblast layers (after TGFβ1 activation) covering these nano landscapes [22, 23]. Ex vivo precision cut lung slices considerably raise the complexity of the model as the lung tissue architecture of various cell types and ECM is greatly conserved [24]. Airways of healthy or diseased lung of human or animal origin are infused with agarose solution followed by subsequent sectioning into uniformly thick tissue slices (200–1000 μm) and placing in culture. Anti-fibrotic drugs may be tested directly on diseased or healthy lung slices after initiation of fibrosis incubating the tissue in a cocktail containing TGFβ1 [25–27]. Precision cut lung slices are widely used and considered a bridging element between in vitro and in vivo studies as they help to reduce animal sacrifices while delivering relevant data.

Here we describe the Scar-in-a-Jar, an in vitro fibrosis model that combines (a) a highly accelerated deposition of ECM by fibroblastic cells such as WI-38 lung fibroblasts through macromolecular crowding (MMC) culture conditions and simultaneous fibrotic stimuli such as TGFβ1 and (b) optical high-content screening to quantify deposited collagen circumventing the need to solubilize ECM and losing material along the way [28]. MMC, the addition of preferably carbohydrate-based macromolecules (50–400 kDa) to culture medium, causes the excluded volume effect. This effect supports protein folding and enzymatic transition complex stabilization by limiting available extracellular space for biomolecule action. Collagen I deposition is boosted on three levels by macromolecular crowding. Firstly, by means of an accelerated conversion rate from procollagen I to collagen I, secondly, an enhanced supramolecular assembly of collagen triple helices to collagen fibers, and thirdly, by an increased stabilization of ECM components by lysyl oxidase- and transglutaminase-mediated cross-linking [29]. In fact, this "crowded state" creates an ECM micro-environment that resembles the in vivo situation more closely than the highly dilute aqueous standard cell culture condition. This is why MMC has been growing in popularity as research discovery tool to unmask processes or functions that may otherwise would have been overlooked [30]. In the Scar-in-a-Jar model, MMC allowed us to deposit significant amounts of ECM (6 day of MMC in a 7 day assay) while keeping the cell density at a subconfluent state [28]. This makes optimal collagen secretion and formation of an ultra-flat 3D model possible, while preventing multilayer formation as in the long-term fibroplasia model [17]. This in turn enabled optical high-content screening by software-assisted determination of immunofluorescent type I collagen area and normalization by a nuclear count [28]. The first adopter of the Scar-in-a-Jar after its publication in 2009 was the pharmaceutical industry, and the assay has since been improved toward medium-throughput (96-well) applications. The Scar-in-a-Jar has stood the test of time as relevant fibrosis model using lung fibroblasts from patients with idiopathic pulmonary fibrosis (IPF) for drug screening (adaptions to the method, *see* **Note 1**) [31]. Recently, an extended (12 day of MMC instead of 6 day) Scar-in-a-Jar assay has been suggested, in combination with clinically validated biochemical markers of ECM synthesis to better monitor long-term effects of anti-fibrotic drugs using ELISA to analyze culture media [32]. These biomarkers are the soluble C-propeptides of collagens I, III, IV–VI and they result from successful cleavage indicating matrix formation. C-propeptides of collagen III and VI have previously been reported to be increased in serum of advanced IPF patients [33]. Of note, α-SMA, a hallmark of myofibroblasts in fibrotic tissue, has recently been detected in serum of patients with IPF [34], and therefore was included in the immunochemical analyses, too. Interestingly, C-propeptides of collagens I, IV-VI were elevated in the first 4 day, while collagen III C-propeptide and α-SMA levels occurred only after 8 days and increased strongly toward day 12 [32]. While this shows that IPF pathology can be emulated in vitro, it also emphasizes the role of the fourth dimension in assay design.

2 Materials

2.1 Cell Culture and Scar-in-a-Jar Procedure

1. Fibroblastic cells such as normal human lung fibroblasts (WI-38) (*see* **Note 2**).
2. Tissue culture plastic ware: 24-well plates, standard cell culture consumables.
3. Proliferation medium: Dulbecco's modified Eagle's medium (DMEM) supplemented with 10% fetal bovine serum (FBS).
4. Assay medium: DMEM supplemented with 0.5% FBS, 100 μM L-ascorbic acid 2-phosphate (*see* **Note 3**), a mixture of 37.5 mg·mL^{-1} Ficoll™ (Fc) 70 with 25 mg·mL^{-1} Fc 400 (*see* **Note 4**), 5 ng·mL^{-1} TGFβ1. This medium would contain any drugs used for fibrosis assessment (*see* **Note 1**).
5. Ciclopiroxolamine (CPX) solution (positive control, *see* **Note 5**): Prepare a stock of 0.5 M CPX in methanol, dilute to a working concentration of 1 mM CPX in DMEM (prior to cell culture addition) and add to cell culture at a final concentration of 8 μM.
6. Humidified incubator at 37 °C and 5% CO_2.
7. Potential anti-fibrotic drugs for assessment (*see* **Note 1**).

2.2 Immunocytochemistry

1. Phosphate-buffered saline (PBS): 137 mM NaCl, 2.7 mM KCl, 6.5 mM Na_2HPO_4, 1.5 mM KH_2PO_4 in apyrogenic ultrapure water; pH 7.4.
2. Fixative: Methanol, ice-cold (*see* **Note 6**).
3. Blocking solution: 3% bovine serum albumin in PBS.

4. Antibodies and nuclei staining: Primary type I collagen and compatible secondary detection antibodies. For nuclei staining, 4′,6-diamidino-2-phenylindoldilactate (DAPI) is suitable (*see* **Note 7**).

2.3 Optical High-Content Screening and Analysis of Data

1. Epifluorescence microscope, microscopic plate reader, or similar device equipped with appropriate filter sets (at least two, preferably one for DAPI) that allows fluorescence image acquisition (*see* **Note 8**).
2. Image analysis software that allows enumeration of nuclei and type I collagen area determination (*see* **Note 9**).

3 Methods

3.1 Cell Culture and Scar-in-a-Jar Procedure

1. Cultivate WI-38 in proliferation medium and passage cells (*see* **Note 10**) when still subconfluent.
2. Plate cells on 24-well plates (*see* **Note 11**) at 50,000 cells per well in 0.5–1 mL of proliferation medium. Plan to have enough experimental wells ready to hold all samples (including desired concentrations and solvent control "vehicle") plus positive control (8 μM CPX in DMEM) and incorporate at least two biological repeats ($n = 2$) per sample and control.
3. Incubate in humidified incubator at 37 °C and 5% CO_2 for 16 h.
4. Change to assay medium (1 mL per well) containing desired drugs (*see* **Note 1**) and controls.
5. Incubate in humidified incubator at 37 °C and 5% CO_2 for 6 day to ensure sufficient ECM deposition.

3.2 Immunocytochemistry

1. Fix cells with ice-cold methanol for 10 min at −20 °C.
2. Remove methanol and wash cells three times with PBS.
3. Add blocking solution and block at room temperature for 30 min.
4. Dilute type I collagen primary antibody in PBS, add to cells, and incubate at room temperature for 90 min.
5. Remove primary antibody and wash cells three times with PBS.
6. From here on work in a dark environment to minimize photobleaching.
7. Prepare a mixture of secondary antibody and DAPI in PBS.
8. Remove primary antibody solution from cells, do **not** wash.
9. Add secondary antibody and DAPI solution to cells and incubate at room temperature for 30 min.

10. Wash cells three times with PBS and leave 0.5 mL of PBS on top of the samples to prevent cells from getting dry.
11. Analyze samples immediately or store at 4 °C in the dark for not longer than 48 h prior to analysis.

3.3 Optical High-Content Screening and Analysis of Data

1. Take multiple images per well (*see* **Note 12**). Look for the strongest signal throughout all experimental samples and controls make sure to adjust exposure times for each channel (nuclei + collagen). Acquire all images using identical settings (*see* **Note 13**).
2. Export and store images in a way that allows analysis of channels individually.
3. Quantify total number of nuclei and corresponding type I collagen content per imaging site (*see* **Note 14**).
4. Normalize type I collagen content by number of nuclei to get area of type I collagen per nuclei per image field in μm^2. Determine mean of each well, calculate fold changes of relevant controls and do statistics on combined biological replicates.

4 Notes

1. GlaxoSmithKline have successfully implemented the Scar-in-a-Jar method to their anti-fibrotic drug discovery and made some updates [31] to our original method [28]: Patient-derived IPF fibroblasts were seeded into black-walled 96-well imaging plates at 10,000 cells per well in DMEM supplemented with 0.4% FBS and 4 mM L-glutamine and incubated for 24 h at 37 °C and 10% CO_2 to reach confluence. Medium was changed to assay medium supplemented with 0.4% FBS, 50 μg·mL^{-1} L-ascorbic acid and Ficoll™ crowders (as described above). Anti-fibrotic drugs or vehicle were pre-incubated for 3 h prior to the addition of 1 ng·mL^{-1} TGFβ1 into the medium. Treated fibroblasts were then cultured for 72 h at 37 °C and 10% CO_2. SB-525334 (ALK5 inhibitor) and CZ415 (mTOR inhibitor) were used as positive controls.
2. Initially, IMR-90 and WI-38 fibroblasts were assessed for their ability to deposit substantial amounts of collagen. In comparison, WI-38 showed a slower proliferation rate which was beneficial for our purpose to remain the cell culture subconfluent after 6 day assay duration.
3. The 2-phosphate derivate was previously found to be more stable than ascorbate [35]. We used it as the magnesium salt hexahydrate.
4. In our original publication [28] we further describe the use of a 500 kDa dextran sulfate molecular crowder at 100 μg·mL^{-1} as an alternative to Fc 70/400. The use of dextran sulfate even

resulted in a quicker method (2 day instead of 6 day); however, type I collagen deposition was rather granular that may result in an under-appreciation of a possible drug effect in image analysis.

5. Since our publication in 2009 [28] the FDA has approved Pirfenidone and Nintedanib for the treatment of IPF in 2014. These drugs may be used as positive controls and benchmark compounds. Further examples of positive controls are described above (*see* **Note 1**).
6. In comparison to 4% paraformaldehyde and live staining, methanol fixation yielded the greatest quantified area of collagen.
7. We used mouse anti-human type I collagen monoclonal primary antibody from Sigma-Aldrich (St. Louis, MO, USA) and Alexa Fluor 594 goat anti-mouse secondary antibody as well as DAPI from Molecular Probes (Eugene, OR, USA).
8. Our data were generated using a Nikon TE600 fluorescence microscope equipped with a Xenon light source, an automated Ludl stage and a Photometrics CoolSNAP high-sensitivity camera.
9. To quantify nuclei and type I collagen content we used Metamorph® Imaging System software.
10. We routinely cultured at low passage (passages 3–8).
11. We found higher well-to-well variations when using 96-well format. Efforts to account for well-to-well variations and port the Scar-in-a-Jar assay to a 96-well plate format for medium-throughput has since been established [31].
12. We acquired nine separate images per well (3×3 grid) covering a total area of 1.35 cm^2 which represents approximately 70% of the 24-well growth area.
13. When using automated image acquisition, it becomes necessary to either turn on autofocus or to define a focus plane by setting focus points in DAPI channel.
14. Using the Metamorph® Imaging System software we accounted for positive DAPI signals when (a) the pixel intensity value of an area was 15 above that of the background and (b) long axis of nucleus measured between 10–15 μm. Suitable fluorescent intensity thresholds for type I collagen were defined according to intensity histograms while fluorescent signals below a defined pixel intensity value were discounted for (background). Corner images of the 3×3 grid were close to the well border. To circumvent illumination variances and interference with auto-fluorescent well borders, triangle-shaped masks (excludes image analysis) were added to these images to account for it.

15. The research group of Ruud Bank recently revisited the Scar-in-a-Jar sucessfully, and found additional value in using PVP40 as an alternative crowder to the Ficoll system that is worth exploring [36].

References

1. Sundarakrishnan A, Chen Y, Black LD, Aldridge BB, Kaplan DL (2018) Engineered cell and tissue models of pulmonary fibrosis. Adv Drug Deliv Rev 129:78–94. https://doi.org/10.1016/j.addr.2017.12.013
2. Neuman RE, Logan MA (1950) The Determination of Hydroxyproline. J Biol Chem 184:299–306
3. Woessner JF (1961) The determination of hydroxyproline in tissue and protein samples containing small proportions of this imino acid. Arch Biochem Biophys 93:440–447. https://doi.org/10.1016/0003-9861(61)90291-0
4. Hofman K, Hall B, Cleaver H, Marshall S (2011) High-throughput quantification of hydroxyproline for determination of collagen. Anal Biochem 417:289–291. https://doi.org/10.1016/j.ab.2011.06.019
5. da Silva CML, Spinelli E, Rodrigues SV (2015) Fast and sensitive collagen quantification by alkaline hydrolysis/hydroxyproline assay. Food Chem 173:619–623. https://doi.org/10.1016/j.foodchem.2014.10.073
6. Kindt E, Gueneva-Boucheva K, Rekhter MD, Humphries J, Hallak H (2003) Determination of hydroxyproline in plasma and tissue using electrospray mass spectrometry. J Pharm Biomed Anal 33:1081–1092. https://doi.org/10.1016/S0731-7085(03)00359-5
7. Junqueira LCU, Bignolas G, Brentani RR (1979) Picrosirius staining plus polarization microscopy, a specific method for collagen detection in tissue sections. Histochem J 11:447–455. https://doi.org/10.1007/BF01002772
8. Lareu RR, Zeugolis DI, Abu-Rub M, Pandit A, Raghunath M (2010) Essential modification of the Sircol Collagen Assay for the accurate quantification of collagen content in complex protein solutions. Acta Biomater 6:3146–3151. https://doi.org/10.1016/j.actbio.2010.02.004
9. Huizen NA, Ijzermans JNM, Burgers PC, Luider TM (2020) Collagen analysis with mass spectrometry. Mass Spectrom Rev 39(4):309–335. https://doi.org/10.1002/mas.21600
10. Seo W-Y, Kim J-H, Baek D-S, Kim S-J, Kang S, Yang WS, Song J-A, Lee M-S, Kim S, Kim Y-S (2017) Production of recombinant human procollagen type I C-terminal propeptide and establishment of a sandwich ELISA for quantification. Sci Rep 7:15946. https://doi.org/10.1038/s41598-017-16290-9
11. Lareu RR, Subramhanya KH, Peng Y, Benny P, Chen C, Wang Z, Rajagopalan R, Raghunath M (2007) Collagen matrix deposition is dramatically enhanced in vitro when crowded with charged macromolecules: the biological relevance of the excluded volume effect. FEBS Lett 581:2709–2714. https://doi.org/10.1016/j.febslet.2007.05.020
12. Clemons TD, Bradshaw M, Toshniwal P, Chaudhari N, Stevenson AW, Lynch J, Fear MW, Wood FM, Iyer KS (2018) Coherency image analysis to quantify collagen architecture: implications in scar assessment. RSC Adv 8:9661–9669. https://doi.org/10.1039/C7RA12693J
13. Bernstein DM, Toth B, Rogers RA, Kling DE, Kunzendorf P, Phillips JI, Ernst H (2020) Evaluation of the dose-response and fate in the lung and pleura of chrysotile-containing brake dust compared to TiO2, chrysotile, crocidolite or amosite asbestos in a 90-day quantitative inhalation toxicology study—Interim results Part 2: histopathological examination, Confocal microscopy and collagen quantification of the lung and pleural cavity. Toxicol Appl Pharmacol 387:114847. https://doi.org/10.1016/j.taap.2019.114847
14. Khorasani H, Zheng Z, Nguyen C, Zara J, Zhang X, Wang J, Ting K, Soo C (2011) A quantitative approach to scar analysis. Am J Pathol 178:621–628. https://doi.org/10.1016/j.ajpath.2010.10.019
15. Stoller P, Celliers PM, Reiser KM, Rubenchik AM (2003) Quantitative second-harmonic generation microscopy in collagen. Appl Opt 42:5209. https://doi.org/10.1364/AO.42.005209
16. Mostaço-Guidolin L, Rosin NL, Hackett T-L (2017) Imaging collagen in scar tissue: developments in second harmonic generation microscopy for biomedical applications. Int J

Mol Sci 18:1772. https://doi.org/10.3390/ijms18081772
17. Clark RAF, McCoy GA, Folkvord JM, McPherson JM (1997) TGF-β1 stimulates cultured human fibroblasts to proliferate and produce tissue-like fibroplasia: a fibronectin matrix-dependent event. J Cell Physiol 170:69–80. https://doi.org/10.1002/(SICI)1097-4652(199701)170:1<69::AID-JCP8>3.0.CO;2-J
18. Pohlers D, Brenmoehl J, Löffler I, Müller CK, Leipner C, Schultze-Mosgau S, Stallmach A, Kinne RW, Wolf G (2009) TGF-β and fibrosis in different organs—molecular pathway imprints. Biochim Biophys Acta BBA Mol Basis Dis 1792:746–756. https://doi.org/10.1016/j.bbadis.2009.06.004
19. Xu Q, Norman JT, Shrivastav S, Lucio-Cazana-J, Kopp JB (2007) In vitro models of TGF-β-induced fibrosis suitable for high-throughput screening of antifibrotic agents. Am J Physiol-Ren Physiol 293:F631–F640. https://doi.org/10.1152/ajprenal.00379.2006
20. Strikoudis A, Cieślak A, Loffredo L, Chen Y-W, Patel N, Saqi A, Lederer DJ, Snoeck H-W (2019) Modeling of fibrotic lung disease using 3D organoids derived from human pluripotent stem cells. Cell Rep 27:3709–3723.e5. https://doi.org/10.1016/j.celrep.2019.05.077
21. Wilkinson DC, Alva-Ornelas JA, Sucre JMS, Vijayaraj P, Durra A, Richardson W, Jonas SJ, Paul MK, Karumbayaram S, Dunn B, Gomperts BN (2017) Development of a three-dimensional bioengineering technology to generate lung tissue for personalized disease modeling. Stem Cells Transl Med 6:622–633. https://doi.org/10.5966/sctm.2016-0192
22. Mastikhina O, Moon B-U, Williams K, Hatkar R, Gustafson D, Mourad O, Sun X, Koo M, Lam AYL, Sun Y, Fish JE, Young EWK, Nunes SS (2020) Human cardiac fibrosis-on-a-chip model recapitulates disease hallmarks and can serve as a platform for drug testing. Biomaterials 233:119741. https://doi.org/10.1016/j.biomaterials.2019.119741
23. Asmani M, Velumani S, Li Y, Wawrzyniak N, Hsia I, Chen Z, Hinz B, Zhao R (2018) Fibrotic microtissue array to predict anti-fibrosis drug efficacy. Nat Commun 9:2066. https://doi.org/10.1038/s41467-018-04336-z
24. Neuhaus V, Danov O, Konzok S, Obernolte H, Dehmel S, Braubach P, Jonigk D, Fieguth H-G, Zardo P, Warnecke G, Martin C, Braun A, Sewald K (2018) Assessment of the cytotoxic and immunomodulatory effects of substances in human precision-cut lung slices. J Vis Exp 135:57042. https://doi.org/10.3791/57042
25. Cedilak M, Banjanac M, Belamarić D, Paravić Radičević A, Faraho I, Ilić K, Čužić S, Glojnarić I, Eraković Haber V, Bosnar M (2019) Precision-cut lung slices from bleomycin treated animals as a model for testing potential therapies for idiopathic pulmonary fibrosis. Pulm Pharmacol Ther 55:75–83. https://doi.org/10.1016/j.pupt.2019.02.005
26. Huang X, Li L, Ammar R, Zhang Y, Wang Y, Ravi K, Thompson J, Jarai G (2018) Molecular characterization of a precision-cut rat lung slice model for the evaluation of antifibrotic drugs. Am J Physiol Lung Cell Mol Physiol 316: L348–L357. https://doi.org/10.1152/ajplung.00339.2018
27. Lehmann M, Buhl L, Alsafadi HN, Klee S, Hermann S, Mutze K, Ota C, Lindner M, Behr J, Hilgendorff A, Wagner DE, Königshoff M (2018) Differential effects of Nintedanib and Pirfenidone on lung alveolar epithelial cell function in ex vivo murine and human lung tissue cultures of pulmonary fibrosis. Respir Res 19:175. https://doi.org/10.1186/s12931-018-0876-y
28. Chen C, Peng Y, Wang Z, Fish P, Kaar J, Koepsel R, Russell A, Lareu R, Raghunath M (2009) The scar-in-a-Jar: studying potential antifibrotic compounds from the epigenetic to extracellular level in a single well: scar-in-a-Jar to assess potential antifibrotics. Br J Pharmacol 158:1196–1209. https://doi.org/10.1111/j.1476-5381.2009.00387.x
29. Chen C, Loe F, Blocki A, Peng Y, Raghunath M (2011) Applying macromolecular crowding to enhance extracellular matrix deposition and its remodeling in vitro for tissue engineering and cell-based therapies. Adv Drug Deliv Rev 63:277–290. https://doi.org/10.1016/j.addr.2011.03.003
30. Benny P, Raghunath M (2017) Making microenvironments: a look into incorporating macromolecular crowding into in vitro experiments, to generate biomimetic microenvironments which are capable of directing cell function for tissue engineering applications. J Tissue Eng 8:2041731417730467. https://doi.org/10.1177/2041731417730467
31. Good RB, Eley JD, Gower E, Butt G, Blanchard AD, Fisher AJ, Nanthakumar CB (2019) A high content, phenotypic 'scar-in-a-jar' assay for rapid quantification of collagen fibrillogenesis using disease-derived pulmonary

fibroblasts. BMC Biomed Eng 1:14. https://doi.org/10.1186/s42490-019-0014-z

32. Rønnow SR, Dabbagh RQ, Genovese F, Nanthakumar CB, Barrett VJ, Good RB, Brockbank S, Cruwys S, Jessen H, Sorensen GL, Karsdal MA, Leeming DJ, Sand JMB (2020) Prolonged Scar-in-a-Jar: an in vitro screening tool for anti-fibrotic therapies using biomarkers of extracellular matrix synthesis. Respir Res 21:108. https://doi.org/10.1186/s12931-020-01369-1

33. Organ LA, Duggan A-MR, Oballa E, Taggart SC, Simpson JK, Kang'ombe AR, Braybrooke R, Molyneaux PL, North B, Karkera Y, Leeming DJ, Karsdal MA, Nanthakumar CB, Fahy WA, Marshall RP, Jenkins RG, Maher TM (2019) Biomarkers of collagen synthesis predict progression in the PROFILE idiopathic pulmonary fibrosis cohort. Respir Res 20:148. https://doi.org/10.1186/s12931-019-1118-7

34. Holm Nielsen S, Willumsen N, Leeming DJ, Daniels SJ, Brix S, Karsdal MA, Genovese F, Nielsen MJ (2019) Serological assessment of activated fibroblasts by alpha-smooth muscle actin (α-SMA): a noninvasive biomarker of activated fibroblasts in lung disorders. Transl Oncol 12:368–374. https://doi.org/10.1016/j.tranon.2018.11.004

35. Hata R-I, Senoo H (1989) L-ascorbic acid 2-phosphate stimulates collagen accumulation, cell proliferation, and formation of a three-dimensional tissuelike substance by skin fibroblasts. J Cell Physiol 138:8–16. https://doi.org/10.1002/jcp.1041380103

36. Puerta Cavanzo N, Bigaeva E, Boersema M, Olinga P, Bank RA (2021) Macromolecular crowding as a tool to screen anti-fibrotic drugs: the Scar-in-a-Jar system revisited. Front Med (Lausanne) 7:615774. https://doi.org/10.3389/fmed.2020.615774. PMID: 33521022; PMCID: PMC7841046

Part II

Myofibroblast Mechanobiology

Chapter 12

Why Stress Matters: An Introduction

Daniel J. Tschumperlin

Abstract

Fibroblasts and myofibroblasts are found throughout mechanically loaded tissues, where they take primary responsibility for generating and maintaining the extracellular matrix scaffold upon which organ structure and function depends. They are thus tasked with creating the appropriate mechanical environment in which cells and tissues function optimally, and constantly adapting this environment as needed in response to changing environmental cues. To carry out these functions, fibroblasts must not only deposit and resorb the extracellular matrix, they must adhere to and sense its physical characteristics, and exert the forces necessary to shape, distort, and remodel it as desired. It is thus only through a constant reciprocal sensing and exertion of stress that fibroblasts can carry out their key functions. This introductory chapter will introduce these aspects of fibroblast stress sensing and matrix remodeling during tissue homeostasis, wound repair and fibrotic disease as a lead in to the detailed method chapters to follow on myofibroblast mechanobiology.

Key words Stiffness, Traction, Collagen, Extracellular matrix, Mechanosensing, Integrin, Myofibroblast, Fibrosis

1 Introduction

Mechanobiology is the field of study dedicated to understanding the reciprocal interactions between cells and their mechanical environment. Fibroblasts and myofibroblasts, by virtue of their close relationship with the extracellular matrix (ECM), provide an illustrative example of the richness and complexity of mechanobiology. The ECM itself provides the structural scaffold upon which all adherent cells (including fibroblasts) grow, differentiate, organize, and function [1]. Fibroblasts and myofibroblasts are tasked with the initial deposition of the ECM during organismal development, as well as maintaining the ECM throughout an organism's life, and remodeling it in response to injury [2]. The ECM thus functions both as an instructive set of physical and chemical cues (i.e., the matrisome [3]), as well as a physical support that allows organs and tissues to withstand stresses and function optimally in the ever-changing dynamic mechanical environments inherent to life. Stresses and strains, whether generated by external forces (e.g., gravity,

Boris Hinz and David Lagares (eds.), *Myofibroblasts: Methods and Protocols*, Methods in Molecular Biology, vol. 2299,
https://doi.org/10.1007/978-1-0716-1382-5_12,

muscular contraction) or by internal forces (e.g., cytoskeletal tension), are propagated along the ECM, with these loads largely born by the ECM scaffold to protect tissue integrity and function [4, 5]. Fibroblast and myofibroblasts are themselves embedded within the ECM, and are therefore in constant contact and communication with the ECM. These cells sense changes in their physical environment, both its composition and mechanical properties, in order to respond to the needs for altered ECM composition, organization, quantity, and physical properties [5]. Myofibroblast functions are thus both shaped by, and in turn shape, the mechanical properties of the ECM. Deciphering the rich interactions between myofibroblasts and the mechanical state of the ECM requires a detailed understanding of the ECM itself as a scaffold and signaling structure that influences myofibroblasts, the matrix- and stress-sensing capacity of myofibroblasts, and the stress-generating capacity of myofibroblasts that dynamically reshapes the ECM. This introductory chapter briefly reviews our growing understanding of these essential pillars of myofibroblast mechanobiology.

2 Extracellular Matrix as Scaffold, Signal, and Stress-Bearing Structure

To optimally fulfill its scaffold function, the ECM varies in composition and organization across the different organs, tissues, and compartments of the organism [3]. ECM scaffolds can be thin, compliant, flexible, and permeable to support endothelial or epithelial barriers, such as typical basement membrane scaffolds found in organs like the gut, lung, kidney, and liver, and underlying the vasculature [6]. Alternatively, ECM scaffolds can be dense, stiff, highly aligned, and strong, as in orthopedic tissues such as tendon and cartilage [7], supporting the high loads necessary for proper function of these tissues. Throughout these scaffolds are distributed the ECM and ECM-associated proteins that collectively define key physical characteristics of the ECM as well its capacity to support and signal to specific cell types [3].

To support the varied needs for tissue-specific ECM, the cells of the mesenchyme must respond to local organ and tissue-specific cues in order to deposit, maintain, and remodel the appropriate extracellular matrix that serves local functional needs, both physical and biochemical. The local cues that regulate fibroblast function can also be provided by other cells in the local environment. For example, epithelial cell types specific to organs and interorgan compartments provide signals that instruct fibroblast programs [8, 9]. In a reciprocal fashion, fibroblasts, acting through the ECM they deposit and the soluble cues they generate, can create tissue-specific niches that support progenitor and resident cell functions in an organ- and tissue-specific fashion [8, 10, 11].

In disease process, disruption of these cellular interactions can promote alterations in ECM abundance, organization, and physical properties. For instance, in fibrotic pathologies, persistent or recurring injury to epithelial compartments leads to persistent and inappropriate activation of a wound healing myofibroblast functional program that ultimately results in aberrant ECM deposition and stiffening [12]. Similarly, oncogenic transformation in epithelial tumors leads to generation of cancer-associated fibroblasts which alter tissue organization and physical properties, often supporting tumor invasion and dissemination [13]. The dramatic increases in tissue stiffness in both of these contexts are linked to fibrillar collagen deposition and cross-linking [14, 15]. Notably, mimicking these limited aspects of the changes in physical environment in culture systems by controlling stiffness alone is sufficient to recapitulate cellular phenotypes of fibrotic and cancer-associated fibroblasts, including increased extracellular matrix gene expression, protein production, and deposition [16–21], as well as proliferation, apoptosis-resistance, contractility, and expression of an invasive phenotype [18, 19, 21–23]. These findings emphasize the potentially major role the physical environment plays in influencing fibroblast function during disease progression [24–29]. While these studies have been largely based on simple 2D hydrogel models, tissue ECM and its surrounding cellular environments are inherently more complex, and more sophisticated models, including those that incorporate true 3D geometry [30, 31], as well as relevant neighboring cell types [32], are likely to provide key insights into how fibroblasts integrate local signals from the mechanical environment and neighboring cells in order to control ECM organization and remodeling.

Cues to remodel the ECM scaffold can come not just from neighboring cells, but also directly from altered physical demands in the tissue [5]. Stresses can be generated locally, by ECM-embedded cells, or at a distance, such as by increased gravitational, muscular or cardiovascular forces. In the absence of an appropriate cellular response and adaptation, the tissue is at risk of chronic dysfunction or even catastrophic failure, as physical limits of the individual ECM fibers and the composite structure of the tissue can be exceeded [33, 34]. Thus, it is paramount that fibroblasts sense and respond quickly and appropriately to alterations in physical demands. For example, stretch promotes fibroblast matrix production [35], which reinforces the tissue and ultimately shields resident cells from further stretch in an attempt to restore mechanical homeostasis [5]. Similarly, in dermal wound repair, rapid deposition and stiffening of the nonlinear provisional ECM promotes myofibroblast activation, supporting wound contraction [36, 37]. Effective tissue repair ultimately relaxes the stresses in the wound bed, allowing activated myofibroblasts to undergo apoptosis or return to a more quiescent state [37, 38]. The key role of mechanical stresses in this cycle is illustrated by wound splinting, which prolongs the mechanical loading of the wound ECM and resident myofibroblasts, impeding healing, and enhancing scar formation. In contrast, physically shielding the wound from stress by using bandages that take on part of the load improves healing and reduces scarring [39, 40]. In a similar fashion, reductions in tissue stiffness during the resolution of fibrosis in the lung and liver trigger myofibroblasts to undergo apoptosis or revert to a more quiescent state [41], and these changes can be triggered by restoring the physical properties of the ECM to normal physiological stiffness levels [42–44]. Such healing and repair processes promote the normalization of the tissue mechanical properties and restoration of homeostasis. It is thus clear that understanding myofibroblast regulation and function must account for the sensing of physical cues provided through its interactions with the ECM.

3 ECM and Mechanosensing

Fibroblasts sense and respond to matrix and mechanical cues through specific adhesion molecules and structures. Chief among these are integrin-based adhesions, which link the ECM to the intracellular cytoskeleton, and to a variety of intracellular signaling pathways [45]. Integrins are heterodimers of α and β subunits that bind to specific polypeptide sequences in extracellular matrix proteins. Assembly and stability of integrin-based adhesions are modulated by the physical force environment. For example, fibroblast sensing of matrix stiffness requires the coordinated interactions of integrins with fibronectin outside the cell, and with cytoskeleton-associated talin and actin inside the cell. This complex forms clutch mechanisms, wherein the balance of force-dependent adhesion and contraction enables the assembly, growth, and maintenance of integrin-based cell-matrix adhesion complexes [46]. Adhesion complexes assemble and disassemble on time scales comparable to dynamic changes in internal or externally applied forces [47]. While integrins themselves have little direct signaling capacity, >500 individual proteins have been identified to assemble into these dynamic cell-ECM adhesion complexes, many with their own inherent signaling capacity [48, 49]. For example, the focal adhesion kinase (FAK) can be activated by stretching of fibroblasts, leading to activation of its catalytic domain and phosphorylation of downstream targets [50]. Force-dependent activation of FAK provides one example of how mechanical information from the cell's microenvironment can activate biochemical signaling. The activation of FAK by mechanical stimuli has been widely linked to myofibroblast activation and persistence, demonstrating its critical role in these processes [51–56]. More broadly, force-induced formation of cell-ECM adhesions can ignite signaling processes that have profound influence on the underlying state of the fibroblast in response to alterations in its physical environment. In addition, as discussed below the formation of stable adhesions to the ECM provides a platform for assembly of a highly organized and strongly contractile actomyosin machinery, enabling fibroblasts to apply stress to their surroundings, leading to remodeling of the ECM and inside-out activation of further signaling responses.

Downstream of cell-matrix adhesions, mechanosensing is integrated into transcriptional and epigenetic mechanisms controlling programs of gene expression, leading to relatively stable alterations in cell state [57]. In the case of fibroblasts, these mechano-responsive epigenetic mechanisms include transcriptional regulators such as YAP/TAZ [58–63] and MRTF [16, 64, 65], microRNAs such as miR-21 [66], and histone alterations such as H3K9 di- and tri-methylation [67]. Together these epigenetic and transcriptional mechanisms modulate dynamic, and in some cases long-lasting changes in gene transcription, resulting in engagement of fibroblast proliferative, contractile, and matrix synthetic states in response to alterations in the physical environment. Further elucidation of these programmatic stress responses, including both the upstream mechanosensitive signaling and downstream epigenetic mechanisms, may open new avenues for therapeutic intervention [26], as manipulating the activation state of myofibroblasts through controlled "reprogramming" will likely be key to remodeling the ECM in support of tissue repair and regeneration.

4 Stress Generation by Myofibroblasts and Its Consequences

Once important consequence of fibroblast activation into a myofibroblast state is the coordinate upregulation of contractile and matrix-adhesive proteins, hence the prefix "myo," meaning muscle, resulting in myofibroblasts gaining a dramatically enhanced contractile capacity [68]. The formation of myofibroblasts is influenced both by chemical (e.g., TGF-β1) and mechanical cues, representing a point of convergence between stress sensing and soluble cues associated with tissue injury and fibrosis [18, 37, 69]. The transition from the fibroblast to myofibroblast state has profound implications on the stress state in the cell's local environment. Myofibroblasts remodel 3D collagen gels in model systems [70], and similarly are responsible for contraction of wounds [71] by engaging in ECM adhesion, actomyosin-mediate force generation, and ECM displacement and deformation. These physical processes alter the compaction and alignment of extracellular matrix proteins such as collagen [70, 72], modulating the tissue-scale physical properties of the ECM [73] as well as the cell's mechanical microenvironment [74].

Myofibroblast-mediated ECM contraction can be beneficial in the context of wound healing, but if excessive in nature can contribute to pathological tissue remodeling as in fibrotic conditions [75]. Excessive contraction and stiffening of the ECM in fibrosis distorts normal tissue architecture, propagates mechanical signaling, and results in grossly impaired function [75]. Hence, interrupting this feedback loop has emerged as an important target in anti-fibrosis therapeutic development [26]. Similarly, in many tumor types, cancer-associated fibroblasts generate and remodel the extracellular matrix and take primary responsibility for determining the stiffness of the local environment via extracellular matrix deposition, contraction, and cross-linking [15, 76–78]. Increases in collagen density, cross-linking, stiffness, and alignment all have been shown to promote cancer cell activation and migration, and these physical features of the ECM correlate in many cancer types with pathologic prognosis [79–83], providing another prominent example of how the stress generating and ECM modulating effects of fibroblasts broadly impact on tissue organization and function.

Beyond such well-studied effects, dynamic contraction of the ECM by fibroblasts has revealed additional and unexpected consequences. For example, forces applied to the ECM can be transmitted over long distances, potentially exerting rapid and directed effects [84] on distant cells via 3D physical communication. Recent work shows that fibroblast dynamic contractions of ECM can "attract" macrophages [85], potentially by providing a mechanical signal that propagates along the ECM to distant cells and modulates their migratory capacity and polarity.

Coming full circle, the stresses applied by myofibroblasts themselves are capable of signaling in an autocrine and paracrine fashion via force-dependent release of growth factors. Activation of TGF-β1, which is stored in a latent complex bound to the extracellular matrix, can be accomplished through integrin binding and requires cell-mediated force generation [86, 87]. The mechanical activation of TGF-β1 can thus be promoted by the enhanced stress-generative capacity of fibroblasts, which are themselves activated to a contractile myofibroblast state by a stiffened ECM, giving rise to a positive feedback loop of amplified TGF-β1 signaling, matrix deposition and stiffening and myofibroblast contractile force generation [68, 88].

5 Emerging Methods to Study Myofibroblast Mechanobiology

As our understanding of myofibroblast mechanobiology continues to progress, continued methodological and technological innovations are essential to pave the way toward greater insight. For example, recent efforts have dramatically enhanced the capacity to control the 2D and 3D mechanical environment in which myofibroblasts are studied, including methods that allow simultaneous control of stiffness and microfibrillar architecture [89], as well as control of viscoelasticity and spatiotemporal patterning of mechanical properties [90]. These methods provide greater fidelity in matching the native tissue mechanical environments in which myofibroblasts typically reside and adapt. To further enhance fidelity, native ECM can be decellularized and used alone [91] or combined with polymer scaffolds in hybrid systems that allow dynamic control of mechanical properties along with presentation of ECM ligands found in normal and fibrotic environments [92]. Even further along the fidelity spectrum, intact normal and disease organs and tissues can be used to generate decellularized scaffolds for studying myofibroblast biology in these environments [93–95]. Ultimately assessing myofibroblast function and cellular and ECM mechanics within native tissue environments remains the long term, and very challenging goal. Approaches using atomic force microscopy offer the current gold standard for sensitivity and spatial resolution in mechanical assessments, and this modality continues to be refined for assessing microanatomical variations in cellular and ECM mechanical properties [96]. However, all contact methods, including AFM, require a free boundary, which limits sampling of most native tissue architectures to surfaces, or requires invasive cutting. Non-contact methods to assess cellular and ECM mechanics continue to be challenging to implement but offer promise [97, 98], and may ultimately provide opportunities for less invasive mechanobiological characterizations. Similarly, in situ assessments of myofibroblast biology are being enabled in animal models through the use of constitutive or inducible markers of gene expression, such as with fluorescence reporters driven by Collagen I or Alpha-smooth muscle actin promoters [99, 100], allowing cells to be visualized in vivo or in situ, or analyzed immediately after tissue dissociation by fluorescence based cell sorting [101]. Continued innovation and thoughtful combination of reductionist and integrative models will be essential as we seek to understand myofibroblast mechanobiology in health and disease.

6 Conclusion

Taken together the concepts introduced in this chapter, some now well-known and others new and unexpected, make it clear that we are still just beginning to understand the broader influence of myofibroblast stress-generating capacity on the state of the ECM, and how this capacity alters the state and function of the cells and the tissues which share the same scaffold. Unlocking these mysteries, not of why, but of how stress matters, will be essential if we are to harness the capacity of myofibroblasts and the ECM to modulate wound repair, tissue healing, and regeneration. The following chapters set the stage for exploration in this area by detailing some of the key concepts, tools, and techniques that will be needed to pave the way toward the long-term goal of understanding myofibroblast mechanobiology.

References

1. Frantz C, Stewart KM, Weaver VM (2010) The extracellular matrix at a glance. J Cell Sci 123(Pt 24):4195–4200
2. Lu P et al (2011) Extracellular matrix degradation and remodeling in development and disease. Cold Spring Harb Perspect Biol 3 (12):a005058
3. Naba A et al (2016) The extracellular matrix: tools and insights for the "omics" era. Matrix Biol 49:10–24
4. Fredberg JJ, Kamm RD (2006) Stress transmission in the lung: pathways from organ to molecule. Annu Rev Physiol 68:507–541
5. Humphrey JD, Dufresne ER, Schwartz MA (2014) Mechanotransduction and extracellular matrix homeostasis. Nat Rev Mol Cell Biol 15(12):802–812
6. Kruegel J, Miosge N (2010) Basement membrane components are key players in specialized extracellular matrices. Cell Mol Life Sci 67(17):2879–2895
7. Screen HR et al (2015) Tendon functional extracellular matrix. J Orthop Res 33 (6):793–799
8. Shannon JM, Hyatt BA (2004) Epithelial-mesenchymal interactions in the developing lung. Annu Rev Physiol 66:625–645
9. Lynch MD, Watt FM (2018) Fibroblast heterogeneity: implications for human disease. J Clin Invest 128(1):26–35
10. Sharpe PM, Ferguson MW (1988) Mesenchymal influences on epithelial differentiation in developing systems. J Cell Sci Suppl 10:195–230
11. Zepp JA et al (2017) Distinct mesenchymal lineages and niches promote epithelial self-renewal and myofibrogenesis in the lung. Cell 170(6):1134–1148. e10
12. Wynn TA (2007) Common and unique mechanisms regulate fibrosis in various fibroproliferative diseases. J Clin Invest 117 (3):524–529
13. LeBleu VS, Kalluri R (2018) A peek into cancer-associated fibroblasts: origins, functions and translational impact. Dis Model Mech 11:4
14. Wells RG (2013) Tissue mechanics and fibrosis. Biochim Biophys Acta 1832(7):884–890
15. Lu P, Weaver VM, Werb Z (2012) The extracellular matrix: a dynamic niche in cancer progression. J Cell Biol 196(4):395–406
16. Huang X et al (2012) Matrix stiffness-induced myofibroblast differentiation is mediated by intrinsic mechanotransduction. Am J Respir Cell Mol Biol 47(3):340–348
17. Li Z et al (2007) Transforming growth factor-beta and substrate stiffness regulate portal fibroblast activation in culture. Hepatology 46(4):1246–1256
18. Liu F et al (2010) Feedback amplification of fibrosis through matrix stiffening and COX-2 suppression. J Cell Biol 190(4):693–706
19. Marinkovic A et al (2012) Improved throughput traction microscopy reveals pivotal role for matrix stiffness in fibroblast contractility and TGF-beta responsiveness. Am J Phys Lung Cell Mol Phys 303(3):L169–L180
20. Olsen AL et al (2011) Hepatic stellate cells require a stiff environment for myofibroblastic differentiation. Am J Physiol Gastrointest Liver Physiol 301(1):G110–G118
21. Calvo F et al (2013) Mechanotransduction and YAP-dependent matrix remodelling is required for the generation and maintenance of cancer-associated fibroblasts. Nat Cell Biol 15(6):637–646
22. Chen H et al (2016) Mechanosensing by the alpha6-integrin confers an invasive fibroblast phenotype and mediates lung fibrosis. Nat Commun 7:12564
23. Lagares D et al (2017) Targeted apoptosis of myofibroblasts with the BH3 mimetic ABT-263 reverses established fibrosis. Sci Transl Med 9:420
24. Parker MW et al (2014) Fibrotic extracellular matrix activates a profibrotic positive feedback loop. J Clin Invest 124(4):1622–1635
25. Southern BD et al (2016) Matrix-driven myosin II mediates the pro-fibrotic fibroblast phenotype. J Biol Chem 291(12):6083–6095
26. Tschumperlin DJ, Lagares D (2020) Mechano-therapeutics: targeting mechanical signaling in fibrosis and tumor stroma. Pharmacol Ther 212:107575
27. Tschumperlin DJ, Liu F, Tager AM (2013) Biomechanical regulation of mesenchymal cell

function. Curr Opin Rheumatol 25 (1):92–100
28. Hinz B, Lagares D (2020) Evasion of apoptosis by myofibroblasts: a hallmark of fibrotic diseases. Nat Rev Rheumatol 16(1):11–31
29. Santos A, Lagares D (2018) Matrix stiffness: the conductor of organ fibrosis. Curr Rheumatol Rep 20(1):2
30. Baker BM et al (2015) Cell-mediated fibre recruitment drives extracellular matrix mechanosensing in engineered fibrillar microenvironments. Nat Mater 14(12):1262–1268
31. Dingal PC et al (2015) Fractal heterogeneity in minimal matrix models of scars modulates stiff-niche stem-cell responses via nuclear exit of a mechanorepressor. Nat Mater 14 (9):951–960
32. Mastikhina O et al (2020) Human cardiac fibrosis-on-a-chip model recapitulates disease hallmarks and can serve as a platform for drug testing. Biomaterials 233:119741
33. Black LD et al (2008) Mechanical and failure properties of extracellular matrix sheets as a function of structural protein composition. Biophys J 94(5):1916–1929
34. Kjaer M (2004) Role of extracellular matrix in adaptation of tendon and skeletal muscle to mechanical loading. Physiol Rev 84 (2):649–698
35. Herum KM et al (2017) Mechanical regulation of cardiac fibroblast pro-fibrotic phenotypes. Mol Biol Cell 28(14):1871–1882
36. Chester D, Brown AC (2017) The role of biophysical properties of provisional matrix proteins in wound repair. Matrix Biol 60–61:124–140
37. Hinz B et al (2001) Mechanical tension controls granulation tissue contractile activity and myofibroblast differentiation. Am J Pathol 159(3):1009–1020
38. Desmouliere A et al (1995) Apoptosis mediates the decrease in cellularity during the transition between granulation tissue and scar. Am J Pathol 146(1):56–66
39. Aarabi S et al (2007) Mechanical load initiates hypertrophic scar formation through decreased cellular apoptosis. FASEB J 21 (12):3250–3261
40. Gurtner GC et al (2011) Improving cutaneous scar formation by controlling the mechanical environment: large animal and phase I studies. Ann Surg 254(2):217–225
41. Kisseleva T et al (2012) Myofibroblasts revert to an inactive phenotype during regression of liver fibrosis. Proc Natl Acad Sci U S A 109 (24):9448–9453
42. Caliari SR et al (2016) Gradually softening hydrogels for modeling hepatic stellate cell behavior during fibrosis regression. Integr Biol (Camb) 8(6):720–728
43. Caliari SR et al (2016) Stiffening hydrogels for investigating the dynamics of hepatic stellate cell mechanotransduction during myofibroblast activation. Sci Rep 6:21387
44. Marinkovic A, Liu F, Tschumperlin DJ (2013) Matrices of physiologic stiffness potently inactivate idiopathic pulmonary fibrosis fibroblasts. Am J Respir Cell Mol Biol 48(4):422–430
45. Hynes RO (2002) Integrins: bidirectional, allosteric signaling machines. Cell 110 (6):673–687
46. Elosegui-Artola A et al (2016) Mechanical regulation of a molecular clutch defines force transmission and transduction in response to matrix rigidity. Nat Cell Biol 18(5):540–548
47. Balaban NQ et al (2001) Force and focal adhesion assembly: a close relationship studied using elastic micropatterned substrates. Nat Cell Biol 3(5):466–472
48. Kuo JC et al (2011) Analysis of the myosin-II-responsive focal adhesion proteome reveals a role for beta-Pix in negative regulation of focal adhesion maturation. Nat Cell Biol 13 (4):383–393
49. Schiller HB et al (2011) Quantitative proteomics of the integrin adhesome show a myosin II-dependent recruitment of LIM domain proteins. EMBO Rep 12(3):259–266
50. Frame MC et al (2010) The FERM domain: organizing the structure and function of FAK. Nat Rev Mol Cell Biol 11(11):802–814
51. Lagares D et al (2012) Inhibition of focal adhesion kinase prevents experimental lung fibrosis and myofibroblast formation. Arthritis Rheum 64(5):1653–1664
52. Horowitz JC et al (2007) Combinatorial activation of FAK and AKT by transforming growth factor-beta1 confers an anoikis-resistant phenotype to myofibroblasts. Cell Signal 19(4):761–771
53. Horowitz JC et al (2012) Survivin expression induced by endothelin-1 promotes myofibroblast resistance to apoptosis. Int J Biochem Cell Biol 44(1):158–169
54. Lagares D et al (2012) Adenoviral gene transfer of endothelin-1 in the lung induces pulmonary fibrosis through the activation of focal adhesion kinase. Am J Respir Cell Mol Biol 47(6):834–842
55. Jiang H et al (2016) Targeting focal adhesion kinase renders pancreatic cancers responsive

to checkpoint immunotherapy. Nat Med 22 (8):851–860
56. Wong VW et al (2011) Focal adhesion kinase links mechanical force to skin fibrosis via inflammatory signaling. Nat Med 18 (1):148–152
57. Tschumperlin DJ et al (2018) Mechanosensing and fibrosis. J Clin Invest 128(1):74–84
58. Mannaerts I et al (2015) The Hippo pathway effector YAP controls mouse hepatic stellate cell activation. J Hepatol 63(3):679–688
59. Piersma B et al (2015) YAP1 is a driver of myofibroblast differentiation in normal and diseased fibroblasts. Am J Pathol 185 (12):3326–3337
60. Szeto SG et al (2016) YAP/TAZ are mechanoregulators of TGF-beta-smad signaling and renal fibrogenesis. J Am Soc Nephrol 27 (10):3117–3128
61. Martin K et al (2016) PAK proteins and YAP-1 signalling downstream of integrin beta-1 in myofibroblasts promote liver fibrosis. Nat Commun 7:12502
62. Liu F et al (2015) Mechanosignaling through YAP and TAZ drives fibroblast activation and fibrosis. Am J Phys Lung Cell Mol Phys 308 (4):L344–L357
63. Noguchi S et al (2017) TAZ contributes to pulmonary fibrosis by activating profibrotic functions of lung fibroblasts. Sci Rep 7:42595
64. Haak AJ et al (2014) Targeting the myofibroblast genetic switch: inhibitors of myocardin-related transcription factor/serum response factor-regulated gene transcription prevent fibrosis in a murine model of skin injury. J Pharmacol Exp Ther 349(3):480–486
65. Scharenberg MA et al (2014) TGF-beta-induced differentiation into myofibroblasts involves specific regulation of two MKL1 isoforms. J Cell Sci 127(Pt 5):1079–1091
66. Li CX et al (2017) MicroRNA-21 preserves the fibrotic mechanical memory of mesenchymal stem cells. Nat Mater 16(3):379–389
67. Ligresti G et al (2019) CBX5/G9a/H3K9me-mediated gene repression is essential to fibroblast activation during lung fibrosis. JCI Insight 5:e127111
68. Wipff PJ et al (2007) Myofibroblast contraction activates latent TGF-beta1 from the extracellular matrix. J Cell Biol 179 (6):1311–1323
69. Hinz B (2010) The myofibroblast: paradigm for a mechanically active cell. J Biomech 43 (1):146–155
70. Sawhney RK, Howard J (2002) Slow local movements of collagen fibers by fibroblasts drive the rapid global self-organization of collagen gels. J Cell Biol 157(6):1083–1091
71. Grinnell F (1994) Fibroblasts, myofibroblasts, and wound contraction. J Cell Biol 124(4):401–404
72. Leung LY et al (2007) A new microrheometric approach reveals individual and cooperative roles for TGF-beta1 and IL-1beta in fibroblast-mediated stiffening of collagen gels. FASEB J 21(9):2064–2073
73. Taufalele PV et al (2019) Fiber alignment drives changes in architectural and mechanical features in collagen matrices. PLoS One 14 (5):e0216537
74. Feng Z et al (2014) The mechanisms of fibroblast-mediated compaction of collagen gels and the mechanical niche around individual fibroblasts. Biomaterials 35 (28):8078–8091
75. Van De Water L, Varney S, Tomasek JJ (2013) Mechanoregulation of the myofibroblast in wound contraction, scarring, and fibrosis: opportunities for new therapeutic intervention. Adv Wound Care (New Rochelle) 2 (4):122–141
76. Gascard P, Tlsty TD (2016) Carcinoma-associated fibroblasts: orchestrating the composition of malignancy. Genes Dev 30 (9):1002–1019
77. Kalluri R (2016) The biology and function of fibroblasts in cancer. Nat Rev Cancer 16 (9):582–598
78. Yamauchi M et al (2018) The fibrotic tumor stroma. J Clin Invest 128(1):16–25
79. Conklin MW et al (2011) Aligned collagen is a prognostic signature for survival in human breast carcinoma. Am J Pathol 178 (3):1221–1232
80. Provenzano PP et al (2009) Matrix density-induced mechanoregulation of breast cell phenotype, signaling and gene expression through a FAK-ERK linkage. Oncogene 28 (49):4326–4343
81. Provenzano PP et al (2008) Collagen density promotes mammary tumor initiation and progression. BMC Med 6:11
82. Venning FA, Wullkopf L, Erler JT (2015) Targeting ECM disrupts cancer progression. Front Oncol 5:224
83. Levental KR et al (2009) Matrix crosslinking forces tumor progression by enhancing integrin signaling. Cell 139(5):891–906
84. Wang H et al (2014) Long-range force transmission in fibrous matrices enabled by tension-driven alignment of fibers. Biophys J 107(11):2592–2603

85. Pakshir P et al (2019) Dynamic fibroblast contractions attract remote macrophages in fibrillar collagen matrix. Nat Commun 10 (1):1850
86. Giacomini MM et al (2012) Epithelial cells utilize cortical actin/myosin to activate latent TGF-beta through integrin alpha(v)beta(6)-dependent physical force. Exp Cell Res 318 (6):716–722
87. Shi M et al (2011) Latent TGF-beta structure and activation. Nature 474(7351):343–349
88. Marinkovic A et al (2012) Improved throughput traction microscopy reveals pivotal role for matrix stiffness in fibroblast contractility and TGF-beta responsiveness. Am J Phys Lung Cell Mol Phys 303(3):L169–L180
89. Davidson CD et al (2020) Myofibroblast activation in synthetic fibrous matrices composed of dextran vinyl sulfone. Acta Biomater 105:78–86
90. Hui E et al (2019) Spatiotemporal control of viscoelasticity in phototunable hyaluronic acid hydrogels. Biomacromolecules 20 (11):4126–4134
91. Pouliot RA et al (2020) Porcine lung-derived extracellular matrix hydrogel properties are dependent on pepsin digestion time. Tissue Eng Part C Methods 26(6):332–346
92. Petrou CL et al (2020) Clickable decellularized extracellular matrix as a new tool for building hybrid-hydrogels to model chronic fibrotic diseases in vitro. J Mater Chem B 8 (31):6814–6826
93. Sun H et al (2016) Netrin-1 regulates fibrocyte accumulation in the decellularized fibrotic sclerodermatous lung microenvironment and in bleomycin-induced pulmonary fibrosis. Arthritis Rheum 68(5):1251–1261
94. Elowsson Rendin L et al (2019) Matrisome properties of scaffolds direct fibroblasts in idiopathic pulmonary fibrosis. Int J Mol Sci 20 (16):4013
95. Tjin G et al (2017) Lysyl oxidases regulate fibrillar collagen remodelling in idiopathic pulmonary fibrosis. Dis Model Mech 10 (11):1301–1312
96. Viji Babu PK et al (2019) Nano-mechanical mapping of interdependent cell and ECM mechanics by AFM force spectroscopy. Sci Rep 9(1):12317
97. Plekhanov AA et al (2020) Histological validation of in vivo assessment of cancer tissue inhomogeneity and automated morphological segmentation enabled by Optical Coherence Elastography. Sci Rep 10(1):11781
98. Scarcelli G et al (2015) Noncontact three-dimensional mapping of intracellular hydromechanical properties by Brillouin microscopy. Nat Methods 12(12):1132–1134
99. Sun KH et al (2016) alpha-Smooth muscle actin is an inconsistent marker of fibroblasts responsible for force-dependent TGFbeta activation or collagen production across multiple models of organ fibrosis. Am J Phys Lung Cell Mol Phys 310(9):L824–L836
100. Yamaguchi T et al (2018) Visualization of specific collagen-producing cells by Col1-GFP transgenic mice revealed novel type I collagen-producing cells other than fibroblasts in systemic organs/tissues. Biochem Biophys Res Commun 505(1):267–273
101. Tsukui T et al (2020) Collagen-producing lung cell atlas identifies multiple subsets with distinct localization and relevance to fibrosis. Nat Commun 11(1):1920

Chapter 13

Soft Substrate Culture to Mechanically Control Cardiac Myofibroblast Activation

Natalie M. Landry, Sunil G. Rattan, and Ian M. C. Dixon

Abstract

Two-dimensional cell culture is the primary method employed for proof-of-concept studies in most molecular biology labs. While immortalized cell lines are convenient and easy to maintain for extended periods in vitro, their inability to accurately represent genuine cell physiology—or pathophysiology—presents a challenge for drug discovery, as most results are not viable for the transition to clinical trial. The use of primary cells is a more biologically relevant approach to this issue; however, simulating in vitro what is observed in vivo is exigent at best. Primary cardiac fibroblasts are particularly difficult to maintain in a quiescent state, due to their innate phenotypic plasticity, and sensitivity to mechanical and biochemical stimulus. As conventional cell culture methods do not consider these factors, here we describe a method that limits environmental input (i.e., mechanical, nutritional, hormonal) to extend the physiological cardiac fibroblast phenotype in vitro.

Key words Soft substrate, Elastic substrate, Cell culture, Cardiac fibroblast, Myofibroblast activation, Cardiac fibrosis

1 Introduction

Recent optimizations to conventional cell culture methods have enabled researchers to better mimic the native cellular environment of various tissues, in vitro. While many of these techniques involve three-dimensional matrices or hydrogels, they are not amenable to everyday use in most molecular laboratories—especially for proof-of-concept studies—due to the significant technical and material requirements for generating appropriate tissue scaffolds [1, 2]. Thus, a modified two-dimensional culture system is a more feasible method to study isolated primary cells.

Fibroblasts, regardless of their native tissue of origin, are especially sensitive to mechanical input [3]. Rather than having a defined phenotype, fibroblasts exist on a phenotypic spectrum, whose limits are defined on one end as "quiescent" or "resting fibroblasts," and as pro-fibrotic "myofibroblasts" or "activated

Boris Hinz and David Lagares (eds.), *Myofibroblasts: Methods and Protocols*, Methods in Molecular Biology, vol. 2299, https://doi.org/10.1007/978-1-0716-1382-5_13, © Springer Science+Business Media, LLC, part of Springer Nature 2021

fibroblasts" on the other [4]. In most tissues, resting fibroblasts are highly motile and maintain tissue homeostasis via the balanced deposition and breakdown of extracellular matrix components. Conversely, activated fibroblasts are highly contractile and hyper-secretory for fibrillar collagens [5]. Due to their significant mechanical loading and violent, pulsatile environment within the contracting myocardium, native cardiac fibroblasts likely exist somewhere between these two extremes.

Primary cardiac fibroblasts, irrespective of the condition of the myocardium from which they are isolated, take on the pro-fibrotic myofibroblast phenotype within hours of plating in conventional cell culture conditions (i.e., on stiff plastic substrata). Here, we describe methods for better maintaining cardiac fibroblasts of murine origin in two-dimensional culture for several days, without significant activation of the myofibroblast phenotype. While we do observe some expression of alpha-smooth muscle actin (α-SMA), it is generally excluded from the actin cytoskeleton and does not promote cell contractility in these conditions [6]. The capacity to maintain primary cardiac fibroblasts in a resting state in vitro for several days enables increasingly sensitive means by which to study their response to various stimuli and can facilitate more accurate and physiologically relevant studies to study cardiac fibrosis.

2 Materials

Prepare solutions using sterile, ultrapure deionized water, and store at 4 °C when not in use.

2.1 Elastic Cell Culture Surfaces

1. ExCellness® PrimeCoat elastic culture surfaces, 5 or 10 kPa (*see* **Notes 1–3**).

2.2 Culture Surface Coating

1. Phosphate-buffered saline (1×; PBS): 137 mM NaCl, 2.7 mM KCl, 10 mM Na_2HPO_4, 1.8 mM KH_2PO_4. Dissolve reagents in 800 mL deionized water and adjust pH to 7.4. Filter sterilize (0.2 μm) the solution in a biosafety cabinet or sterilize by autoclaving for 20 min at 121 °C, 15 psi.
2. Gelatin stock solution (1% w/v): 1 g Porcine Gelatin Type A, 100 mL ultrapure, sterile water. Pre-warm water to 37 °C, then add the gelatin. Allow to dissolve, and sterilize by autoclaving at 121 °C, 15 psi, for 30 min. Cool slightly (~50 °C) and make appropriate aliquots under sterile conditions. Store the stock solution indefinitely at 4 °C until use.
3. Isopropanol (Reagent grade is sufficient).

2.3 Primary Rat Cardiac Fibroblast Isolation and Maintenance

1. Dulbecco's Modified Eagle Medium: Nutrient Mixture F12 (DMEM/F12).
2. Minimum Essential Medium, Spinner's Modification (S-MEM).
3. Ham's F-10 Nutrient Mix (F-10). It is strongly recommended to use culture media supplemented with 25 mM HEPES buffer.
4. Ketamine-Xylazine solution: Combine 100 mg/mL sterile veterinary-grade ketamine hydrochloride and 20 mg/mL sterile veterinary-grade xylazine to a final ratio of 10:0.5 (ketamine: xylazine), at an appropriate stock concentration to minimize the volume injected into each animal. Maintain the sterility of the solution by combining in a sterile medication vial while working in a biosafety cabinet or laminar flow hood; store at 4 °C for up to 1 month.
5. 6 mg/mL Heparin solution: Dissolve 60 mg of Heparin Sodium in 10 mL of sterile water. Filter sterilize and store in a sterile, amber medication vial at 4 °C for up to 3 months. Protect from light.
6. Surgical suture, 3-0 gauge.
7. 640 U/mL Collagenase Type II solution: Dissolve 32,000 units Collagenase Type II in 50 mL S-MEM. Mix well while avoiding the formation of bubbles—do not vortex. 50 mL of this solution are required for each rat heart.
8. Fetal Bovine Serum (FBS), sterile.
9. Penicillin-Streptomycin (Cell culture grade, 100× stock solution).
10. Phosphate-Buffered Saline (1×).
11. Langendorff apparatus with thermostat controller, reservoirs, and tubing for recirculation of 50 mL of liquid.
12. 27-gauge × ½ inch (Tuberculin) needles.
13. 1 mL Luer lock syringes.
14. Surgical tools suitable for small animals: forceps, scissors, scalpel.
15. 10 cm polystyrene tissue culture dishes.
16. Serological pipettes (10 mL, 25 mL), sterile.
17. 40 μm cell strainers, sterile.
18. Conical tubes suitable for centrifugation (50 mL).

2.4 Primary Mouse Cardiac Fibroblast Isolation

1. Isoflurane gas (3% is used for this protocol).
2. Mouse compatible gas induction chamber and face mask.
3. A fume hood.

4. Dulbecco's Modified Eagle Medium: Nutrient Mixture F12 (DMEM/F12); Minimum Essential Medium, Spinner's Modification (S-MEM).
5. Ham's F-10 Nutrient Mix (F-10). It is strongly recommended to use culture media supplemented with 25 mM HEPES buffer.
6. Collagenase Type II solution (330 U/mL): Dissolve 8250 units Collagenase Type II in 25 mL S-MEM. Mix well while avoiding the formation of bubbles—do not vortex. 25 mL of this solution are required for each mouse heart.
7. EDTA solution: in 400 mL sterile water, dissolve 5 mM EDTA, 130 mM NaCl, 5 mM KCl, 500 nM NaH_2PO_4, 10 mM HEPES, 10 mM glucose, 10 mM 2,3-Butanedione 2-monoxime, and 10 mM Taurine. Adjust the pH to 7.8 with NaOH, bring volume up to 500 mL with water, and filter sterilize. Store solution at 4 °C for up to 6 months.
8. Perfusion buffer: in 400 mL sterile water, dissolve 1 mM $MgCl_2$, 130 mM NaCl, 5 mM KCl, 500 nM NaH_2PO_4, 10 mM HEPES, 10 mM glucose, 10 mM 2,3-Butanedione 2-monoxime, and 10 mM Taurine. Adjust the pH to 7.8 with NaOH and bring up to 500 mL with water. Filter sterilize and store at 4 °C for up to 6 months.
9. Fetal Bovine Serum (FBS), sterile.
10. Penicillin-Streptomycin (Cell culture grade, 100× stock solution).
11. Ascorbic Acid (100 mM): Dissolve 0.88 g ascorbic acid in 50 mL sterile ultrapure water. Filter sterilize (0.2 μm) and store in a dark or opaque vial. Protect from light and store at 4 °C. Do not use if solution becomes oxidized (i.e., turns yellow/orange in color).
12. Insulin-Transferrin-Selenium-Sodium Pyruvate cell culture supplement (100× solution): 1.00 g/L insulin, 0.55 g/L transferrin, 3.87 μM sodium selenite, 100 mM sodium pyruvate. Dissolve all components in Earle's Balanced Salt Solution (5.3 mM KCl, 26 mM $NaHCO_3$, 117 mM NaCl, 1.0 mM NaH_2PO_4, 5.55 mM D-glucose), filter sterilize, and store at 4 °C for up to 18 months.
13. Phosphate-Buffered Saline (1×; PBS).
14. 10 cm Polystyrene tissue culture dishes.
15. 27-gauge × ½ inch needles.
16. 20 mL Luer lock syringes.
17. Surgical tools suitable for small animals: forceps, scissors, hemostat, scalpel.

18. Serological pipettes (10 mL), sterile.
19. 40 μm cell strainers, sterile.
20. Conical tubes suitable for centrifugation (50 mL).

3 Methods

Cell culture surface preparation must be conducted in a sterile environment, such as a laminar flow hood or biosafety cabinet, using proper aseptic technique. Because this method involves the use of primary cells, all handling of cell cultures should be done in a Class II biosafety cabinet.

We have included fibroblast isolation and culture techniques for both rat and mouse, as the two differ quite significantly. The protocols described here have been optimized for fibroblasts isolated from young animals, as cells originating from older animals tend to senesce too quickly and isolated yields are much lower. It is expected that all experiments involving animals are to be approved by the appropriate ethics committee and are conducted according to the local and federal governing bodies associated with your institute (e.g., Canadian Council for Animal Care, United States Animal Welfare Act, and NIH guidelines).

3.1 Coating Elastic Tissue Culture Surfaces

1. The day prior to isolating primary cardiac fibroblasts, sterilize elastic culture surfaces using several washes of 100% isopropanol—two or three rounds of vigorous swirling, both on the dish and in the lid, is sufficient. Allow the plates to dry thoroughly in a sterile environment.
2. While the plates are drying, dilute the gelatin in pre-warmed (37 °C) 1× PBS to a final concentration of 10 μg/mL. An excess volume (250 μL/cm^2) is typically required to properly cover the entire hydrophobic silicone surface, and so the following volumes are recommended:

Culture vessel	Volume
35 mm dish or 6-well plate	2 mL/dish or well
10 cm dish	15 mL/dish
12-well plate	1 mL/well
24-well plate	0.5 mL/well

3. Immerse the elastic surfaces in the diluted gelatin solution. Allow the gelatin to coat the culture surfaces overnight at 37 °C (*see* **Note 4**).

3.2 Primary Rat Cardiac Fibroblast Isolation and Culture

1. Anesthetize young rats (101–125 g in mass) using ketamine-xylazine cocktail (90 mg/kg ketamine; 10 mg/kg xylazine), by intraperitoneal injection.
2. Upon loss of limb reflexes, make a small incision in the skin of the mid-thigh to reveal the great arteries of the leg. Administer 6 mg/kg heparin sodium via the saphenous or femoral artery.
3. Excise hearts and promptly rinse in DMEM/F12 to remove excess blood but preserve cardiac contraction.
4. Cannulate the heart onto the pre-warmed (37 °C) Langendorff apparatus via the aorta and secure with surgical suture.
5. Perfuse the hearts in retrograde for 5 min with DMEM/F12.
6. Perfuse for another 5 min with S-MEM to halt cardiac contraction and promote cell dissociation.
7. Briefly stop the apparatus and drain the S-MEM from the tubing. Change the mode to recirculation, add 40 mL of the Collagenase Type II solution to the recirculation vessels, and resume perfusion for 25 min, or until the tissue has been completely digested (*see* **Note 5**).
8. While the tissue is digesting, dilute the remaining 10 mL of Collagenase solution 1:1 with S-MEM, and keep it warm in a 10 cm tissue culture dish at 37 °C.
9. Remove the hearts from the apparatus by cutting below the atria to isolate only the ventricles. Place the digested tissue into the dish of diluted Collagenase solution from **step 8**. Gently tease the tissue apart with forceps, and incubate at 37 °C, 5% CO_2 for 10 min.
10. **Steps 10 through 18 are to be performed in a Class II biosafety cabinet.** Using a 10 mL serological pipette, further dissociate the cells by triturating (pipetting up and down), being careful not to introduce bubbles into the suspension.
11. Neutralize the Collagenase by adding 10 mL DMEM/F12 supplemented with 2% FBS.
12. Allow the cell suspension to pass through a cell strainer to remove debris and undigested tissue matrix.
13. Pellet the cells by centrifugation at 200 × *g* for 7 min.
14. Aspirate the supernatant and resuspend the cells in an appropriate volume of F-10 medium supplemented with 2% FBS, 100 Units penicillin-streptomycin (*see* **Note 6**).
15. Remove the gelatin solution from the prepared elastic culture surfaces and immediately seed the isolated cells (do not wash off any remaining gelatin solution).
16. Allow fibroblasts to adhere for 2.5–4 h at 37 °C, 5% CO_2.

17. Briefly wash adhered cells twice with pre-warmed (37 °C) PBS supplemented with penicillin-streptomycin, then replace the culture medium.
18. Maintain cells at 37 °C, 5% CO_2, in a humidified incubator.
19. Each subsequent day, repeat **step 17**, until cells are at the optimum confluency for in vitro assays (*see* **Note 7**).

3.3 Primary Mouse Cardiac Fibroblast Isolation and Culture

1. Anesthetize young mice (<12 weeks of age) using 3% isoflurane in an induction chamber.
2. Upon loss of limb reflexes, transfer the animal to a fume hood and maintain anesthesia using a face mask.
3. Open the chest and excise the heart while keeping a portion of the ascending aorta attached.
4. In a 10 cm tissue culture dish, immediately flush the heart with 10 mL pre-warmed (37 °C) EDTA buffer by injecting into the ventricles using a 27-gauge needle inserted via the apex.
5. In another 10 cm dish, flush the heart with 10 mL pre-warmed (37 °C) perfusion buffer.
6. Transfer the heart to yet another 10 cm dish. While still injecting through the apex, clamp the aorta with a hemostat, and perfuse the heart with 50 mL of Collagenase solution (i.e., by injecting 25 mL twice), while collecting the solution in the dish.
7. Once the heart is sufficiently digested (*see* **Note 5**), gently pull the tissue apart into small pieces using forceps. Allow the tissue to digest further at 37 °C, 5% CO_2 for an additional 10 min.
8. **Steps 8 through 12 are to be performed in a Class II biosafety cabinet.** Using a 10 mL serological pipette, triturate the tissue until no large pieces are apparent.
9. Neutralize the Collagenase by adding 10 mL DMEM/F12 supplemented with 10% FBS.
10. Allow the cell suspension to pass through a sterile cell filter to remove any undigested tissue and debris.
11. Pellet the cells by centrifugation at 200 × *g* for 7 min.
12. Remove the supernatant and resuspend the cells in 40 mL DMEM/F12 supplemented with 10% FBS, 1 μM ascorbic acid, and 100 U/mL penicillin-streptomycin.
13. Remove the gelatin solution from the prepared elastic surfaces.
14. Seed the cells according to the vessel sizes prepared (*see* **Note 8**) and allow them to adhere for 3 h.
15. Gently wash adhered cells twice with pre-warmed (37 °C) PBS supplemented with penicillin-streptomycin and replace with fresh DMEM/F12 supplemented with 10% FBS, ascorbic acid, and antibiotics.

16. Maintain the cells at 37 °C, 5% CO_2, in a humidified incubator.
17. Repeat **step 15** every 24 h until the cells develop a spindle-shaped morphology; this typically takes 3–4 days post-isolation.
18. Change the culture medium to F-10 supplemented with 2% FBS, 1× Insulin-Transferrin-Selenium, and penicillin-streptomycin.
19. Change the culture medium every subsequent day, until the cells reach the desired confluency for in vitro assays (*see* **Note 9**).

4 Notes

1. While acrylamide hydrogels are often used to create soft cell culture surfaces, this method employs the use of commercially available silicone soft substrates. The myocardium has an elastic (E) of approximately 7 kPa, and so we recommend using surfaces with $E = 5$ to 10 kPa for optimal results. For a list of commercial silicone substrates and recommended moduli, *see* **Note 3**.
2. We have observed a severe reduction in cell attachment and survival with both commercially available and lab-made acrylamide soft-well hydrogels, regardless of the coating used to promote cell attachment (e.g., collagen, gelatin, fibronectin). There is speculation that the techniques used to attach the hydrogel to the dish/well generates reactive oxygen species (ROS), to which primary cells are particularly sensitive [7].
3. At the time of this methodology's publication, the following silicone elastic substrates are commercially available for primary cardiac fibroblasts: ExCellness® PrimeCoat 10 cm dishes, cover slips, and multi-well plates ($E = 5$ kPa or 10 kPa); Advanced BioMatrix Cytosoft® imaging plates ($E = 8$ kPa).
4. Silicone surfaces may be coated the same day as cardiac fibroblast cell isolation; however, we have found that primary cells adhere much better to gelatin if coating is carried out for a minimum of 12 h.
5. Hearts are fully digested when they are pale in color, and the cells are pulling away from the decellularized epicardium. Testing of compliance with forceps should show little to no resistance.
6. We have found that resuspending one rat heart in 45 mL of culture medium provides the best seeding confluency. For 10 cm dishes, we typically use 3 mL of cell suspension at plating; for coverslips (6-well or 35 mm dishes), we use 0.5 mL per well/dish.

7. Primary rat cardiac fibroblasts cultured under these conditions will typically remain quiescent and subconfluent (<70%) for at least 4 days post-plating.
8. For freshly isolated mouse cardiac fibroblasts, we find that one heart can be divided into up to four 10 cm dishes, or two 6-well dishes (12 × 35 mm dishes). Typically, we are able to seed the cells at ~10% confluency.
9. From our experience with mouse cardiac fibroblasts, they can be maintained at subconfluent (<70%) density on elastic surfaces in a quiescent state for up to 10 days post-isolation.

Acknowledgments

This work was funded by grants to IMCD by the Heart & Stroke Foundation of Canada (#G-17-0018631) and from the Canadian Institutes for Health Research (PJT—162163). Operating support and facilities were also provided by the St. Boniface Hospital Research Foundation and Albrechtsen Research Centre.

References

1. Fang Y, Eglen RM (2017) Three-dimensional cell cultures in drug discovery and development. SLAS Discov 22:456–472. https://doi.org/10.1177/1087057117696795
2. Breslin S, O'Driscoll L (2013) Three-dimensional cell culture: the missing link in drug discovery. Drug Discov Today 18:240–249. https://doi.org/10.1016/j.drudis.2012.10.003
3. Solon J, Levental I, Sengupta K, Georges PC, Janmey PA (2007) Fibroblast adaptation and stiffness matching to soft elastic substrates. Biophys J 93:4453–4461
4. Ivey MJ, Tallquist MD (2016) Defining the cardiac fibroblast. Circ J 80:2269–2276. https://doi.org/10.1253/circj.CJ-16-1003
5. Hinz B (2010) The myofibroblast: paradigm for a mechanically active cell. J Biomech 43:146–155. https://doi.org/10.1016/j.jbiomech.2009.09.020
6. Landry NM, Rattan SG, Dixon IMC (2019) An improved method of maintaining primary murine cardiac fibroblasts in two-dimensional cell culture. Sci Rep 9:12889. https://doi.org/10.1038/s41598-019-49285-9
7. Caliari SR, Burdick JA (2016) A practical guide to hydrogels for cell culture. Nat Methods 13:405–414. https://doi.org/10.1038/nmeth.3839

Chapter 14

Quantitative Analysis of Myofibroblast Contraction by Traction Force Microscopy

Shuying Yang, Fernando R. Valencia, Benedikt Sabass, and Sergey V. Plotnikov

Abstract

Myofibroblasts play important roles in physiological processes such as wound healing and tissue repair. While high contractile forces generated by the actomyosin network enable myofibroblasts to physically contract the wound and bring together injured tissue, prolonged and elevated levels of contraction also drive the progression of fibrosis and cancer. However, quantitative mapping of these forces has been difficult due to their extremely low magnitude ranging from 100 pN/μm^2 to 2 nN/μm^2. Here, we provide a protocol to measure cellular forces exerted on two-dimensional compliant elastic hydrogels. We describe the fabrication of polyacrylamide hydrogels labeled with fluorescent fiducial markers, functionalization of substrates with ECM proteins, setting up the experiment, and imaging procedures. We demonstrate the application of this technique for quantitative analysis of traction forces exerted by myofibroblasts.

Key words Traction force microscopy, Cellular forces, Contractility, Myofibroblasts, Actomyosin cytoskeleton

1 Introduction

Myofibroblasts play important roles in physiological processes such as wound healing and tissue repair. High contractile forces generated by the actomyosin network enable myofibroblasts to physically contract the wound and bring together injured tissue [1]. However, prolonged and elevated levels of contraction also drive the progression of fibrosis and cancer [2–4]. Therefore, the ability to measure contractile forces generated by myofibroblasts can substantially increase our understanding of the cellular and molecular mechanisms that underlie the function of myofibroblasts in both healthy and diseased tissue. Traction force microscopy (TFM) is an established technique that quantitatively measures contractile forces of

Shuying Yang and Fernando R. Valencia contributed equally to this work.

Boris Hinz and David Lagares (eds.), *Myofibroblasts: Methods and Protocols*, Methods in Molecular Biology, vol. 2299, https://doi.org/10.1007/978-1-0716-1382-5_14, © Springer Science+Business Media, LLC, part of Springer Nature 2021

adherent cells to the nanonewton scale. Such contractile forces generated by the actomyosin network are transmitted to the extracellular matrix through cell adhesion sites called focal adhesions—integrin-based protein complexes that physically connect the cell's cytoskeleton to the underlying matrix [5]. TFM takes advantage of this mechanical linkage by using deformation of the extracellular matrix as an indicator of the magnitude of contractile force. Harris, Wild and Stopak (1980) pioneered the idea of using elastic substrates to study forces exerted by individual cells by plating chicken heart fibroblasts on a thin layer of polymerized silicone weak enough to be wrinkled when fibroblasts contract [6]. By manually applying forces of known magnitudes, Harris and colleagues distorted the silicone surface to similar levels to determine the magnitude of traction force produced by individual cells. However, difficulties due to the nonlinear elasticity of silicone and low spatial resolution have hindered researchers' efforts in force quantification using this method [7, 8].

More recently, continuous, linearly elastic materials such as polydimethylsiloxane, polyethylene glycol, and polyacrylamide have replaced silicone substrates for TFM measurements [9–11]. Polyacrylamide hydrogels (PAAG) in particular, introduced by Pelham and Wang (1997), have become the most widely used material. The superior optical transparency of PAAG allows researchers to simultaneously visualize cytoskeletal proteins and analyze focal adhesions dynamics that complement traction force measurements [12]. In addition, the stiffness of PAAG can be readily adjusted by varying the concentration of acrylamide and the cross-linker bis-acrylamide to mimic the mechanical properties of various biological tissues ranging from soft brain to stiff bone [13]. Moreover, PAAG can be covalently cross-linked to different ECM proteins such as fibronectin, collagen, and laminin to control the activation of specific classes of adhesion receptors, thereby mimicking the biochemical properties of the microenvironment of various cell types.

Alongside developments in TFM-optimized elastic substrates, advances in substrate preparation techniques have enabled higher spatial resolution for traction force measurements with sub-micrometer resolution. Instead of quantifying large-scale deformations such as that of wrinkled silicone or tilted PDMS pillars, deformations in the elastic substrate are detected by capturing the displacement of fluorescent bead markers embedded within the substrate [14, 15]. First, the position of fluorescent beads in substrates strained by contractile cells is captured (Fig. 1). Then, this position is compared to that in substrates where cells have undergone enzymatic or chemical detachment. Two-dimensional vector maps capturing both the magnitude and direction of displacement are extracted through mathematical modeling based on

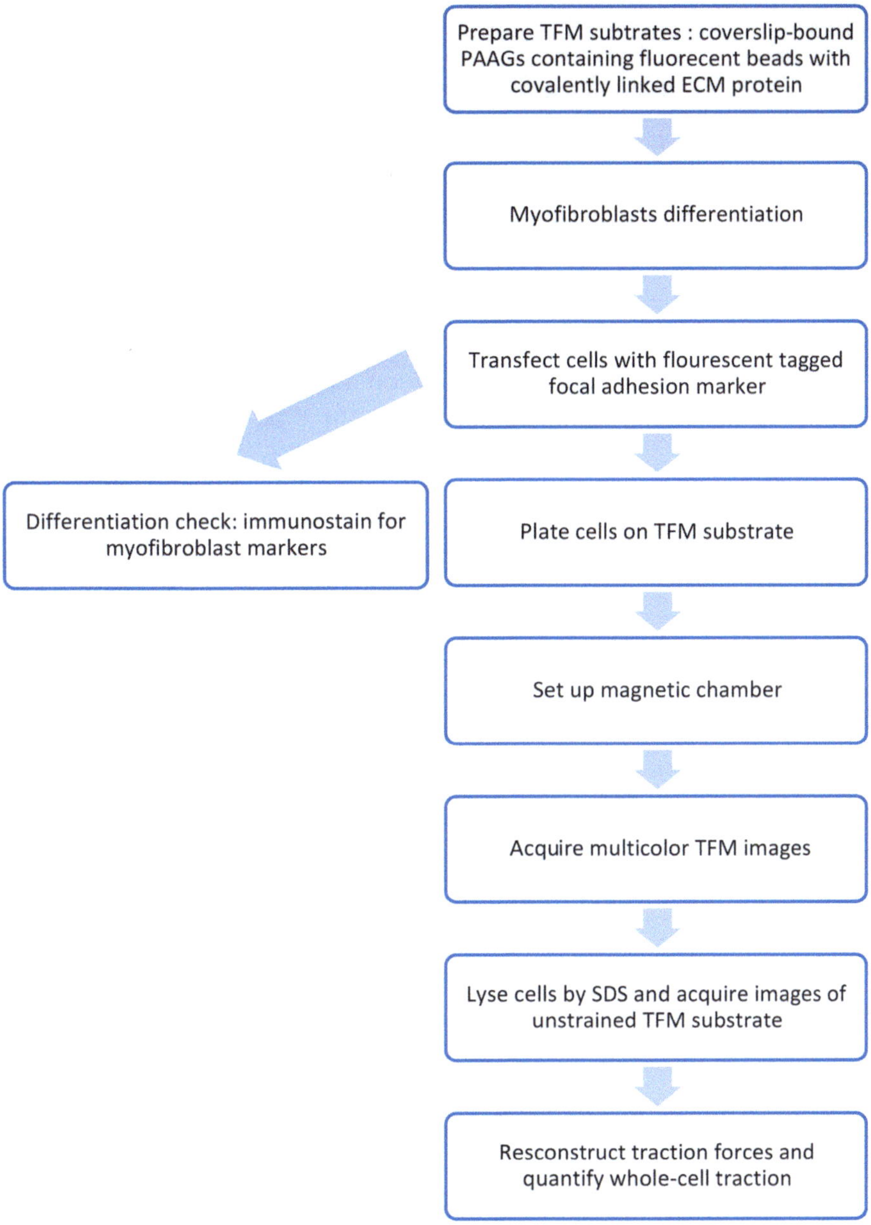

Fig. 1 Flowchart for performing traction force microscopy (TFM) on a compliant polyacrylamide substrate

the elastic mechanics of the substrate to generate traction force maps. Fluorescent beads of one single color are commonly used if acquisition of additional fluorescent proteins is required to address the biological question. Although using fluorescent beads of two colors increases spatial resolution by the square root of two, it is not required to measure forces exerted by the entire cell.

2 Materials

2.1 Activation of Coverslips

1. "Squeaky clean" 22×22 mm coverslips (*see* **Note 1**).
2. 15 cm diameter glass Petri dish.
3. (3-aminopropyl)trimethoxysilane (APTMS) solution: Mix equal parts of APTMS and ultrapure water to make 0.5 mL of 50% APTMS solution for each coverslip. To ensure reliable binding of the PAAG to the glass surface, this solution should be made on the day of coverslip activation. Note that the dilution of APTMS in water is exothermic and produces a lot of heat — the solution may start to boil.
4. 0.5% Glutaraldehyde: add 1 mL of 25% electron microscopy grade glutaraldehyde into 49 mL of ultrapure water.

2.2 Preparation of Fibronectin-Coated TFM Substrates

1. Microscope slides with a hydrophobic surface.
2. Rain-X (e.g., ITW Global Brands, available at automotive parts suppliers).
3. KimWipe tissue.
4. Ultrasonic bath.
5. Razor blade.
6. Fluorescently labeled polystyrene microspheres. We use 40 nm red-orange (565/580 nm) fluorescent microspheres supplied by Life Technologies.
7. Master mix stock solution of acrylamide/bis-acrylamide: Mix 40% acrylamide and 2% bis-acrylamide stock solution as per Table 1. Stock solutions can be kept for at least 1 year at 4 °C.
8. 10% Ammonium persulfate (APS): Dissolve 10 mg APS in 100 μL ultrapure water. Prepare fresh APS solution on the day of the experiment. Alternatively, APS solution can be stored at −20 °C for up to a month.
9. *N*,*N*,*N′*,*N′*-Tetramethylethylenediamine (TEMED).
10. *N*-Sulfosuccinimidyl-6-(4′-azido-2′-nitrophenylamino)hexanoate (sulfo-SANPAH) (*see* **Note 2**).
11. Ultraviolet cross-linker (e.g., Fisher Scientific) equipped with 365 nm light tubes.
12. Parafilm.
13. Deionized water.
14. Native fibronectin purified from human plasma (1 mg/mL) (*see* **Note 3**).

Table 1
Composition of Acrylamide/Bis-acrylamide Stock and Working Solutions Required for Preparation of 2.3–55 kPa PAAGs

	Polyacrylamide gel stiffness (G)					
	2.3 kPa	4.1 kPa	8.6 kPa	16.3 kPa	30 kPa	55 kPa
PAAG stock solution (can be stored up to a year at 4 °C)						
40% Acrylamide (mL)	2.75	2.34	2.34	3.00	3.00	2.25
2% Bis-acrylamide (mL)	0.75	0.94	1.88	0.75	1.40	2.25
Ultrapure water (mL)	0.50	1.72	0.78	1.25	0.60	0.50
Total volume (mL)	5.00	5.00	5.00	5.00	5.00	5.00
PAAG working solution (use immediately)						
PAAG stock solution (μL)	125	200	200	250	250	333
Red fluorescent beads (μL)	7.5	7.5	7.5	7.5	7.5	7.5
10% APS (μL)	2.5	2.5	2.5	2.5	2.5	2.5
TEMED (μL)	0.75	0.75	0.75	0.75	0.75	0.75
Ultrapure water (μL)	364	289	289	239	239	170
Total Volume (μL)	500	500	500	500	500	500

Use indicated volumes of stock solution (upper half of table) to make the working solution (lower half of table) when preparing TFM substrates of the desired stiffness. Note that working solutions should be used immediately after adding 10% APS and *N,N,N′,N′*-tetramethylethylenediamine (TEMED) as these chemicals induce rapid polymerization of acrylamide. In contrast, the stock solutions can be kept for at least a year if stored at 4 °C

2.3 Differentiation and Culturing of Myofibroblasts

Preparation of mammalian cells for TFM requires basic equipment for sterile tissue culture. For this chapter, we assume that the reader is familiar with tissue culture techniques and has access to an equipped tissue culture facility. Myofibroblast cell cultures differentiate spontaneously or upon treatment with growth factors. For example, we use immortalized mouse embryo fibroblasts differentiated into myofibroblasts upon treatment with transforming growth factor beta (TGF-β).

1. TGF-β. We use human TGF-β from Cell Signaling Technologies. Make a stock solution of TGF-β by adding 20 mM citric acid, pH 3.0 to a final concentration of 50 μg/mL. Store small aliquots of the stock solution at −20 °C.
2. Fibroblasts (e.g., mouse embryonic fibroblasts).
3. Standard culture medium (e.g., DMEM supplemented with 10% fetal bovine serum, 2 mM L-glutamine, 100 U/mL penicillin/streptomycin, and nonessential amino acids).
4. Incubator with humid atmosphere of 95% air and 5% CO_2 at 37 °C.

5. Transfection reagents and apparatus. We perform electroporation with a Lonza Nucleofector 2b system and Nucleofector Kit V. However, lipid-based transfection reagents such as Lipofectamine or FuGENE 6 can also be used.
6. Expression construct(s) to visualize focal adhesions in live cells. For most experiments, we use a C-terminal fusion of eGFP to paxillin since overexpression of this construct has no significant effect on focal adhesion dynamics [16]. The expression constructs should be endotoxin free.

2.4 Immunofluorescence Staining of Differentiation Markers

1. Phosphate-buffered saline (PBS): 137 mM NaCl, 2.7 mM KCl, 6.5 mM Na_2HPO_4, 1.5 mM KH_2PO_4 in apyrogenic ultrapure water, pH 7.4.
2. 4% Paraformaldehyde in PBS.
3. 0.1% Triton-X100 in PBS.
4. 5% BSA in PBS.
5. Alpha smooth muscle actin (αSMA) antibody. We use αSMA antibody from Cell Signaling Technology.
6. Secondary antibody. We use anti-rabbit-IgG secondary antibody from Cell Signaling Technology.
7. Fluorescently labeled phalloidin.
8. DAPI (4′,6-diamidino-2-phenylindole).
9. Microscopy glass slides.
10. Mounting medium.
11. Standard confocal or widefield microscope.

2.5 Live-Cell Imaging

1. A computer-controlled research-grade inverted epifluorescence microscope equipped with a spinning disk confocal scanning head. For our experiments, we use a Nikon Eclipse Ti2 microscope equipped with a CSU-X1 scanning head (Yokogawa), iChrome MLE laser combiner (TOPTICA Photonics), and Photometrics CoolSNAP Myo CCD camera (*see* **Note 4**).
2. Highly corrected, high-numerical aperture water-immersion objective lens. For our experiments, we use a Nikon 60× Plan Apo NA 1.2 WI objective.
3. Stage-top incubator and an objective heater (e.g., Tokai Hit).
4. Live-cell imaging chamber. We use Chamlide CMS 35 mm dish type magnetic chambers.
5. Imaging media. For our experiments, we use phenol-red-free DMEM media supplemented with 10% fetal bovine serum, 2 mM L-glutamine, 100 U/mL penicillin/streptomycin and nonessential amino acids.
6. Sodium dodecyl sulfate (SDS): Dissolve 2 g of SDS in 10 mL of ultrapure water.

2.6 Quantification of TFM Substrate Deformation and Calculation of Traction Force

1. Custom MATLAB script to track TFM substrate deformations (https://github.com/CellMicroMechanics/Measure_2d_displacements_for_TFM).
2. TFM package developed by the Sabass group (https://github.com/CellMicroMechanics/Easy-to-use_TFM_package) [17].

3 Methods

3.1 Activation of Coverslips

1. Dry "squeaky clean" coverslips and arrange them in a 15 cm glass Petri dish so that they are evenly spaced and not touching each other (Fig. 1).
2. Add ~1 mL of 50% APTMS solution to each individual coverslip such that the solution domes on the coverslip.
3. Incubate the coverslips for 10 min at room temperature.
4. Add ultrapure water to the Petri dish to cover the coverslips. Ensure that no coverslips are stuck to the dish. Do not flip any coverslips as only the side exposed with 50% APTMS will be activated.
5. Place the Petri dish with coverslips on an orbital shaker for 30 min.
6. Rinse the coverslips 3–4 times with deionized water.
7. Add 50 mL of 0.5% glutaraldehyde solution to the Petri dish and agitate for 30 min.
8. Rinse the coverslips three times with deionized water, dry, and store for up to 2 weeks in a vacuum desiccator. Note which side is activated.

3.2 Preparation of Fibronectin-Coated TFM Substrates

1. Use KimWipe tissue to coat microscope slides (one slide for up to two TFM substrates) with Rain-X (ITW Global Brands, available at automotive parts suppliers). Allow the slides to dry for 5–10 min, remove excess Rain-X with a KimWipe tissue, wash extensively with deionized water, and rinse several times with ethanol (*see* **Note 5**).
2. Sonicate the desired fluorescent microspheres in an ultrasonic bath for 30 s.
3. Prepare acrylamide/bis-acrylamide working solution by mixing the desired volume of the stock solution, ultrapure water, and fluorescent microspheres as per Table 1 (*see* **Note 6**).
4. Add 0.75 μL of TEMED and 2.5 μL of 10% APS to initiate polymerization of the acrylamide/bis-acrylamide working solution. Mix the solution by pipetting up and down several times. Avoid introducing air bubbles. Do not vortex.

5. Apply 10 μL of acrylamide/bis-acrylamide working solution containing TEMED and APS on a hydrophobic slide and cover with an activated coverslip, ensuring that the activated surface faces the solution (*see* **Note 7**).
6. Allow the PAAG to polymerize for 20–30 min, then lift the corner of the coverslip with a razor blade to detach the gel from the hydrophobic slide and immerse the gel attached to the coverslip in ultrapure water. Hydrated PAAG can be stored at 4 °C for up to 2 weeks.
7. Take a 40 μL aliquot of sulfo-SANPAH stock solution out of −80 °C freezer and thaw at room temperature for several minutes. One aliquot of sulfo-SANPAH is sufficient to functionalize two 22×22 mm PAAGs.
8. While thawing sulfo-SANPAH, remove excess water from the PAAGs with KimWipe tissue. Be careful not to touch the gel surface in the center of the coverslip. Make sure that gels do not dry.
9. Add 0.96 mL of ultrapure water to the sulfo-SANPAH aliquot, mix by pipetting, and apply 0.5 mL on top of each PAAG. Activate sulfo-SANPAH by exposing gels to long-wavelength UV light (365 nm) for 2 min. During activation, the color of the sulfo-SANPAH solution will change from orange to brown.
10. Place a piece of Parafilm into a 15 cm Petri dish and pipet 50 μL of fibronectin solution (1 mg/mL) on the Parafilm for each PAAG.
11. Quickly wash PAAGs with deionized water. Remove excess water with KimWipe tissue, invert, and place the gel surface on the drop of fibronectin solution.
12. Incubate at room temperature for 4 h, and then wash gels multiple times with DPBS with agitation on a shaker. TFM substrates coated with fibronectin can be stored at 4 °C for up to 1 week.

3.3 Differentiation and Culturing of Myofibroblasts

1. Culture the desired cell line in the appropriate medium and desired environmental conditions.
2. Two days prior to measuring traction forces, plate cells to a confluency desirable for your cell type, keeping in mind that the cells will go under division for 48 h. We plate mouse embryonic fibroblasts at 40% confluency in DMEM supplemented with 10 ng/mL TGF-β to induce myofibroblast differentiation.
3. One day prior to measuring traction forces, transfect cells with a pDNA expression construct to visualize focal adhesions (eGFP-paxillin). After transfection, culture cells for ~24 h in

DMEM supplemented with 10 ng/mL of TGF-β. To check if cells have differentiated as expected, proceed to Subheading 3.4.

4. On the day of traction force measurement, preincubate fibronectin-coated TFM substrates in live-cell imaging media for 30 min at 37 °C to equilibrate ion/nutrient concentration. Harvest transfected cells by trypsinization, resuspend in live-cell imaging media, and plate on TFM substrates for 2–6 h. In our experiments, we plate the cells for just enough time for them to fully spread and polarize. This decreases matrix deposition and remodeling by cells and improves the reproducibility of TFM measurements.

3.4 Immunofluorescence Staining of Differentiation Markers

Although immunofluorescence staining is not a part of traction force measurement, it is useful to confirm that myofibroblasts have differentiated as expected before proceeding with TFM.

1. After treating cells with 10 ng/mL TGF-β for 48 h, wash cells once with warm PBS to stabilize cytoskeletal structures that would otherwise destabilize upon washing with cold PBS.
2. Fix cells with 4% paraformaldehyde in PBS for 15 min at room temperature. Wash cells three times with PBS.
3. Permeabilize the cell membrane with 0.1% Triton-X100 in PBS for 10 min at room temperature. Rinse three times with PBS for 5 min each.
4. Incubate cells in blocking solution that is appropriate for the primary antibody. We block cells in 5% BSA in PBS for 1 h at room temperature.
5. Dilute αSMA antibody in the appropriate blocking solution as recommended by the antibody manufacturer. Apply 50 μL of working solution to a piece of parafilm and cover with the glass coverslip. Incubate cells at room temperature for 1 h. Alternatively, incubate cells overnight at 4 °C in a humidified chamber. Wash three times with PBS for 5 min each to remove excess primary antibody.
6. Dilute secondary antibody, DAPI, and phalloidin as recommended by the manufacturer in the appropriate blocking solution. Apply 50 μL of working solution to a piece of parafilm and cover with the glass coverslip. Incubate cells at room temperature for 1 h. Wash three times with PBS for 5 min each.
7. Apply a small droplet of mounting media to a glass slide and cover with coverslip. Allow mounting media to dry before sealing with nail polish.
8. In confocal or widefield mode, acquire multicolor images of αSMA, the transfected focal adhesion marker, actin, and DAPI. Compare αSMA intensity between control and TGF-β-treated

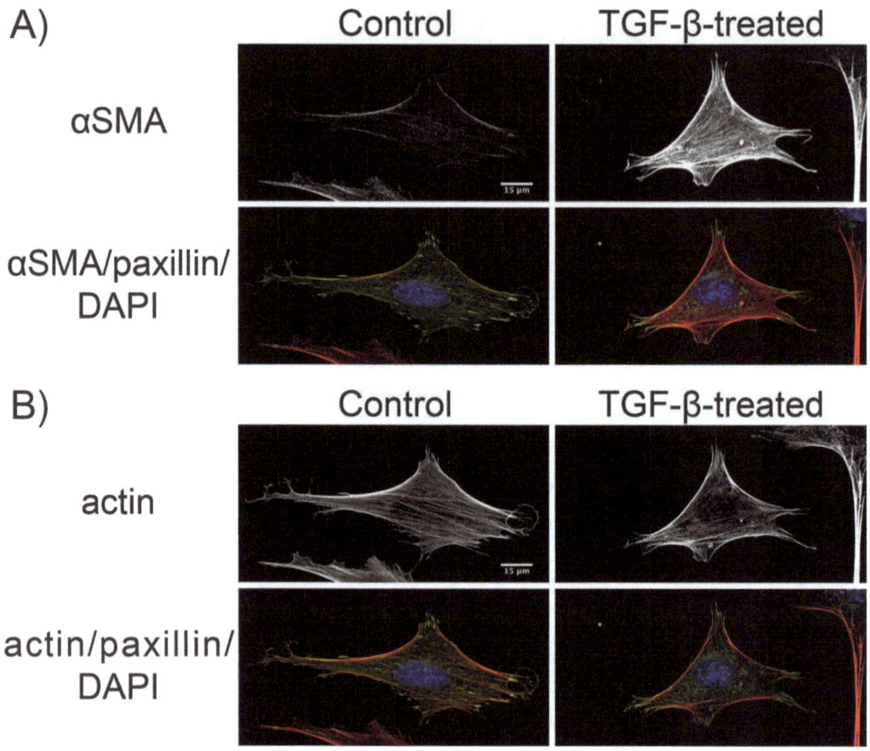

Fig. 2 Validation of myofibroblast differentiation. Mouse embryonic fibroblasts (NIH3T3) were treated with 10 ng/mL TGF-β for 48 h and transfected with eGFP-paxillin to label focal adhesions. (**a**) Confocal images of a representative control cell and TGF-β-treated cell stained for the differentiation marker αSMA. αSMA intensity in TGF-β-treated cells is significantly higher than in control cells, suggesting that cells have differentiated into myofibroblasts. (**b**) Confocal images of actin staining in the control cell and TGF-β-treated cell

cells to determine whether cells have differentiated into myofibroblasts. *See* Fig. 2 for representative images of control and TGF-β-treated cells.

9. After verifying that cells have differentiated as expected, proceed to the following steps for TFM measurements.

3.5 Acquisition of TFM Images

1. Assemble a magnetic chamber according to the manufacturer's manual. For a Chamlide 35 mm chamber, transfer a TFM substrate with cells attached to the bottom plate of the chamber and place a silicon gasket over the coverslip. Install the main body (the main body will attach to the bottom plate by magnetic force). Add ~1.5 mL of warm live-cell imaging media into the chamber. Do not completely fill up the chamber with media.
2. Use water and ethanol to clean the glass surface of the coverslip with the TFM substrate attached. Confirm that the chamber is watertight before mounting it on the microscope and make sure that the chamber is locked on the stage.

3. Focus the 60× lens on the cells adhered to the surface of the PAAG by transmitted light mode, and then locate a candidate cell expressing moderate levels of fluorescently tagged focal adhesion marker. Use transmitted light to make sure that the candidate cell is isolated with no other cells within 10–20 μm from the cell edge.
4. In confocal imaging mode focus the microscope on focal adhesions. In the case of a GFP-tagged focal adhesion marker, the focal adhesions can be observed in the green fluorescence channel.
5. Acquire a multicolor image of GFP labeled focal adhesions and red microspheres. If you have automated focus control, you may image the focal adhesions at their precise focal plane and image the beads at ~0.5 μm deeper into the substrate, which will give higher contrast images of the beads. If the imaging system is equipped with an automated microscope stage, memorize X and Y coordinates and repeat measurements for several cells.
6. Remove cells from the TFM substrate. Following imaging of the cells and the strained substrate, add 500 μL of 20% SDS directly into imaging media in the magnetic chamber, taking care to avoid moving the magnetic chamber or touching the substrate. Wait for 5 min for the cells to lyse and the substrate to relax.
7. Image beads in the unstrained TFM substrate. Move the stage to every memorized position, focus on the uppermost layer of the PAAG, and acquire Z-stacks of bead images in 100 nm increments. Later, one of these image planes will be selected as a reference (unstrained) image to quantify substrate deformation.

3.6 Quantification of TFM Substrate Deformation and Calculation of Traction Force

1. Find reference optical planes in a Z-stack of images from the unstrained gel. The drift of the microscope makes it necessary to find a best-matching reference plane for each cell. Align the images of the deformed and unstrained substrate to correct for microscope stage drift. This alignment can be done in Nikon Elements by using the Manual Channel Alignment tool.
2. Remove out-of-focus fluorescence and smooth each reference image by subtracting a median-filtered image from the original image.
3. Find local maxima in reference images by locating pixels that are brighter than their neighbors and brighter than the average intensity of the whole image.
4. Remove maxima that are too close to each other to avoid tracking similar features multiple times. We exclude the peaks that are less than 1 μm away from other peaks.

5. Track displacement of fiducial markers. In our experiments, we track TFM substrate deformations by running a custom MATLAB script (https://github.com/CellMicroMechanics/Measure_2d_displacements_for_TFM). Here we describe the flow of the algorithm tracking the displacement of individual beads with subpixel accuracy. Use the unfiltered bead image for tracking. Place windows $W_{ref}(m_1, m_2)$ on top of each recognized bead in the reference images. The size of the windows is usually ~1 μm^2. Calculate the cross correlation between all reference windows $W_{ref}(m_1, m_2)$ and corresponding shifted windows. Obtain a subpixel estimate of the position of the bead by fitting a two-dimensional Gaussian to the maximum of the correlation matrix. The formulas for fitting two-dimensional Gaussians can be found elsewhere [18].
6. Reconstruct traction forces by using the TFM package developed by the Sabass group (https://github.com/CellMicroMechanics/Easy-to-use_TFM_package) [17]. Run the script "TF reconstruction" providing the following input data: fiducial marker displacements, Young's modulus of the substrate, and pixel size of the images. Analyze the data with the default setting for the mesh size. Choose the option "Bayesian regularization" for force reconstruction. Choose "Manual noise selection" and use the mouse pointer to select a region far away from the cell where the systematic displacements caused by the cell is negligible. The subsequently generated output file will contain the substrate deformation vectors projected on a square grid and traction force vectors in units of Pascals (N/m^2).
7. Usually, traction forces exerted by cells on a substrate are represented as a heat map of traction magnitude (Fig. 3). For quantitative comparison, whole-cell traction should be quantified through the median of traction magnitude. Bar graphs of traction magnitude yield clear visual representation of contractile forces. An alternative measure of the overall strength of the cell is the total strain energy, which is one half of the discretized surface integral of the product of traction and surface deformation.

4 Notes

1. In our experiments we use borosilicate coverslips supplied by Corning Inc. (#2850-22) due to their cleanliness, reliable optical properties, and low thermal expansion. To ensure tight and uniform gel attachment to the glass surface, we clean the coverslips according to the protocol developed by the Salmon lab (http://labs.bio.unc.edu/Salmon/protocolscoverslippreps.html). Briefly, coverslips are sonicated for 20 min in hot water

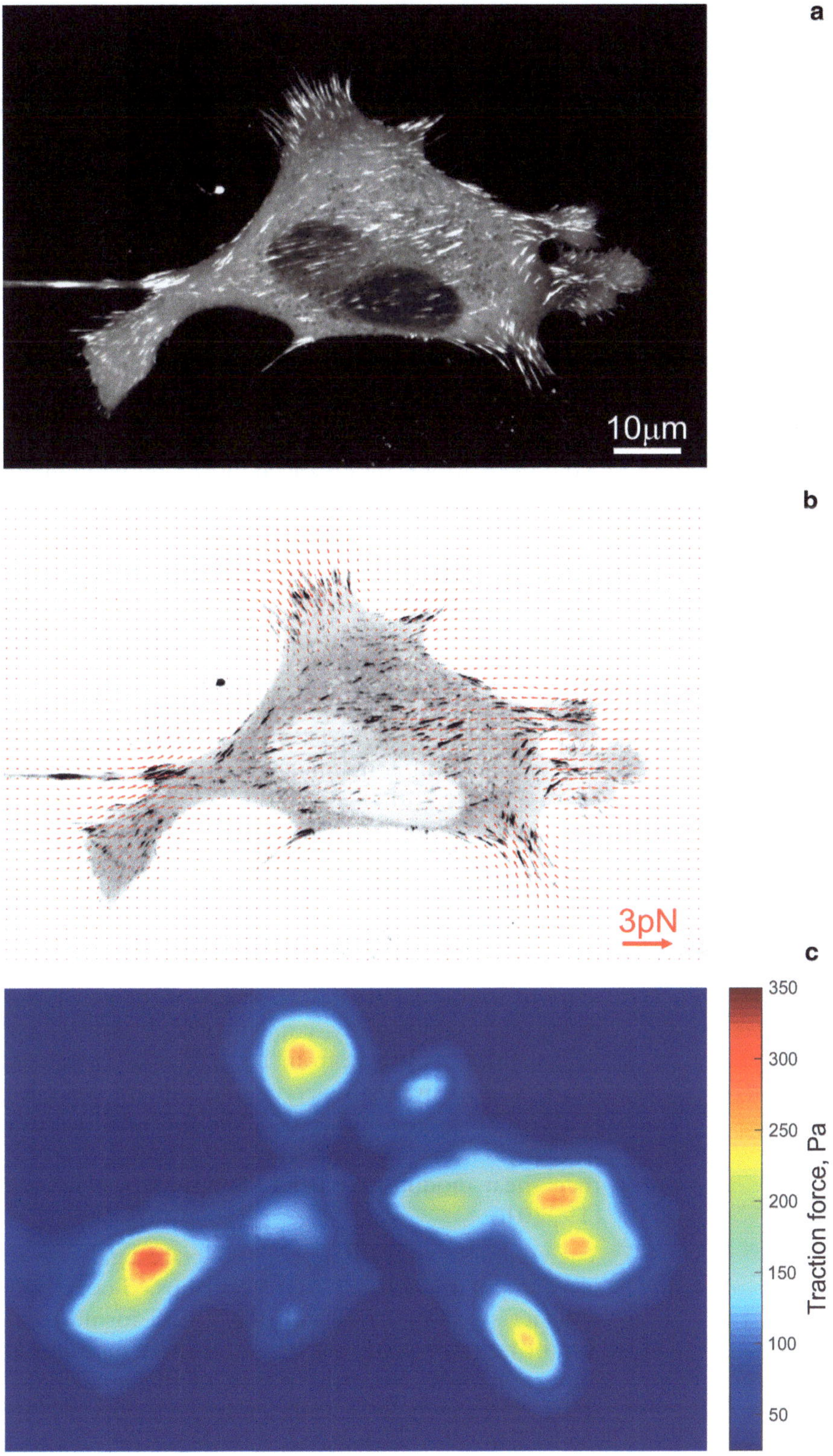

Fig. 3 Quantification of traction forces exerted by a myofibroblast on TFM substrate. Myofibroblasts differentiated from mouse embryonic fibroblasts (NIH3T3) were transfected with eGFP-paxillin and plated onto a 32 kPa TFM substrate embedded with red fluorescent microspheres. (**a**) Confocal image of focal adhesions labeled with eGFP-paxillin. (**b**) Traction force vector field overlaid on the inverted-contrast image of eGFP-paxillin. (**c**) Color-coded heat map of reconstructed traction force exerted on the ECM

containing Contrad 70 detergent, rinsed several times with tap water, and then sonicated for 30 min in each of the following solutions in the following order: tap water, distilled water, 1 mM EDTA, 70% ethanol, and 100% ethanol. Clean coverslips are stored at room temperature submerged in 100% ethanol.

2. Note that sulfo-SANPAH undergoes rapid hydrolysis when dissolved in water. Prepare stock solution of sulfo-SANPAH by dissolving 50 mg of sulfo-SANPAH powder in 2 mL of anhydrous DMSO, aliquot in Eppendorf tubes (40 μL per tube), and freeze by immersing the tubes into liquid nitrogen. We store stock solutions of sulfo-SANPAH at −80 °C for several months.
3. Note that fibronectin dissolved in Tris-buffered saline must be dialyzed prior to cross-linking to the TFM substrate since the amine groups of Tris will react with the cross-linker.
4. It is critical that the imaging system acquires images that satisfy the Nyquist sampling criterion. This can be achieved either by using a camera with small pixels (4.45 μm for 60× objective lens) or by adding an intermediate magnification lens (1.5–2×) in the optical path of the microscope.
5. Make sure to remove all debris from the slides since any remaining particles will contaminate the PAAG surface. We usually prepare slides with hydrophobic surfaces in bulk and store them in a vacuum desiccator for several months.
6. The mechanical properties of PAAG are determined by the concentration of acrylamide and *N*,*N*-′-methylenebisacrylamide, and thus can be adjusted to mimic the stiffness of most biological tissues (100 Pa to 100 kPa) without losing the elastic properties of the gel. For cells exhibiting high contractility, e.g., myofibroblasts, we use PAAGs with stiffnesses ranging from 16 to 55 kPa. Different formulations of acrylamide/bisacrylamide can be used to produce PAAG with similar stiffnesses. The formulations described in this chapter have been adapted from Yeung et al. (2005) and their shear moduli have been confirmed by the Gardel group [19, 20].
7. The volumes described herein will create a gel with a thickness of ~30 μm on 22×22 mm coverslips. Since the extent of polymer swelling varies with the PAAG formulation [8] and cannot be easily predicted based on shear modulus alone, it is important to measure the height of the resulting TFM substrate. The gel must be sufficiently thick such that the gel can freely deform in response to cellular forces without the influence of the underlying glass [21].

Acknowledgments

Shuying Yang and Fernando R. Valencia contributed equally to this work.

References

1. Gabbiani G, Hirschel BJ, Ryan GB et al (1972) Granulation tissue as a contractile organ: a study of structure and function. J Exp Med 135:719–734
2. Forbes SJ, Rosenthal N (2014) Preparing the ground for tissue regeneration: from mechanism to therapy. Nat Med 20:857–869
3. Duffield JS (2014) Cellular and molecular mechanisms in kidney fibrosis. J Clin Invest 124:2299–2306
4. Otranto M, Sarrazy V, Bonté F et al (2012) The role of the myofibroblast in tumor stroma remodeling. Cell Adhes Migr 6:203–219
5. Kanchanawong P, Shtengel G, Pasapera AM et al (2010) Nanoscale architecture of integrin-based cell adhesions. Nature 468:580–584
6. Harris AK, Wild P, Stopak D (1980) Silicone rubber substrata: a new wrinkle in the study of cell locomotion. Science 208:177–179
7. Beningo KA, Wang YL (2002) Flexible substrate for the detection of cellular traction forces. Trends Cell Biol 12:79–84
8. Kraning-Rush CM, Carey SP, Califano JP et al (2012) Quantifying traction stresses in adherent cells. Methods Cell Biol 110:139–178
9. Yoshie H, Koushki N, Kaviani R et al (2018) Traction force screening enabled by compliant PDMS elastomers. Biophys J 114:2194–2199
10. Herrick WG, Nguyen TV, Sleiman M et al (2013) PEG-phosphorylcholine hydrogels as tunable and versatile platforms for mechanobiology. Biomacromolecules 14:2294–2304
11. Pelham RJ, Wang YL (1997) Cell locomotion and focal adhesions are regulated by substrate flexibility. Proc Natl Acad Sci U S A 94:13661–13665
12. Gardel ML, Sabass B, Ji L et al (2008) Traction stress in focal adhesions correlates biphasically with actin retrograde flow speed. J Cell Biol 183:999–1005
13. Pasqualini FS, Agarwal A (2018) Traction force microscopy of engineered cardiac tissues. PLoS One 13(3):e0194706. https://doi.org/10.1371/journal.pone.0194706
14. Sabass B, Gardel ML, Waterman CM et al (2008) High resolution traction force microscopy based on experimental and computational advances. Biophys J 94:207–220
15. Plotnikov SV, Pasapera AM, Sabass B et al (2012) Force fluctuations within focal adhesions mediate ECM-rigidity sensing to guide directed cell migration. Cell 151:1513–1527
16. Webb DJ, Donais K, Whitmore LA et al (2004) FAK-Src signalling through paxillin, ERK and MLCK regulates adhesion disassembly. Nat Cell Biol 6:154–161
17. Huang Y, Gompper G, Sabass B (2020) A Bayesian traction force microscopy method with automated denoising in a user-friendly software package. Comput Phys Commun 256:107313. https://doi.org/10.1016/j.cpc.2020.107313
18. Nobach H, Honkanen M (2005) Two-dimensional Gaussian regression for sub-pixel displacement estimation in particle image velocimetry or particle position estimation in particle tracking velocimetry. Exp Fluids 38:511–515
19. Yeung T, Georges PC, Flanagan LA et al (2005) Effects of substrate stiffness on cell morphology, cytoskeletal structure, and adhesion. Cell Motil Cytoskeleton 60:24–34
20. Aratyn-Schaus Y, Gardel ML (2010) Transient frictional slip between integrin and the ECM in focal adhesions under myosin II tension. Curr Biol 20:1145–1153
21. Sen S, Engler AJ, Discher DE (2009) Matrix strains induced by cells: computing how far cells can feel. Cell Mol Bioeng 2:39–48

Chapter 15

Nucleocytoplasmic Shuttling of the Mechanosensitive Transcription Factors MRTF and YAP/TAZ

Michael Kofler and András Kapus

Abstract

Myocardin-related transcription factor (MRTF) and the paralogous Hippo pathway effectors Yes-associated protein (YAP) and transcriptional co-activator with PDZ-binding motif (TAZ) are transcriptional co-activators that play pivotal roles in myofibroblast generation and activation, and thus the pathogenesis of organ fibrosis. They are regulated by a variety of chemical and mechanical fibrogenic stimuli, primarily at the level of their nucleocytoplasmic shuttling. In this chapter we describe the tools and protocols that allow for exact, quantitative, and automated determination and analysis of the nucleocytoplasmic distribution of endogenous or heterologously expressed MRTF and YAP/TAZ, measured in large cell populations. Dynamic monitoring of nucleocytoplasmic ratios of transcription factors is a novel and important approach, suitable to address both the structural requirements and the regulatory mechanisms underlying transcription factor traffic and the consequent reprogramming of gene expression during fibrogenesis.

Key words Nucleocytoplasmic shuttling, Transcription factors, Organ fibrosis, MRTF, YAP/TAZ, Nuclear import, CRM1-dependent export, Fluorescence imaging, Automated nucleocytoplasmic ratio determination

1 Introduction

1.1 Nucleocytoplasmic Traffic of Transcriptional Regulators: Key Role in Fibrosis

Fibrosis in general and the formation of myofibroblasts in particular are manifestations of major phenotypic reprogramming, affecting both mesenchymal and epithelial cells. In response to fibrogenic stimuli fibroblasts or other mesenchymal cells transform into myofibroblasts (fibroblast-myofibroblast transition, FMT) [1, 2], while epithelial cells either undergo epithelial-mesenchymal transition (EMT) or, more often, acquire a profibrotic epithelial phenotype (PEP), characterized by a partial loss of epithelial features and the production of fibrogenic cytokines and other mediators [3–5]. These products then promote fibroblast proliferation and FMT. All these transitions are brought about by large-scale changes in gene expression, mainly due to altered transcription mediated by a key set of transcription factors/transcriptional co-activators (here

Boris Hinz and David Lagares (eds.), *Myofibroblasts: Methods and Protocols*, Methods in Molecular Biology, vol. 2299,
https://doi.org/10.1007/978-1-0716-1382-5_15,

collectively abbreviated as TFs). For example, members of the Smad family (e.g., Smad2 and 3) are phosphorylated by the activated TGFβ receptor which results in their nuclear translocation/accumulation and the consequent initiation of the fibrogenic program [6–8]. Altered nuclear TF traffic, a mechanism not restricted to Smads, has recently emerged as a main theme in the molecular pathogenesis of fibrosis. Since, beside chemical mediators such as TGFβ [9], mechanical inputs (e.g., increased matrix stiffness, stretch, contractility, altered cell density, wounding) are also key triggers of phenotype shifts in fibrosis [2, 10–15], recent research has focused on the characterization of nucleocytoplasmic shuttling of those TFs, which (1) are regulated primarily at the level of their nuclear accumulation, and (2) are responsive to chemical and/or mechanical fibrogenic stimuli. This short overview concentrates on two such families: myocardin-related transcription factors (MRTF) [16–18] and Yes-associated protein (YAP)/transcriptional co-activator with PDZ-binding motif (TAZ) [19–22] . Both MRTF [5, 10, 13, 23–27] and YAP/TAZ [5, 10, 28–31] have been implicated in FMT, EMT, and PEP, as well as in fibrogenesis in multiple organs, including the heart, lung, liver, kidney, and skin [5, 24, 30, 32–48]. Importantly, genetic or pharmacological interference with MRTF or YAP/TAZ were shown to mitigate fibrogenesis (*see* the cited studies above). Consequently, selective and clinically relevant pharmacological inhibitors for MRTF [49, 50] and YAP/TAZ are being developed [51, 52].

1.2 Quantitation of Nuclear Traffic of MRTF and YAP/TAZ: Importance and Challenges

Unraveling the basic mechanisms and regulation of nucleocytoplasmic traffic of TFs is essential for understanding the pathobiology of fibrosis. However, the overwhelming majority of previous studies assessing MRTF or YAP/TAZ distribution used binary (in/out, i.e., nuclear versus cytosolic) or tripartite (in/even/out) categorization. These relatively crude approaches are often neither accurate nor sufficient. First, these methods are essentially subjective. Second, they are obviously inadequate for the detection of subtle changes that may have major biological significance (e.g., the nuclear content may significantly increase yet may remain within the same category). Third, YAP/TAZ distribution is notoriously heterogeneous within a cell population. This is understandable, given that many local factors (e.g., cell density/confluence, altered contractility) that impact TAZ localization differ in various areas within a single monolayer. Thus, detection of the effect of a treatment or comparing the localization of different mutants may be very challenging. Fourth, while qualitative snapshots are informative, quantitative and dynamic monitoring of TF traffic gains increasingly recognized importance; experimental observations and mathematical models show that the kinetics of the changes in the nuclear concentration of TFs, as well as various response patterns (amplitude, time-integrated concentrations, oscillations, etc.)

are important determinants of the ensuing transcriptional response and can lead to different outcomes [53–56]. Clearly, reliable conclusions about the impact of stimuli, the involved regulatory circuits, the structural requirements of import and export, and the underlying molecular interactions can only be drawn by quantitative characterization and analysis of nucleocytoplasmic distributions of endogenous or heterologously expressed TFs, measured in large (hundreds to thousands of cells) populations. Therefore, quantitative methods are needed to determine the spatial distributions across a population and their dynamic changes.

1.3 The Nuclear Shuttling of MRTF: Overview and Open Questions

MRTF is a G-actin binding protein which, depending on cell type, localizes partly or fully in the cytosol in the resting state (Fig. 2a). Upon actin polymerization (in response to chemical mediators and/or mechanical stress), G-actin dissociates from MRTF and thereby liberates its nuclear localization signal (NLS) [16, 57, 58] . In addition, nuclear actin polymerization reduces nuclear export of MRTF [59, 60] presumably because intranuclear G-actin binding supports nuclear efflux via a nuclear export signal (NES) [61]. The molecular details underlying efflux regulation have not been fully elucidated. Moreover, nuclear traffic and/or activity of MRTF is modified by phosphorylation of multiple residues (e.g., 26 serines and threonines in MRTF-A)[61], the impact of which warrants further characterization. Once in the nucleus, MRTF binds to its cognate transcription factor, serum response factor (SRF), and the complex drives gene expression through CC(A/T-rich)$_6$GG cis-elements called CArG boxes [62]. A large number of MRTF/SRF target genes are relevant for fibrosis, including extracellular matrix proteins, integrins, cytokines, many components of the acto-myosin cytoskeleton, and TFs [63] including TAZ [10, 37]. In addition, MRTF plays a critical role in "mechanical memory," the capacity of cells to "remember" the stiffness of their previous environments, determining their fibrogenic potential for extended time periods [64]. Our protocol will describe how to quantify changes in the nucleocytoplasmic distribution of endogenous MRTF in large cell populations. In this example, net nuclear uptake of MRTF is initiated by uncoupling of intercellular contacts (a potent inducer of RhoGTPase-mediated actin polymerization) using low calcium medium (LCM) [13, 65, 66].

1.4 The Nuclear Shuttling of YAP and TAZ: Overview and Open Questions

YAP and TAZ are paralogous transcriptional coactivators and downstream effectors of the Hippo pathway, a critical regulator of cell/organ size, cell proliferation, contact inhibition, stemness, and differentiation [22, 67–71]. The Hippo pathway is composed of the serine kinases MST1/2 (along with their adaptor SAV) and their downstream targets LATS1/2 kinases (along with their adaptors MOB1/2), which in turn phosphorylate YAP and TAZ at 5 and 4 serine residues, respectively [69, 71]. Phosphorylation of these sites (predominantly S127 in YAP and S89 in TAZ) results in their cytosolic retention due to binding to cytosolic adaptors, such as 14-3-3 proteins [21]. Accordingly, phosphorylation-incompetent mutants (e.g., TAZ 4SA) are constitutively nuclear (Fig. 3b, c). The Hippo pathway exerts tonic inhibition on the nuclear uptake of YAP/TAZ. Cell contact disruption, for example by using LCM (Fig. 2d), inactivates the Hippo pathway and allows YAP/TAZ to enter the nucleus [10, 72]. LCM also leads to actin polymerization, which is another key inducer of the nuclear accumulation of YAP/TAZ by promoting entry and/or inhibiting egress through poorly understood but partly Hippo-independent mechanisms [73–76]. Indeed, most mechanical factors (stiffness, stretch), local architectural clues (low cell density, edge effects), as well as soluble ligands (activating heterotrimeric G-proteins) are believed to foster nuclear accumulation of YAP/TAZ through cytoskeletal reorganization (actin polymerization and myosin-mediated contraction) [12, 73, 77–80] . Once in the nucleus YAP/TAZ bind to transcription factors, predominantly members of TEAD family which drive a large array of genes involved in the abovementioned functions via 5′-g/aCATTCCa/t-3′ cis-elements called TEAD-binding elements (TBE) [81, 82]. Importantly, YAP/TAZ collaborate with Smad2/3 and MRTF to orchestrate fibrotic gene expression, and their direct interactions impact their localization and actions [10, 37, 83–87].

While the basic mechanisms regulating nuclear traffic of fibrogenic TFs have been established, fundamental questions remained open. In fact, the nucleocytoplasmic shuttling of YAP/TAZ and its regulation is much less understood than that of MRTF. First, the mechanism whereby "free" cytoplasmic YAP/TAZ enters the nucleus was unknown. It was not even clear whether these molecules are taken up by passive diffusion or by mediated import across the nuclear pore. Assuming the latter, the responsible sequences were unknown, as YAP/TAZ do not contain a classic NLS. Second, no putative NES have been identified that could account for their CRM1-mediated nuclear export. Further, the regulation of the import or export processes per se (as opposed to cytosolic or nuclear retention) remained elusive. The approaches described in the next section and detailed in our protocol were essential to answer some of these questions.

1.5 New Tools and Strategies for Studying Nucleocytoplasmic Traffic

To study the basic mechanism, regulation, and dynamics of TAZ nuclear import, we developed a molecular toolkit (elements of which will be presented in our protocols, *see* Table 1) [88]. These approaches can easily be adapted to other TFs. To test the existence of active transport, we fused wild-type TAZ or its Hippo pathway-independent mutant (TAZ 4SA) to a large tag (5 mCitrine, 5C), which precludes passive diffusion through the nuclear pore (Fig. 3a–c). Inhibition of the classic (CRM1-dependent) nuclear

Table 1
All constructs are based on pcDNA3.1(-) [88]

Name	Details	Comments
1C-control	mCitrine-mCherry fusion	small-tagged control
1C-TAZ 4SA	mCitrine-TAZ 4SA fusion	small-tagged TAZ fusion protein, with 4 serine-to-alanine mutations that destroy Lats-phosphorylation sites
5C-control	(5XmCitrine)-mCherry fusion	diffusion-limited, large-tagged control
5C-TAZ 4SA	(5XmCitrine)-TAZ 4SA fusion	diffusion-limited, large-tagged TAZ 4SA fusion protein
5C-anchorless	(5XmCitrine)-TAZ 4SA F52A fusion	diffusion-limited, large-tagged TAZ 4SA fusion protein, F52A mutation disrupts TEAD binding
5C-R5A	(5XmCitrine)-R5A fusion	diffusion-limited, large-tagged construct comprising the R5A mutation within SV40-NLS (residues 126–132 of the large T antigene)
H2B-2xFKBP-mCherry	H2B-FKBP-FKBP-mCherry fusion	H2B serves as anchor to stably localize the FKBP domains in the nucleus; anchor-construct for the rapamycin nuclear sequestration system
1C-NES-FRB	mCitrine-NES-FRB fusion	mCitrine-FRB fusion protein, comprising a NES; control for import studies using the rapamycin nuclear sequestration system
1C-NES-FRB-(290–345)	mCitrine-NES-FRB-TAZ(290–345) fusion	TAZ fragment 290–345, fused to mCitrine-NES-FRB; for the rapamycin nuclear sequestration system

export pathway by leptomycin B (LMB) induced nuclear accumulation of the constructs, allowing us to study the unmasked nuclear import (Fig. 3d–f). To provide positive controls, the 5C tag was also fused to known "classic" nuclear localization signals (e.g., the NLS from simian virus 40 (SV40) or its less potent mutant R5A [89]). We also generated an "anchorless" mutant of TAZ (4SA/F52A), which binds neither the major cytoplasmic (14-3-3 proteins) nor the main nuclear (TEADs) retention factors so that the import can be studied independent of retention (not shown). The nucleocytoplasmic ratios of these constructs (Fig. 3c) as well as that of endogenous YAP, TAZ, and MRTF (Fig. 2d) were determined in cell populations utilizing the ImageXpress Micro platform with integrated MetaXpress Imaging and Analysis software (as described in Subheading 3.1). To monitor stimulus-induced changes in the distribution of these factors, we used three approaches. These provide well-defined starting points from which the change can be followed, enabling kinetic analysis. (1) We uncoupled intercellular contacts (a known stimulus both for MRTF and YAP/TAZ translocation) by incubating cells in LCM (Fig. 2c, d). (2) We inhibited the classic (CRM1-dependent) nuclear export pathway by LMB [90], thereby unmasking import (Fig. 3e, f). (3) We developed a drug-inducible (rapamycin-triggered) protein-protein interaction system, termed Rapamycin-induced sequestration in the nucleus (RIS'N) (Fig. 4). In order to better approximate physiological transport rates, we utilized single mCitrine (1C) fusions for these experiments. The principle of this method is that rapamycin induces high-affinity heterodimerization between the FK506 binding protein (FKBP) and the FKBP-rapamycin-binding (FRB) domain of mTOR [91]. Accordingly, we fused FKBP to a red fluorescence tag (mCherry) along with Histone-2B as a nuclear entrapment element (H2B-2xFKPB-mCherry). The construct was then co-expressed with 1C-NES-FRB fused to full-length TAZ or fragments thereof (Fig. 4). Addition of rapamycin eliminates TAZ export by nuclear retention (via H2B-2xFKPB-mCherry), uncovering net nuclear import of the labeled TAZ molecules. These movements were followed by live imaging using the VivaView incubator-housed fluorescence microscope system.

Using these and other tools, we were able to prove that TAZ shuttles by mediated transport via a novel and unusual (negatively charged) NLS in the C-terminal half of the molecule, and an N-terminal NES, which is masked by TEAD binding [88]. These studies also started to unravel the regulation of these processes. Moreover, these approaches called attention to a broader context, namely the absolute necessity of using highly quantitative and sensitive methods to monitor the nuclear shuttling of TFs. This chapter provides basic protocols for such measurements.

2 Materials

Follow standard tissue culture procedures and waste disposal regulations to maintain sterility and insure biosafety. We recommend the use of the specific antibodies noted in the protocol.

2.1 Cell Culture

1. LLC-PK1 cells.
2. Cell culture medium: Dulbecco's modified Eagle medium (D-MEM), with 1 g/L D-glucose, L-glutamine, and 110 mg/L sodium pyruvate, supplemented with 10% fetal bovine serum (FBS, sterile-filtered), 100 units of penicillin, and 100 μg/L of streptomycin.
3. Commercially available sterile phosphate-buffered saline (PBS, pH 7.4).
4. Trypsin-EDTA (0.05% Trypsin with EDTA 4Na).

5. Sterile tissue culture (TC) vessels: 75 cm^2 flasks (250 mL, vented, polystyrene TC-treated), 96-well clear flat-bottom polystyrene TC-treated microplates with lid, 35 mm glass-bottom dishes (uncoated, sterile, No. 1.5 coverslip thickness).

2.2 Cell Transfection

1. Purified plasmids (*see* Table 1 for details): 1 μg/μL in 10 mM Tris (pH 8.0) containing 1 mM EDTA.
2. jetPRIME® buffer and transfection reagent.

2.3 Cell Stimulation

1. LCM: D-MEM with 4.5 g/L D-glucose, without L-glutamine, sodium pyruvate, and calcium chloride.
2. Rapamycin medium: cell culture medium with 21 nM rapamycin.

2.4 Immunofluorescence

1. Paraformaldehyde (PFA), 4% in PBS.
2. Neutralization buffer: 100 mM glycine in PBS.
3. Permeabilization buffer: PBS with 0.2% Triton X-100 (*see* **Note 1**).
4. Washing buffer: PBS.
5. Blocking buffer: 5% (w/v) bovine serum albumin (BSA, fraction V, heat shock isolation) in PBS.
6. Antibody incubation buffer: 1% (w/v) BSA in PBS.
7. Primary antibodies: anti-TAZ (BD Biosciences, 560,235) mouse monoclonal antibody, anti-MRTF-B rabbit polyclonal antibody (Santa Cruz, sc- 98989).
8. Secondary antibodies (e.g., Alexa Fluor® 555 anti-mouse and Alexa Fluor® 555 anti-rabbit antibody).

2.5 Imaging Equipment and Software

1. Micro automated wide-field high-content imaging platform, e.g., ImageXpress with integrated MetaXpress Imaging and Analysis software (Molecular Devices LLC).
2. VivaView FL Incubator Microscope (e.g., Olympus) with MetaMorph imaging software (Molecular Devices).
3. Software: Fiji (*see* [92]), Microsoft Excel, and GraphPad Prism.

3 Methods

3.1 Automated Acquisition and Analysis Using Fixed Cells in 96-Well Plates

(A) refers to nuclear accumulation of endogenous MRTF and TAZ upon disruption of intercellular contacts; (B) refers to nucleocytoplasmic distribution of five mCitrine (5C) constructs.

3.1.1 Cell Culturing and Stimulation

1. Grow LLC-PK1 cells in cell culture medium in a humidified incubator at 37 °C and 5% CO_2. Maintain growth by regular passaging.

   Detach cells with Trypsin-EDTA and seed 10,000 cells into each well of a 96-well plate, with a total of 100 μL cell culture medium/well.

   (a) Low calcium stimulation: 48 h after seeding, when cells are 100% confluent. Incubate cells for 0, 30, 60, and 120 min with LCM. To eliminate trace amounts of calcium, wash cells five times with 100 μL LCM before finally putting them into LCM. Ensure all incubations end at the same time as the entire plate will be fixed simultaneously.

   (b) JetPrime transfection: 24 h later, when cells are 50–100% confluent: Per well, mix 0.1 μg plasmid with 10 μL jetPrime buffer and 0.2 μL jetPrime reagent, incubate for 10 min and add to well. Incubate for 24 h.

3.1.2 Cell fixation and staining

1. Remove medium and fix cell with 50 μL/well 4% PFA for 20 min at room temperature in a chemical hood. Collect and dispose PFA solution appropriately.
2. Incubate cells with 150 μL/well neutralization buffer for 20 min at room temperature in a chemical hood. For (B), skip **steps 3**–**7**.
3. Incubate cells with 100 μL/well permeabilization buffer for 20 min at room temperature in a chemical hood (*see* **Note 1**).
4. Wash cells three times 5 min with PBS.
5. Block with 100 μL blocking buffer for 1 h at room temperature.
6. Wash as in **step 4**.
7. Incubate cells over night at 4 °C with 40 μL/well primary antibody, diluted 1:50 in antibody incubation buffer.
8. Wash as in **step 4**.

   (a) Incubate for 1 h at room temperature with 40 μL/well secondary antibody, diluted 1:500 in antibody incubation buffer, containing 2 μg/mL 4′,6-diamidino-2-phenylindole (DAPI). Protect plate from light.

   (b) Incubate for 1 h at room temperature with 40 μL/well PBS, containing 2 μg/mL DAPI (*see* **Note 2**). Protect plate from light.
9. Wash as in **step 4**.

3.1.3 Image Acquisition and Analysis Using the ImageXpress Micro Platform with Integrated MetaXpress Imaging and Analysis Software (Figs. 1, 2, and 3)

1. Use the 10× objective and select wavelengths DAPI and Texas Red, adjust focus and acquisition times and record multiple images per well (*see* **Notes 3–5**).
2. Set up the inbuilt "Multi Wavelength Translocation Module" for image analysis:
3. Select DAPI as "Source Image" and optimize "Approximate minimum width," "Approximate maximum width," and "Intensity above local background" settings to detect most cell nuclei (Fig. 1). The created nuclei masks are the bases for defining nuclear and cytoplasmic compartments (*see* **Note 6**).
4. Using the Texas Red channel as source, use 0.4 μm as value for both "inner distance from boarder" and "outer distance from boarder" and define the outer compartment width as 2 μm (Fig. 1; *see* **Notes** 7 and **8**).
5. In the analysis module, choose recording of the following parameters for each cell: "Cell: Compartment Stained Area (MultiWaveTransData)"—(i.e., the DAPI stained area), "Cell: Mean Inner Intensity W2 (MultiWaveTransData)"—(i.e., the mean fluorescent signal in the nucleus), "Cell: Mean Outer Intensity W2 (MultiWaveTransData)"—(i.e., the mean fluorescent signal in the cytoplasm) (Fig. 1).

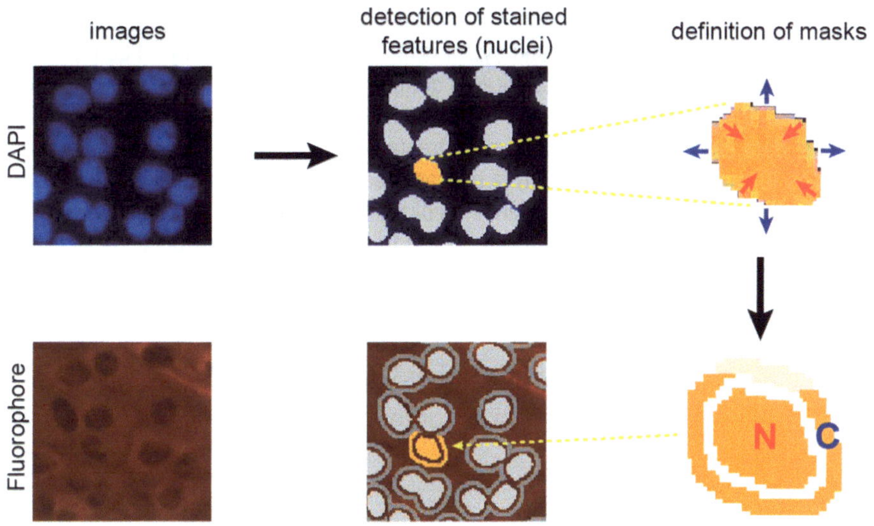

Fig. 1 Analysis of images by the MetaXpress Imaging and Analysis software. The nuclear DAPI staining is automatically detected and serves to define masks (orange and gray). The nuclear region (N) is deduced by deflation of the mask boundaries, thereby excluding potential bleed-through from the cytoplasm. For the cytoplasmic region (C), expanded mask boundaries are used to define a band around the nucleus. Areas which cannot be unambiguously assigned to one cell are removed from the cytoplasmic region mask (light orange). N and C masks are then used to measure the individual mean fluorescence intensity values in the nucleus and cytoplasm, respectively

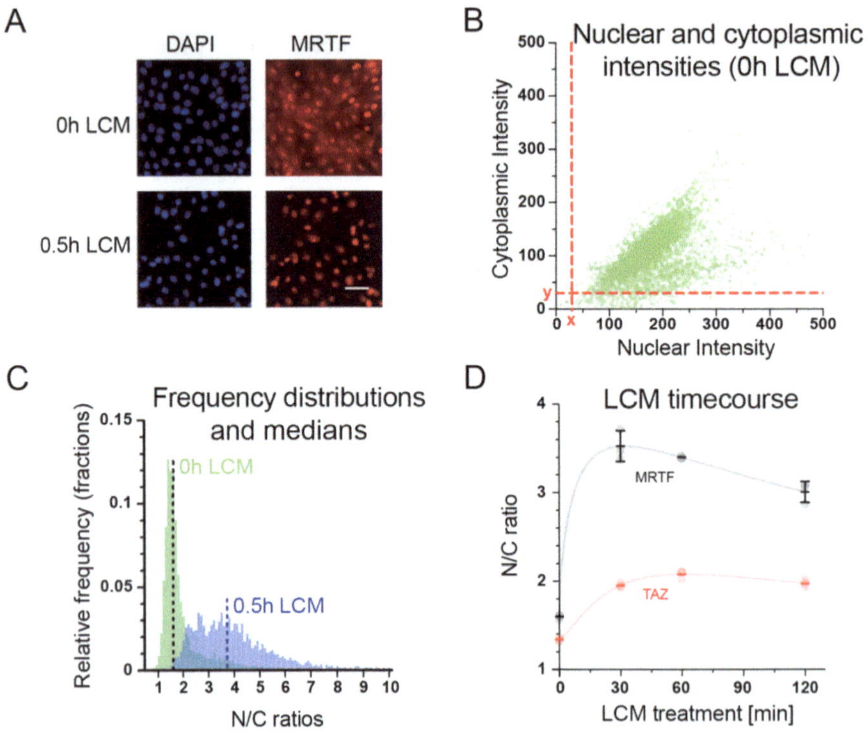

Fig. 2 Depiction and processing of acquired localization data. (**a**) Representative images of LLC-PK1 cells without treatment or 0.5 h LCM treatment. DAPI staining (blue) and MRFT-staining (red) are shown. The scale bar represents 50 μm. (**b**) Plot of nuclear and cytoplasmic mean MRTF intensity value pairs from untreated cells (one well of the 96-well plate, 5022 cells). Red dotted lines and values (*x*, *y*) are thresholds to filter out very low intensities. (**c**) Frequency histograms of nucleocytoplasmic fluorescence (N/C) ratios for untreated cells (green bars, data calculated from **a**) and cells treated for 0.5 h with LCM (blue bars). Location of the median N/C ratios (0 h LCM and 0.5 h LCM) is indicated as black and blue dotted lines, respectively. The medians serve as robust descriptors of the MRTF localization in these cell populations. (**d**) Time-course of MRTF (black) and TAZ (red) median N/C ratios upon LCM treatment. Each data point is calculated from more than 1000 cells and plotted against the LCM-treatment time. Mean and standard deviation (SD) from three repetitions are indicated

6. Run the analysis module (*see* **Note 9**).
7. Export data as comma separated values text files (*see* **Note 10**).

3.1.4 Data Analysis

1. We calculate the median nucleocytoplasmic ratio value using Excel (*see* **Notes 11–13**).
2. For histograms, graphical depiction of median values, and statistical analysis, we use the software GraphPad Prism (Fig. 2 and 3).

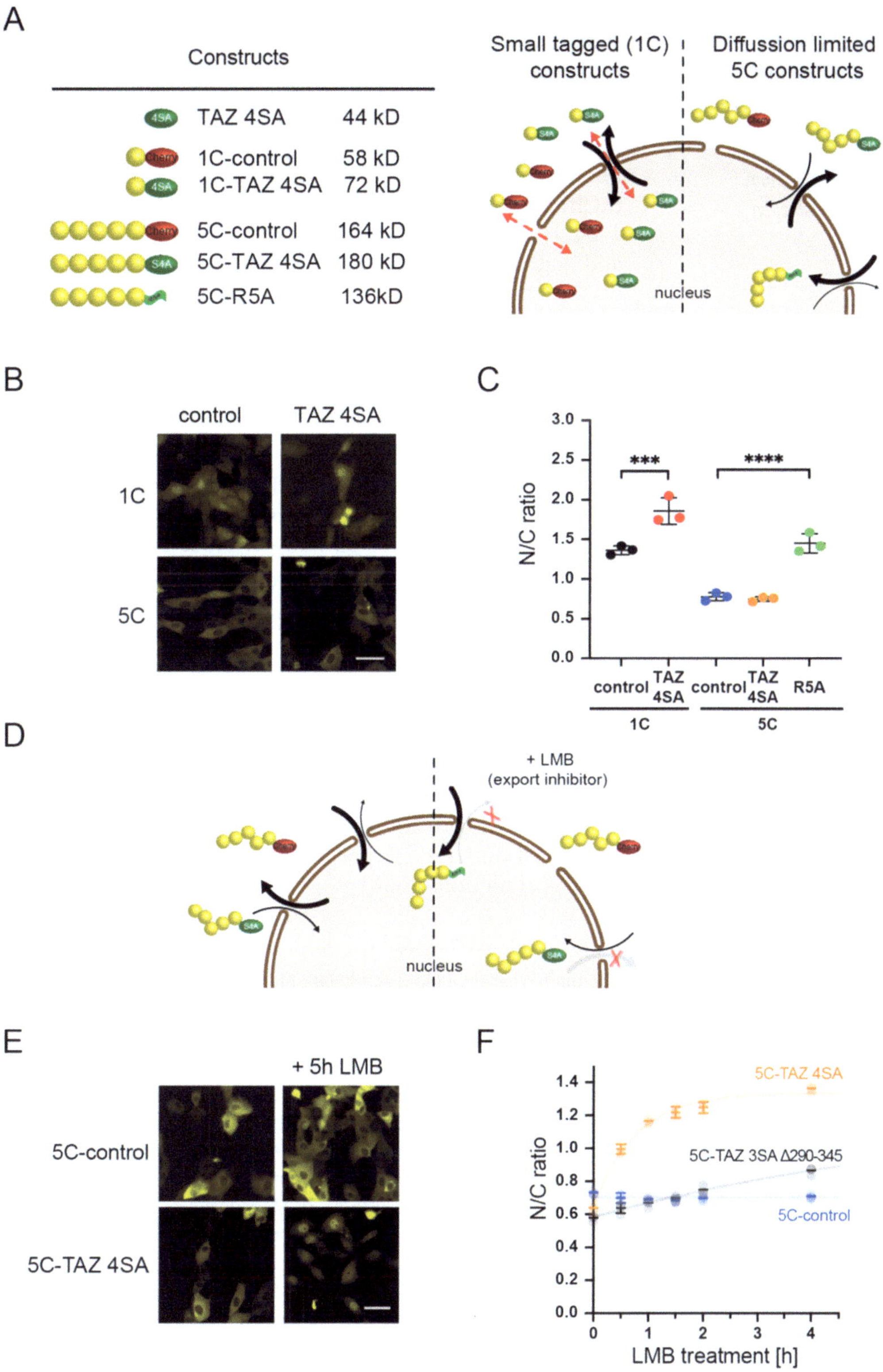

Fig. 3 Localization data of overexpressed constructs (**a**): (Left) Single mCitrine 1C and five mCitrine (5C)-control constructs comprise the respective tag, fused to mCherry as unrelated protein of similar size to TAZ.

3.2 Live Cell Imaging: Kinetics of Nuclear Import Using the Rapamycin Sequestration System

3.2.1 Cell Culturing and Cell Transfection

1. Maintain LLC-PK1 cells as described in Subheading 3.1.1, **step 1**.
2. Seed 400,000 cells into glass-bottom dishes, in a total of 2 mL cell culture medium (*see* Subheading 3.1.1, **step 2**).
3. jetPrime transfection 24 h later, when cells are 50–100% confluent: Per well, mix 2 μg plasmids (1 μg H2B-2xFKBP-mCherry and either 1 μg single mCitrine (1C)-NES-FRB or 1 μg 1C-NES-FRB-(290–345)) with 100 μL jetPrime buffer and 4 μL jetPrime reagent, incubate for 10 min, and add to dishes. Incubate for 24 h.
4. Replace cell culture medium (2 mL).

3.2.2 Cell Stimulation and Live-Cell Imaging

1. Transfer dishes into VivaView FL Incubator and let them equilibrate for about 30 min (*see* **Note 14**).
2. Using the 10× objective, select wavelengths GFP and Texas Red, adjust light intensities and focuses and set up Z-stack acquisition (*see* **Notes 15**). Set up time-lapse imaging to record a Z-stack every 30 s, for 15 min (*see* **Note 16**).
3. Start acquisition and add quickly 100 μL rapamycin medium before the second time-point (*see* **Note 17**).

3.2.3 Data Analysis

1. Using the multi-dimensional analysis module, select the best in-focus GFP image for each time point and export them as a stack.
2. For each image stack, manually draw a background region of interest (ROI). For each cell in the image stack, draw an additional ROI in the cytoplasm and one in the nucleus (*see* **Notes 18** and **19**). Export the mean intensity values for the regions at all time-points.

Fig. 3 (continued) 5C-R5A comprises the 5C tag fused to a weak variant of the minimal SV40-NLS (residues 126–132) labeled as R5A. Molecular weights of the fusion proteins are shown. (Right) Effect of tag-size on the cellular distribution of indicated constructs. The small tagged construct 1C-control can diffuse into and out of the nucleus (red dotted arrow) while 1C-TAZ 4SA is also subject to mediated import and export (black arrows). Diffusion limited 5C constructs cannot enter the nucleus by passive diffusion, resulting in cytoplasmic confinement of 5C-control. Efficient import leads to nuclear accumulation of 5C-R5A. The balance between slow nuclear import and faster export of 5C-TAZ 4SA results in an apparent nuclear exclusion. (**b**) Images of LLC-PK1 cells overexpressing 1C and 5C constructs for 24 h. The scale bar represents 50 μm. (**c**) Median N/C ratios of indicated 1C and 5C constructs, overexpressed for 24 h. (**d**) Mechanism of LMB induced nuclear accumulation of shuttling constructs. LMB is a specific export inhibitor which prevents CRM1-mediated nuclear export. Since shuttling proteins continue to be imported, they accumulate in the nucleus upon LMB treatment. (**e**) Images of LLC-PK1 cells overexpressing 5C constructs for 24 h, in the absence and presence of LMB for 5 h. The scale bar represents 50 μm. (**f**) Median N/C ratios of 5C constructs during a LMB time-course. Individual data points and means ± SD are depicted. ***$p < 0.001$, ****$p < 0.0001$; ANOVA with multiple comparisons. Figure 3a, b, c, e, and f are modified from [88] Figs. 1 and 2. Material is available under Public License and disclaimer of warranties http:/creativecommons.org/licenses/by/4.0/

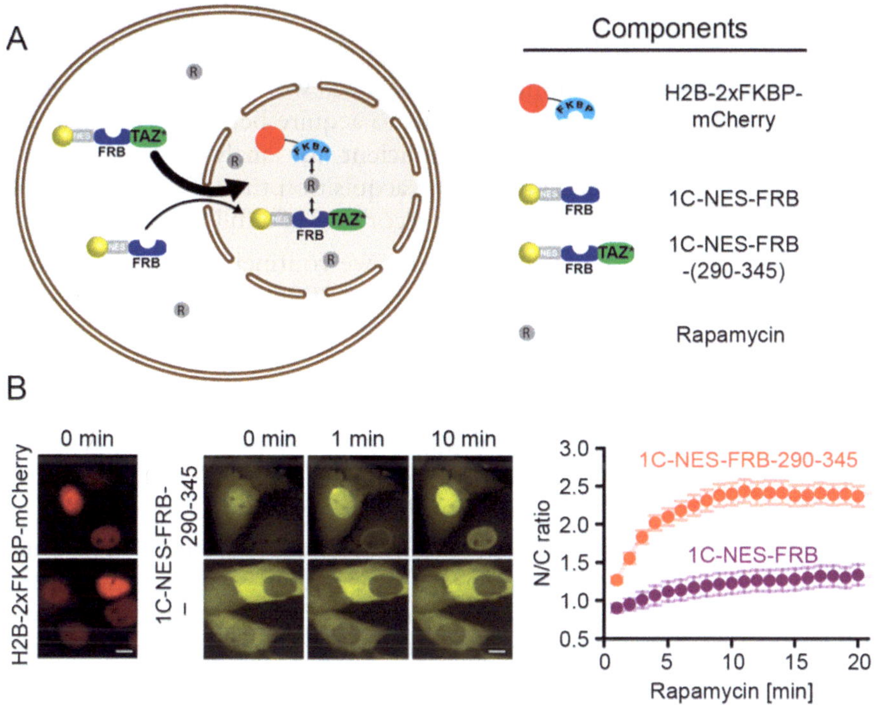

Fig. 4 Rapamycin-induced sequestration in the nucleus (RIS'N) system. (**a**) Model showing rapamycin-induced nuclear entrapment of 1C-NES-FRB or 1C-NES-FRB-(290–345) by H2B-2xFKBP-mCherry (*see* also Subheading 1.5). Right: Description of components and their purpose. (**b**) Live-cell RIS'N import assay. Fluorescence images were recorded using a VivaView system and show co-expression of H2B-2xFKBP-mCherry (red) and 1C-NES-FRB or 1C-NES-FRB-290–345 (yellow) at the indicated time-points (left). Scale bar: 10 μm. Average N/C values ± SEM were plotted against time of rapamycin treatment. 1C-NES-FRB, $n = 40$ cells; 1C-NES-FRB-290–345, $n = 80$ cells. Differences between the two probes are significant at all time-points ($p < 0.01$, not indicated). Curves are fits to an exponential rise to maximum function. Figure 4 is modified from [88] Fig. 4. Material is available under Public License and disclaimer of warranties http:/creativecommons.org/licenses/by/4.0/

3. For each time-point, subtract the background intensity from the fluorescence intensity in the nucleus and the cytoplasm of each cell and calculate the mean nucleocytoplasmic ratio (*see* **Note 20**).
4. Plot graphs and perform statistical analysis with appropriate software such as Excel or Prism (Fig. 4).

4 Notes

1. Neutralization and permeabilization can be done in one step: 100 mM glycine in PBS with 0.2% Triton X-100.
2. DAPI staining can be reduced to 5–10 min if cells are permeabilized.

3. Clean the plate bottom to remove auto-fluorescent particles that could affect image acquisition and quality.
4. For immunofluorescence, we routinely record five images/well. This allows us to acquire between 1000 and 6000 cells/well which is sufficient for analysis with robust statistics. Depending on the acquisition time, five images/well can take up to 90 min for recording of a full plate.
5. For transfected cells, we routinely record 25 images/well to collect between 100 and 2000 transfected cells/well. Recording takes usually around 90–120 min due to the short acquisition times.
6. Avoid fragmentation of individual nuclei. Masks comprising multiple overlapping or clustered nuclei can be sorted out at a later point based on their "Cell: Compartment Stained Area (MultiWaveTransData)" values.
7. Inner and outer distances are chosen to (1) minimize cross contamination of fluorescence signals between the defined nuclear and cytoplasmic compartments, (2) maximize the nuclear compartment, and (3) maximize the overlap of the defined cytoplasmic compartment with the actual cytoplasm.
8. Maximizing the overlap of the defined cytoplasmic compartment with the actual cytoplasm becomes an issue for cells that are very spindle shaped. LLC PK1 or other epithelial cells have a large cytoplasm and are hence ideal for studying the cellular distribution of proteins.
9. Analyses using inbuilt modules are usually faster than customized analysis modules. The inbuilt "Multi Wavelength Translocation Module" is fast, but will take up to several hours for a 96-well plate. We usually run analyses in batches overnight.
10. We exclude cells from data export that either have "Cell: Compartment Stained Area (MultiWaveTransData)" values smaller than 50 (i.e., DAPI precipitate, nuclear debris or nuclei in mitosis) or that have mean fluorescence values below 30 (i.e., very weak staining or just background signal).
11. The aggregate function allows simultaneous filtering of data for multiple criteria and calculation of the 50th percentile (median) of the nucleocytoplasmic ratios.
12. We usually set the threshold number of cells for the calculation of medians to a minimum of 100/well.
13. We optimize minimum and maximum threshold levels for nuclear and cytoplasmic mean intensities, especially for overexpressed proteins since both very low and very high expression can result in nonphysiological distribution of the constructs.

14. Remove lids to facilitate quick administration of rapamycin.
15. Administration of the drug requires opening of the incubator which will result in a focus drift. Recording Z-stacks of 11 images ensures in-focus acquisition.
16. Multi-position imaging or shorter acquisition intervals are difficult to achieve using Z-stacks.
17. The final Rapamycin concentration will be 1 nM. The relatively large volume added (1/20th of the total) minimizes the diffusional equilibration time of the drug in the dish.
18. Whenever possible, choose regions in the stack that do not change due to cell movements or changes in focus.
19. We frequently use Fiji and its "Multi Measure" plugin (Wayne Rasband, Bob Dougherty, Tony Collins, Audrey Karperian and Ulrik Stervbo) for drawing of ROIs and measuring mean intensity values.
20. We use the mean nucleocytoplasmic ratio instead of the median for data from small cell populations (less than 100).

Acknowledgments

The authors' own experimental work was supported by grants from the Canadian Institute of Health Research (CIHR), Natural Sciences and Engineering Council of Canada (NSERC) and the Kidney Foundation of Canada (KFoC) to AK.

References

1. Hinz B, Phan SH, Thannickal VJ, Galli A, Bochaton-Piallat ML, Gabbiani G (2007) The myofibroblast: one function, multiple origins. Am J Pathol 170:1807–1816
2. Hinz B, McCulloch CA, Coelho NM (2019) Mechanical regulation of myofibroblast phenoconversion and collagen contraction. Exp Cell Res 379:119–128. https://doi.org/10.1016/j.yexcr.2019.03.027
3. Lovisa S, LeBleu VS, Tampe B, Sugimoto H, Vadnagara K, Carstens JL, Wu CC, Hagos Y, Burckhardt BC, Pentcheva-Hoang T, Nischal H, Allison JP, Zeisberg M, Kalluri R (2015) Epithelial-to-mesenchymal transition induces cell cycle arrest and parenchymal damage in renal fibrosis. Nat Med 21:998–1009. https://doi.org/10.1038/nm.3902
4. Grande MT, Sanchez-Laorden B, Lopez-Blau C, De Frutos CA, Boutet A, Arevalo M, Rowe RG, Weiss SJ, Lopez-Novoa JM, Nieto MA (2015) Snail1-induced partial epithelial-to-mesenchymal transition drives renal fibrosis in mice and can be targeted to reverse established disease. Nat Med 21:989–997. https://doi.org/10.1038/nm.3901
5. Bialik JF, Ding M, Speight P, Dan Q, Miranda MZ, Di Ciano-Oliveira C, Kofler MM, Rotstein OD, Pedersen SF, Szaszi K, Kapus A (2019) Profibrotic epithelial phenotype: a central role for MRTF and TAZ. Sci Rep 9:4323. https://doi.org/10.1038/s41598-019-40764-7
6. Derynck R, Budi EH (2019) Specificity, versatility, and control of TGF-beta family signaling. Sci Signal 12:eaav5183. https://doi.org/10.1126/scisignal.aav5183
7. Hill CS (2009) Nucleocytoplasmic shuttling of Smad proteins. Cell Res 19:36–46. https://doi.org/10.1038/cr.2008.325
8. Hill CS (2016) Transcriptional control by the SMADs. Cold Spring Harb Perspect Biol 8:a022079. https://doi.org/10.1101/cshperspect.a022079

9. Lodyga M, Hinz B (2019) TGF-beta1-a truly transforming growth factor in fibrosis and immunity. Semin Cell Dev Biol 101:123–139. https://doi.org/10.1016/j.semcdb.2019.12.010
10. Speight P, Kofler M, Szaszi K, Kapus A (2016) Context-dependent switch in chemo/mechanotransduction via multilevel crosstalk among cytoskeleton-regulated MRTF and TAZ and TGFbeta-regulated Smad3. Nat Commun 7:11642. https://doi.org/10.1038/ncomms11642
11. Dasgupta I, McCollum D (2019) Control of cellular responses to mechanical cues through YAP/TAZ regulation. J Biol Chem 294:17693–17706. https://doi.org/10.1074/jbc.REV119.007963
12. Dupont S (2019) Regulation of YAP/TAZ activity by mechanical cues: an experimental overview. Methods Mol Biol 1893:183–202. https://doi.org/10.1007/978-1-4939-8910-2_15
13. Fan L, Sebe A, Peterfi Z, Masszi A, Thirone AC, Rotstein OD, Nakano H, McCulloch CA, Szaszi K, Mucsi I, Kapus A (2007) Cell contact-dependent regulation of epithelial-myofibroblast transition via the rho-rho kinase-phospho-myosin pathway. Mol Biol Cell 18:1083–1097. https://doi.org/10.1091/mbc.E06-07-0602
14. Janmey PA, Wells RG, Assoian RK, McCulloch CA (2013) From tissue mechanics to transcription factors. Differentiation 86:112–120. https://doi.org/10.1016/j.diff.2013.07.004
15. Gomez EW, Chen QK, Gjorevski N, Nelson CM (2010) Tissue geometry patterns epithelial-mesenchymal transition via intercellular mechanotransduction. J Cell Biochem 110:44–51. https://doi.org/10.1002/jcb.22545
16. Miralles F, Posern G, Zaromytidou AI, Treisman R (2003) Actin dynamics control SRF activity by regulation of its coactivator MAL. Cell 113:329–342
17. Wang DZ, Li S, Hockemeyer D, Sutherland L, Wang Z, Schratt G, Richardson JA, Nordheim A, Olson EN (2002) Potentiation of serum response factor activity by a family of myocardin-related transcription factors. Proc Natl Acad Sci U S A 99:14855–14860. https://doi.org/10.1073/pnas.222561499
18. Parmacek MS (2007) Myocardin-related transcription factors: critical coactivators regulating cardiovascular development and adaptation. Circ Res 100:633–644. https://doi.org/10.1161/01.RES.0000259563.61091.e8
19. Sudol M (1994) Yes-associated protein (YAP65) is a proline-rich phosphoprotein that binds to the SH3 domain of the Yes proto-oncogene product. Oncogene 9:2145–2152
20. Yagi R, Chen LF, Shigesada K, Murakami Y, Ito Y (1999) A WW domain-containing yes-associated protein (YAP) is a novel transcriptional co-activator. EMBO J 18:2551–2562. https://doi.org/10.1093/emboj/18.9.2551
21. Kanai F, Marignani PA, Sarbassova D, Yagi R, Hall RA, Donowitz M, Hisaminato A, Fujiwara T, Ito Y, Cantley LC, Yaffe MB (2000) TAZ: a novel transcriptional co-activator regulated by interactions with 14-3-3 and PDZ domain proteins. EMBO J 19:6778–6791. https://doi.org/10.1093/emboj/19.24.6778
22. Pocaterra A, Romani P, Dupont S (2020) YAP/TAZ functions and their regulation at a glance. J Cell Sci 133:jcs230425. https://doi.org/10.1242/jcs.230425
23. Masszi A, Speight P, Charbonney E, Lodyga M, Nakano H, Szaszi K, Kapus A (2010) Fate-determining mechanisms in epithelial-myofibroblast transition: major inhibitory role for Smad3. J Cell Biol 188:383–399. https://doi.org/10.1083/jcb.200906155
24. Small EM, Thatcher JE, Sutherland LB, Kinoshita H, Gerard RD, Richardson JA, Dimaio JM, Sadek H, Kuwahara K, Olson EN (2010) Myocardin-related transcription factor-a controls myofibroblast activation and fibrosis in response to myocardial infarction. Circ Res 107:294–304. https://doi.org/10.1161/CIRCRESAHA.110.223172
25. Crider BJ, Risinger GM Jr, Haaksma CJ, Howard EW, Tomasek JJ (2011) Myocardin-related transcription factors A and B are key regulators of TGF-beta1-induced fibroblast to myofibroblast differentiation. J Invest Dermatol 131:2378–2385. https://doi.org/10.1038/jid.2011.219
26. Small EM (2012) The actin-MRTF-SRF gene regulatory axis and myofibroblast differentiation. J Cardiovasc Transl Res 5:794–804. https://doi.org/10.1007/s12265-012-9397-0
27. Morita T, Mayanagi T, Sobue K (2007) Dual roles of myocardin-related transcription factors in epithelial mesenchymal transition via slug induction and actin remodeling. J Cell Biol 179:1027–1042. https://doi.org/10.1083/jcb.200708174
28. Speight P, Nakano H, Kelley TJ, Hinz B, Kapus A (2013) Differential topical susceptibility to TGFbeta in intact and injured regions of the epithelium: key role in myofibroblast transition. Mol Biol Cell 24:3326–3336. https://doi.org/10.1091/mbc.E13-04-0220

29. Lei QY, Zhang H, Zhao B, Zha ZY, Bai F, Pei XH, Zhao S, Xiong Y, Guan KL (2008) TAZ promotes cell proliferation and epithelial-mesenchymal transition and is inhibited by the hippo pathway. Mol Cell Biol 28:2426–2436. https://doi.org/10.1128/MCB.01874-07
30. Liu F, Lagares D, Choi KM, Stopfer L, Marinkovic A, Vrbanac V, Probst CK, Hiemer SE, Sisson TH, Horowitz JC, Rosas IO, Fredenburgh LE, Feghali-Bostwick C, Varelas X, Tager AM, Tschumperlin DJ (2015) Mechanosignaling through YAP and TAZ drives fibroblast activation and fibrosis. Am J Phys Lung Cell Mol Phys 308:L344–L357. https://doi.org/10.1152/ajplung.00300.2014
31. Gui Y, Li J, Lu Q, Feng Y, Wang M, He W, Yang J, Dai C (2018) Yap/Taz mediates mTORC2-stimulated fibroblast activation and kidney fibrosis. J Biol Chem 293:16364–16375. https://doi.org/10.1074/jbc.RA118.004073
32. Rahaman SO, Grove LM, Paruchuri S, Southern BD, Abraham S, Niese KA, Scheraga RG, Ghosh S, Thodeti CK, Zhang DX, Moran MM, Schilling WP, Tschumperlin DJ, Olman MA (2014) TRPV4 mediates myofibroblast differentiation and pulmonary fibrosis in mice. J Clin Invest 124:5225–5238. https://doi.org/10.1172/JCI75331
33. Sisson TH, Ajayi IO, Subbotina N, Dodi AE, Rodansky ES, Chibucos LN, Kim KK, Keshamouni VG, White ES, Zhou Y, Higgins PD, Larsen SD, Neubig RR, Horowitz JC (2015) Inhibition of myocardin-related transcription factor/serum response factor signaling decreases lung fibrosis and promotes mesenchymal cell apoptosis. Am J Pathol 185:969–986. https://doi.org/10.1016/j.ajpath.2014.12.005
34. Tian W, Hao C, Fan Z, Weng X, Qin H, Wu X, Fang M, Chen Q, Shen A, Xu Y (2015) Myocardin related transcription factor A programs epigenetic activation of hepatic stellate cells. J Hepatol 62:165–174. https://doi.org/10.1016/j.jhep.2014.07.029
35. Rozycki M, Bialik JF, Speight P, Dan Q, Knudsen TE, Szeto SG, Yuen DA, Szaszi K, Pedersen SF, Kapus A (2016) Myocardin-related transcription factor regulates Nox4 protein expression: linking cytoskeletal organization to redox state. J Biol Chem 291:227–243. https://doi.org/10.1074/jbc.M115.674606
36. Xu H, Wu X, Qin H, Tian W, Chen J, Sun L, Fang M, Xu Y (2015) Myocardin-related transcription factor a epigenetically regulates renal fibrosis in diabetic nephropathy. J Am Soc Nephrol 26:1648–1660. https://doi.org/10.1681/ASN.2014070678
37. Miranda MZ, Bialik JF, Speight P, Dan Q, Yeung T, Szaszi K, Pedersen SF, Kapus A (2017) TGF-beta1 regulates the expression and transcriptional activity of TAZ protein via a Smad3-independent, myocardin-related transcription factor-mediated mechanism. J Biol Chem 292:14902–14920. https://doi.org/10.1074/jbc.M117.780502
38. Wu P, Liu Z, Zhao T, Xia F, Gong L, Zheng Z, Chen Z, Yang T, Duan Q (2019) Lovastatin attenuates angiotensin II induced cardiovascular fibrosis through the suppression of YAP/-TAZ signaling. Biochem Biophys Res Commun 512:736–741. https://doi.org/10.1016/j.bbrc.2019.03.158
39. Haak AJ, Kostallari E, Sicard D, Ligresti G, Choi KM, Caporarello N, Jones DL, Tan Q, Meridew J, Diaz Espinosa AM, Aravamudhan A, Maiers JL, Britt RD Jr, Roden AC, Pabelick CM, Prakash YS, Nouraie SM, Li X, Zhang Y, Kass DJ, Lagares D, Tager AM, Varelas X, Shah VH, Tschumperlin DJ (2019) Selective YAP/TAZ inhibition in fibroblasts via dopamine receptor D1 agonism reverses fibrosis. Sci Transl Med 11:eaau6296. https://doi.org/10.1126/scitranslmed.aau6296
40. Martin K, Pritchett J, Llewellyn J, Mullan AF, Athwal VS, Dobie R, Harvey E, Zeef L, Farrow S, Streuli C, Henderson NC, Friedman SL, Hanley NA, Piper Hanley K (2016) PAK proteins and YAP-1 signalling downstream of integrin beta-1 in myofibroblasts promote liver fibrosis. Nat Commun 7:12502. https://doi.org/10.1038/ncomms12502
41. Mannaerts I, Leite SB, Verhulst S, Claerhout S, Eysackers N, Thoen LF, Hoorens A, Reynaert H, Halder G, van Grunsven LA (2015) The Hippo pathway effector YAP controls mouse hepatic stellate cell activation. J Hepatol 63:679–688. https://doi.org/10.1016/j.jhep.2015.04.011
42. Szeto SG, Narimatsu M, Lu M, He X, Sidiqi AM, Tolosa MF, Chan L, De Freitas K, Bialik JF, Majumder S, Boo S, Hinz B, Dan Q, Advani A, John R, Wrana JL, Kapus A, Yuen DA (2016) YAP/TAZ are mechanoregulators of TGF-beta-Smad signaling and renal fibrogenesis. J Am Soc Nephrol 27:3117–3128. https://doi.org/10.1681/ASN.2015050499
43. Liang M, Yu M, Xia R, Song K, Wang J, Luo J, Chen G, Cheng J (2017) Yap/Taz deletion in Gli+ cell-derived myofibroblasts attenuates fibrosis. J Am Soc Nephrol 28:3278–3290. https://doi.org/10.1681/ASN.2015121354

44. Anorga S, Overstreet JM, Falke LL, Tang J, Goldschmeding RG, Higgins PJ, Samarakoon R (2018) Deregulation of Hippo-TAZ pathway during renal injury confers a fibrotic maladaptive phenotype. FASEB J 32 (5):2644–2657. https://doi.org/10.1096/fj.201700722R
45. Rinschen MM, Grahammer F, Hoppe AK, Kohli P, Hagmann H, Kretz O, Bertsch S, Hohne M, Gobel H, Bartram MP, Gandhirajan RK, Kruger M, Brinkkoetter PT, Huber TB, Kann M, Wickstrom SA, Benzing T, Schermer B (2017) YAP-mediated mechanotransduction determines the podocyte's response to damage. Sci Signal 10:aaf8165. https://doi.org/10.1126/scisignal.aaf8165
46. Haak AJ, Tsou PS, Amin MA, Ruth JH, Campbell P, Fox DA, Khanna D, Larsen SD, Neubig RR (2014) Targeting the myofibroblast genetic switch: inhibitors of myocardin-related transcription factor/serum response factor-regulated gene transcription prevent fibrosis in a murine model of skin injury. J Pharmacol Exp Ther 349:480–486. https://doi.org/10.1124/jpet.114.213520
47. Shiwen X, Stratton R, Nikitorowicz-Buniak J, Ahmed-Abdi B, Ponticos M, Denton C, Abraham D, Takahashi A, Suki B, Layne MD, Lafyatis R, Smith BD (2015) A role of myocardin related transcription factor-A (MRTF-A) in scleroderma related fibrosis. PLoS One 10: e0126015. https://doi.org/10.1371/journal.pone.0126015
48. Toyama T, Looney AP, Baker BM, Stawski L, Haines P, Simms R, Szymaniak AD, Varelas X, Trojanowska M (2018) Therapeutic targeting of TAZ and YAP by dimethyl fumarate in systemic sclerosis fibrosis. J Invest Dermatol 138:78–88. https://doi.org/10.1016/j.jid.2017.08.024
49. Hutchings KM, Lisabeth EM, Rajeswaran W, Wilson MW, Sorenson RJ, Campbell PL, Ruth JH, Amin A, Tsou PS, Leipprandt JR, Olson SR, Wen B, Zhao T, Sun D, Khanna D, Fox DA, Neubig RR, Larsen SD (2017) Pharmacokinetic optimitzation of CCG-203971: Novel inhibitors of the Rho/MRTF/SRF transcriptional pathway as potential antifibrotic therapeutics for systemic scleroderma. Bioorg Med Chem Lett 27:1744–1749. https://doi.org/10.1016/j.bmcl.2017.02.070
50. Yu-Wai-Man C, Spencer-Dene B, Lee RMH, Hutchings K, Lisabeth EM, Treisman R, Bailly M, Larsen SD, Neubig RR, Khaw PT (2017) Local delivery of novel MRTF/SRF inhibitors prevents scar tissue formation in a preclinical model of fibrosis. Sci Rep 7:518. https://doi.org/10.1038/s41598-017-00212-w
51. Crawford JJ, Bronner SM, Zbieg JR (2018) Hippo pathway inhibition by blocking the YAP/TAZ-TEAD interface: a patent review. Expert Opin Ther Pat 28:867–873. https://doi.org/10.1080/13543776.2018.1549226
52. Pobbati AV, Hong W (2020) A combat with the YAP/TAZ-TEAD oncoproteins for cancer therapy. Theranostics 10:3622–3635. https://doi.org/10.7150/thno.40889
53. Schmierer B, Tournier AL, Bates PA, Hill CS (2008) Mathematical modeling identifies Smad nucleocytoplasmic shuttling as a dynamic signal-interpreting system. Proc Natl Acad Sci U S A 105:6608–6613. https://doi.org/10.1073/pnas.0710134105
54. Cheong R, Hoffmann A, Levchenko A (2008) Understanding NF-kappaB signaling via mathematical modeling. Mol Syst Biol 4:192. https://doi.org/10.1038/msb.2008.30
55. Hansen AS, O'Shea EK (2013) Promoter decoding of transcription factor dynamics involves a trade-off between noise and control of gene expression. Mol Syst Biol 9:704. https://doi.org/10.1038/msb.2013.56
56. Ay A, Arnosti DN (2011) Mathematical modeling of gene expression: a guide for the perplexed biologist. Crit Rev Biochem Mol Biol 46:137–151. https://doi.org/10.3109/10409238.2011.556597
57. Olson EN, Nordheim A (2010) Linking actin dynamics and gene transcription to drive cellular motile functions. Nat Rev Mol Cell Biol 11:353–365. https://doi.org/10.1038/nrm2890
58. Mouilleron S, Langer CA, Guettler S, McDonald NQ, Treisman R (2011) Structure of a pentavalent G-actin*MRTF-A complex reveals how G-actin controls nucleocytoplasmic shuttling of a transcriptional coactivator. Sci Signal 4:ra40. https://doi.org/10.1126/scisignal.2001750
59. Vartiainen MK, Guettler S, Larijani B, Treisman R (2007) Nuclear actin regulates dynamic subcellular localization and activity of the SRF cofactor MAL. Science 316:1749–1752. https://doi.org/10.1126/science.1141084
60. Staus DP, Weise-Cross L, Mangum KD, Medlin MD, Mangiante L, Taylor JM, Mack CP (2014) Nuclear RhoA signaling regulates MRTF-dependent SMC-specific transcription. Am J Physiol Heart Circ Physiol 307: H379–H390. https://doi.org/10.1152/ajpheart.01002.2013
61. Panayiotou R, Miralles F, Pawlowski R, Diring J, Flynn HR, Skehel M, Treisman R (2016) Phosphorylation acts positively and negatively to regulate MRTF-A subcellular

localisation and activity. elife 5:e15460. https://doi.org/10.7554/eLife.15460
62. Sun Q, Chen G, Streb JW, Long X, Yang Y, Stoeckert CJ Jr, Miano JM (2006) Defining the mammalian CArGome. Genome Res 16:197–207. https://doi.org/10.1101/gr.4108706
63. Esnault C, Stewart A, Gualdrini F, East P, Horswell S, Matthews N, Treisman R (2014) Rho-actin signaling to the MRTF coactivators dominates the immediate transcriptional response to serum in fibroblasts. Genes Dev 28:943–958. https://doi.org/10.1101/gad.239327.114
64. Li CX, Talele NP, Boo S, Koehler A, Knee-Walden E, Balestrini JL, Speight P, Kapus A, Hinz B (2017) MicroRNA-21 preserves the fibrotic mechanical memory of mesenchymal stem cells. Nat Mater 16:379–389. https://doi.org/10.1038/nmat4780
65. Busche S, Descot A, Julien S, Genth H, Posern G (2008) Epithelial cell-cell contacts regulate SRF-mediated transcription via Rac-actin-MAL signalling. J Cell Sci 121:1025–1035. https://doi.org/10.1242/jcs.014456
66. Sebe A, Masszi A, Zulys M, Yeung T, Speight P, Rotstein OD, Nakano H, Mucsi I, Szaszi K, Kapus A (2008) Rac, PAK and p38 regulate cell contact-dependent nuclear translocation of myocardin-related transcription factor. FEBS Lett 582:291–298. https://doi.org/10.1016/j.febslet.2007.12.021
67. Piccolo S, Dupont S, Cordenonsi M (2014) The biology of YAP/TAZ: hippo signaling and beyond. Physiol Rev 94:1287–1312. https://doi.org/10.1152/physrev.00005.2014
68. Totaro A, Panciera T, Piccolo S (2018) YAP/-TAZ upstream signals and downstream responses. Nat Cell Biol 20:888–899. https://doi.org/10.1038/s41556-018-0142-z
69. Fu V, Plouffe SW, Guan KL (2017) The Hippo pathway in organ development, homeostasis, and regeneration. Curr Opin Cell Biol 49:99–107. https://doi.org/10.1016/j.ceb.2017.12.012
70. Hansen CG, Moroishi T, Guan KL (2015) YAP and TAZ: a nexus for Hippo signaling and beyond. Trends Cell Biol 25:499–513. https://doi.org/10.1016/j.tcb.2015.05.002
71. Ma S, Meng Z, Chen R, Guan KL (2019) The Hippo Pathway: Biology and Pathophysiology. Annu Rev Biochem 88:577–604. https://doi.org/10.1146/annurev-biochem-013118-111829
72. Varelas X, Samavarchi-Tehrani P, Narimatsu M, Weiss A, Cockburn K, Larsen BG, Rossant J, Wrana JL (2010) The Crumbs complex couples cell density sensing to Hippo-dependent control of the TGF-beta-SMAD pathway. Dev Cell 19:831–844. https://doi.org/10.1016/j.devcel.2010.11.012
73. Dupont S, Morsut L, Aragona M, Enzo E, Giulitti S, Cordenonsi M, Zanconato F, Le Digabel J, Forcato M, Bicciato S, Elvassore N, Piccolo S (2011) Role of YAP/TAZ in mechanotransduction. Nature 474:179–183. https://doi.org/10.1038/nature10137
74. Aragona M, Panciera T, Manfrin A, Giulitti S, Michielin F, Elvassore N, Dupont S, Piccolo S (2013) A mechanical checkpoint controls multicellular growth through YAP/TAZ regulation by actin-processing factors. Cell 154:1047–1059. https://doi.org/10.1016/j.cell.2013.07.042
75. Mana-Capelli S, Paramasivam M, Dutta S, McCollum D (2014) Angiomotins link F-actin architecture to Hippo pathway signaling. Mol Biol Cell 25:1676–1685. https://doi.org/10.1091/mbc.E13-11-0701
76. Ege N, Dowbaj AM, Jiang M, Howell M, Hooper S, Foster C, Jenkins RP, Sahai E (2018) Quantitative analysis reveals that actin and src-family kinases regulate nuclear YAP1 and its export. Cell Syst 6:692–708. e613. https://doi.org/10.1016/j.cels.2018.05.006
77. Halder G, Dupont S, Piccolo S (2012) Transduction of mechanical and cytoskeletal cues by YAP and TAZ. Nat Rev Mol Cell Biol 13:591–600. https://doi.org/10.1038/nrm3416
78. Panciera T, Azzolin L, Cordenonsi M, Piccolo S (2017) Mechanobiology of YAP and TAZ in physiology and disease. Nat Rev Mol Cell Biol 18(12):758–770. https://doi.org/10.1038/nrm.2017.87
79. Zmajkovicova K, Bauer Y, Menyhart K, Schnoebelen M, Freti D, Boucher M, Renault B, Studer R, Birker-Robaczewska M, Klenk A, Nayler O, Gatfield J (2020) GPCR-induced YAP activation sensitizes fibroblasts to profibrotic activity of TGFbeta1. PLoS One 15:e0228195. https://doi.org/10.1371/journal.pone.0228195
80. Yasuda D, Kobayashi D, Akahoshi N, Ohto-Nakanishi T, Yoshioka K, Takuwa Y, Mizuno S, Takahashi S, Ishii S (2019) Lysophosphatidic acid-induced YAP/TAZ activation promotes developmental angiogenesis by repressing Notch ligand Dll4. J Clin Invest 130:4332–4349. https://doi.org/10.1172/JCI121955
81. Zhao B, Ye X, Yu J, Li L, Li W, Li S, Yu J, Lin JD, Wang CY, Chinnaiyan AM, Lai ZC, Guan

KL (2008) TEAD mediates YAP-dependent gene induction and growth control. Genes Dev 22:1962–1971. https://doi.org/10.1101/gad.1664408

82. Lin KC, Park HW, Guan KL (2017) Regulation of the hippo pathway transcription factor TEAD. Trends Biochem Sci 42:862–872. https://doi.org/10.1016/j.tibs.2017.09.003

83. Varelas X, Sakuma R, Samavarchi-Tehrani P, Peerani R, Rao BM, Dembowy J, Yaffe MB, Zandstra PW, Wrana JL (2008) TAZ controls Smad nucleocytoplasmic shuttling and regulates human embryonic stem-cell self-renewal. Nat Cell Biol 10:837–848. https://doi.org/10.1038/ncb1748

84. Beyer TA, Weiss A, Khomchuk Y, Huang K, Ogunjimi AA, Varelas X, Wrana JL (2013) Switch enhancers interpret TGF-beta and Hippo signaling to control cell fate in human embryonic stem cells. Cell Rep 5:1611–1624. https://doi.org/10.1016/j.celrep.2013.11.021

85. Yu OM, Miyamoto S, Brown JH (2016) Myocardin-related transcription factor A and yes-associated protein exert dual control in G protein-coupled receptor- and RhoA-mediated transcriptional regulation and cell proliferation. Mol Cell Biol 36:39–49. https://doi.org/10.1128/MCB.00772-15

86. Kim T, Hwang D, Lee D, Kim JH, Kim SY, Lim DS (2017) MRTF potentiates TEAD-YAP transcriptional activity causing metastasis. EMBO J 36:520–535. https://doi.org/10.15252/embj.201695137

87. Foster CT, Gualdrini F, Treisman R (2017) Mutual dependence of the MRTF-SRF and YAP-TEAD pathways in cancer-associated fibroblasts is indirect and mediated by cytoskeletal dynamics. Genes Dev 31:2361–2375. https://doi.org/10.1101/gad.304501.117

88. Kofler M, Speight P, Little D, Di Ciano-Oliveira C, Szaszi K, Kapus A (2018) Mediated nuclear import and export of TAZ and the underlying molecular requirements. Nat Commun 9:4966. https://doi.org/10.1038/s41467-018-07450-0

89. Hodel AE, Harreman MT, Pulliam KF, Harben ME, Holmes JS, Hodel MR, Berland KM, Corbett AH (2006) Nuclear localization signal receptor affinity correlates with in vivo localization in Saccharomyces cerevisiae. J Biol Chem 281:23545–23556. https://doi.org/10.1074/jbc.M601718200

90. Fung HY, Chook YM (2014) Atomic basis of CRM1-cargo recognition, release and inhibition. Semin Cancer Biol 27:52–61. https://doi.org/10.1016/j.semcancer.2014.03.002

91. Putyrski M, Schultz C (2012) Protein translocation as a tool: the current rapamycin story. FEBS Lett 586:2097–2105. https://doi.org/10.1016/j.febslet.2012.04.061

92. Schindelin J, Arganda-Carreras I, Frise E, Kaynig V, Longair M, Pietzsch T, Preibisch S, Rueden C, Saalfeld S, Schmid B, Tinevez JY, White DJ, Hartenstein V, Eliceiri K, Tomancak P, Cardona A (2012) Fiji: an open-source platform for biological-image analysis. Nat Methods 9:676–682. https://doi.org/10.1038/nmeth.2019

Chapter 16

Atomic Force Microscopy for Live-Cell and Hydrogel Measurement

Alexander J. Whitehead, Natalie J. Kirkland, and Adam J. Engler

Abstract

Atomic force microscopy (AFM) has emerged as a popular method for determining the mechanical properties of cells, their components, and biomaterials. Here, we describe AFM setup and application to obtain stiffness measurements from single indentations for hydrogels and myofibroblasts.

Key words Fibroblast, Atomic force microscopy, Force-curve, Stiffness, Young's modulus, Hydrogels, Live-cell measurement

1 Introduction

Cell rheological properties are determined by the combined influence of the extracellular matrix, adhesions, cytoskeleton, membrane proteins and nuclear components. The mechanical properties of each can influence cell function and define whether a cell is normal or diseased. For example, cellular stiffness can influence processes including differentiation [1], aging [2–4], and cancer [5]. Atomic force microscopy (AFM) has emerged as a useful tool to examine cell mechanics, and as a method to characterize mechanical differences in disease states [3].

Standard AFM technique for measuring cell mechanics relies on producing nanometer indentations, via a flexible cantilever with a pyramidal or spherical probe (tip), onto a cell or substrate surface. As force is exerted by the tip onto the cell surface, the deformation of the cantilever is proportional to the deformation of the substrate. To detect the cantilever deformation, a small laser is focused onto the back of the cantilever and the reflection is detected by a photodiode detector (Fig. 1). When the cantilever deforms, the light is deflected away from the point of incidence. As the tip reactively

Alexander J. Whitehead and Natalie J. Kirkland contributed equally to this work.

Boris Hinz and David Lagares (eds.), *Myofibroblasts: Methods and Protocols*, Methods in Molecular Biology, vol. 2299, https://doi.org/10.1007/978-1-0716-1382-5_16, © Springer Science+Business Media, LLC, part of Springer Nature 2021

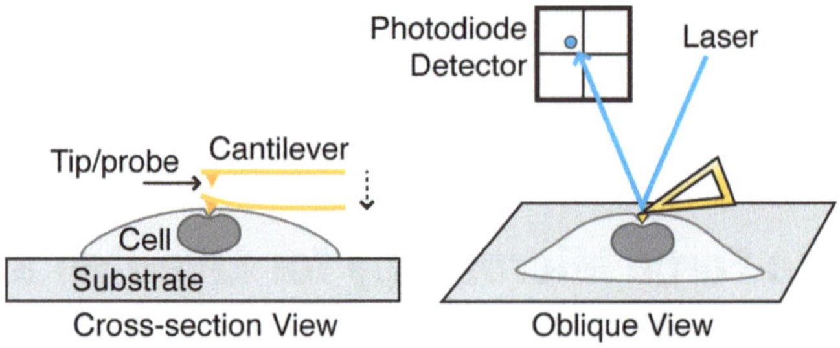

Fig. 1 *Schematic of atomic force microscopy principle.* Precise indentation can be made apically on cells or biomaterials. A laser directed onto the cantilever is deflected and detected by a photodiode so that the cantilever deflection at the surface can be measured

deforms at the surface during indentation and retraction, the deflection is recorded against the distance from the surface to produce a force-indentation curve (which accounts for the deflection against an infinitely rigid substrate, e.g., glass, where deflection is equivalent to indentation; Fig. 2a–c). The force-indentation curves can then be fitted to several proposed models in order to determine the Elastic/Young's modulus, i.e., elastic/stiffness properties, with the Hertz model being one of the most commonly used [6, 7]. The standard Hertz model assumes that the indenter is spherical and the sample being indented is significantly thicker than the indentation depth, i.e., an infinite half-space. When a pyramidal tip is used, the Sneddon's cone indenter variation is used [8] (Table 1). However, stiffness measurements have been reported to vary widely depending on the model applied. Therefore, rather than using absolute stiffness values, it is better practice to compare values relatively between samples using the same indenter geometry [9, 10]. More recently, models are emerging to obtain greater information from force-curves, including cell viscoelastic properties [11]. For simplicity, the protocol here specifically reflect data obtained for elastic interactions, though for viscoelastic properties, additional indentation parameters can be added such as dwell time or force clamps such that creep, stress relaxation, etc. can be measured.

In addition to variance from indenter geometry and model type, cell stiffness measurements from AFM can vary as a result of position within the cell, as well as cell type; such variation can range from 1 to 100 kPa [10, 12]. As measured herein, we show how these data can vary as the tip interacts with regions of the cell (Fig. 2d, e). Thus, we describe the basic procedure for calibrating the AFM (Subheading 1) and characterizing stiffness of biomaterials (Subheading 2) and myofibroblast cells (Subheading 3). Note that we describe the procedure to obtain classic force curves from

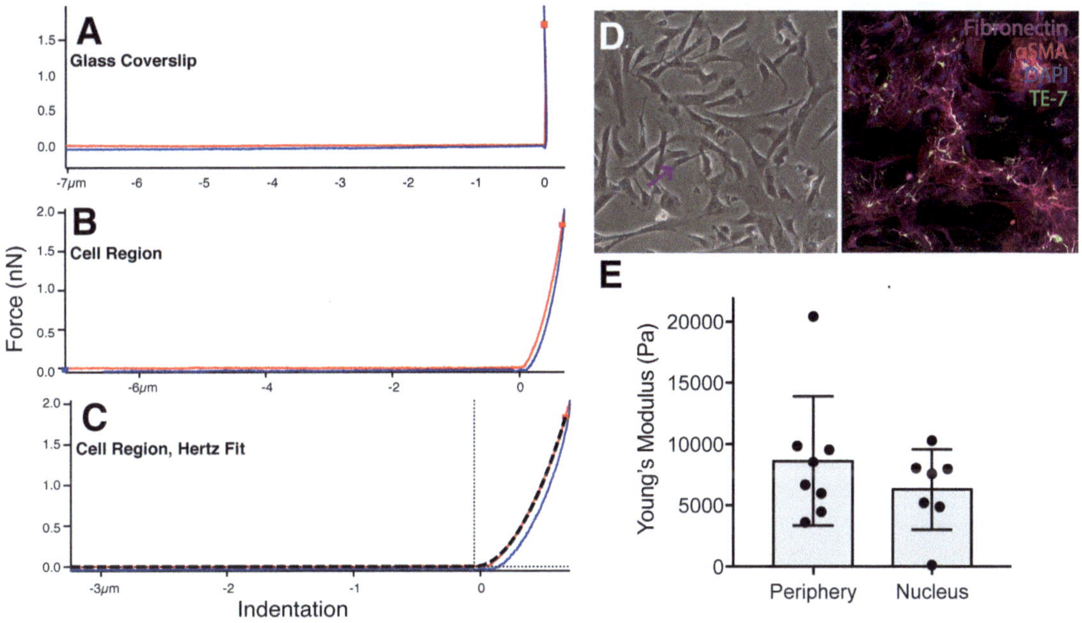

Fig. 2 *Measurement of surface stiffness.* Force-curve generated on (**a**) glass, (**b**) cellular region, and (**c**) Force curve generated Hertz fit for calculating Young's Modulus. (**d**) Cardiac fibroblasts cultured on glass coverslips, imaged live with phase contrast (left) and fixed, confocal (right) imaging. Right panel, cells are stained with DAPI (blue), Fibronectin (purple), alpha-SMA (red), TE-7 (green). The purple and blue arrows refer to the cell periphery and above nucleus, respectively, that upon AFM would generate force-curves as shown in panels **b–c** and cell stiffness measurements as in panel **e**. The black arrow refers to the glass substrate that when measured with AFM would generate curves as in panel **a**. (**e**) Example of cell stiffness measurements for cardiac fibroblast cellular regions. Each point represents average Young's Modulus for a 5 × 4 force map covering a 3–5 μm^2 area. $n = 8$ (cell periphery) and 7 (above nucleus) cells. For independent *t*-test, $p = 0.3316$

Table 1
Equations used for determining the Young's modulus

Determining Force on Cantilever: h = depth of the indentation E^* = the effective modulus of a system tip-sample	$F(h) = \frac{4\sqrt{R}}{3} E^* h^{3/2}$
Hertz equation (determination of E^*): E = Young's Modulus ν = Poisson ratio	$\frac{1}{E^*} = \frac{1-\nu_{tip}^2}{E_{tip}} + \frac{1-\nu_{sample}^2}{E_{sample}}$
Sneddon's variation of Hertz: (for cone tip AFM) α = the half-opening angle of the AFM tip	$F(h) = \frac{2}{\pi} \tan\alpha \frac{E_{sample}}{1-\nu_{sample}^2} h^2$

single indentations and extract stiffness measurements from the Asylum MFP3D software though we and others typically use more complicated analysis methods for complicated, multilayer materials.

2 Materials

2.1 Hardware

1. Asylum MFP-3D Atomic Force Microscope.
2. Epifluorescent Bulb with Blue Filter.
3. Fluorescent Bulb for Brightfield.
4. Asylum Live cell chamber.
5. Flat pointed forceps.

2.2 Disposables for General AFM

1. Vacuum Grease.
2. 25 mm glass coverslips.
3. Syringe (18 Gauge, Smaller gauge is better).
4. Kimwipes.
5. Plastic 5 mL eye droppers.
6. 1× Phosphate Buffered Saline Solution (PBS).
7. Clean, non-charged glass slides.
8. Optional: insect pins.
9. Optional: 0.1% Neutrad solution in DI Water.
10. Optional: 1% Bovine Serum Albumin in DI Water.

2.3 Disposables for AFM on Gels

1. Hydrogels, either bonded to methacrylated coverslips or poly-d-lysine treated coverslips, in 6-well plate and covered in PBS.

2.4 Disposables for AFM with Live Cells

1. Pre-warmed cell culture media.
2. 25 mm coverslips.

2.5 Software

1. Asylum Research 13, Igor Pro 6.34A (or above).

3 Methods

3.1 AFM General Setup

1. Turn on the AFM and brightfield bulb (set to ~25%), and start the Asylum Research software using Standard, Force, Force-ContactMode. Display the Live camera panel, Master panel, and Heater panel (if doing live-cell measurements).
2. In the Master panel, Main tab, set the integral gain to 0.5–1 and Set Point to 1 V. Set your saving path to automatically save force curves.
3. Under the Force tab, set the trigger to a force of 2nN (*see* **Note 1**). Select to keep the Approach Velocity constant at 2 μm/s and set the Force Distance to 6 μm. Before calibrating a new tip, under the Calibration tab, set the spring constant to 50 pN/nm and the Defl Invols to 50 nm/V.

4. In the Thermal tab, set the frequency center and width according to the cantilever being used (*see* **Note 2**); we will be using Nanoworld PNP-TR tips, which are set to 2 and 10 kHz, respectively.
5. Invols and spring constant will need to be experimentally determined for each tip being used, as these values should converge when alternating between thermal calibration (determines spring constant) and hitting glass (determines inVols) (*see* **Note 3**). Ensure the Spring Constant is "unlocked."
6. Set the Cantilever holder into the holder stand on a benchtop. Use forceps to grab the long edges of the tip and place on the Cantilever holder (*see* **Note 4**). Use a flat-tipped screwdriver to lift the metal overhang from the side and slide the cantilever underneath. Seat the tip in the center and remove the screwdriver, allowing the overhang to contact the tip. Gently tighten the center screw until the tip is flush against the overhang and Cantilever holder, then tighten each side screw, and finally re-tighten the center screw (*see* **Note 5**).
7. Return the Cantilever holder to the AFM head.
8. Use a 1.5 mL Pasteur pipette to add PBS (*see* **Note 6**) on the cantilever holder to pre-wet the tip. Start on one side of the cantilever and work around it, being careful not to add fluid directly on top of the cantilever or knock the cantilever with the Pasteur pipette (*see* **Note 7**). Aspirate PBS before continuing.
9. Seat a glass slide on the AFM stage, securing with magnets at the edges of the slide. Add PBS, or the solution being used with your samples, onto the glass slide to form a domed droplet with a height of approximately 5 mm (*see* **Note 8**).
10. Flip the AFM head onto the stage and lower the unit using the front dial until the surface tension of the bubble is broken (*see* **Note 9**). The cantilever should now be visible on the live video screen (*see* **Note 10**). Using the camera adjustment knobs, align the camera with the tip while lowering.
11. Move the laser onto the tip using the rear and right-facing dials, maximizing the sum value in the Sum and Deflection Meter (*see* **Note 11**).
12. Allow the deflection to stabilize, waiting approximately 5 min.
13. Lower the tip until just after the appearance of diffraction patterns in the live video, but so that the tip is still above the surface of the glass.
14. Set the deflection to 0, close the isolation doors, and click "Capture Thermal Data" (Master → Thermal → Capture Thermal Data).

15. Allow frequency spectrum to stabilize over at least 1 minute, then repetitively click "Fit Thermal Data" until the fitting is constant. Check that the obtained spring constant is consistent with your cantilever lot or literature values (here, we expect a value of 20–50 pN/nm).
16. Open the isolation doors, set the deflection to less than 0, and click "Engage" in the Sum and Deflection Meter. Ensure that your computer's sound is turned on and is loud enough to hear the force trigger noise.
17. Slowly lower the tip until the force trigger is reached (Z voltage 70–100, in Sum and Deflection Meter), and then press "Withdraw" on the Sum and Deflection Meter.
18. Under the Force tab, click "Single Force."
19. On the resulting displacement vs. deflection graph (Fig. 2a), zoom in on the top-most part of the curve (the linear region) by clicking and dragging a box around the region, right-click the end of the curve, and calculate InVols. Check the inVols are consistent with your tip (here, we expect values from 40 to 50 nm/V).
20. Repeat the thermal calibration and InVols measurement until the values remain relatively constant (*see* **Note 12**).
21. Using the front dial, raise the head to the maximum height before removing the head. The AFM is now calibrated and ready for samples to be mounted.

3.2 Hydrogel Analysis Using AFM

1. Synthesize gels according to established protocols [13–15] on coverslips. Gels should be in the calibration buffer at room temperature before measurement.
2. Take a fresh glass slide and add vacuum grease to the center of the slide with a Q-tip, using a small circular motion to spread the grease and cover an area approximately the size of the coverslip.
3. Use a bent syringe to pry the coverslips (and overlying gels) from the bottom of the well with one hand, grabbing the elevated edge with forceps with the other hand.
4. Place the gel-laden coverslip into the greased portion of the slide and push down along the edges of the gel to ensure level seating on the slide.
5. Place the slide on the AFM stage and secure with magnets.
6. Add buffer solution to the gel, creating a large, domed droplet.
7. Restore the AFM head on top of the stage and lower until the surface tension is broken.
8. Allow 5 minutes for the deflection to stabilize, then adjust to below 0 using the left dial.

9. Engage the tip and lower until the trigger is reached.
10. Make note of the front dial position at which the trigger is reached, then raise the head by one thumb-stroke to the left, followed by returning it to the original height by aligning the dial with the guide marks on the AFM head (*see* **Note 13**).
11. Press "single force" (*see* **Note 14**).
12. In the Master Force panel, within the Display tab, load the directory in which the curves are saved. Navigate to the Elastic tab and change the Poisson's Ratio to 0.5, and select the tip geometry and material according to your cantilever (*see* **Note 15**).
13. Due to intra-sample topological variances and biological heterogeneity, it is customary to compile a Force Map of several points in the $X–Y$ plane at each point of measurement. This also rapidly increases the rate of data acquisition versus individual points and allows an average stiffness to be calculated per region. To create a force map, go to the FMap tab on the Master Panel. Input the scan size and the amount of points in the x (points) and y (lines) directions. You can also alter the scan speed. Set an appropriate base name and saving path for the force map.
14. Select "Do Fmap."
15. To review the Force Map, select a single point in the Map view or the Force Plots list (Master Force Panel → Display), then move to the Entire Force Map tab (under the Elastic tab) and click Fit. You will observe a curve fit as shown in Fig. 2b, c.
16. To output the Young's Moduli for the entire force map, click the Analyze (A) button in the Map view, then under the histogram tab of the Analyze panel, click "Make New Histogram," and then edit. Individual values can then be pasted into spreadsheets for downstream analysis (Fig. 2e).
17. To obtain additional measurements, raise the AFM head using the front dial before moving the stage so the tip does not break by tugging on raised regions, then repeat **steps 9–16** in Subheading 3.2.
18. When finished, remove the AFM head from the stage, flip it over, and set on the rest. If planning to use the tip for the next experiment, rinse with DI water. Otherwise, dispose of the tip or replace in the tip box, making note of the date it was used.

3.3 Live Cell Analysis Using AFM

1. Culture cells of interest on sterilized glass coverslips.
2. Transfer 50 mL of growth media or calibration buffer to a mL conical tube and heat to 37 °C and remove from the incubator just before mounting cells.

3. Transfer cell-seeded coverslip into the live-cell chamber, using ample vacuum grease and ensuring a water-tight seal between the glass and chamber.
4. Add just enough warmed growth media to reach the heater ring on the live-cell chamber and secure the chamber on the AFM stand with the electrical cable resting to the right side of the AFM stand.
5. Place the AFM head over the live-cell chamber and lower until the silicone forms a seal on the live-cell chamber (*see* **Note 16**), and turn on the feedback and Target Temp in the heater panel.
6. Lower the cantilever until the surface tension is broken (*see* **Note 17**).
7. Wait for the temperature to reach the Target Temp and for the deflection to stabilize, usually a minimum of 5 min.
8. Conduct force measurements as described in the previous section (*see* **Note 18**).
9. Turn off the heater and set point before removing the AFM head (*see* **Note 19**).

4 Notes

1. Soft gels may require lower indentation forces, gel may get deposited on tip and result in biphasic response. Test tips by indenting glass, and switch as necessary. To clean organic material from a cantilever, submerge in 0.1% Neutrad solution. Several rinses may be required.
2. The frequency will be significantly less than what is listed on the packaging due to the tip being submerged in liquid [16, 17].
3. As mentioned above, the spring constant in water-based buffers will also be less than the value printed on the packaging (approximately a third in our experience, but dependent on buffer used).
4. It may be helpful to check the integrity of the cantilever using a magnifying glass and ensure that tips were not damaged or broken in transit.
5. Only tighten until finger-tight, over-tightening can result in snapping the cantilever—this renders measured values inaccurate and can destroy the opposite end of the cantilever.
6. PBS can be replaced with whichever buffer the hydrogel or cells are in.
7. Dropping fluid directly on the cantilever can result in bent or broken tips—either preventing data collection or providing inaccurate measurements. The entire cantilever should be

submerged before proceeding, which may require pipetting off PBS and reapplying until the surface is covered. Adding too much PBS at once may cause it to flow over the side and damage the head of the AFM. If small air bubbles appear under the cantilever, Insect pins can be used to scrape the bubbles toward the edge of the platform where they are easier to burst.

8. If performing live-cell AFM, calibration should be done on a glass coverslip in the live-cell incubator with the buffer heated to the desired temperature.
9. This should be readily apparent when looking at the front of the AFM—if you are unsure if you have broken the surface tension, check at the live video feed. The cantilever should be in focus once it is submerged.
10. The AFM head should be level at the Z-height of the sample. If the head is not level, the angle of approach of the cantilever will not be perpendicular to the surface, resulting in erroneous measurements.
11. For PNP-TR tips, sum on the long cantilever should be 8+, and 4–5 for the short cantilever.
12. After identifying standard values for your tips of choice, we typically repeat the thermal and glass force measurement three times before proceeding. For PNP-TR tips, acceptable values are typically $k = 20$–50 pN/nm and InVols = 40–50 nm/V.
13. This is important for soft gels and somewhat "stickier" biological materials that cause adhesion with the tip. This will ensure that the tip is no longer in contact with the sample when the force measurement is taken, resulting in a horizontally flat approach curve.
14. A smaller indentation distance will provide more accurate and quicker measurements, though a larger distance is recommended for force maps to minimize the amount of failed trigger points.
15. PNP-TR tips are pyramidal silicon nitride, though generally the tip material and half angle minimally affect the calculated Young's Modulus. Biphasic curves usually indicate that you have indented through the material and are likely hitting the glass. To prevent excessive adhesion, sample can be submerged in a 1% BSA solution for 5 minutes.
16. The x-y directions of the AFM stage may need to be adjusted to seat the silicone in the live-cell chamber. If the silicone is not sealed well, the media or buffer will evaporate more rapidly and may result in an oblique cantilever approach, providing erroneous results.

17. This is best visualized using the live video feed, as the silicone ring and chamber will impede direct observation.
18. Force map areas and lateral movement of the AFM stage should be minimized to preserve the silicone gasket seal. Cells may contract, peel off the substrate, and die after several indentations.
19. Failure to do this can result in the heating element overheating and burning sugars from the media. When using a single sample for a long time, additional media may be required to compensate for the evaporated volume. If you smell a sweet bread-like aroma, the sample may have run dry, causing remaining media on the heating element to caramelize the sugars.

References

1. Engler AJ, Sen S, Sweeney HL, Discher DE (2006) Matrix elasticity directs stem cell lineage specification. Cell 126:677–689
2. Lieber SC, Aubry N, Pain J, Diaz G, Kim SJ, Vatner SF (2004) Aging increases stiffness of cardiac myocytes measured by atomic force microscopy nanoindentation. Am J Physiol Heart Circ Physiol 287:H645–H651
3. Jin H, Xing X, Zhao H, Chen Y, Huang X, Ma S, Ye H, Cai J (2010) Detection of erythrocytes influenced by aging and type 2 diabetes using atomic force microscope. Biochem Biophys Res Commun 391:1698–1702
4. Starodubtseva MN (2011) Mechanical properties of cells and ageing. Ageing Res Rev 10:16–25
5. Suresh S (2007) Biomechanics and biophysics of cancer cells. Acta Biomater 3:413–438
6. Hertz H (1881) Ueber den kontakt elastischer koerper. J fuer die Reine Angewandte Mathematik 92:156
7. Johnson KL (1985) Contact mechanics. Cambridge University Press, Cambridge
8. Laurent VM, Kasas S, Yersin A, Schaffer TE, Catsicas S, Dietler G, Verkhovsky AB, Meister JJ (2005) Gradient of rigidity in the lamellipodia of migrating cells revealed by atomic force microscopy. Biophys J 89:667–675
9. Ohashi T, Ishii Y, Ishikawa Y, Matsumoto T, Sato M (2002) Experimental and numerical analyses of local mechanical properties measured by atomic force microscopy for sheared endothelial cells. Biomed Mater Eng 12:319–327
10. Kuznetsova TG, Starodubtseva MN, Yegorenkov NI, Chizhik SA, Zhdanov RI (2007) Atomic force microscopy probing of cell elasticity. Micron 38:824–833
11. Efremov YM, Wang WH, Hardy SD, Geahlen RL, Raman A (2017) Measuring nanoscale viscoelastic parameters of cells directly from AFM force-displacement curves. Sci Rep 7:1541
12. Thomas G, Burnham NA, Camesano TA, Wen Q (2013) Measuring the mechanical properties of living cells using atomic force microscopy. J Vis Exp (76):50497
13. Ondeck MG, Engler AJ (2016) Mechanical characterization of a dynamic and tunable methacrylated hyaluronic acid hydrogel. J Biomech Eng 138:0210031–0210031
14. Tse JR, Engler AJ (2010) Preparation of hydrogel substrates with tunable mechanical properties. Curr Protoc Cell Biol 47:10.16.11–10.16.16. John Wiley & Sons, Ltd.
15. Benton JA, Deforest CA, Vivekanandan V, Anseth KS (2009) Photocrosslinking of gelatin macromers to synthesize porous hydrogels that promote valvular interstitial cell function. Tissue Eng A 15:3221–3230
16. Vančura C, Dufour I, Heinrich SM, Josse F, Hierlemann A (2008) Analysis of resonating microcantilevers operating in a viscous liquid environment. Sensors Actuators A Phys 141:43–51
17. Habibnejad Korayem M, Jiryaei Sharahi H, Habibnejad Korayem A (2012) Comparison of frequency response of atomic force microscopy cantilevers under tip-sample interaction in air and liquids. Sci Iran 19:106–112

Chapter 17

Method for Investigating Fibroblast Durotaxis

Hossam Kadry, David Lagares, and Taslim A. Al-Hilal

Abstract

Durotaxis is the phenomena of directed cell migration driven by gradients of extracellular matrix stiffness. Durotaxis has been recently involved in the development of fibrosis by promoting the recruitment of pathological fibroblasts to areas of established fibrosis, thus amplifying the fibrotic response. Here, we describe the fabrication of mechanically patterned hydrogels that can be used to investigate molecular mechanisms controlling durotaxis of fibroblasts and other cells with mechanosensing properties. This method effectively creates a stiffness gradient of 275 Pa/μm, mimicking the natural spatial stiffness variations we observed in fibrotic tissues from mouse models of fibrosis and human fibrotic diseases.

Key words Durotaxis, Hydrogel, Fibrosis, Myofibroblasts, ECM

1 Introduction

Directed cell migration has critical effects on tissue development, homeostasis, and disease progression [1]. Directed migration requires continuous sampling and analyzing environmental cues, followed by processing the results to produce directional cell movement [2]. Cell migration can be driven by gradients of diffusible soluble chemicals (chemotaxis), substrate-bound biomolecules (haptotaxis), or physical cues such as electric fields or extracellular matrix stiffness (ECM) [3]. The directed migration of cells from regions of lower-to-higher stiffness that occurs in the absence of chemoattractant gradients is known as durotaxis, which has been recently involved in cancer progression and metastasis [4], atherosclerosis [5], liver fibrosis [6], and innate immunity as well as nervous system development [7]. Cell durotaxis is triggered by spatial gradients of matrix stiffness, which are formed by localized cross-linking of ECM proteins including collagen and elastin [8–10]. In most fibrotic diseases, collagen deposition and cross-linking result in increased matrix stiffness [11]. For instance, healthy human lung tissue possesses a Young's modulus of 1.96 ± 0.13 kPa, while diseased lung tissue from patients with

Boris Hinz and David Lagares (eds.), *Myofibroblasts: Methods and Protocols*, Methods in Molecular Biology, vol. 2299,
https://doi.org/10.1007/978-1-0716-1382-5_17,

idiopathic pulmonary fibrosis exhibits increased Young's modulus of 16.52 ± 2.25 kPa, with peaks up to 50 kPa. In addition, our recent studies show that fibrotic tissues display striking differences in the spatial distribution of matrix stiffness compared to healthy tissues. While healthy organs are characterized by uniform homogenous distribution of matrix stiffness, fibrotic tissues are characterized by focal "peaks" and "valleys" of matrix stiffness, which rise and fall between spatial gradients of stiffness. The slope of these stiffness gradients (average slope for lung fibrosis: 500 Pa/μm, skin fibrosis: 433 Pa/μm and kidney fibrosis 115 Pa/μm) is consistent with the notion that cell durotaxis may be implicated in the development of organ fibrosis [12, 13]. In this regard, our laboratory has recently shown that fibroblast durotaxis is a major disease mechanism in the pathogenesis of lung fibrosis in which stiffness gradients promote the recruitment of fibroblasts to stiff fibrotic tissues, thus amplifying the fibrotic response. Our data demonstrate that targeting mechanosensing pathways abrogates fibroblast durotaxis in vitro and mitigates lung fibrosis in vivo in preclinical models.

The study of cell durotaxis in fibrotic diseases and the molecular pathways controlling matrix rigidity sensing has been hindered by the lack of methods to assess stiffness-directed migration in vitro. Here, we describe the fabrication of mechanically patterned hydrogels that can be used to study durotaxis of fibroblasts or other cells with mechanosensing properties in fibrosis. The system is based on micropatterned hydrogels that aligns 100-μm-wide "stiff" stripes interspersed with 200-μm-wide "soft" stripes corresponding to an average elastic modulus of 40 and 4 kPa, respectively [adapted from 14]. This method effectively creates a stiffness gradient of 275 Pa/μ m, mimicking the natural spatial stiffness variations we observed in vivo in fibrotic tissues.

2 Materials

Prepare all solutions using ultrapure water (double-distilled deionized water to attain a sensitivity of 18 MΩ-cm at 25 °C) and analytical grade reagents. Prepare and store all reagents at room temperature unless otherwise indicated. Please follow all waste disposal regulations when disposing waste materials.

2.1 Reagents

1. Dimethyldichlorosilane (DMDCS).
2. Absolute ethanol.
3. Glacial acetic acid.
4. Methyl methacrylate.
5. Glass coverslip 25 mm.

6. Glass coverslip 18 mm.
7. 2% bis-acrylamide.
8. 40% Acrylamide.
9. 2-hydroxy-4′-(2-hydroxyethoxy)-2-methylpropiophenone (Irgacure®).
10. Phosphate-buffered saline (PBS): 0.138 M NaCl, 0.0027 M KCl, pH 7.4.
11. Sulfo-SANPAH.
12. HEPES buffer (pH 8.0).
13. IrgaCure®.
14. Fibronectin.
15. Rat tail collagen Type I (rColT1).
16. Dulbecco's modified Eagle medium (DMEM).
17. Fetal bovine serum (FBS), aliquoted and stored at −20 °C.
18. Paraformaldehyde (PFA) solution 4% in PBS, stored at 4 °C.
19. Human primary lung fibroblasts.
20. Antifade mounting media with DAPI.
21. Florescent Phalloidin for F-Actin Stain.

2.2 Deactivating the Surface of Coverslip (25 mm)

Dilute 25 mL of dimethyldichlorosilane (DMDCS) in the bottle to 500 mL with toluene, yielding a 5% solution of DMDCS (*see* **Note 1**). Then soak the coverslip in 5% DMDCS solution for 15 min, rinse the coverslip twice with toluene, rinse again the coverslip twice in methanol for 5 min each, and finally dry the coverslip with high-purity nitrogen (*see* **Note 2**).

2.3 Preparation of 1:10 (v/v) Acetic Acid Solution

Add 1 mL glacial acetic acid to 9 mL of deionized water in a 15 mL conical tube and vortex to mix well (*see* **Note 3**).

2.4 Preparation of Methacrylate Solution

Add 20 mL of absolute ethanol, 600 μL of 1:10 (v/v) acetic acid solution, and 100 μL of methacrylate in a 50 mL conical tube and vortex to mix well (*see* **Notes 3** and **4**).

2.5 Preparation of 5% IrgaCure® Solution

Take a clean spatula and weigh 5 mg IrgaCure® in a 1.5 mL Eppendorf tube. Add 100 μL of absolute ethanol and vortex well until all the granules disappear (*see* **Note 3**).

2.6 Preparation of High-Resolution Photomask

A design consisting of 200 μm black stripes separated by 100 μm transparent stripes was prepared using AutoCAD (Autodesk) and the patterns were transferred to chrome covered quartz transparency mask using high-resolution printing.

2.7 Preparation of Sulfo-SANPAH Solution

Prepare a 1 mg/mL working solution (1×) of sulfo-SANPAH in HEPES buffer (pH 8.0). Aliquot and keep it in the freezer until use. The sulfo-SANPAH must be thawed at 37 °C before use.

2.8 Fibronectin/Collagen Coating Solution

Prepare a 10 μg/mL of bovine fibronectin plus 40 μg/mL of Rat tail collagen Type 1 (rColT1) solution in sterile PBS from stock on a daily basis. To make a 10 mL of Fibronectin/Collagen coating solution, add 100 μL of fibronectin and 100 μL of rColT1 (assuming that fibronectin and rColT1 stock concentration is 1 and 4 mg/mL, respectively) with 9.8 mL of sterile PBS (*see* **Note 3**).

2.9 Equipment

1. Plastic square petri dishes, sterile.
2. 15 mL Conical sterile polypropylene centrifuge tubes.
3. 50 mL Conical sterile polypropylene centrifuge tubes.
4. Pipettes for volumes 10–1000 μL.
5. Pipettes for volumes 10–100 μL.
6. UVP transilluminator.
7. Dry bath.
8. Surgical razor.
9. Forceps.
10. Glass staining jar.
11. Centrifuge for cell culture.
12. Laminar flow hood.
13. CO_2 gas incubator.
14. Cell culture petri dishes, sterile.
15. Cell counting chamber Neubauer hemocytometer.
16. Microscope.

3 Methods

3.1 Design and Fabrication of Mechanically Patterned Soft-Stiff Hydrogel Onto a Coverslip

3.1.1 Activation of Glass Coverslips (18 mm)

1. Expose the glass coverslips (18 mm) to UV light using a UVP Transilluminator for 5 min on each side (*see* **Note 5**).
2. Pour the methacrylate solution onto a glass staining dish and soak the UV-treated coverslips for 5 min (*see* **Note 6**).
3. After 5 min, remove the coverslips carefully using a tweezer and rinse twice using absolute ethanol (*see* **Note 7**).
4. Dry the activated coverslips using high-purity nitrogen and keep onto kimwipes until further use (*see* **Note 8**).

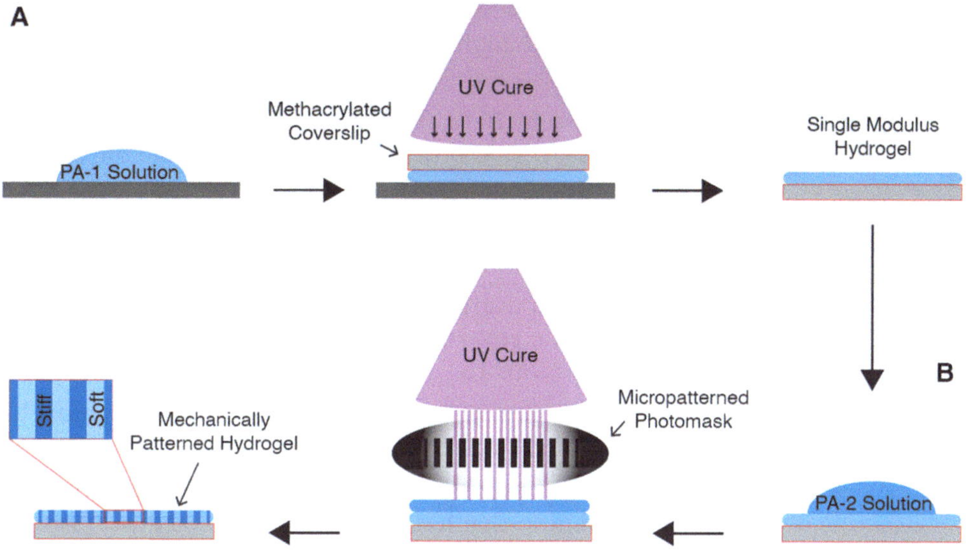

Fig. 1 Fabrication of mechanically patterned hydrogel to analyze the durotactic property of fibroblasts. (**a**) The exposure of the hydrogel to UV light creates a soft layer onto a coverslip. (**b**) The second exposure of UV light that includes a photomask creates the soft-stiff patterned hydrogel

3.1.2 Fabrication of the Soft Layer

1. Mix 200 μL of 2% bis-acrylamide (0.4%), 100 μL of 40% acrylamide (4%), and 700 μL of PBS to prepare 1 mL of pre-polymer solution 1 (PA-1) (*see* **Note 3**).
2. Mix 1 μL of 5% IrgaCure® solution with 100 μL PA-1 solution.
3. Place the DMDCS-treated coverslips (25 mm coverslip) onto the working surface area of UVP Transilluminator.
4. Add an aliquot of PA-1 solution (13–14 μL per 18-mm coverslip) onto a DMDCS-treated coverslip (Fig. 1a).
5. Place the methacrylated 18-mm coverslip onto the drop and allow the PA1 solution to spread evenly between two coverslips (*see* **Note 9**).
6. UV expose the coverslips for 5 min to polymerize the solution (*see* **Note 5**).
7. Remove the PA1-coated glass coverslip from the DMDCS-treated coverslip using the edge of a razor blade.
8. Dehydrate the PA1 gel for an hour onto a hotplate at low heat (~30 °C) (*see* **Note 10**).

3.1.3 Fabrication of the Stiff Layer on Top of the Soft Layer

1. Mix 200 μL of 2% bis-acrylamide (0.4%), 120 μL of 40% acrylamide (4%), and 680 μL of PBS to prepare 1 mL of pre-polymer solution 2 (PA-2) (*see* **Note 3**).
2. Mix 1 μL of 5% IrgaCure® solution with 100 μL PA-2 solution.
3. Place the high-resolution photomask onto the working surface area of UVP transilluminator (*see* **Note 11**).

4. Add an aliquot of PA-2 solution (12–13 μL per 18-mm coverslip) onto the high-resolution photomask with stripe patterns (Fig. 1b).
5. Place the dehydrated PA1-gel-coated coverslips onto the PA-2 solution, gel side down, taking care to avoid air bubbles (*see* **Note 12**).
6. UV expose the solution for 5 min to polymerize the PA-2 solution (*see* **Note 5**).
7. After UV exposure, add 1 mL of PBS on top of the coverslip and peel the coverslip off from the photomask using a razor blade (*see* **Note 13**).
8. After detaching, immediately place the coverslips into a 12-well dish and add 1 mL of sterile PBS.
9. Wash the gel three times using sterile PBS to remove any excess PA-2 solution that has not been polymerized (*see* **Note 14**).
10. Observe under the microscope to make sure that the stripes are intact and not distorted.
11. Store the coverslips at 4 °C until seeding in an air-tight bag to avoid dryness (*see* **Note 15**).

3.2 Seeding of Lung Fibroblasts in Mechanically Patterned Hydrogel

3.2.1 Surface Activation and Coating of the Soft-Stiff Patterned Hydrogel Surface

1. Remove the 12-well plate containing patterned hydrogel from refrigerator and thaw it in room temperature for 20–30 min.
2. Wash each well two times using sterile PBS.
3. Add 1.0 mL of thawed 1x sulfo-SANPAH working solution to each well containing patterned hydrogel.
4. UV expose the plate for 15 min (*see* **Note 16**).
5. Wash three times with sterile PBS, aspirate, and add 1 mL of sterile PBS (*see* **Note 17**).
6. Remove the coverslip and put into a new well of 12-well dish (*see* **Note 18**).
7. Wash once with sterile PBS and aspirate (*see* **Note 17**).
8. Add 1.5 mL of Fibronectin/Collagen coating solution to each well of the patterned hydrogel.
9. Keep the plate overnight in a cell incubator at 37 °C (*see* **Note 19**).
10. Next day, seed the cells at an appropriate density.

3.2.2 Seeding of Lung Fibroblasts Onto the Patterned Hydrogel

1. Calculate the required number of cells, based on the number of experimental groups used in the specific experiment.
2. Aspirate fibroblast growth medium of a 100 mm fibroblast culture plate of 70–80% confluency.
3. Wash with sterile PBS and aspirate.
4. Repeat once.

5. Add 1 mL trypsin/EDTA solution and incubate at 37 °C for 3 min.
6. Stop the reaction by adding 3 mL PBS containing 10% FBS (v/v) on the cells.
7. Pipette the cells into a 15 mL centrifuge tube and centrifuge them for 5 min at 150 × *g*, RT.
8. Discard the supernatant and resuspend the cells in 2–3 mL of growth medium.
9. Count the cells. This step can be performed through an automated cell counter or manually through a cell counting chamber.
10. Take the calculated number of cells in a 50 mL centrifuge tube and spin them, followed by resuspending them by adding growth medium to prepare a final suspension of 5000 or 10,000 cells per mL.
11. Add 1.0 mL of cells suspension to each well of patterned hydrogel, then gently shake 12-well plate in order to disperse cells homogeneously (*see* **Note 20**).
12. Under a light microscope, confirm that fibroblasts are equally distributed on both soft and stiff stripes (*see* **Note 21**).
13. Culture cells for 4, 8, 12, or 24 h.
14. After growing the cells, remove the media very carefully and wash each well two times with sterile PBS (*see* **Note 22**).
15. Fix cells by adding 1 mL of 4% PFA to each well for 20 min at room temperature. Wash out 4% PFA, leave fixed cells on 1 mL of PBS (*see* **Note 22**).

3.3 Assessment of Durotactic Index

1. Using a light microscope, take bright-field images of fixed fibroblasts on micropatterned gels, then determine their "durotactic index." This is defined by the number of cells accumulated on stiff over soft stripes. For instance, if the durotactic index is 1, it indicates there was one cell on stiff stripes per cell found on soft stripes, meaning 50% distribution and therefore no durotactic capacity (Fig. 2). However, if the durotactic index is 5, it means there were five cells on stiff stripes per cell found on soft stripes, indicating durotactic capacity of the cell line under investigation (Fig. 2). Typical durotactic index of fibroblast lines ranges from 4 to 8.
2. Alternatively, fibroblasts can be stained with phalloidin and mounted with antifade media with DAPI for studying cell morphology of durotaxing cells. Staining can be performed from 4 h after fixation to a few days later. The samples should be kept at 4 °C and liquid should be regularly added to avoid dryness.

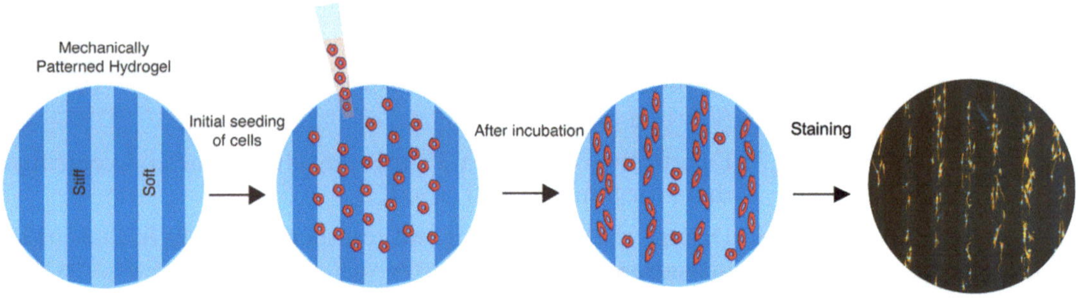

Fig. 2 Durotaxis assay. Upon homogeneous seeding of fibroblasts onto mechanically patterned, mesenchymal cells durotax from soft to stiff stripes at 24 h after plating. Immunostaining with DAPI and phalloidin enables assessment of durotactic index and study cell morphology

3.4 Mechanistic Studies Investigating Molecular Control of Durotaxis

1. Our platform is suitable for mechanistic studies aimed at investigating mechanosensing and mechanotransduction pathways involved in cell durotaxis. Gain- and loss-of-function studies using pharmacological or genetic tools can be easily incorporated to assess their effect of fibroblast's durotactic index.
2. Immunofluorescence on fixed cells and live cell imaging can be utilized to visualize molecular pathways on durotaxing cells.
3. Since stiff matrices have been shown to promote cell proliferation, inhibition of cell cycle is advised if working with fast-cycling cells. We recommend treating cells with mitomycin C at 10 mg/mL for 3 hours and rinsed three times with media before plating.

4 Notes

1. This solution is flammable, and moisture will reduce its effectiveness. Store in brown glass at room temperature.
2. We find that a deactivated glass slide or coverslip can be used up to 15–20 times before re-silanizing them repeating the same procedure.
3. The solution can be stored at 4 °C up to months but it is preferred to prepare it on a daily basis.
4. To ensure the maximum activity of the solution, we find that it is best to add the required amount of methacrylate before using it.
5. Use high UV exposure.
6. Do not allow the coverslips to stick to each other.
7. Prepare two 50 mL beakers each containing 40 mL of absolute ethanol for rinsing.

8. Do not keep the coverslips on top of each other.
9. Caution must be taken to avoid the trapping of bubbles inside.
10. Make sure that the temperature is set at 30 °C and cover the hot plate with aluminum foil to avoid the infiltration of dust particles.
11. Make sure the ink/printed side (black stripes) is facing up.
12. Remember to place the PA-1-coated gel side, not the uncoated glass side, onto the solution. Tap it slowly to make sure that PA-2 solution spread all over the coverslip; be very careful to avoid big air bubbles.
13. Peel off the cover slip parallel to the stripes of the photomask to avoid the damage to the stiff layer.
14. Clean the photomask using sterile PBS and soak the liquid using kimwipes.
15. The mechanically patterned hydrogels can be stored at 4 °C up to 3 months.
16. The plates can be exposed to UV light under the cell culture hood until the solution of the color turns into blackish from red.
17. Do not touch the surface of the hydrogel during aspiration.
18. Use a 200 μL pipette tip to peel off the coverslip from the surface of wells and then hold the edge of the coverslip using forceps.
19. We may also coat it for ~4 h. We can also use other protein of choice, such as laminin.
20. Make sure to disperse the cells homogeneously by shaking the plates gently. Check under the microscope whether the cells are dispersed well; if necessary, gently shake more.
21. Initially, the fibroblasts should attach all over the patterned hydrogels, but as time goes on fibroblasts will durotax and move to the stiff pattern.
22. Do not aspirate the media. Remove and wash with the media using a micropipette very slowly.

Acknowledgments

D.L. gratefully acknowledges funding support from the NIH (grant R01 HL147059-01), the Start-up Package from Massachusetts General Hospital, the Scleroderma Foundation New Investigator Grant, the Scleroderma Research Foundation Investigator-Initiated Research Grant, the American Thoracic Society Foundation/Pulmonary Fibrosis Foundation Research Grant and Sponsored Research Grants from Boehringer Ingelheim, Unity

Biotechnology and Indalo Therapeutics. Dr. Lagares has a financial interest in Mediar Therapeutics. The company is developing treatments for organ fibrosis. Dr. Lagares's interests were reviewed and are managed by MGH and Partners HealthCare in accordance with their conflict of interest policies.

References

1. Handorf AM, Zhou Y, Halanski MA, Li WJ (2015) Tissue stiffness dictates development, homeostasis, and disease progression. Organogenesis 11(1):1–15. https://doi.org/10.1080/15476278.2015.1019687
2. Janmey PA, McCulloch CA (2007) Cell mechanics: integrating cell responses to mechanical stimuli. Annu Rev Biomed Eng 9:1–34. https://doi.org/10.1146/annurev.bioeng.9.060906.151927
3. Petrie RJ, Doyle AD, Yamada KM (2009) Random versus directionally persistent cell migration. Nat Rev Mol Cell Biol 10(8):538–549. https://doi.org/10.1038/nrm2729
4. Ulrich TA, de Juan Pardo EM, Kumar S (2009) The mechanical rigidity of the extracellular matrix regulates the structure, motility, and proliferation of glioma cells. Cancer Res 69(10):4167–4174. https://doi.org/10.1158/0008-5472.CAN-08-4859
5. Isenberg BC, Dimilla PA, Walker M, Kim S, Wong JY (2009) Vascular smooth muscle cell durotaxis depends on substrate stiffness gradient strength. Biophys J 97(5):1313–1322. https://doi.org/10.1016/j.bpj.2009.06.021
6. Georges PC, Hui JJ, Gombos Z, McCormick ME, Wang AY, Uemura M, Mick R, Janmey PA, Furth EE, Wells RG (2007) Increased stiffness of the rat liver precedes matrix deposition: implications for fibrosis. Am J Physiol Gastrointest Liver Physiol 293(6):G1147–G1154. https://doi.org/10.1152/ajpgi.00032.2007
7. Mandeville JT, Lawson MA, Maxfield FR (1997) Dynamic imaging of neutrophil migration in three dimensions: mechanical interactions between cells and matrix. J Leukoc Biol 61(2):188–200. https://doi.org/10.1002/jlb.61.2.188
8. Gosline J, Lillie M, Carrington E, Guerette P, Ortlepp C, Savage K (2002) Elastic proteins: biological roles and mechanical properties. Philos Trans R Soc Lond Ser B Biol Sci 357(1418):121–132. https://doi.org/10.1098/rstb.2001.1022
9. Yue B (2014) Biology of the extracellular matrix: an overview. J Glaucoma 23(8 Suppl 1):S20–S23. https://doi.org/10.1097/IJG.0000000000000108
10. Sherratt MJ (2009) Tissue elasticity and the ageing elastic fibre. Age (Dordr) 31(4):305–325. https://doi.org/10.1007/s11357-009-9103-6
11. Fenner J, Stacer AC, Winterroth F, Johnson TD, Luker KE, Luker GD (2014) Macroscopic stiffness of breast tumors predicts metastasis. Sci Rep 4:5512. https://doi.org/10.1038/srep05512
12. Liu F, Mih JD, Shea BS, Kho AT, Sharif AS, Tager AM, Tschumperlin DJ (2010) Feedback amplification of fibrosis through matrix stiffening and COX-2 suppression. J Cell Biol 190(4):693–706. https://doi.org/10.1083/jcb.201004082
13. Bülow RD, Boor P (2019) Extracellular matrix in kidney fibrosis: more than just a scaffold. J Histochem Cytochem 67(9):643–661. https://doi.org/10.1369/0022155419849388
14. Happe CL, Tenerelli KP, Gromova AK, Kolb F, Engler AJ (2017) Mechanically patterned neuromuscular junctions-in-a-dish have improved functional maturation. Mol Biol Cell 28(14):1950–1958. https://doi.org/10.1091/mbc.E17-01-0046

Chapter 18

Decellularized Extracellular Matrix (ECM) as a Model to Study Fibrotic ECM Mechanobiology

Chiuan-Ren Yeh, Grace C. Bingham, Jagathpala Shetty, Ping Hu, and Thomas H. Barker

Abstract

Aberrant deposition of the extracellular matrix (ECM) causes fibrosis and leads to ECM stiffening. This fibrotic ECM provides biological and biophysical stimulations to alter cell activity and drive progression of fibrosis. As an emerging discipline, mechanobiology aims to access the impact of both these cues on cell behavior and relates the reciprocity of mechanical and biological interactions; it incorporates concepts from different fields, like biology and physics, to help study the mechanical and biological facets of fibrosis extensively. A useful experimental platform in mechanobiology is decellularized ECM (dECM), which mimics the native microenvironment more accurately than standard 2D culture techniques as its composition includes similar ECM protein components and stiffness. dECM, therefore, generates more reliable results that better recapitulate in vivo fibrosis.

Key words Decellularized extracellular matrix (dECM), Fibrosis, Bleomycin, Atomic Force Microscopy (AFM), Mechanobiology, Myofibroblasts

1 Introduction

The majority of the extracellular matrix (ECM) is composed of cell-secreted molecules, such as fibrous proteins and glycosaminoglycans (GAGs), that offer biochemical and structural support to cells. The fibrillar components, including collagens, elastin, tenascins, fibronectin, and laminin, interconnect to form a complex three-dimensional network that provides structure and determines the mechanical properties of the ECM [1]. There are four primary types of GAGs, including hyaluronic acid, keratan sulfate, chondroitin sulfate, and heparan sulfate, which can form proteoglycans when assembling with core proteins. These proteoglycans are negatively charged, attracting water, which enables it to act as a lubricant and keeps tissues hydrated [2]. Meanwhile, proteoglycan can

Boris Hinz and David Lagares (eds.), *Myofibroblasts: Methods and Protocols*, Methods in Molecular Biology, vol. 2299, https://doi.org/10.1007/978-1-0716-1382-5_18,

also interact with other matrix components, growth factors, and cell surface receptors to elicit cell signaling and biological processes that alter cell activities, matrix organization and mediate disease progression. Cells are capable of sensing both ECM components and rigidity through integrin-mediated interactions with the matrix [3].

Integrins are transmembrane proteinsthat serve as the primary ECM receptors responsible for transducing mechanical inputs from the ECM into intracellular biochemical signals. Furthermore, which cellular pathways are altered is dependent upon the abundance of binding sites available on specific ECM proteins (i.e., the composition of ECM), allowing ECM to influence migration, proliferation, differentiation, inflammation, and apoptosis [4, 5]. For example, stiff matrix drives fibroblast to myofibroblast differentiation and the presence of specialized matrix proteins, such as the splice variant of fibronectin (extra domain A), drive the differentiation of fibroblasts to myofibroblasts [6].

Fibrosis is characterized by aberrant deposition of ECM proteins such as collagens, fibronectin, glycoproteins, and proteoglycans. Excessive secretion of these ECM proteins by activated fibroblasts contribute to disease progression by providing both biochemical and biomechanical signaling cues that alter cell activity and increase tissue stiffness, which in turn reduces organ function [7]. For example, idiopathic pulmonary fibrosis (IPF) is characterized by excessive assembly of ECM in the lung interstitium by activated fibroblasts, in a dysfunctional form of wound healing after tissue damage. After micro-injuries cause epithelial cells to undergo apoptosis, vascular permeability is increased for proteins (e.g., fibrinogen and fibronectin) that cause the formation of a provisional matrix (wound clot). In this progressive abnormal lung remodeling, myofibroblasts secrete excessive amounts of extracellular matrix proteins, mainly fibrillar collagens that change the ECM stiffness. Booth et al. confirmed that the stiffness of the normal human lung and IPF tissue was 1.96 ± 0.13 kPa and 16.52 ± 2.25 kPa by atomic force microscopy (AFM), respectively [8]. Also, they showed the difference in ECM protein composition between healthy and IPF lung; several proteins, like tenascin, vitronectin, collagens I, V, and XV, were only expressed in IPF but not normal lungs [8]. ECM protein variants, such as the splice variant of fibronectin (extra domain A), also changes cell activity and accelerates fibrosis progression [6].

In the fibrotic ECM, the result of cross-talking of ECM component proteins, rigidity, and cells drives disease progression, but the mechanism is still poorly understood. Mechanobiology is an emerging discipline that relates to the reciprocity of mechanical and biological interactions. However, it more precisely considers the effects of mechanical forces on cellular processes. It provides the concepts to approach the study of fibrosis from a physical, biological, and material perspective that mimics the real microenvironment. The most commonly used in vitro method for biological studies is 2D tissue culture on a plastic culture dish. However, tissue culture plastic possesses Young's moduli ranging from 2 to 4 GPa [9], suggesting that cells will experience a much stiffer surface when cultured on plastic dishes in vitro than their physiological origin. Hydrogels can recapitulate physiologically relevant stiffness and add an additional degree of complexity by coating them with a protein of interest such as collagen. Although these techniques and others help investigate specific mechanobiology-driven hypotheses, it is in an isolated manner, and these techniques do not capture the full complexity of in vivo ECM. For example, Fiore et al. demonstrated that fibronectin rich, soft matrices promote fibroblast to myofibroblast differentiation similarly to stiff matrices, whereas soft, provisional matrix rich in type 1 collagen could not induce myofibroblast differentiation [10]. In contrast to 2D methods, decellularization, the detergent-mediated removal of cells yielding a complex ECM-derived material from native tissues (dECM), has been successfully used to recapitulate the instructive biochemical and biophysical cues of native microenvironmental niches.

The dECM is a platform that preserves both the structure and composition of the native ECM, making it a useful platform for probing mechanotransduction. It may be that decellularization results in ECM relaxation, making it slightly less stiff but still significantly stiffer than normal tissue. To recapitulate the biophysical and biochemical features of the mechanical microenvironment experiments investigating cell functions heavily influenced by mechanics, we propose the use of dECM to gain further insight into the interplay between ECM and cells during disease progression. dECM is a 3-D scaffold that closely mimics the fibrotic milieu and can be purposed to test cellular contribution in ECM deposition by decellularizing in vitro cultures. The impact of ECM on cell behavior can also be probed through dECM by decellularizing healthy and disease tissues and seeding naive cells on them. In this chapter, we will introduce how to generate decellularized ECM from human and rodent's lung, prepare dECM for mass spectrometry analysis, make stiffness adjustable hydrogels coated with dECM, and use atomic force microscopy (AFM) to evaluate dECM.

2 Materials

2.1 Human Lung Decellularization

1. Vibratome.
2. Phosphate-buffered saline (PBS): 137 mM NaCl, 2.7 mM KCl, 10 mM Na_2HPO_4, 1.8 mM KH_2PO_4 pH 7.4. Weigh 8 g of NaCl, 0.2 g of KCl, 1.44 g of Na_2HPO_4, and 0.24 g of

KH_2PO_4. Then transfer all of the salts listed above to a 1000 mL beaker and add 800 mL of distilled water. Adjust the pH to 7.4 with HCl and add distilled water to a total volume of 1 L.

3. Wash solution: prepare 5× and 1× penicillin/streptomycin (PS) stock solutions dissolved in PBS. The concentration of 5× PS contains 500 U/mL penicillin and 500μg/mL streptomycin. In the 1× PS, adjust the concentration to 100 U/mL and 100μg/mL.
4. Lysis buffer: 0.1% Triton X-100 and 2% sodium deoxycholate (SDC) (w/v).
5. 1 M NaCl: weigh 58.44 g of NaCl and dissolve it in diH_2O (deionized water) until it reaches a final volume of 1 L.
6. DNAse reagent (30μg/mL) in 1.3 mM $MgSO_4$ and 2 mM $CaCl_2$.
7. Peracetic acid (PAA) solution containing 0.18% peracetic acid (w/v) and 4.8% ethanol (v/v).
8. 2% UltraPure Low Melting Point agarose: weigh 1 g of UltraPure Low Melting Point agarose in 50 mL of diH_2O and heat it to 65 °C by hotplate stirrer with stirring. After agarose powder completely dissolved, keep the bottle in 37 °C water bath so that it remains a fluid.

2.2 Mouse Lung Decellularization After Bleomycin Treatment

1. Adult C57BL/6 mice (usually 8–10 weeks of age).
2. Anesthesia: a mixture of medetomidine (0.5 mg/kg mouse body weight), midazolam (5.0 mg/kg mouse body weight), and fentanyl (0.05 mg/kg mouse body weight) (MMF).
3. Anesthesia antagonist: a mixture of atipamezole (2.5 mg/kg mouse), flumazenil (0.5 mg/kg mouse), and naloxone (1.2 mg/kg mouse).
4. Bleomycin: Bleomycin (1 U/kg mouse) prepared in 50μL of sterile 1× PBS.
5. Mouse intubation kit.
6. Blunted 21 gauge cannula: cutdown the sharp end of the 21 gauge and use smooth sandpaper to polish the cutting edge.
7. 2% UltraPure Low Melting Point: the same recipe described in Subheading 2.1.

2.3 Decellularizing In Vitro Cell Generated ECM

1. PDMS (Polydimethylsiloxane) sheets (2 × 8 cm^2).
2. Cell culture petri dishes (150 mm).
3. 70% EtOH (ethanol).
4. 0.1 N NaOH in diH_2O: weigh 0.4 g of NaOH and dissolve it in 100 mL of diH_2O.

5. 1% Glutaraldehyde in diH_2O (w/v).
6. 3% 3-Aminopropyltriethoxysilane in diH2O (APTES) (w/v).
7. Fibronectin (10μg/mL). Prepare Fn in sterile PBS containing 1% PS (v/v).
8. 1% BSA: weigh 0.1 g of BSA dissolved into diH_2O. Make the final volume of the solution 10 mL.
9. 2μM Latrunculin B reagent: dilute Latrunculin B in 50 mM EDTA to a final concentration of 2μM.
10. DOC Lysis buffer: make a 2% (w/v) solution by dissolving 1 g of sodium deoxycholate (SDC) in 50 mL of 20 mM Tris–Cl (pH 7.6) solution containing 2 mM PMSF (phenylmethylsulfonyl fluoride) and 2 mM EDTA.
11. DNase reagent: add 40μL of 500 U/mL DNase into PBS, including 2 mM $CaCl_2$ and 1 mM $MgSO_4$ to reach a final volume of 20 mL.

2.4 Fabrication of Hydrogels with Varying Stiffness

1. 35 mm Coverslips.
2. (3-Aminopropyl) triethoxysilane (APTES): 5% APTES in absolute ethanol (v/v).
3. 0.5% Glutaraldehyde in diH2O (v/v).
4. PBS.
5. Dichlorodimethylsilane (DCDMS).
6. Polyacrylamide (PAA): 40% Acrylamide solution, 2% Bis-acrylamide solution, TEMED, 10% ammonium persulfate (APS) (w/v).
7. Sulfo-SANPAH: the stock solution is 25 mg/mL. Add 2 mL molecular-grade DMSO directly into the bottle.
8. 50 mM HEPES buffer, pH 8.5. Add 1.1915 g HEPES (free acid) to a suitable container and make up to 80 mL with distilled water. Adjust the pH to 8.5 with NaOH and add diH_2O to a final volume of 100 mL.

2.5 In-Solution Digestion of ECM Proteins to Peptides for Mass Spectrometry

1. 8 M urea: 8 M urea in 100 mM ammonium bicarbonate.
2. 2 M urea: Dilute 8 M urea to 2 M urea within the 100 mM ammonium bicarbonate pH 8.0.
3. Lodoacetamide solution in HPLC-grade water (make 500 mM stock).
4. 1 M DTT (Dithiothreitol).
5. PNGaseF (commercial solution of PNGaseF is at 500 U/μL, used at a final concentration of 1μL of PNGase F/2–6 mg of proteins).
6. Lys-C (Endoproteinase LysC, mass spectrometry grade). Make a stock solution of 0.5μg/μL in HPLC-grade water.

7. Trypsin (mass spectrometry grade, commercial solution at 0.5μg/μL).
8. 50% Trifluoroacetic acid (TFA) (v/v): dilute the TFA (≥99%) in MilliQ water to 50%.
9. 60% Acetonitrile and 0.1% trifluoroacetic acid mixture. Mix 600μL acetonitrile, 10μL 10%TFA, in 390μL MilliQ water.
10. 3% Acetonitrile with 0.1% trifluoroacetic acid. Mix 30μL acetonitrile and 10μL 10%TFA in 960μL MilliQ water.

2.6 In-Gel Digestion of Proteins for Mass Spectrometry Analysis

1. 2× Laemmli sample buffer and separation by SDS–PAGE. Samples are allowed to migrate approximately half-way down a 4–12% acrylamide gel.
2. Coomassie staining solution [0.025% (w/v) Coomassie Brilliant Blue R-250, 10% (v/v) acetic acid, 25% (v/v) propanol].
3. Odyssey IR imaging system or any other available imaging system.
4. V-bottomed 96-well plate (Proxeon, Odense, Denmark).
5. Acetonitrile (ACN).
6. Vacuum centrifuge.
7. NH_4HCO_3 stock: to make NH_4HCO_3 stock in 100 mM, weigh 0.791 g of NH_4HCO_3, and dissolve in 100 mL of water.
8. Reduction buffer (10 mM dithiothreitol, 25 mM NH_4HCO_3): to make the reducing buffer, the ratio of 1 M DTT:100 mM NH_4HCO_3:H_2O is 1:40:59.
9. Alkylation buffer (55 mM iodoacetamide, 25 mM NH_4HCO_3): dissolve 0.046 g of iodoacetamide in 500μl H_2O to make iodoacetamide stock in 500 mM; dilute iodoacetamide and NH_4HCO_3 to final concentration 55 and 25 mM.
10. Digestion buffer (25 mM NH_4HCO_3) containing 12.5 ng/μL sequencing-grade modified trypsin (Promega, Southampton, UK).
11. V-bottomed 96-well storage plate (Thermo Fisher Scientific, Waltham, MA, USA).
12. 5% (v/v) formic acid in 50% (v/v) ACN.

2.7 AFM

1. **Cantilever:** When purchasing cantilevers, it is important to select one with a spring constant comparable to the sample of interest. The spring constant required to measure each dECM will be dependent upon the tissue of origin, but a standard range for biological probe stiffness is 0.01–0.60 N/m [11] (*see*

Note 1). For force mapping, we recommend silicon nitride triangle-shaped cantilevers that are tipless and have used MLCT-O10 cantilevers from Bruker for this method.

2. **Tips:**
   (a) 5 or 25μm glass beads. Size of bead is dependent upon the resolution of force measurements from user.
   (b) Epoxy-resin for attaching bead to cantilever.

3 Methods

3.1 Human Lung Decellularization

Using the following procedures, process the lung tissue in a sterile environment.

1. On Day 1: profuse lung tissue with warmed 2% UltraPure Low Melting Point agarose (SeaPrep; Lonza, Rockland, ME), (*see* **Note 2**). Then cool the perfused lung at 4 °C until the agarose solidifies. After gelation, cut the tissue into 1000μm thick sections on a standard Vibratome. Later, gently warm sections to remove the agarose surrounding and within the tissue sections.
2. Agitate lung samples three times in sterile PBS containing 5× PS.
3. Subsequently, rinse lung sections with cold 0.1% Triton X-100, discard, then add new solution and store in cold 0.1% Triton X-100 at 4 °C for 24 h while shaking.
4. On Day 2: discard the 0.1% Triton X-100. Wash sections three times with cold PBS containing 5× PS, followed by a wash with cold 2% sodium deoxycholate (SDC) (w/v). After washing with SDC reagent, keep sections in 2% SDC at 4 °C and shaking 24 h at 4 °C, subsequently.
5. On Day 3: after SDC lysis, wash sections three times with cold 1× PS PBS, then wash sections one time with 1 M NaCl. Discard 1 M NaCl and continue to incubate sections in 1 M NaCl while shaking for 1 h at RT to lyse residual nuclei. Follow incubation by washing the sections three times with cold 1× PS PBS. Incubate tissues with DNAse (30μg/mL) reagent diluted in a 1.3 mM MgSO4 and 2 mM $CaCl_2$ solution.
6. Finally, wash lung tissues three times with sterile PBS, then sterilize the sections in 0.18% peracetic acid and 4.8% ethanol solution for 20 min (*see* **Note 3**).
7. Later, wash tissues three times with sterile PBS and store at 4 °C (*see* **Note 4**).

3.2 Decellularizing Mouse Lung After Bleomycin Treatment

3.2.1 Bleomycin Intratracheal Administration in Mice

1. 8- to 10-week-old C57BL/6 mice are preferred for the following protocol.
2. Anesthetize the mouse according to local IACUC protocols.
3. Intratracheal bleomycin instillation. Administer 1 U/kg bleomycin by intubating mouse and injecting 50μL of bleomycin solution in the trachea.
4. Allow the mouse to recover from anesthesia or antagonize the narcosis.
5. Monitor mouse weight and health during the next 14 days.

3.2.2 Mouse Lung Harvest

On Day 14 post bleomycin instillation:

1. Perform the following steps in a sterile environment.
2. Euthanize the rodents according to the approved local IACUC protocols.
3. Open the chest of a mouse to expose the lung and heart; then dissect the throat with scissors to expose the trachea and cut both SCMs (sternocleidomastoid muscles); to ensure easy access, dissect away strap muscles on top of the trachea.
4. Clear blood from the lung vasculature by injecting 3 mL PBS into the right ventricle, twice.
5. Cut the top of the trachea with scissors *carefully* and cannulate the trachea with a blunted 21 gauge cannula; secure the trachea and cannula with the suture.
6. Lavage the lungs three times with 1 mL of 1× PBS; airway T Cells/macrophages should be removed by this step.
7. Inflate lungs through the intratracheal cannulae with warm 2% UltraPure low melting point agarose around 2 ml (*see* **Note 2**), then cool at 4 °C to solidify in 1 min by dropping ice on top until the agarose solidifies.
8. Cut 1000μm thick lung sections on a standard Vibratome.
9. Proceed to the decellularization protocol described above in Subheading 3.1.

3.3 Decellularizing In Vitro Cell Generated ECM

3.3.1 Surface Treatment of PDMS Sheets

1. Submerge PDMS sheets (2 × 8 cm^2) in a 50 mL conical tube filled with 70% EtOH. Sonicate in bath sonicator at max power for 15 min.
2. Transfer all PDMS sheets to the underside of a petri dish lid (diameter: 150 mm), and cover with 0.1 N NaOH for 30 min, then lift PDMS sheets so that they float on top of NaOH solution to treat the underside of the sheets for another 30 min (*see* **Note 5**).
3. Transfer sheets from the lid of the petri dish to its respective deep dish counter part to wash four times with diH_2O.

4. Cover sheets with 3% APTES solution for 20 min, and make 50 mL of 1% glutaraldehyde in the meantime for efficiency. After APTES incubation, wash PDMS sheets four times with diH_2O.
5. Cover sheets with 1% glutaraldehyde for 30 min, then wash two times with diH_2O, followed by two more washes with PBS containing 2% PS.
6. Transfer PDMS sheets to sterile cell culture dishes (diameter: 150 mm).
7. Cover each PDMS strip with 1 mL of fibronectin (10μg/mL) and incubate for 1 h in a 37 °C cell incubator.
8. After removing the fibronectin solution and subsequently washing the sheets three times with sterile PBS, blocked with 1% BSA for 1 h at RT.
9. Wash three times with sterile PBS.
10. Plating and culturing to form matrix:
    (a) Suspend desired cells in 1 mL cell culture media containing FBS (FBS percentage is user-dependent and will vary based on the cell type being used and spread on the strip).
    (b) After cells firmly adhere, add media to cover the whole dish.
    (c) When using HFF fibroblast, seed cells at a density of 10,000 cells/cm^2.
    - For other cell types, seeding density will have to be determined by the user as cell density is dependent on cell size.
11. Place plated cells in the incubator (5% CO_2), and change media every 48 h for at least 8 days.

3.3.2 Decellularized Cell-Based ECM

1. Make EDTA with Latrunculin B solution and DOC lysis, warm to 37 °C.
2. Aspirate media from the growth surface of PDMS sheet; wash 1× with sterile PBS.
3. Remove PBS, add Latrunculin B solution, and incubate for 1 h at 37 °C. Latrunculin B prevents actin from polymerizing and actin cytoskeleton disassembly that causes by osmotic pressure.
4. Wash 1× with PBS.
5. Add warmed DOC lysis, incubate at 37 °C for 5 min.
6. Gently aspirate DOC lysis, rinse with 20 mL PBS, and wash three times.
7. Incubate sheets with DNase I solution for 30 min, after cell lysis. Then wash two times with PBS to remove remaining cell debris and reagents.

8. Keep matrix in PBS with 2% PS containing proteinase inhibitors. Observe the lysis result under a microscope. The matrix should be visible.

3.4 Fabrication of Polyacrylamide Hydrogels with Varying Stiffness

The basic concept of the following protocol is to generate uniform PAA hydrogels adhered to coverslips by "sandwiching" the PAA hydrogel between a silane-treated coverslips (to attach the PAA) and DCDMS-treated slides (to repel the PAA for easy removal), and thus both the coverslips and the slides need to be pretreated ahead of the experiment.

3.4.1 Coverslip Preparation

1. *Coverslip treatment:* Sonicate coverslips (35 mm diameter) submerged in 70% EtOH for 10 min and air-dry. Afterward, use a plasma cleaner to treat the cleaned coverslips (*see* **Note 6**). After plasma treatment, immediately add 360μL of 5% APTES in absolute ethanol to the coverslips (*see* **Note** 7). Incubate for 10 min under ambient temperature. Remove APTES and *thoroughly* wash the coverslips with excessive diH_2O three times. Incubate coverslips in 0.5% glutaraldehyde in diH_2O for 30 min at ambient temperatures by either adding 500μL of 0.5% glutaraldehyde solution per coverslip or completely submerging coverslips in a 6-well plate. Remove glutaraldehyde solution. Wash three times with diH_2O and keep the coverslip in diH_2O until use.
2. *DCDMS coverslide treatment*: Treat each slide with 360μL DCDMS in a ventilation hood for 5 min, make sure the whole surface is covered. Remove excessive DCDMS and rinse slides with diH_2O three times, then air dry.

3.4.2 Solubilized dECM Preparation

1. Remove lids from frozen tubes containing decellularized lung tissue and place filter paper over the tube opening and secure it with a rubber band. In the meantime, the tube and contents should still in frozen status otherwise place at −80 °C until frozen.
2. Lyophilize the tissue until all excess liquid is gone, using a freeze dryer according to manufacturer directions. Store at −80 °C until ready to mill.
3. Before beginning, add liquid nitrogen to freezer mill to cool insulation and internal components to working conditions.
4. Remove magnetic mill bar from mill tube and add tissue to cover the bottom.
5. Replace mill bar and add tissue block. The mill bar should still move freely.
6. Close the mill tube and place it in the freezer mill.
7. Fill liquid nitrogen to max fill line.

8. Freezer mill all tissue into a fine powder (approximately 5 min, impaction rate of 600 min^{-1}), in a polycarbonate cylinder with a stainless steel impactor as well as stainless steel end plugs using the freezer mill according to manufacturer directions. Store at −80 °C until ready for use.
9. To dissolve dECM powder, take the fine dECM powder into the tube, then add 0.01 M HCl that contains 0.1% (w/v) of pepsin to reach a final concentration of dECM is 1% (w/v), under constant agitation at room temperature, for 48 h.
10. After digestion for 48 h, place the solution and reagents on ice for 5 min.
11. Using refrigerated 10% (v/v) 0.1 M NaOH, and 11.11% (v/v) 10× PBS (to bring the entire solution to 1× concentration), bring the digested protein solution to physiologic pH of 7.4. The dECM solution can now be stored at 4 °C for up to 1 week.

3.4.3 Hydrogel Preparation

1. Prepare the acrylamide/bis-acrylamide mixture according to Table 1 (10 mL total volume): Mix the components thoroughly and degas for 15 min to expel resolved air.
2. Hydrogel generation*:* for a 35 mm coverslip, mix 360μL of desired PAA mixture solution with 3.6μL 10% APS and 0.36μL TEMED (*see* **Note 8**).
3. Immediately vortex the tube and pipette 43.2μL of the mixture onto DCDMS-treated slide. Place no more than four drops at a time and immediately place the treated coverslips (treated face DOWN) onto the drops.
4. Discard the remaining PAA hydrogel solution.
5. Wait 2 h to allow the hydrogel to finish polymerizing (*see* **Note 9**).
6. Carefully peel off the coverslips with fine tip forceps and place them in a multi-well plate filled with PBS (hydrogel side up). Keep hydrogels covered at 4 °C for 2–3 months.
7. *Hydrogel f unctionalization and storage:* For each 35 mm slip, mix 496μL 50 mM HEPES buffer (pH 8.5) with 4μL Sulfo-SANPAH stock (*see* **Note 10**); thoroughly mix and add onto

Table 1
The optimized conditions for polyacrylamide hydrogels preparation

Stiffness (kPa)	Acrylamide %	Acry volume (mL)	Bis-acry %	Bis volume (mL)	PBS volume (mL)
1.8	5%	1.25	0.06%	0.3	8.45
4.5	5%	1.25	0.15%	0.75	8
8.4	5%	1.25	0.23%	1.125	7.63
19.6	8%	2	0.26%	1.32	6.68

the hydrogel surface. UV treat (*see* **Note 11**) for 10 min (9.9 min is the max for our UV oven). Afterward, tap off the excessive Sulfo-SANPAH solution; quickly wash with sterile PBS two times and coat the functionalized hydrogel with solubilized dECM for 1 h at 37 °C (*see* **Note 12**).

3.5 In-Solution Digestion of dECM Proteins to Peptides for Mass Spectrometry Analysis

The procedure described here for the in-solution digestion of the ECM proteins for proteomic analysis is adapted from the method developed by Naba et al. [12].

The pellet obtained after the decellularization procedure (Subheading 3.1) is highly enriched with insoluble ECM proteins. For further probing by mass spectrometry and proteomic analysis, these proteins need to be digested into peptides. As a consequence of ECM protein insolubility, it is not possible at this step to measure the protein concentration of the sample. For this mass spectrometry procedure, freshly prepare all solutions in HPLC-grade water.

1. Reduction and alkylation. The purpose of this step is to reduce disulfide bonds and alkylate the sulfhydryl groups of cysteine side. It makes the Cys-containing peptides to be identified during database searching. Resuspend the fine dECM powder by adding the appropriate volume (50μL to 1 mm thick pellet) of 8 M urea to the ECM-enriched pellet and add DTT at a final concentration of 10 mM. Continuously agitate at 1400 rpm while incubating for 2 h at 37 °C (*see* **Note 13**).

   Cool the sample to RT and add the iodoacetamide to a final concentration of 25 mM. For complete alkylation, the DTT: iodoacetamide ratio should be between 1:2.5 and 1:3. Incubate in the dark for 30 min at RT.

2. Deglycosylation: Glycopeptides have several characters, like lower detection sensitivity and also reduce digestion efficiency, that cause the inadequate spectral data collection during MS analysis. Deglycosylation of the glycopeptides increases protein sequence coverage and improves protein identification.

   Dilute urea to 2 M with 100 mM ammonium bicarbonate (pH 8.0) and add the appropriate amount of PNGaseF at a final concentration of 1μL of PNGase F/2–6 mg of proteins. Incubate with continuous agitation at 1400 rpm for 2 h at 37 °C (*see* **Note 14**).

3. Digestion. Add Lys-C (Endoproteinase LysC, mass spectrometry grade, make a stock solution of 0.5μg/μL in HPLC-grade water, use 1μg) and incubate with continuous agitation at 1400 rpm for 2 h at 37 °C. Add trypsin (mass spectrometry grade, commercial solution at 0.5μg/μL, use 3μg) and incubate with continuous agitation at 1400 rpm O/N at 37 °C (*see* **Note 15**). Add another aliquot of trypsin (1.5μg) to the sample

and incubate with continuous agitation at 1400 rpm for an additional 2 h at 37 °C.

4. Acidification. Upon completion of the digestion, inactivate the trypsin by acidifying the sample with freshly prepared 50% trifluoroacetic acid (TFA). The sample should reach pH <2. Add 1–1.5μL of 50% TFA at a time and use 1μL of the peptide solution to measure the pH of the solution using pH paper. Centrifuge the acidified sample at 16,000 × *g* for 5 min at RT. Collect the supernatant in a clean low-retention tube. At this point, the peptide solution can be stored at −20 °C.

5. Desalting. Before the proteomics analysis, desalt peptides eluted with freshly prepared 60% acetonitrile, 0.1% trifluoroacetic acid, and the sample. Then concentrate desalted samples in a vacuum concentrator. Resuspend the peptides in freshly prepared 3% acetonitrile, 0.1% trifluoroacetic acid (*see* **Note 16**).
   (a) Now analyze the sample by mass spectrometry, according to the optimal procedures of the mass spectrometry facility.

3.6 In-Gel Digestion of Proteins for Mass Spectrometry Analysis

The procedure described here for the processing and in-gel digestion of the dECM proteins for proteomic analysis is adapted from the method developed by Robertson et al. [13, 14] and Shevchenko et al. [15].

1. For proteomic analyses, the protein in solution is mixed with 2× Laemmli sample buffer and separated by SDS–PAGE. Samples are allowed to migrate approximately half-way down a 4–12% acrylamide gel.
2. Total protein in the gel is visualized by incubating gels in Coomassie staining solution [0.025% (w/v) Coomassie Brilliant Blue R-250, 10% (v/v) acetic acid, 25% (v/v) propanol] for 1 h at room temperature.
3. Destain the gel by washing in water overnight at 4 °C, scan the gel using Odyssey IR imaging system, or any other available imaging system.
4. Wash the gel in water, excise, and divide each lane into 15–30 slices. Process each slice using an in-gel tryptic digestion procedure in 96-well plates as below.
5. Chop each slice of the gel into ~1 mm cubes and transfer to separate wells of a perforated V-bottomed 96-well plate (Proxeon, Odense, Denmark) and shrink the gel pieces by dehydration in acetonitrile (ACN) for 5 min.
6. Remove ACN by centrifugation (289 × *g*, 1 min) and dry the gel pieces in a vacuum centrifuge.
7. Rehydrate the gel pieces in reduction buffer (10 mM dithiothreitol, 25 mM NH4HCO3) for 1 h at 56 °C.

8. Remove the supernatant by centrifugation, and inoculate with alkylation buffer (55 mM iodoacetamide, 25 mM NH_4HCO_3) for 45 min at room temperature in the dark.
9. Wash the gel pieces with 25 mM NH_4HCO_3 for 10 min. Gel pieces are then shrunk in ACN (5 min), washed with 25 mM NH_4HCO_3 (5 min), and shrunk again in ACN (5 min).
10. Dry the gel pieces in a vacuum centrifuge before rehydrating with digestion buffer (25 mM NH_4HCO_3) containing 12.5 ng/μL sequencing-grade modified trypsin (Promega, Southampton, UK) for 45 min at 4 °C.
11. Incubate the sample for 16 h at 37 °C. Collect the peptides from the digestion buffer by centrifugation into a V-bottomed 96-well storage plate (Thermo Fisher Scientific, Waltham, MA, USA).
12. Extract the remaining peptides from the gel matrix with 20 mM NH_4HCO_3 (20 min) and collect by centrifugation. Repeat the peptide extraction two more times with 5% (v/v) formic acid in 50% (v/v) ACN.
13. Concentrate pooled extracts to 20μL and remove ACN by vacuum centrifugation. Store the samples at −20 °C until analysis by liquid chromatography-tandem mass spectrometry (LC-MS/MS).

The methods presented above are based on the review of the contributions in the field of proteomics regarding the extracellular matrix and adhesion mediated signaling events. There may be areas of limitation of the techniques presented. For example, the stringent washing steps in the procedure may result in the loss of a few key readily soluble proteins like growth factors or remodeling enzymes and attempt to fully profile the composition of matrisome-associated proteins that may be compromised.

3.7 AFM Calibration and Experiment Preparation

Here, we will be describing how to use the nanoindentation technique to measure the elasticity, in terms of Young's modulus (E), of dECM through the AFM mechanical mapping mode. Note that methods pertaining to AFM in this chapter are specific to the Asylum Research MFP-3D-BIO AFM as we are most familiar with this AFM model and this protocol may need to be modified for other user models. Experimental parameters of AFM must be predetermined ahead of the actual assay. Most critical parameters, including force-distance (piezo traveling distance), trigger point, and velocity, are case dependent and should be determined by the user themselves.

3.7.1 Laser Alignment

Fill a 30 mm petri dish with the same medium that will be used with the dECM sample during data collection. This can be PBS, media, or even air, but should not contain a sample yet.

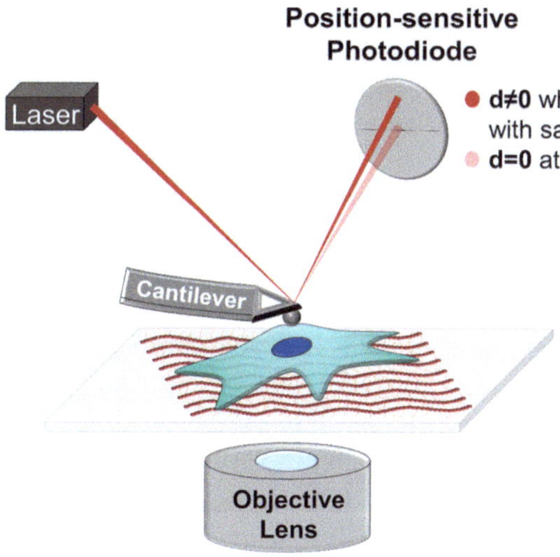

Fig. 1 Basic principles of AFM. The AFM operates by measuring the deflection of a cantilever of predetermined stiffness (i.e., spring constant $= k$) as it indents a given sample upon contact. These small deflections are measured with high sensitivity via the optical lever method. A laser beam is reflected onto the back of the cantilever, and bending of the cantilever induces an amplified movement of the laser. A position-sensitive photodiode detects that optical deflection toward or away from the sensor to infer the deflection of the cantilever (x) which is then used to determine force via Hooke's law (Eq. (1): $F = kx$) (*see* **Note 17**)

1. Load cantilever into the AFM apparatus, so that the tip of the cantilever is parallel to the surface of the piezo crystal. Tighten screw and ensure cantilever is fixed into apparatus.
2. Add the solution from the petri dish to the AFM tip to prevent undesired damage caused by surface tension when the tip enters the solution. Then mount AFM head onto a petri dish.
3. Locate the laser, then direct the beam to the edge of the cantilever tip. Place the laser at the center of the tips edge as depicted in Fig. 1.
4. Adjust the deflection by rotating the deflection wheel adjustment until deflection is approximately zero.
   (a) The deflection (d) adjustment changes the position of the photodiode used to detect the movement of the laser beam reflecting off the back of the cantilever. It is essential that $d = 0$ for proper force measurement.

3.7.2 Cantilever Calibration

Prior to data acquisition, the cantilever should be calibrated to validate both the spring constant (k) and its deflection sensitivity (the conversion factor between voltage change on the photodiode and laser movement from the cantilever bending) (*see* **Note 18**). These two calibrations are covered below as these two measurements are essential for accurately determining force using Hooke's law (Eq. (1)).

1. *Deflection Calibration:* The following will instruct you on how to calibrate the system to obtain accurate measurements of cantilever bending. Here, you will collect a force measurement on a flat, rigid surface (such as a plastic 30 mm petri dish) to produce a linear graph displaying the relationship between a known change in voltage on the photodiode and the actual deflection of the cantilever for calibration [11, 16]. On this voltage-deflection graph, the slope can be used to calculate the cantilever sensitivity as we know the precise distance the cantilever is lowered in nm via a piezo, and on a stiff substrate, this distance will be equivalent to cantilever deflection (*see* **Note 19**).
   (a) Begin with setting the imaging mode to "contact."
   (b) Press engage and begin lowering the cantilever toward the dish using the engagement wheel adjustment until the program indicates contact.
      - Ensure that $d = 0$ while the Z Voltage is still 150.0 V (value based on Asylum system).
      - Then s*lowly* lower the cantilever into the sample till the Z voltage = ~70–75 V (value based on Asylum system).
   (c) Under the force tab:
      - Select the start distance option.
      - Set the force-distance. This value can vary but should not exceed 1 μm.
      - Set trigger channel to deflection or similar setting.
        - Acquire a single force curve and align sensitivity with the linear portion of the approaching curve (*see* Fig. 2B) to calibrate according to the Hertz model.
        - Under the calibration tab set sensitivity to deflection.
2. *Spring Constant Calibration*: Use the thermal resonance frequency method (also referred to as the thermal noise or thermal fluctuation method), as it is simple, offered with most software packages, and widely employed [11, 17–19].
   (a) After calibrating the deflection, raise the cantilever tip, so it is no longer in contact with the surface (~3 rotations of the engagement wheel for MPF-3D Bio AFM).
   (b) Make sure there is sufficient room for the AFM to fluctuate the tip without it contacting the surface, then reset deflection to zero.
   (c) Under the thermal tab, commence the thermal frequency method. Data will be continuously collected. Stop the analysis after ~120 inputs. Fit the curve and evaluate the spring constant.

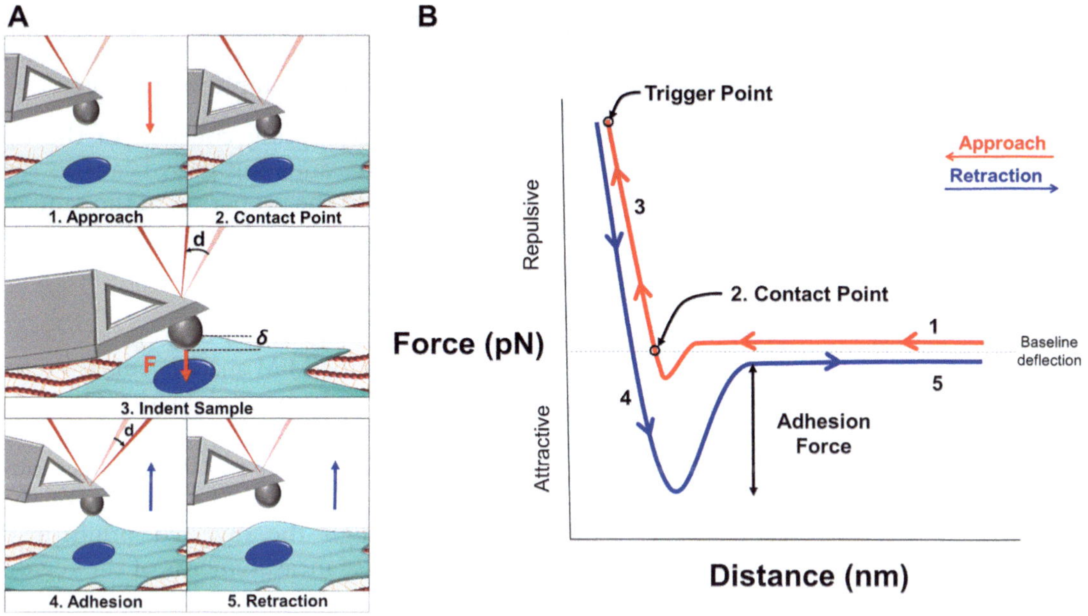

Fig. 2 (**a**) Schematic demonstrating different steps of cantilever approach and retraction on a single cell. (**b**) Cartoon depiction of the force-distance curve, where numbers related to steps shown in (**a**)

(d) Spring constants will vary based on the cantilever manufacturer, but for nanoindentation of normal (~2 kPa) and fibrotic (~15–20 kPa) human lung, our lab has used k values ranging 0.03–0.08 N/m [10, 20].

3.7.3 Tip Preparation

Tips, also known as probes, fixed at the end of the cantilever are used to probe the sample and can modulate the resolution of the AFM measurements depending on their shape and size.

1. Sharp, pyramidal tips are useful for performing topography mapping due to their ability to acquire high-resolution images (0.25μm at 0.5μm indention) [17]. However, these force measurements are typically confined to a small region within a heterogeneous tissue. They therefore require extrapolation to account for the mechanics of the entire sample, leading to inaccuracies [21].
2. Spherical beads fixed to the underside of a tipless cantilever provide greater contact area (e.g., ~1.22μm resolution with a 6μm diameter sphere at 0.5μm indention), ergo averaging the measurement of the mechanical stress (stress: $\sigma = F/A$) used to determine the young's modulus (E) over a larger area within the sample for more accurate measurements [16] (*see* **Note 20**).
   (a) We use a water-resistant, two-part epoxy-resin to attach the bead to the bottom of a v-shaped tip. Allow epoxy-resin to dry for at least 24 h prior to re-evaluating the k of the tip.

- Use epoxy-resin sparingly when adhering beads to ensure that the mixture does not leak around the bead and interfere with bead-sample interaction.

(b) It should be noted that the spring constant of the tip should be re-evaluated after adhering glass beads to the cantilever as you have altered the properties of the cantilever, but the *k* should be re-evaluated before every experiment regardless.

3.8 dECM Sample Preparation

Here we will briefly describe how to select an appropriate tissue section thickness for dECM samples for AFM analysis but also refer readers to a more thorough review by Jorba et al. [16].

3.8.1 Tissue Thickness

Thin sections (~10–50μm in thickness and up to 100μm as demonstrated by our lab [22]) are preferred so that the position of the cantilever can be visualized through the inverted bright-field microscope in order to obtain precise measurements of the region of interest (Fig. 1).

1. Although force mapping can be performed on thicker samples (e.g., 1000μm [8]—2000μm [23]), it is difficult to confirm precisely where the measurement is being acquired in the tissue [16].
2. Thicker samples can also be associated with more complex geometries, and some tissues, such as lungs, which must be inflated [22], may need to be altered for analysis. This can lead to errors when the cantilever engages a region of the tissue that is higher than the previous, and the setup must be re-calibrated.

3.8.2 Mounting dECM

Before proceeding with mechanical characterization, we advise that the dECM be mounted to a surface. Immobilizing the sample prevents it from moving or floating during analysis, which otherwise introduces error into the system and impedes, if not entirely prevents, data acquisition.

1. Many methods, such as administering Cell-Tak to adhere dECM, super gluing dECM [23], mounting dECM sections to super frost glass slides [24], and chemically binding dECM to poly-L-lysine-coated coverslips [25], have been used to immobilize samples, but researchers should evaluate if these methods contribute to differences in mechanical measurements with their own workflow and system.
2. Once the dECM has been sectioned to the desired thickness and has been mounted, it is beneficial to immediately re-hydrate the dECM in PBS. This step can be used to:

(a) Clean the sample of detergents from the decellularization protocol or storage reagents such as OCT.

(b) Prevent drying.

(c) Sterilize the dECM prior to reseeding cells if antibiotics are also added to the solution.

3.9 AFM Settings/Parameters for Obtaining Force Curves

3.9.1 Ramping Rate

In general, we would recommend a ramping rate (oscillation frequency) of 1 Hz as it has been successfully applied to dECM extensively [23, 26–28] and is comparable in efficiency to setting the deflection set point (maximum force input) to 1–2 nN at a velocity of ~22μm/s [23, 29]. These optimal inputs for each experiment, however, will have to be determined by the user based on AFM setup, a sample of interest, and other experimental constraints (*see* **Note 19**).

3.9.2 Loading Rate and Velocity

While measuring dECM mechanical properties, users should particularly be aware of nonlinear changes with the loading rate as well as how the timescale for these force measurements alter the evaluated elasticity (17, 30, 31) (*see* **Note 21**).

3.9.3 Probing Area

Biological samples are also heterogeneous, and therefore force maps should be acquired in various regions within the sample to accurately portray the mechanical characteristics of the matrices. We recommend obtaining two-dimensional force maps in 5–12 different positions.

3.9.4 Tip Indentation Depth

Another critical factor for accurately calculating the Young's modulus is the depth at which the cantilever tip indents the sample (δ) (Eq. (4)).

1. We recommend to initially set indentation depths to >500 nm [16] to avoid inaccuracies in determining the contact point [30]. The real experimental parameter, however, has to be determined by the users.
2. Restricting indentation to ~10% of the sample's thickness limits the mechanical contributions from the substrate beneath the sample, which would otherwise lead to overestimations of E [31].
3. Meanwhile, greater indentation depth allows the stiffness to be averaged over a larger contact area for reliable measurement of E (*see* **Notes 22** and **23**).

3.10 Force Curve Interpretation and Data Analysis

At every measuring point, the AFM outputs what is known as a force-distance curve (FD curve), which can be used to estimate the stiffness of dECM. These curves display the distance moved by the cantilever toward the sample on the x-axis (read right to left during the approach and left to right during the retraction of the cantilever) as seen in Fig. 2b, and the *y*-axis denotes the deflection of the cantilever or compliance of the sample in response to a force.

3.10.1 Basic Principles of FD Curve

A FD curve consists of two parts: a red approaching curve and a blue retraction curve. The approaching curve encompasses the measurements acquired while the cantilever descends, contacts, and indents the sample (Fig. 2a). The straight region is known as the noncontact curve, which is generated during tip approaching but not yet in contact with the sample (red on Fig. 2b). When the cantilever is close enough to a sample, attractive van der Waals forces will bend the cantilever toward the sample (negative force), but the cantilever will eventually be deflected upward by repulsive electrostatic forces (positive force) once it is lowered further [32], creating a sudden dip in the approaching curve before the contact point (Fig. 2b). The region to the left of the contact point represents when the tip is indenting the sample and are used to determine the sample deformation. Values to the right of the contact point represent the separation between the tip and sample. The contact point provides the input required for accurately calculating the indentation depth (Eq. (4)), thus choosing the correct contact point is essential for providing reliable measurements for the E of the sample (Eq. (2)) [11, 33, 34].

From the contact point onward, the cantilever is indenting the sample, bending in response to the sample, and deflecting the laser, which results in an upward slope as more force is required to lower the cantilever further into the sample. Once the cantilever reaches the predetermined trigger point or the maximum force-distance, the indentation ends. The slope of the approaching curve (from the contact point to the trigger point) represents the stiffness or relative hardness of the sample. Therefore, when the same contact force is reached by the cantilever traveling a smaller distance to deform the sample, the slope is steeper, implying the sample is stiffer and its Young's modulus higher.

3.10.2 FD Analysis

To obtain the E from the FD curve, the slope of the contact region (between the contact point and trigger point) is fitted to a contact model. While there are many contact models reviewed here [11, 21], the Hertz model is one of the most commonly used contact models and has been specifically designed for spherical tips, as shown by the Eq. (2) [35]. We will not cover the details of how to fit and acquire E from a given FD curve as the exact procedure differs across different AFM systems.

$$E = \frac{3F\left(1 - \upsilon^2\right)}{\left(4\sqrt{R}\right)\left(\delta^{3/2}\right)} \tag{2}$$

[36, 37] where E is the Young's modulus, F is the force the cantilever exerts on the sample as determined by cantilever deflection, R is the radius of the sphere probe attached to the cantilever tip, δ is the indention depth as determined by Eq. (4) (*see* **Note 22**), and υ is the Poisson ratio of the sample (normally assumed to equal 0.5 for an incompressible material, but values as low as 0.3 have been used.

On the other hand, the blue force curve in Fig. 2b represents the retraction of the cantilever and ascent away from the sample. The force decreases as the cantilever is risen off the sample, but there are usually adhesive forces that bind the cantilever to the sample which require positive force from the cantilever to break the bond as depicted in Fig. 2a.4 and the dip in the retraction curve between 4 and 5 on Fig. 2b.

4 Notes

1. This ensures that the measurement sensitivity is preserved as the cantilever will be stiff enough to indent the sample, while also being soft enough to generate sufficient bending in the cantilever to accurately measure the deflection while deforming the sample [].
2. The UltraPure Low Melting Point agarose should not overheat, and temperature should allow you to hold the bottle. Keeping bottle in 37 °C water bath and syringe on heat pad can prevent the solidity of agarose.
3. The residue of detergents, like, Triton X-100, SDC, NaCl, and DNAse, will damage cells when reseeding cells back. Try to profuse lung with washing buffer at least time every step after detergent applied.
4. A sample of the decellularized matrix can be frozen in Tissue Tek O.C.T. freezing medium, sectioned into 10μm slices, and stained with hematoxylin and eosin (H&E) to confirm the absence of cells.
5. While treating PDMS with NaOH, make 50 mL of 3% (APTES) to allow for hydrolysis.
6. With the following protocol: vacuuming for the 60s; followed with 90s plasma treatment at high power level.
7. Make sure to add sufficient APTES (around 40μL/cm^2) to cover the whole glass surface, but the excessive amount of APTES may make the coverslips very dirty later on.
8. It is recommended to make a larger volume of PAA mixture with TEMED (e.g., 500μL with 0.5μL TEMED) ahead and divide the mixture into an appropriate amount (e.g., 360μL per

tube) before adding the appropriate volume of 10% APS. Add APS only when ready to make hydrogels.

9. If everything has done correctly, you should be able to see a circular hydrogel formed between the coverslip and the slide with a "ring" like a gap between the edge of the hydrogel and the edge of the coverslip.
10. Prepare 25 mg/mL Sulfo-SANPAH stock solution (Thermofisher #22589) by adding 2 mL molecular-grade DMSO directly into the bottle (balance to ambient temperature before the step). Vortex the bottle a few times to ensure all Sulfo-SANPAH is resolved. Aliquot the stock solution and store at −80 C and protect from light. Sulfo-SANPAH is very sensitive to water, especially under low pH, and will be hydrolyzed quickly. Therefore repeated freeze-thaw should be avoided, and any unused stock solution after three times the thaw-freeze cycle should be discarded.
11. When UV oven is available, keep the UV oven inside the large cell culture hood to maintain the sterile condition.
12. dECM-coated hydrogel should be used within a week to ensure the best quality. Unfunctionalized/coated hydrogel, however, can be kept in 1% P.S. containing PBS for up to a month (or even longer).
13. The ECM proteins will not be fully dissolved, and the large visible protein particles should not be discarded by centrifugation or filtration. The suspension is expected to clear upon deglycosylation and digestion.
14. Deglycosylation is needed to remove carbohydrate side chains that interfere with the identification of peptides modified by N-linked glycosylation.
15. The ECM-rich suspension that began cloudy upon initial reconstitution in 8 M urea appears clear after O/N digestion.
16. After desalting, the concentration of the peptide solution should be measured by spectrophotometry.
17. In Eq. (1), the bending of the cantilever (i.e., the movement of the laser beam expressed as a change in voltage on the photodetector) is the distance (x) and the stiffness of the cantilever is the spring constant (k).
18. Deflection sensitivity varies based on how the laser beam is reflected, whether it be due to the initial positioning of the laser on the cantilever at the beginning of each experiment, effectively changing the trajectory of the laser beam's reflection onto the photodetector or the variability in light transmission through different media [].

19. The ramping rate and velocity control the offset between the approach and retraction force curves, which can be used to estimate the viscosity of the sample (Fig. 2b); by decreasing the ramping rate, we can decrease the hysteresis between these force curves [].
    (a) Using a low ramping rate can also decrease the hysteresis between the approach and retraction force curves [11].
20. The further the sphere is lowered into the sample, the greater the contact area between the sphere and the sample. This expands the force applied by the cantilever to the sample across a wider area, averaging the stress force (σ) measurement for more effective data acquisition []. If we consider the following equation for E in its simplest form, it is easy to see why averaging the force over a larger area would lead to more effective data acquisition.

$$E = \frac{\sigma}{\varepsilon} \tag{3}$$

    where the stress force is $\sigma = {}^{F}\!/_{A}$, the strain is ε, and the Young's modules is E.

    Here strain and elastic stress have a linear dependence on the Young's modulus, and in order to maintain this relationship, the sample should not be indented more than ~10% of its thickness [21, 31, 40, 41].
21. Krieg et al., in their review of AFM-based mechanobiology, explain the importance of specifying the loading rate and velocity for AFM experiments, as these inputs can define the mechanical properties of biological samples which often have viscoelastic behaviors determined by the forces exerted upon them (17).
22. The indention (δ) of the sample can be calculated as:

$$\delta = (z - z_c) - (d - d_{\text{off}}) \tag{4}$$

    where z is the position of the cantilever, d is the position of the laser on the photodiode, and d_{off} is the offset of the laser from its initial position. The subscript c represents z at the point of contact with the sample where $d - d_{\text{off}} = 0$ because the cantilever has not begun to bend yet. It should also be noted that $d - d_{\text{off}}$ is x in Hooke's law (Eq. (1)).
    (a) All of this should be kept in mind whenever deciding the thickness of one's sections.
23. Deeper indentions increase the complexity of the estimated effects the probe has on the system.

References

1. Frantz C, Stewart KM, Weaver VM (2010) The extracellular matrix at a glance. J Cell Sci 123:4195–4200
2. Yue B (2014) Biology of the extracellular matrix: an overview. J Glaucoma 23:S20–S23
3. Plotnikov SV, Pasapera AM, Sabass B, Waterman CM (2012) Force fluctuations within focal adhesions mediate ECM-rigidity sensing to guide directed cell migration. Cell 151:1513–1527
4. Campbell ID, Humphries MJ (2011) Integrin structure, activation, and interactions. Cold Spring Harb Perspect Biol 3
5. Lee DY et al (2008) Integrin-mediated expression of bone formation-related genes in osteoblast-like cells in response to fluid shear stress: roles of extracellular matrix, Shc, and mitogen-activated protein kinase. J Bone Miner Res 23:1140–1149
6. Hinz B et al (2007) The myofibroblast: one function, multiple origins. Am J Pathol 170:1807–1816
7. Bjoraker JA et al (1998) Prognostic significance of histopathologic subsets in idiopathic pulmonary fibrosis. Am J Respir Crit Care Med 157:199–203
8. Booth AJ et al (2012) Acellular normal and fibrotic human lung matrices as a culture system for in vitro investigation. Am J Respir Crit Care Med 186:866–876
9. Butcher DT, Alliston T, Weaver VM (2009) A tense situation: forcing tumour progression. Nat Rev Cancer 9:108–122
10. Fiore VF et al (2018) αvβ3 integrin drives fibroblast contraction and strain stiffening of soft provisional matrix during progressive fibrosis. JCI Insight 3
11. Gavara N (2017) A beginner's guide to atomic force microscopy probing for cell mechanics. Microsc Res Tech 80:75–84
12. Naba A, Clauser KR, Hynes RO (2015) Enrichment of extracellular matrix proteins from tissues and digestion into peptides for mass spectrometry analysis. J Vis Exp:e53057
13. Robertson J et al (2017) Characterization of the phospho-adhesome by mass spectrometry-based proteomics. Methods Mol Biol 1636:235–251
14. Humphries JD et al (2009) Proteomic analysis of integrin-associated complexes identifies RCC2 as a dual regulator of Rac1 and Arf6. Sci Signal 2:ra51
15. Shevchenko A, Tomas H, Havlis J, Olsen JV, Mann M (2006) In-gel digestion for mass spectrometric characterization of proteins and proteomes. Nat Protoc 1:2856–2860
16. Jorba I, Uriarte JJ, Campillo N, Farré R, Navajas D (2017) Probing micromechanical properties of the extracellular matrix of soft tissues by atomic force microscopy. J Cell Physiol 232:19–26
17. Hutter JL, Bechhoefer J (1993) Calibration of atomic-force microscope tips. Rev Sci Instrum 64:1868–1873
18. Butt H-J, Jaschke M (1995) Calculation of thermal noise in atomic force microscopy. Nanotechnology 6(1):1
19. Neumeister JM, Ducker WA (1994) Lateral, normal, and longitudinal spring constants of atomic force microscopy cantilevers. Rev Sci Instrum 65:2527–2531
20. Brown AC, Fiore VF, Sulchek TA, Barker TH (2013) Physical and chemical microenvironmental cues orthogonally control the degree and duration of fibrosis-associated epithelial-to-mesenchymal transitions. J Pathol 229:25–35
21. Krieg M et al (2019) Atomic force microscopy-based mechanobiology. Nat Rev Phys 1:41–57
22. Brown AC, Fiore VF, Sulchek TA, Barker TH (2013) Physical and chemical microenvironmental cues orthogonally control the degree and duration of fibrosis-associated epithelial-to-mesenchymal transitions. J Pathol 229:25–35
23. Viji Babu PK, Rianna C, Mirastschijski U, Radmacher M (2019) Nano-mechanical mapping of interdependent cell and ECM mechanics by AFM force spectroscopy. Sci Rep 9:12317
24. Achterberg VF et al (2014) The nano-scale mechanical properties of the extracellular matrix regulate dermal fibroblast function. J Invest Dermatol 134:1862–1872
25. Liu F et al (2010) Feedback amplification of fibrosis through matrix stiffening and COX-2 suppression. J Cell Biol 190:693–706
26. Melo E et al (2014) Effects of the decellularization method on the local stiffness of acellular lungs. Tissue Eng Part C Methods 20:412–422
27. Perea-Gil I et al (2015) In vitro comparative study of two decellularization protocols in search of an optimal myocardial scaffold for recellularization. Am J Transl Res 7:558–573
28. Luque T et al (2013) Local micromechanical properties of decellularized lung scaffolds measured with atomic force microscopy. Acta Biomater 9:6852–6859

29. Fiore VF et al (2015) Conformational coupling of integrin and Thy-1 regulates Fyn priming and fibroblast mechanotransduction. J Cell Biol 211:173–190
30. Rigato A, Miyagi A, Scheuring S, Rico F (2017) High-frequency microrheology reveals cytoskeleton dynamics in living cells. Nat Phys 13:771–775
31. Dimitriadis EK, Horkay F, Maresca J, Kachar B, Chadwick RS (2002) Determination of elastic moduli of thin layers of soft material using the atomic force microscope. Biophys J 82:2798–2810
32. Wegmann S, Medalsy I, Mandelkow E, Müller D (2012) The fuzzy coat of pathological human tau fibrils is a two-layered polyelectrolyte brush. Proc Natl Acad Sci U S A 110
33. Gavara N (2016) Combined strategies for optimal detection of the contact point in AFM force-indentation curves obtained on thin samples and adherent cells. Sci Rep 6:21267
34. Crick SL, Yin FCP (2007) Assessing micromechanical properties of cells with atomic force microscopy: importance of the contact point. Biomech Model Mechanobiol 6:199–210
35. Heinrich H (1882) Ueber die Berührung fester elastischer Körper. Journal für die reine und angewandte Mathematik 1882:156–171
36. Johnson KL (1987) Contact mechanics. Cambridge University Press, New York
37. Frey MT, Engler A, Discher DE, Lee J, Wang YL (2007) Microscopic methods for measuring the elasticity of gel substrates for cell culture: microspheres, microindenters, and atomic force microscopy. Methods Cell Biol 83:47–65
38. Medalsy ID, Muller DJ (2013) Nanomechanical properties of proteins and membranes depend on loading rate and electrostatic interactions. ACS Nano 7:2642–2650
39. Fabry B et al (2001) Scaling the microrheology of living cells. Phys Rev Lett 87:148102
40. Gavara N, Chadwick RS (2012) Determination of the elastic moduli of thin samples and adherent cells using conical AFM tips. Nat Nanotechnol 7:733–736
41. Cross SE, Jin YS, Rao J, Gimzewski JK (2007) Nanomechanical analysis of cells from cancer patients. Nat Nanotechnol 2:780–783

Chapter 19

Fibrosis on a Chip for Screening of Anti-Fibrosis Drugs

Mohammadnabi Asmani and Ruogang Zhao

Abstract

Idiopathic pulmonary fibrosis (IPF) is a chronic pathological disorder that targets alveoli interstitial tissues and is characterized by the progressive stiffening of alveolar membrane. The median survival rate of the patients with IPF is less than 5 years. Currently, IPF has no cure and there are few options to alleviate the progress of this disease. A critical roadblock in developing new anti-fibrosis therapies is the absence of reliable cell based in vitro models that can recapitulate the progressive features of this disease. Here a novel fibrotic microtissue on a chip system is created to model the fibrotic transition of the lung interstitial tissue and the effect of anti-fibrosis drugs on such transitions. This system will not only help to expedite the efficacy analysis of anti-fibrotic therapies but also help to unveil their potential mode of action.

Key words Pulmonary fibrosis, Microtissue array, Lung on a chip, Anti-fibrosis therapy, Drug screening, Tissue mechanics, Stiffness, and contractile force, Pirfenidone, Nintedanib

1 Introduction

Idiopathic pulmonary fibrosis (IPF) leads to lung failure and death and currently has no cure [1, 2]. IPF is characterized by the stiffening of the membranous tissue that makes up the alveolar air sacs. The accumulation of highly contractile and collagen-producing myofibroblasts predominantly contributes to the stiffening of the alveolar tissue [3–5]. Conventional 2D cell culture-based models do not capture the biomechanical and physiological complexities of the fibrotic alveolar tissue. Although candidates for anti-fibrotic therapies such as inhibitors directed against transforming growth factor-β (TGF-β) and lysyl oxidase-like 2 (LoxL2) have been studied [2, 6], their development is hindered by the high cost and the need for large sample sizes associated with animal studies and clinical trials. In vitro screening assays that can justify the development of new therapies in prolonged and expensive clinical trials are highly desired.

To model the physio-pathological characteristics of lung fibrosis in vitro, different experimental systems have been developed.

Boris Hinz and David Lagares (eds.), *Myofibroblasts: Methods and Protocols*, Methods in Molecular Biology, vol. 2299, https://doi.org/10.1007/978-1-0716-1382-5_19,

Conventional 2D culture-based assays are very limited in examining tissue mechanics. On the other hand, centimeter-scale, cell-embedded tissue models are not suitable for drug screening due to the diffusion limitation in culture and the low reproducibility of functional read outs [7–12]. Recently, microfabricated culture systems such as organ-on-chip systems have been studied as a potential high content screening platform [13–15]. The goal of these novel micro-systems is to replicate in vivo-like physiological conditions. For instance, in a recently published lung-on-chip device, authors built two separate chambers to mimic gas exchange in lung. An alveolar epithelial cell-lined air chamber and a capillary endothelial cell-lined fluidic chamber were separated by a porous silicone membrane [13]. With modeling pulmonary edema in sight, this system fails to provide a suitable model for IPF since extracellular matrix (ECM) components are missing which are critical to the tissue remodeling aspect of fibrosis [16]. Inability to assess the biomechanical characteristics of the alveoli wall such as its stiffness and contractility is a significant obstacle for any cell-based model that aims to mimic soft tissue fibrosis.

The system discussed here consists of arrays of membranous 3D microtissues formed through cell-mediated ECM remodeling. During their formation, microtissues undergo morphological and structural changes that are guided by poly (dimethyl siloxane) (PDMS) micropillar-defined mechanical boundary conditions. These membranous lung microtissues recapitulate key biomechanical properties of both healthy and fibrotic lung alveolar tissues. Moreover, pathological transition of this originally compliant, membranous lung interstitial microtissue to fibrotic stiff tissue can be achieved in this system. Proof of principle studies have been carried out to test the efficacy of two FDA approved anti-fibrosis drugs, Pirfenidone and Nintedanib [4, 17, 18].

2 Materials

2.1 Cell Culture

1. Primary human lung fibroblasts (NHLFs, *see* **Note 1**) (Lonza), maintained in recommended growth medium (FGM-2 Bullet-Kit, Lonza).
2. Freezing media: 10% DMSO, 45% media, and 45% fetal bovine serum (FBS).

2.2 Micropillar Array Device and Collagen Prepolymer

1. SU-8 photoresist (Microchem), silicon wafer (University Wafer), laser plotted transparency mask (CAD/Art Services Inc.) for multilayer photolithography. PDMS (Sylgard 184, DowCorning) for soft lithography of micropillar array device. P35 petri dishes and 12-well plate (Fig. 1a, b).

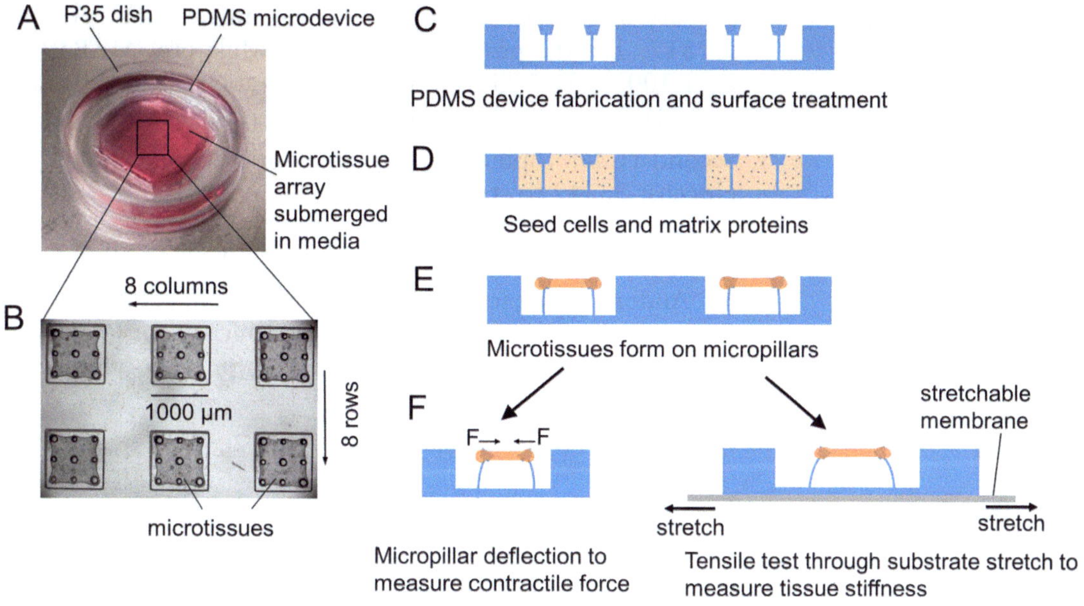

Fig. 1 (**a**) PDMS microdevice is casted in a P35 dish. Arrays of microtissues are submerged in the culture media. (**b**) Enlarged view of a portion of the microtissue array. (**c–f**) Flowchart shows the steps of microtissue device preparation, microtissue seeding, and tissue mechanical property testing

2. Transparent stretchable silicone substrate (SMI silicone sheeting, 0.01-inch NRV G/G 40D, Saginaw, MI) for stretchable microdevices.
3. 70% Ethanol and Pluronic F-127 solution (2% w/v).
4. Neutralized Collagen solution: Collagen type-I solution (Rat tail, Corning®) (*see* **Note 2**), 1 M NaOH, PBS or distilled water, 250 mM HEPES, M199 10× solution, 5% $NaHCO_3$ (w/v).
5. Lung fibroblast growth medium (FGM-2 BulletKit, Lonza).

2.3 Pharmacological Treatment

1. Anti-fibrotic drugs Pirfenidone (TCI America) and Nintedanib (Selleckchem). The lyophilized compounds were dissolved in DMSO per manufacturer description.
2. TGF-β1. The lyophilized TGF-β1 was dissolved in PBS to create stock solution of 1 μg/mL.
3. F-12k medium supplemented with 2% FBS, 100 U/mL penicillin, and 100 μg/mL streptomycin as a base medium and negative control for NHLF-populated microtissues.

2.4 Immunofluorescence

1. 4% Formaldehyde for fixation and methanol for cell permeabilization (*see* **Note 3**), 3% bovine serum albumin (BSA) for blocking.

2. Primary antibodies against α-smooth muscle actin (Abcam, ab7817, 1:300), pro-collagen type-I (Millipore, MAB1913-C, 1:400), ED-A-fibronectin (Abcam, ab6328, 1:400), and rat collagen type-I (Millipore, AB755P, 1:300).
3. Secondary fluorophore-conjugated, anti-IgG antibodies and Hoechst 33342 to stain nuclei.

2.5 Microscopy and Image Analysis

1. Inverted Nikon Ti-U microscope for bright-field imaging.
2. An Andor Technology DSD2 confocal unit coupled to an Olympus IX-81 motorized inverted microscope equipped with Plan-Apochromat ×10 air objective for confocal imaging of the microtissue.
3. Fiji software for image analysis.

2.6 Centrifuges and Vacuum Chamber

1. Since the working temperature for collagen prepolymer is 4 °C, a refrigerated centrifuge should be used.
2. A vacuum desiccator to degas during microtissue seeding.

3 Methods

3.1 Microtissue Device Fabrication

1. Multilayer microlithography and soft-lithography techniques were employed to fabricate micropillar arrays [19]. SU-8 masters were created through spin coating, alignment, exposure, and baking of multiple layers of SU-8 photoresists. UV exposure was performed through transparency masks printed by laser plotting. PDMS stamps were casted over the SU-8 master at 10:1 mixing ratio. Micropillar devices were then produced through replica molding from stamps in P35 petri dishes or 12-well plate (Fig. 1a, b).
2. To allow mechanical conditioning and stiffness measurement of the microtissues, the PDMS micropillar device was casted on a transparent stretchable silicone substrate. Stretching of the microtissue was achieved by holding one end of the silicone membrane stationary and increasing the displacement of the moving end.
3. Microtissue seeding and culture in microdevices cased either in P35 dishes or on the stretchable membrane follow the same procedure as described in Subheading 3.2.

3.2 Microtissue Seeding and Pharmacological Treatment

1. Sterilize PDMS micropillar device by filling the microdevices to the top with 70% ethanol and pipette up and down with 1 mL pipette to dislodge any trapped air pockets in the molds for 15 min (Fig. 1c) (*see* **Note 4**).
2. Aspirate ethanol and let the device dry in the hood or use nitrogen gun if available.

Table 1
Sample collagen prepolymer compositions. Note that the HEPES, NaHCO3, and M199 amounts are always the same for 1.8 mL of solution regardless of the collagen stock concentration. Regardless of collagen stock concentration the ratio of NaOH/Collagen should always remain the same

Stock collagen concentration of 3.83 mg/mL	Total 1.8 mL (in mL)		
	1 mg/mL collagen	2 mg/mL collagen	2.5 mg/mL collagen
PBS or DI H_2O	1017	532	290
HEPES (250 mM)	83	83	83
$NaHCO_3$ (5% w/v)	14	14	14
M199 10×	200	200	200
NaOH (1 M)	16	31	38
Collagen (3.83 mg/mL)	470	940	1175

3. Pluronic treatment of the device (*see* **Note 5**): Dilute the 2% Pluronic F127 in PBS to 0.2%, then add it into the microdevice, and pipette up and down with 1 mL pipette to dislodge any trapped air pockets. The treatment time depends on the concentration of the collagen prepolymer (*see* **Note 6**).
4. Aspirate F127 solution and rinse once with PBS and let it completely dry. From this step forward, all solutions should be chilled.
5. Mix collagen solution components (1.8 mL is enough for 1 device. Ingredients ratio needs to be adjusted for different collagen lot concentration). Add all the ingredients in Table 1 (except for NaOH and collagen) to a conical tube kept on ice (to cool them down to avoid early polymerization of collagen). After few minutes, add collagen followed by NaOH. Pipette up and down to mix the components and avoid generating too many bubbles. Final solution should be "salmon" colored, or roughly the color of CO_2 equilibrated DMEM (*see* Table 1 and **Note 7**).
6. Add ~0.6 mL of the collagen solution without cells (all on ice) (Fig. 1d) to each microdevice to fully cover the bottom surface of the microdevice with some excess on top. Then proceed immediately to **step 7**.
7. Degas the devices in a chilled vacuum desiccator to allow the collagen solution fill each microwell (*see* **Note 8**).
8. Add an additional 0.4 mL of collagen solution (without cells) to each microdevice (bringing the final amount to 1 mL per microdevice).

9. Mix the collagen solution and cells: Prepare cells by trypsinizing, counting, and spinning down. Aspirate supernatant and resuspend the cells with cold collagen solution from the conical tube. 0.1 to 0.2 mL of cell/collagen mixture solution is needed for seeding one device (for example, if you plan to seed four devices you will need to suspend the cell pellet with at least 0.4 mL of collagen solution). Pipette up and down gently to mix the cell and collagen and avoid making bubbles. While working with collagen, use caution with pipetting to avoid frothing up the collagen solution.
10. Add cell/collagen mixture into the microdevice that is already partially filled with collagen. Mix cell/collagen solution with pre-existing collagen well by gently shaking or pipetting (avoid generating a lot of bubbles). Then, spin down the cells into the microwells at 1150 rpm for 2 min in a temperature-controlled centrifuge at 4 °C (*see* **Note 9**).
11. Remove excess cell/collagen solution using vacuum aspiration. Tilt the microdevice at ~30 degree and start to aspirate the collagen solution from a corner. Run the glass pipette until you reach the bottom of the microdevice. Then slowly work your way around the edge of the microdevice with the glass pipette to aspirate remaining collagen.
12. Polymerize the collagen gels. Add 0.5 of PBS to the lid of each device and invert the microdevice onto the lid (this keeps the collagen from drying out during the polymerization). Place the device (inverted) into an incubator for 9 min. After 9 min take the device out and add growth media (1–2 mL total) very slowly on top of the microdevice surface. While adding media, be cautious and start from a corner of the chamber.
13. Place the device back into the incubator and wait until the constructs form (Fig. 1e).
14. To induce myofibroblast differentiation of the NHLFs, add 5 ng/mL of TGF-β1 to the F12-K base media and maintain for different durations.
15. Pirfenidone at 1.6 and 5.3 mM and Nintedanib at 0.5 and 3 μM are used in two treatment approaches. In preventative treatment, anti-fibrosis drugs are co-administered with TGF-β1 at the beginning of microtissue formation and maintained for 6 days. In therapeutic treatment, TGF-β1-induced fibrosis is allowed to progress for the first 3 days and TGF-β1 is removed, and anti-fibrosis drugs are then administered and remained for another 3 days. The expressions of fibrosis biomarkers and the changes in tissue mechanical properties are assessed by the end of the treatments (Fig. 1f).

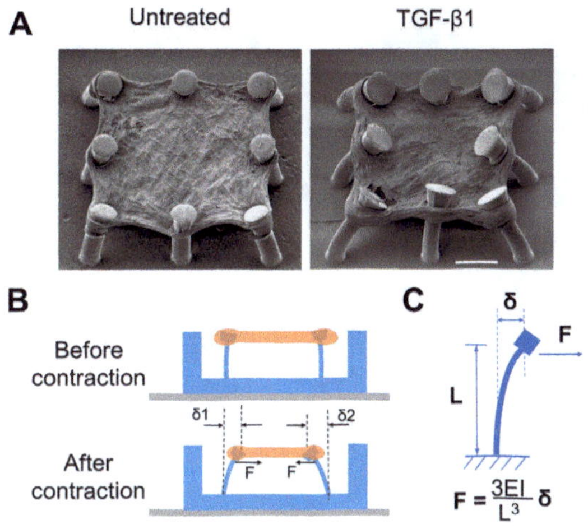

Fig. 2 (**a**) SEM images of an untreated and a TGF-β1-treated microtissue. Significant micropillar deflection caused by elevated tissue contraction can be seen in TGF-β1-treated microtissue. (**b**) Side view of the micropillars shows the measurement of micropillar deflection (*δ*). (**c**) Schematic shows the cantilever bending theory. Scale bar is 200 μm

3.3 Force Measurement

1. Tissue-generated contraction force is calculated from the deflection of micropillars based on cantilever bending theory (Fig. 2) (*see* **Note 10**). Micropillar deflection is measured by image analysis of the images taken by phase contrast microscopy. The deflected position of the centroid of each pillar top to the centroid of its base is measured and tracked during the time period of interest.
2. Microtissue contraction force is measured for each pharmacological condition. Then the traveled distance is multiplied by the spring constant. Micropillar's spring constant, which is 0.66 μN/μm for micropillars in the current study, is determined based on the actual geometry of the micropillar and the elastic modulus of the PDMS.
3. The absolute values of calculated force from each micropillar is added up and divided by 2 to reflect the collective force generated by one microtissue.

3.4 Stiffness Measurement

1. After microtissues are cultured for desired period, the silicone substrate containing the microtissue array is transferred to the stretching platform (Fig. 3a, b). One end of the silicone substrate is kept stationary and the other end can move.
2. The substrate carrying the microtissue array is then lowered to close to the bottom of a P100 petri dish and submerged in the culture media (Fig. 3c). The stretching platform and the P100-

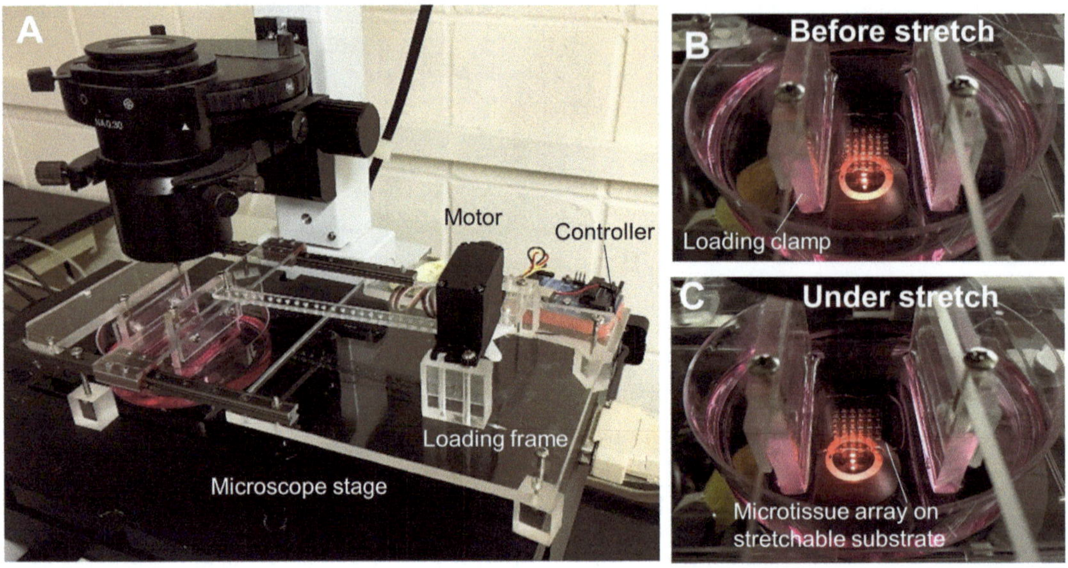

Fig. 3 Microtissue mechanical stretching system setup. (**a**) Custom-made stretching system was mounted on the microscope stage. A transparent stretchable silicone substrate was mounted on the loading frame driven by a DC motor controlled by an Arduino microcontroller. The mounting fixture allows the silicone substrate to be lowered to close to the bottom of a P100 petri-dish containing culture media. (**b**) Microtissue array before stretch. Microtissue array is directly above the objective to allow imaging. (**c**) Microtissue array under stretch

dish are mounted together on a microscope stage to enable real-time measurement of the microtissue deformation and micropillar deflection of multiple microtissues through optical imaging.

3. After stretching the silicone substrate, the tensile strain transfer from the silicone substrate to the microtissue and the tensile force applied to the microtissues can be characterized by imaging (Fig. 4a).
4. Tensile tests are performed on untreated, TGF-β1-treated, and anti-fibrosis drug-treated microtissues to measure their stiffness and compliance (*see* **Note 11**).
5. Microtissue strain was calculated as the tissue extension ($\Delta L1 + \Delta L2$) along the tensile direction divided by original tissue length (L) (Fig. 4b). Stretching-induced tensile force was calculated based on the micropillar deflection caused by stretching ($F = k\,(\delta 1' + \delta 2')$). Summation of the tensile forces from all micropillars was divided by 2 and then divided by the cross-section area of the microtissue to obtain the tensile stress. Elastic modulus (E, stiffness) was calculated as tensile stress divided by tensile strain. The thickness of microtissues was obtained using confocal microscopy. Microtissue compliance could be calculated as stretching-induced tissue extension ($\Delta L1 + \Delta L2$) divided by tensile force (F).

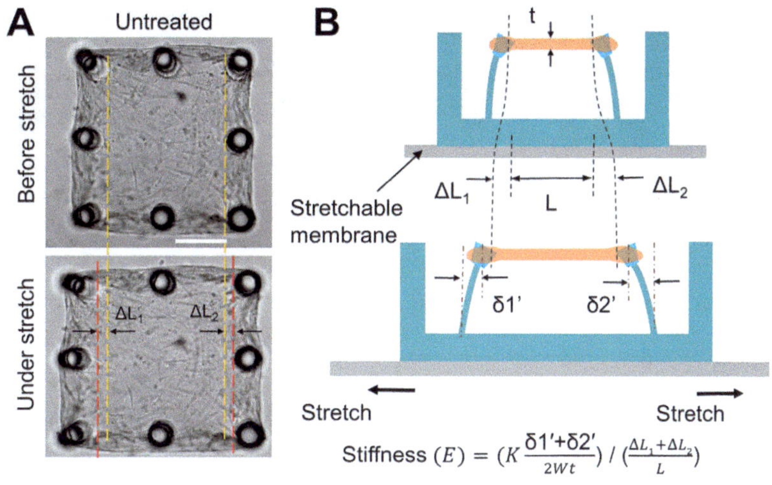

Fig. 4 (**a**) Phase contrast images of microtissues before and under stretching. Stretch-induced extension of microtissue is ($\Delta L1 + \Delta L2$). Scale bar is 200 μm. (**b**) Schematic shows the principle of tissue stiffness measurement

3.5 Immunostaining and Microscopy

Tissue sectioning during immunostaining is not required since the microtissues are within desired imaging thickness, it is also not feasible to cut the tissues since they are suspended over flexible PDMS pillars.

1. Wash microdevice with PBS for a few minutes.
2. Fix for 10 min in 4% PFA.
3. Remove fixative and rinse twice with PBS for 5 min.
4. If staining for intracellular protein, permeabilize with chilled methanol in −20 °C for 3 min.
5. Rinse twice with PBS for 5 min. At this point, fixed and permeabilized cells can be stored in PBS at 4 °C if necessary.
6. Block with 3% BSA for 20–30 min at 37 °C or 60 min at room temperature.
7. Dilute primary antibodies in 3% BSA to working concentration. This depends on the antibody; for instance, for α-SMA antibody add 2–3 μL of primary ab to 600 μL of 3% BSA (~1:300).
8. Remove blocking solution (do not rinse with PBS) and apply primary antibody, 600 μL is enough to cover the surface. Incubate overnight at 4 °C in the fridge.
9. Perform secondary staining in the dark to avoid photobleaching of fluorophores. Remove the primary antibody and wash twice with PBS for 10 min.
10. Block slides with 10% goat serum for 30 min at room temperature.

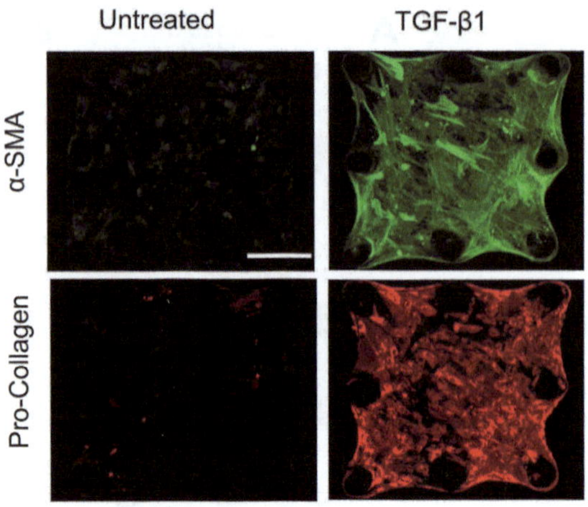

Fig. 5 Representative fluorescent confocal images of untreated and TGF-β1-treated microtissues. TGF-β1 treatment induced strong expressions of α-SMA stress fibers and cytosolic pro-collagen in lung fibroblast-populated microtissue. Scale bar is 200 μm

11. Dilute secondary antibody in 10% goat serum to working concentration (1:400), make sure the secondary matches the host of the primary. For example, if the primary is mouse anti-α-SMA, then the secondary should be goat anti-mouse.
12. Remove blocking solution (do not rinse with PBS) and apply secondary antibody to slides for 1 h at room temperature.
13. Rinse with PBS for 5 min.
14. Add Hoechst 33342 (diluted 1:1000 in PBS) for 5 min to stain nuclei.
15. Rinse with PBS for 5 min, and then keep them in PBS, so they will not dry out until imaging.
16. The microtissue array is then removed from the PDMS base and inverted on a glass slide for imaging. The microtissue array is submerged under a puddle of PBS during imaging (Fig. 5).

4 Notes

1. Primary cells at density of 187,000 cells per cm^2 were plated in T75 flask and 15 mL of media was added. After 4 days the flasks have reached 75–80% confluency and were sub-cultured into four flasks. For cell freezing, primary cells were frozen at passages 1 and 2 in freezing media. For the purpose of this research NHLFs were cultured up to 6 passages with media change every 3 days. After 6 passages the effect of aging was evident in the cells, as their growth rate slows down, and they become less sensitive to TGF-β1 treatment.

2. Final concentration of neutralized collagen is 3 mg/mL in the current research.
3. Methanol should be prechilled in −20 C freezer before adding to the fixed tissues. The permeabilizing process must take place in freezer as well.
4. As an optional step you can sterilize the device under UV light while it is being ethanol treated.
5. The purpose of this step is to make the microwell walls and the pillars non-adhesive so that the cell/collagen mixture can detach from the wall and slide to the top of the pillar during formation.
6. F127 Pluronic treatment duration varies depending on the cell and ECM type and is a major factor affecting the tissue formation. Some typical treatment times are: 1–2 min for 1 mg/ml collagen; 7–8 min for 2.5 mg/mL collagen; 10 min for 3 mg/mL collagen.
7. The amounts/ratios listed in Table 1 is adjusted for 3T3 fibroblasts, human umbilical cord vascular endothelial cells, and bovine pulmonary artery smooth muscle cells. Some cells such as the neonatal rat myocytes require Matrigel (~10%) to be supplemented into the mixture as well.
8. Placing an ice pack in the desiccator works well for maintaining the low temperature so that the collagen will not polymerize during degassing. However, the collagen may freeze if left for a long period of time (10–15 min) under vacuum in cold desiccator.
9. The cell density needs to the optimized for different experiments. For instance, if you spin down 400,000 cells in a microdevice, you should get around 300–400 cells per microtissue construct. The ideal number of cells in each microwell depends on the cell type and the shape/configuration of the micropillar and microwell. If the cells in each microwell are not enough, gently shake the device and spin down again.
10. Micropillar devices used for our studies contain micropillars that can be treated as microcantilevers. They serve not only as mechanical constraints for the cells but also as force sensors to allow the measurement of contractile forces exerted by the cells. Contraction forces of the tissues are calculated from the deflection of the head of the micropillars, based on cantilever bending theory (Fig. 2b, c).
11. After 6 days of culture, microtissue array-mounted silicone substrate was mounted on the stretching platform and was stretched to 120% of initial length. Phase images of the microtissue and the micropillars before and after tensile test were taken.

References

1. Wynn TA, Ramalingam TR (2012) Mechanisms of fibrosis: therapeutic translation for fibrotic disease. Nat Med 18:1028. https://doi.org/10.1038/nm.2807
2. Friedman S, Sheppard D, Duffield J, Violette S (2013) Therapy for fibrotic diseases: nearing the starting line. Sci Transl Med 5:167sr1. https://doi.org/10.1126/scitranslmed.3004700
3. King TE, Pardo A, Selman M (2011) Idiopathic pulmonary fibrosis. Lancet 378:1949. https://doi.org/10.1016/S0140-6736(11)60052-4
4. Hinz B (2012) Mechanical aspects of lung fibrosis: a spotlight on the myofibroblast. Proc Am Thorac Soc 9:137. https://doi.org/10.1513/pats.201202-017AW
5. Ebihara T, Venkatesan N, Tanaka R, Ludwig MS (2000) Changes in extracellular matrix and tissue viscoelasticity in bleomycin-induced lung fibrosis. Temporal aspects. Am J Respir Crit Care Med 162:1569. https://doi.org/10.1164/ajrccm.162.4.9912011
6. Ahluwalia N, Shea BS, Tager AM (2014) New therapeutic targets in idiopathic pulmonary fibrosis. Aiming to rein in runaway wound-healing response. Am J Respir Crit Care Med 190:867. https://doi.org/10.1164/rccm.201403-0509PP
7. Xu Q, Norman JT, Shrivastav S, Lucio-Cazana-J, Kopp JB (2007) In vitro models of TGF-beta-induced fibrosis suitable for high-throughput screening of antifibrotic agents. Am J Physiol Ren Physiol 293:F631. https://doi.org/10.1152/ajprenal.00379.2006
8. Porras AM, Hutson HN, Berger AJ, Masters KS (2016) Engineering approaches to study fibrosis in 3-D in vitro systems. Curr Opin Biotechnol 40:24. https://doi.org/10.1016/j.copbio.2016.02.006
9. Neidert MR, Tranquillo RT (2006) Tissue-engineered valves with commissural alignment. Tissue Eng 12:891. https://doi.org/10.1089/ten.2006.12.891
10. Kimlin L, Kassis J, Virador V (2013) 3D in vitro tissue models and their potential for drug screening. Expert Opin Drug Discov 8:1455. https://doi.org/10.1517/17460441.2013.852181
11. Feng Y, Mitchison TJ, Bender A, Young DW, Tallarico JA (2009) Multi-parameter phenotypic profiling: using cellular effects to characterize small-molecule compounds. Nat Rev Drug Discov 8:567. https://doi.org/10.1038/nrd2876
12. Chen CZ, Raghunath M (2009) Focus on collagen: in vitro systems to study fibrogenesis and antifibrosis state of the art. Fibrogenesis Tissue Repair 2. https://doi.org/10.1186/1755-1536-2-7
13. Huh D (2010) Reconstituting organ-level lung functions on a chip. Science 328:1662. https://doi.org/10.1126/science.1188302
14. Grosberg A, Alford P, McCain M, Parker K (2011) Ensembles of engineered cardiac tissues for physiological and pharmacological study: heart on a chip. Lab Chip 11:4165. https://doi.org/10.1039/c1lc20557a
15. Bhatia SN, Ingber DE (2014) Microfluidic organs-on-chips. Nat Biotechnol 32:760. https://doi.org/10.1038/nbt.2989
16. Huh D (2012) A human disease model of drug toxicity–induced pulmonary edema in a lung-on-a-chip microdevice. Sci Transl Med 4:159ra147. https://doi.org/10.1126/scitranslmed.3004249
17. Piciucchi S (2016) From "traction bronchiectasis" to honeycombing in idiopathic pulmonary fibrosis: a spectrum of bronchiolar remodeling also in radiology? BMC Pulm Med 16. https://doi.org/10.1186/s12890-016-0245-x
18. King TE (2014) A phase 3 trial of pirfenidone in patients with idiopathic pulmonary fibrosis. N Engl J Med 370:2083. https://doi.org/10.1056/NEJMoa1402582
19. Zhao R, Boudou T, Wang WG, Chen CS, Reich DH (2013) Decoupling cell and matrix mechanics in engineered microtissues using magnetically actuated microcantilevers. Adv Mater 25:1699. https://doi.org/10.1002/adma.201203585

Part III

Animal and Human Models to Study Myofibroblast Biology

Chapter 20

Animal and Human Models of Tissue Repair and Fibrosis: An Introduction

David Lagares and Boris Hinz

Abstract

Reductionist cell culture systems are not only convenient but essential to understand molecular mechanisms of myofibroblast activation and action in carefully controlled conditions. However, tissue myofibroblasts do not act in isolation and the complexity of tissue repair and fibrosis in humans cannot be captured even by the most elaborate culture models. Over the past five decades, numerous animal models have been developed to study different aspects of myofibroblast biology and interactions with other cells and extracellular matrix. The underlying principles can be broadly classified into: (1) organ injury by trauma such as prototypical full thickness skin wounds or burns; (2) mechanical challenges, such as pressure overload of the heart by ligature of the aorta or the pulmonary vein; (3) toxic injury, such as administration of bleomycin to lungs and carbon tetrachloride to the liver; (4) organ infection with viruses, bacteria, and parasites, such as nematode infections of liver; (5) cytokine and inflammatory models, including local delivery or viral overexpression of active transforming growth factor beta; (6) "lifestyle" and metabolic models such as high-fat diet; and (7) various genetic models. We will briefly summarize the most widely used mouse models used to study myofibroblasts in tissue repair and fibrosis as well as genetic tools for manipulating myofibroblast repair functions in vivo.

Key words Mouse models, Fibrosis, Myofibroblast, Mechanical overload, Injury, Toxins, Inflammation, Infection

1 Introduction: Different Philosophies to Study Myofibroblasts

It is not uncommon to meet different phenotypes of fibrosis researchers at the respective specialist meetings: Type 1: The "Humanist" frowns upon any data that is not directly derived from human subjects. Frequent arguments of the Humanist are: "Mice are not man" or "A dish is a dish and not a human" or "How is this finding clinically relevant?" or "It is bedside to bench and not vice versa." Type 2: "The Mouse Healer" knocks out every possible murine gene and hopes for a phenotype. Absence of a knockout phenotype is countered by increasing the levels of insult to the mouse. For more general considerations of this approach, the reader is encouraged to consult an excellent essay on knockout

Boris Hinz and David Lagares (eds.), *Myofibroblasts: Methods and Protocols*, Methods in Molecular Biology, vol. 2299,
https://doi.org/10.1007/978-1-0716-1382-5_20,

strategies [1]. Variations of the Mouse Healer are "The Tracer" who labels proteins and cell types to follow their location and fate in the mouse, and "The Counter" who uses surface proteins or RNA codes of single cells to sort them into clusters—which is the scientific term for "box." Type 3: "The Reductionist" lives in their own world—which is typically a petri dish. Humans and mice are merely tolerated as donors for cell material that is then analyzed to the atomic level with neat linear pathways as research outcome. Very precise data obtained from cells sitting on a plastic surface in an isotonic cocktail spiced with cow blood extracts are interpreted with universal statements about tissue fibrosis—often with insinuated relevance for tumor biology. The information age and technical advances created Type 4, "The Analyst," who does not need to know—and often does not care—where data came from: human, mouse, or cells in dishes as long as they are so big that nobody can make sense of them anymore. The Analyst is a good friend of the Cell Counter and together they create esthetically pleasing color-coded diagrams or relationship networks which awe their colleagues but are still too complex to be understood. Of course, in real life, all these phenotypes coexist at different expression levels within every researcher studying myofibroblasts, wound healing, and fibrosis. And only if they collaborate, we stand a chance to understand and eventually cure fibrosis.

2 Advanced Human Culture Models

While the preceding chapters were dedicated to "simple" cell culture models that please the Reductionist, the following contributions will give credit to animal models and elaborate models with human cells and tissues that advanced our understanding of myofibroblasts in their "real" tissue environment. Examples of miniaturized two-dimensional (2D) substrates with and without externally applied strain [2–4], and more complex in vitro environment using myofibroblast-derived extracellular matrix (ECM), either with or without the cells still present [5, 6] have been presented in the preceding chapter. In this chapter, authors present complex heterocellular culture models with human cells, such as organotypic cultures and stem cell-derived organoids [7–9]. If human tissue is available for experimental studies, myofibroblast cultures can be established and studied on human tissue-derived decellularized ECM [10–13], living human tissue slices, precision-cut with a vibratome, or grafting human tissue onto immune-compromised mice, as done with human skin grafts to study hypertrophic scarring [14]. Such models are essential to study mechanisms of human-specific processes and cell interactions or approximate fibrotic tissue conditions that cannot be produced in animal models. Examples are "true" skin hypertrophic scarring (*see* contribution by V. Moulin)

[15, 16] and formation of keloids [17, 18], as well as myofibroblast-driven palmar fascia contractures that are characteristic Dupuytren's disease [19, 20].

3 Mouse Models of Tissue Repair and Fibrosis

Wound healing and fibrosis models exist for different animal species with different potential to translate to the human condition [21]. Like in other areas of biomedical research, mice are the predominant study objects because of low purchase and maintenance cost, fast fibrogenic kinetics, and—most importantly—the ability to be manipulated genetically, e.g., by gene knockouts (KO) (Fig. 1). As discussed by many contributors to this chapter, one animal model is often not sufficient to reproduce a human condition and multiple models are required to illuminate myofibroblasts from different angles, all with their strengths and limitations. We will not be able to capture all organs and approaches in the following chapter. The inclined reader is referred to excellent reviews with technical details provided for animal models of fibrosis of the heart [22, 23], liver [24, 25], skin [26, 27], skeletal muscle [28, 29], kidney [30, 31], intestine [32], lung [33–35], brain [36], and eye [37] (Table 1).

While animals have been used to study repair after organ insult and ensuing fibrosis throughout human history, the existence of myofibroblasts has been first documented by ultrastructural analysis of rat wounds, healing after full thickness excision of skin, physical scarring of liver tissue, and subcutaneous injection of croton oil to elicit an inflammatory reaction in response to its phorbol ester content [38], not unlike current models of bleomycin-induced skin fibrosis. Myofibroblasts are part of the normal body response to injury which becomes dysregulated in fibrosis due to persisting or overwhelming organ damage. Consequently, most experimental models of mouse fibrosis generate conditions of injury and/or inflammation using different principles: physical injury or trauma, acute and chronic mechanical overload, tissue damage by toxins, organ infections, and delivery of pro-inflammatory and pro-fibrotic cytokines (Fig. 1, Table 1).

Fibrosis models based on "physical" tissue damage, such as resections of kidneys, or bowels [30, 32], keratectomy in the eye [37], excisions of full thickness skin and burns [26, 27, 39], implantation or delivery of body foreign materials [26, 27, 33–35], overstretching of vessels by ballooning [40, 41], and gamma irradiation [32, 34, 42]. Related approaches exploit the fact that myofibroblasts are activated by mechanical stress [43], which is discussed in Chap. 2 of this issue. In mechanically induced animal models of fibrosis, enhanced or chronic mechanical load is experimentally generated in skin, kidney, heart, and lungs. In rodent skin,

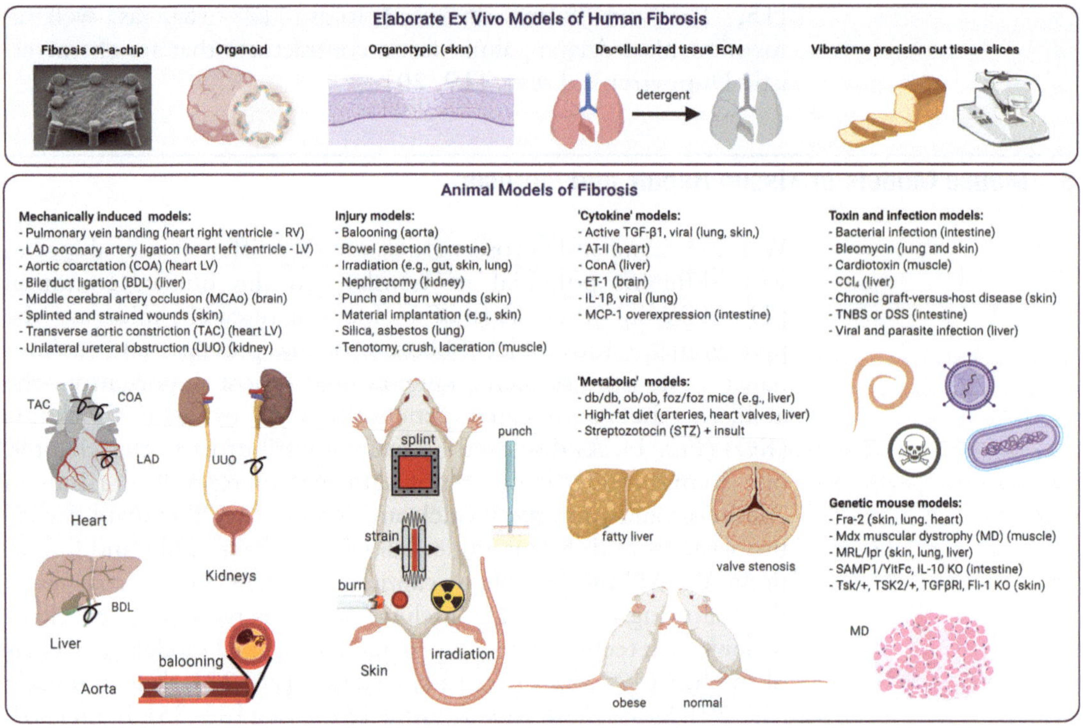

Fig. 1 Human ex vivo and animal models of fibrosis. The figure summarizes different models of fibrosis. *Abbreviations:* TNBS 2,4,6-Trinitrobenzene sulfonic acid, AT-II Angiotensin II, ET-1 Endothelin-1, COA Aortic coarctation, BDL Bile duct ligation, CCl_4 Carbon tetrachloride, ConA Concanavalin A, DSS Dextran sodium sulfate, Fra2 Fos related antigen 2, Fli-1 Friend leukemia integration 1, IL interleukin, IRI Ischemia reperfusion injury, KO Knockout, LAD left anterior descending [95], Left anterior descending coronary artery ligation, LV Left ventricle, MCAo Middle cerebral artery occlusion, MCP-1 Monocyte chemoattractant protein-1; MRL/lpr MRL lymphoproliferation, MD Muscular dystrophy, tight skin [96], TGFβRI TGF-β1 receptor type II, RV Right ventricle, STZ Streptozotocin, TAC Transverse aortic constriction [89], UUO Unilateral ureteral obstruction

splints are used to restrain contraction of full-thickness wounds passively or actively using actuated expander devices sutured to the edges of incisional wounds [44–47]. Tubular organ systems, such as liver bile ducts, urethra, arteries and veins, or the cerebral artery are surgically constricted to increase upstream fluid pressure and/or ischemia in these systems (Fig. 1, Table 1). For instance, aortic coarctation (COA) results in left ventricle (LV) heart fibrosis and ligation of pulmonary veins in right ventricle (RV) fibrosis [22, 23]. Alternatively, vasoconstrictors, such as angiotensin (AT)-II and endothelin (ET)-1, are delivered locally or systemically to generate high blood pressure and organ overload [48]. Although mechanical stress does not directly trigger inflammation, tissue microinjuries often ensue that then cause secondary chronic inflammatory reactions.

Inducing acute or chronic inflammation is another strategy to cause a subsequent wound healing response that involves myofibroblast activation and actions. The most widely used

Table 1
Animal models of fibrosis

Animal model	Effect	Organ	References
Injury models			
Ballooning	Partial constriction of thoracic inferior vena cava; pressure overload, aortic intima damage	Heart, aorta	[40, 41]
Bowel resection	Pressure overload	Intestine	[32]
Nephrectomy	Hyperfiltration of the remnant kidney	Kidney	[30]
Punch and burn wounds	Tissue injury; inflammation	Skin	[26, 27]
Keratectomy	Corneal damage; inflammation	Eye	[37]
Irradiation	Bowel inflammation (intestine), tissue microdamage (skin), alveolar epithelial cell damage (lung)	Intestine, skin, lung	[32, 34, 42]
Material implantation	Tissue injury; inflammation	Skin	[26, 27]
Silica, asbestos	Tissue microinjury, development of fibrotic nodules	Lung	[33–35]
Mechanically induced models			
Splinted and strained wounds	Wound tissue stress	Skin	[26, 27]
Pulmonary vein banding	Blood pressure overload	Heart, RV	[22, 23]
Aortic coarctation (COA)	Blood pressure overload	Heart, LV	[22, 23]
Transverse aortic constriction	Blood pressure overload	Heart, LV	[22, 23, 89, 97]
Left anterior descending coronary artery ligation	Ischemia, infarct	Heart, LV	[22, 23, 98]
Bile duct ligation (BDL)	Bile pressure overload	Liver	[24, 25]
Unilateral ureteral obstruction (UUO)	Urine pressure overload	Kidney	[30, 31]
Ischemia reperfusion injury (IRI)	Acute kidney injury	Kidney	[30, 99]
Middle cerebral artery occlusion (MCAo)	Ischemia, stroke	Brain	[36]
Cytokine models			
Virally expressed active TGF-β1	Fibroblast activation, lung epithelial cell injury	Lung, skin, intestine	[26, 27, 34, 100]
AT-II	Fibroblast proliferation and ECM protein synthesis	Heart	[48]

(continued)

Table 1 (continued)

Animal model	Effect	Organ	References
ConA	Activation and recruitment of T-cells to the liver	Liver	[24, 25, 53]
ET-1	Fibroblast activation	Brain	[48]
Virally expressed IL-1β	Inflammation	Lung	[34, 101, 102]
MCP-1 overexpression	Inflammation	Intestine	[32, 54]
Metabolic models			
db/db, ob/ob, foz/foz mice	Persistent hyperphagia, and are obese and diabetic Hyperglycemia, hyperinsulinemia, high serum leptin levels; hepatic inflammation	Liver	[52, 58]
High-fat diet	Obesity, peripheral insulin resistance; hypertension	Arteries, heart valves, liver	[24, 25, 52, 56]
Streptozotocin (STZ) + insult	persistent hyperphagia, and are obese and diabetic Hyperglycemia	Systemic	[57]
Genetic mouse models			
Fos related antigen 2 (Fra2)	Microvascular injury and fibrosis	Skin, lung, heart	[26, 27, 34, 103, 104]
Mdx muscular dystrophy (MD)	Reduction of muscle fibrosis and fibrotic proteins	Muscle	[28, 29]
MRL/lpr	Spontaneous autoimmune response	Skin, lung, liver	[26, 27, 52, 103]
SAMP1/YitFc, IL-10 KO	Spontaneous inflammation; granulomas in the mucosa and submucosa	Intestine	[32]
Tight skin (TSK)1/+, TSK2/+, TGFβRI, Fli-1 KO	Altered ECM elastic fiber assembly or altered TGF-β bioavailability	Skin	[26, 27, 96, 103, 105]
Toxin and infection models			
Bacterial infection	Inflammation	Intestine	[32]
Bleomycin	Tissue injury and inflammation via direct DNA damage to cells	Lung, skin	[26, 27, 33–35, 51]
Cardiotoxin	Skeletal muscle injury	Muscle	[28, 29]

(continued)

Table 1 (continued)

Animal model	Effect	Organ	References
CCl4	Hepatotoxin; tissue damage	Liver	[24, 25, 52]
Aristolochic acid, folic acid, adenine crystals	Tubular damage	Kidney	[30]
Chronic graft-versus-host disease	Inflammation	Skin	[26, 27, 103]
TNBS or DSS	Epithelial cell injury; inflammation	Intestine	[32]
Viral and parasite infection	Immune response and fibrosis	Liver	[24, 25]

All abbreviations are defined in the text

pro-inflammatory toxic agents to induce animal fibrosis are bleomycin in skin and lung [33, 49–51], carbon tetrachloride (CCl_4) in liver [24, 25, 52], cardiotoxin in muscle [28, 29], and 2,4,6-trinitrobenzene sulfonic acid (TNBS) and dextran sodium sulfate (DSS) in the intestine [32]. In addition, pro-inflammatory and pro-fibrotic cytokines are locally produced upon injection of viral vectors that code the expression of cytokines such as active TGF-β1, interleukin (IL)-1β, or are directly administered like concanavalin A (ConA), and monocyte chemoattractant protein (MCP)-1 [24, 25, 32, 53, 54].

One important limitation of many mouse models of induced fibrosis is spontaneous resolution after removing the insult, whereas human fibrosis is often irreversible. For instance, the lungs of young mice develop severe fibrosis with dramatic histological changes after 14–21 days of repeated intratracheal administration of bleomycin in moderate doses, but reestablish almost normal morphology and function within 35–50 days when the treatment is discontinued [34, 35, 55]. In contrast, human idiopathic pulmonary fibrosis is considered irreversible at the advanced stages typically presenting at first diagnosis [35]. To overcome these limitations, genetic and/or dietary/metabolic "lifestyle" mouse models are used that either develop spontaneous fibrosis and/or sensitize the animal to experimentally induced fibrosis (Table 1, Fig. 1). Subjecting mice to high-fat diet results in spontaneous cardiovascular fibrosis, including development of interstitial valve fibrosis [24, 25, 52, 56]. Furthermore, diabetic or hyperglycemic mice are often more prone to develop organ fibrosis which is studied by killing pancreatic beta-cells with streptozotocin (STZ) [57] or using db/db, ob/ob or foz/foz mice [52, 58]. Other genetic mouse models that develop spontaneous fibrosis are mostly organ-specific and summarized in Table 1 and Fig. 1.

4 Genetic Strategies to Study Myofibroblast Functions In Vivo

Myofibroblasts are in the crosshairs of research on wound repair and fibrotic diseases due to their matrix synthetic and contractile phenotype [59]. These distinct functional characteristics of activated myofibroblasts have served as a basis for developing genetic strategies to manipulate their biology in vivo. The Cre-LoxP system is widely used to knock out genes in myofibroblasts in vivo [60] (Table 2), enabling the study of myofibroblast functions in experimental models. Mouse lines expressing Cre recombinase under the control of alpha smooth muscle actin (α-SMA) and collagen type I promoters (α-SMA-Cre [61], Col1a1-Cre [62] and Col1a2-Cre [63] lines) are commonly used to knock out pro-fibrotic genes in myofibroblasts and in lineage tracing studies in mouse models of tissue repair and fibrosis [60, 64]. In addition, tamoxifen-inducible versions of these Cre lines including Col1a2-Cre^{ERT} [65] and α-SMA-Cre^{ERT} [66] have been generated in order to conditionally disrupt myofibroblast genes in adult mice, albeit recombination at lower frequency has been reported due to lower Cre protein levels and activity. While these genetically engineered mice are valuable research tools and certainly target a major population of collagen-producing cells during wound repair and fibrosis in multiple organs, controversy is always around these tools due to issues with recombination efficiency and specificity. First, it is typically argued that both collagen type I and α-SMA are not specific markers of myofibroblasts, raising doubts about the selectivity of these mouse Cre lines at targeting myofibroblasts in vivo. To solve this problem, alternative tools have been generated, including platelet-derived growth factor receptor (PDGFR)α-Cre [67, 68], PDGFR-β-Cre [69–71], periostin-Cre [72, 73], and FSP1-Cre [74] mouse lines, which intent to target more specific subsets of myofibroblasts. Whereas these Cre lines mostly target collagen-producing myofibroblasts in vivo, recombination has been detected in non-fibroblastic cell types as well.

Underlying this cell specificity issue are fundamental questions in myofibroblasts biology, the origin of myofibroblasts and myofibroblast heterogeneity. Given the multiple cellular sources of myofibroblasts (see Reviews [75, 76]), manipulation of specific progenitor population has been proposed as an alternative approach to target distinct myofibroblast linages in tissue repair and fibrosis. FoxD1-Cre^{ERT} has been used to target a progenitor cell population in the kidney that gives rise to specific α-SMA+ myofibroblasts during the development of kidney fibrosis [77]. Similarly, the early embryonic transcription factor T-box gene 4 (TBX4) has been shown to label progenitor mesenchymal progenitors in the lungs; therefore TBX4-Cre^{ERT} mice have been used to target myofibroblasts in experimental models of lung

Table 2
Mouse Cre lines to study myofibroblast functions in vivo

Cre Mouse Line	Expression in Cell Types	Organs	References
α-SMA-Cre, α-SMA-Cre^{ERT}	Myofibroblasts, smooth muscle cells, pericytes	Lung, skin, kidney, liver, heart	[61, 66, 106, 107]
Col1a1-Cre, Col1a2-Cre, Col1a2-Cre^{ERT}	Myofibroblasts, fibroblasts, hepatic stellate cells, osteoblasts	Lung, skin, kidney, liver, heart, peritoneum	[62, 63, 65, 108–111]
PDGFRα-Cre, PDGFRα-Cre^{ERT}	Myofibroblasts, fibroblasts, glia	Lung, liver, heart, skin, brain	[67, 68, 112, 113]
PDGFRβ-Cre, PDGFRβ-Cre^{ERT}	Myofibroblasts, pericytes, endothelial cells, adipocytes	Kidney, liver, skin, lung, heart	[70, 71, 114]
Periostin-Cre, Periostin-Cre^{ERT}	Myofibroblasts, fibroblasts	Heart	[72, 73]
FSP1-Cre	Myofibroblasts, fibroblasts, macrophages	Lung, kidney, liver, heart	[115–118]
FoxD1-Cre, FoxD1-Cre^{ERT}	Embryonic renal cells Adult myofibroblasts, pericytes	Kidney, lung	[77, 119]
TBX4-Cre	Embryonic lung mesenchymal myofibroblasts, pericytes	Lung	[78]
Gli-Cre^{ERT}	Myofibroblasts, pericytes, smooth muscle cells, neural stem cells	Kidney, lung, liver, heart	[79, 120, 121]
Engrailed-1-Cre	Myofibroblasts	Skin	[81]
Lrat-Cre	Myofibroblasts	Liver	[83]

fibrosis [78]. In addition, the Gli-Cre^{ERT} mouse line has been used to target cardiac myofibroblasts arising from perivascular progenitors in the context of cardiac fibrosis [79]. Linage tracing studies have shown that FoxD1, TBX4, and Gli target myofibroblasts derived from different origins in multiple organs [60]. Myofibroblast heterogeneity is also associated with functional diversity, which may reflect not only distinct myofibroblast lineages but also different activates states. In this regard, two dermal fibroblast populations derived from the same embryonic progenitor have been shown to exert distinct function during tissue repair and fibrosis [80, 81]. Noteworthy, the engrailed-1-Cre line has been shown to specifically label fibrogenic myofibroblasts in the context of skin scarring [82], indicating that targeting myofibroblasts based on their functional selectivity could be a feasible approach. This approach has been also pursued in mouse models of liver fibrosis, where hepatic stellate cells represent a unique tissue-specific myofibroblast population due to their retinoid storage. This characteristic has led to the development of lecithin-retinol acyltransferase (Lrat)-Cre line, which targets over 90% of α-SMA+ myofibroblasts during the development of liver fibrosis [83]. More recently, novel single-cell analysis studies have revealed molecularly distinct (myo)-fibroblast subpopulations in fibrotic tissues [84–94], which may have different functions during the development and progression of tissue fibrosis. Whether these subsets represent unique myofibroblast subsets or different transient activation states remains to be elucidated. The development of novel Cre lines targeting specific myofibroblast subsets will help advance our understanding around myofibroblast heterogeneity and their functions in tissue repair and fibrotic disease. The abovementioned genetic tools in combination with established experimental models of tissue repair and fibrosis described in this chapter have also led to establishing preclinical proof of concept of potential targets for anti-fibrotic therapy, which are typically validated using ex vivo models of fibrotic diseases such as human precision cut tissue slices or human Organ-on-a-Chip. Some of these human and murine models are introduced and described in the upcoming contributions to this chapter.

Acknowledgments

The research of B.H. is supported by a foundation grant from the Canadian Institutes of Health Research (#375597) and support from the John Evans Leadership funds (#36050 and #38861) and innovation funds ("Fibrosis Network, #36349") from the Canada Foundation for Innovation (CFI) and the Ontario Research Fund (ORF). D.L. gratefully acknowledges funding support from the NIH (grant R01 HL147059-01), the Start-up Package from Massachusetts General Hospital, the Scleroderma Foundation New Investigator Grant, the Scleroderma Research Foundation Investigator-Initiated Research Grant, the American Thoracic Society Foundation/Pulmonary Fibrosis Foundation Research Grant and Sponsored Research Grants from Boehringer Ingelheim, Unity Biotechnology and Indalo Therapeutics. Dr. Lagares has a financial interest in Mediar Therapeutics. The company is developing treatments for organ fibrosis. Dr. Lagares's interests were reviewed and are managed by MGH and Partners Health Care in accordance with their conflict of interest policies.

References

1. Lazebnik Y (2002) Can a biologist fix a radio?—or, what I learned while studying apoptosis. Cancer Cell 2(3):179–182
2. MacQueen L, Sun Y, Simmons CA (2013) Mesenchymal stem cell mechanobiology and emerging experimental platforms. J R Soc Interface 10(84):20130179
3. Asmani M et al (2018) Fibrotic microtissue array to predict anti-fibrosis drug efficacy. Nat Commun 9(1):2066
4. Mazza G, Al-Akkad W, Rombouts K (2017) Engineering in vitro models of hepatofibrogenesis. Adv Drug Deliv Rev 121:147–157

5. Benny P, Raghunath M (2017) Making microenvironments: a look into incorporating macromolecular crowding into in vitro experiments, to generate biomimetic microenvironments which are capable of directing cell function for tissue engineering applications. J Tissue Eng 8:2041731417730467
6. Franco-Barraza J et al (2016) Preparation of extracellular matrices produced by cultured and primary fibroblasts. Curr Protoc Cell Biol 71:10 9 1–10 9 34
7. Moulin VJ (2013) Reconstitution of skin fibrosis development using a tissue engineering approach. Methods Mol Biol 961:287–303
8. Sobral-Reyes MF, Lemos DR (2020) Recapitulating human tissue damage, repair, and fibrosis with human pluripotent stem cell-derived organoids. Stem Cells 38(3):318–329
9. Sundarakrishnan A et al (2018) Engineered cell and tissue models of pulmonary fibrosis. Adv Drug Deliv Rev 129:78–94
10. Paish HL et al (2019) A bioreactor technology for modeling fibrosis in human and rodent precision-cut liver slices. Hepatology 70(4):1377–1391
11. Clouzeau-Girard H et al (2006) Effects of bile acids on biliary epithelial cell proliferation and portal fibroblast activation using rat liver slices. Lab Investig 86(3):275–285
12. Parker MW et al (2014) Fibrotic extracellular matrix activates a profibrotic positive feedback loop. J Clin Invest 124(4):1622–1635
13. Uhl FE, Wagner DE, Weiss DJ (2017) Preparation of decellularized lung matrices for cell culture and protein analysis. Methods Mol Biol 1627:253–283
14. Ding J, Tredget EE (2017) Transplanting human skin grafts onto nude mice to model skin scars. Methods Mol Biol 1627:65–80
15. Gauglitz GG et al (2011) Hypertrophic scarring and keloids: pathomechanisms and current and emerging treatment strategies. Mol Med 17(1–2):113–125
16. Kuehlmann B et al (2020) Mechanotransduction in wound healing and fibrosis. J Clin Med 9(5):1423
17. Supp DM (2019) Animal models for studies of keloid scarring. Adv Wound Care (New Rochelle) 8(2):77–89
18. Sharma JR et al (2019) In vitro and ex vivo models for functional testing of therapeutic anti-scarring drug targets in keloids. Adv Wound Care (New Rochelle) 8(12):655–670
19. Hinz B, Gabbiani G (2011) The role of myofibroblasts in Dupuytren's disease: fundamental aspects of contraction and therapeutic perspectives. In: Eaton C et al (eds) Morbus Dupuytren and related hyperproliferative disorders: principles, research, and clinical perspectives. Springer, New York
20. Layton T, Nanchahal J (2019) Recent advances in the understanding of Dupuytren's disease. F1000Res 8. https://doi.org/10.12688/f1000research.17779.1
21. Gordillo GM et al (2013) Preclinical models of wound healing: is man the model? Proceedings of the wound healing society symposium. Adv Wound Care (New Rochelle) 2(1):1–4
22. Rai V et al (2017) Relevance of mouse models of cardiac fibrosis and hypertrophy in cardiac research. Mol Cell Biochem 424 (1–2):123–145
23. Cowling RT et al (2019) Mechanisms of cardiac collagen deposition in experimental models and human disease. Transl Res 209:138–155
24. Reimer KC et al (2020) New drugs for NAFLD: lessons from basic models to the clinic. Hepatol Int 14(1):8–23
25. Klein S et al (2017) Mouse and rat models of induction of hepatic fibrosis and assessment of portal hypertension. Methods Mol Biol 1627:91–116
26. Yue X et al (2018) Recent advances in mouse models for systemic sclerosis. Autoimmun Rev 17(12):1225–1234
27. Barnes LA et al (2018) Mechanical forces in cutaneous wound healing: emerging therapies to minimize scar formation. Adv Wound Care (New Rochelle) 7(2):47–56
28. Mahdy MAA (2019) Skeletal muscle fibrosis: an overview. Cell Tissue Res 375(3):575–588
29. Pessina P et al (2014) Novel and optimized strategies for inducing fibrosis in vivo: focus on Duchenne muscular dystrophy. Skelet Muscle 4:7
30. Nogueira A, Pires MJ, Oliveira PA (2017) Pathophysiological mechanisms of renal fibrosis: a review of animal models and therapeutic strategies. In Vivo 31(1):1–22
31. Zeisberg M, Soubasakos MA, Kalluri R (2005) Animal models of renal fibrosis. Methods Mol Med 117:261–272
32. Rieder F et al (2012) Animal models of intestinal fibrosis: new tools for the understanding of pathogenesis and therapy of human disease. Am J Physiol Gastrointest Liver Physiol 303 (7):G786–G801
33. Liu T, De Los Santos FG, Phan SH (2017) The Bleomycin model of pulmonary fibrosis. Methods Mol Biol 1627:27–42

34. Moore BB et al (2013) Animal models of fibrotic lung disease. Am J Respir Cell Mol Biol 49(2):167–179
35. Jenkins RG et al (2017) An official American Thoracic Society workshop report: use of animal models for the preclinical assessment of potential therapies for pulmonary fibrosis. Am J Respir Cell Mol Biol 56(5):667–679
36. Fluri F, Schuhmann MK, Kleinschnitz C (2015) Animal models of ischemic stroke and their application in clinical research. Drug Des Devel Ther 9:3445–3454
37. Rittie L, Hutcheon AEK, Zieske JD (2017) Mouse models of corneal scarring. Methods Mol Biol 1627:117–122
38. Gabbiani G, Ryan GB, Majno G (1971) Presence of modified fibroblasts in granulation tissue and their possible role in wound contraction. Experientia 27(5):549–550
39. Nabai L, Ghahary A (2017) Hypertrophic scarring in the rabbit ear: a practical model for studying dermal fibrosis. Methods Mol Biol 1627:81–89
40. Cops J et al (2019) Current animal models for the study of congestion in heart failure: an overview. Heart Fail Rev 24(3):387–397
41. Levy M (1972) Effects of acute volume expansion and altered hemodynamics on renal tubular function in chronic caval dogs. J Clin Invest 51(4):922–938
42. Jang WH et al (2016) In vivo characterization of early-stage radiation skin injury in a mouse model by two-photon microscopy. Sci Rep 6:19,216
43. Hinz B, McCulloch CA, Coelho NM (2019) Mechanical regulation of myofibroblast phenoconversion and collagen contraction. Exp Cell Res 379(1):119–128
44. Aarabi S et al (2007) Mechanical load initiates hypertrophic scar formation through decreased cellular apoptosis. FASEB J 21 (12):3250–3261
45. Abercrombie M, James DW, Newcombe JF (1960) Wound contraction in rabbit skin, studied by splinting the wound margins. J Anat 94:170–182
46. Davidson JM, Yu F, Opalenik SR (2013) Splinting strategies to overcome confounding wound contraction in experimental animal models. Adv Wound Care (New Rochelle) 2 (4):142–148
47. Hinz B et al (2001) Mechanical tension controls granulation tissue contractile activity and myofibroblast differentiation. Am J Pathol 159(3):1009–1020
48. Frangogiannis NG (2019) Cardiac fibrosis: cell biological mechanisms, molecular pathways and therapeutic opportunities. Mol Asp Med 65:70–99
49. Do NN, Eming SA (2016) Skin fibrosis: models and mechanisms. Curr Res Transl Med 64 (4):185–193
50. Moeller A et al (2008) The bleomycin animal model: a useful tool to investigate treatment options for idiopathic pulmonary fibrosis? Int J Biochem Cell Biol 40(3):362–382
51. Yamamoto T (2017) Intradermal injections of bleomycin to model skin fibrosis. Methods Mol Biol 1627:43–47
52. Yanguas SC et al (2016) Experimental models of liver fibrosis. Arch Toxicol 90 (5):1025–1048
53. Heymann F et al (2015) The concanavalin a model of acute hepatitis in mice. Lab Anim 49 (1 Suppl):12–20
54. Distler JH et al (2009) Monocyte chemoattractant proteins in the pathogenesis of systemic sclerosis. Rheumatology (Oxford) 48 (2):98–103
55. Schiller HB et al (2015) Time- and compartment-resolved proteome profiling of the extracellular niche in lung injury and repair. Mol Syst Biol 11(7):819
56. Watanabe S et al (2018) A high-fat and high-cholesterol diet induces cardiac fibrosis, vascular endothelial, and left ventricular diastolic dysfunction in SHRSP5/Dmcr rats. J Atheroscler Thromb 25(5):439–453
57. Kitada M, Ogura Y, Koya D (2016) Rodent models of diabetic nephropathy: their utility and limitations. Int J Nephrol Renovasc Dis 9:279–290
58. Lau JK, Zhang X, Yu J (2017) Animal models of non-alcoholic fatty liver disease: current perspectives and recent advances. J Pathol 241(1):36–44
59. Hinz B (2010) The myofibroblast: paradigm for a mechanically active cell. J Biomech 43 (1):146–155
60. Swonger JM et al (2016) Genetic tools for identifying and manipulating fibroblasts in the mouse. Differentiation 92(3):66–83
61. LeBleu VS et al (2013) Origin and function of myofibroblasts in kidney fibrosis. Nat Med 19 (8):1047–1053
62. Kisseleva T et al (2012) Myofibroblasts revert to an inactive phenotype during regression of liver fibrosis. Proc Natl Acad Sci U S A 109 (24):9448–9453
63. Zheng B et al (2002) Ligand-dependent genetic recombination in fibroblasts : a potentially powerful technique for investigating gene function in fibrosis. Am J Pathol 160 (5):1609–1617

64. Greenhalgh SN, Conroy KP, Henderson NC (2015) Cre-ativity in the liver: transgenic approaches to targeting hepatic nonparenchymal cells. Hepatology 61(6):2091–2099
65. Florin L et al (2004) Cre recombinase-mediated gene targeting of mesenchymal cells. Genesis 38(3):139–144
66. Wendling O et al (2009) Efficient temporally-controlled targeted mutagenesis in smooth muscle cells of the adult mouse. Genesis 47 (1):14–18
67. Roesch K et al (2008) The transcriptome of retinal Muller glial cells. J Comp Neurol 509 (2):225–238
68. Kang SH et al (2010) NG2+ CNS glial progenitors remain committed to the oligodendrocyte lineage in postnatal life and following neurodegeneration. Neuron 68(4):668–681
69. Hong KY et al (2015) Perilipin+ embryonic preadipocytes actively proliferate along growing vasculatures for adipose expansion. Development 142(15):2623–2632
70. Foo SS et al (2006) Ephrin-B2 controls cell motility and adhesion during blood-vessel-wall assembly. Cell 124(1):161–173
71. Henderson NC et al (2013) Targeting of alphav integrin identifies a core molecular pathway that regulates fibrosis in several organs. Nat Med 19(12):1617–1624
72. Takeda N et al (2010) Cardiac fibroblasts are essential for the adaptive response of the murine heart to pressure overload. J Clin Invest 120(1):254–265
73. Kaur H et al (2016) Targeted ablation of periostin-expressing activated fibroblasts prevents adverse cardiac remodeling in mice. Circ Res 118(12):1906–1917
74. Bhowmick NA et al (2004) TGF-beta signaling in fibroblasts modulates the oncogenic potential of adjacent epithelia. Science 303 (5659):848–851
75. Hinz B et al (2007) The myofibroblast: one function, multiple origins. Am J Pathol 170 (6):1807–1816
76. Hinz B, Lagares D (2020) Evasion of apoptosis by myofibroblasts: a hallmark of fibrotic diseases. Nat Rev Rheumatol 16(1):11–31
77. Humphreys BD et al (2010) Fate tracing reveals the pericyte and not epithelial origin of myofibroblasts in kidney fibrosis. Am J Pathol 176(1):85–97
78. Xie T et al (2016) Transcription factor TBX4 regulates myofibroblast accumulation and lung fibrosis. J Clin Invest 126(9):3626
79. Kramann R et al (2015) Perivascular Gli1+ progenitors are key contributors to injury-induced organ fibrosis. Cell Stem Cell 16 (1):51–66
80. Driskell RR et al (2013) Distinct fibroblast lineages determine dermal architecture in skin development and repair. Nature 504 (7479):277–281
81. Jiang D et al (2018) Two succeeding fibroblastic lineages drive dermal development and the transition from regeneration to scarring. Nat Cell Biol 20(4):422–431
82. Rinkevich Y et al (2015) Skin fibrosis. Identification and isolation of a dermal lineage with intrinsic fibrogenic potential. Science 348 (6232):aaa2151
83. Mederacke I et al (2013) Fate tracing reveals hepatic stellate cells as dominant contributors to liver fibrosis independent of its aetiology. Nat Commun 4:2823
84. Peyser R et al (2019) Defining the activated fibroblast population in lung fibrosis using single-cell sequencing. Am J Respir Cell Mol Biol 61(1):74–85
85. Guerrero-Juarez CF et al (2019) Single-cell analysis reveals fibroblast heterogeneity and myeloid-derived adipocyte progenitors in murine skin wounds. Nat Commun 10 (1):650
86. Philippeos C et al (2018) Spatial and single-cell transcriptional profiling identifies functionally distinct human dermal fibroblast subpopulations. J Invest Dermatol 138 (4):811–825
87. Reyfman PA et al (2019) Single-cell transcriptomic analysis of human lung provides insights into the pathobiology of pulmonary fibrosis. Am J Respir Crit Care Med 199 (12):1517–1536
88. Xie T et al (2018) Single-cell deconvolution of fibroblast heterogeneity in mouse pulmonary fibrosis. Cell Rep 22(13):3625–3640
89. Krenkel O et al (2019) Single cell RNA sequencing identifies subsets of hepatic stellate cells and myofibroblasts in liver fibrosis. Cell 8(5):503
90. Mizoguchi F et al (2018) Functionally distinct disease-associated fibroblast subsets in rheumatoid arthritis. Nat Commun 9(1):789
91. Dobie R et al (2019) Single-cell transcriptomics uncovers zonation of function in the mesenchyme during liver fibrosis. Cell Rep 29 (7):1832–1847. e8
92. Layton TB et al (2020) Author correction: cellular census of human fibrosis defines functionally distinct stromal cell types and states. Nat Commun 11(1):3275
93. Valenzi E et al (2019) Single-cell analysis reveals fibroblast heterogeneity and myofibroblasts in systemic sclerosis-associated interstitial lung disease. Ann Rheum Dis 78 (10):1379–1387

94. Beisang DJ et al (2020) Single-cell RNA sequencing reveals that lung mesenchymal progenitor cells in IPF exhibit pathological features early in their differentiation trajectory. Sci Rep 10(1):11162
95. Holt DR et al (1992) Effect of age on wound healing in healthy human beings. Surgery 112 (2):293–297; discussion 297-8
96. Pickup M, Novitskiy S, Moses HL (2013) The roles of TGFbeta in the tumour microenvironment. Nat Rev Cancer 13(11):788–799
97. Bosch L et al (2020) The transverse aortic constriction heart failure animal model: a systematic review and meta-analysis. Heart Fail Rev. https://doi.org/10.1007/s10741-020-09960-w
98. Fraccarollo D, Galuppo P, Bauersachs J (2017) Modeling cardiac fibrosis in mice: (Myo)fibroblast phenotype after ischemia. Methods Mol Biol 1627:123–137
99. Sasaki K et al (2015) Phase II evaluation of IPI-926, an oral hedgehog inhibitor, in patients with myelofibrosis. Leuk Lymphoma 56(7):2092–2097
100. Chong SG et al (2019) Fibrocytes and fibroblasts-where are we now. Int J Biochem Cell Biol 116:105595
101. Kolb M, Bonella F, Wollin L (2017) Therapeutic targets in idiopathic pulmonary fibrosis. Respir Med 131:49–57
102. Kolb M et al (2001) Transient expression of IL-1beta induces acute lung injury and chronic repair leading to pulmonary fibrosis. J Clin Invest 107(12):1529–1536
103. Distler JHW et al (2019) Shared and distinct mechanisms of fibrosis. Nat Rev Rheumatol 15(12):705–730
104. Maurer B, Distler JH, Distler O (2013) The Fra-2 transgenic mouse model of systemic sclerosis. Vasc Pharmacol 58(3):194–201
105. Smith GP, Chan ES (2010) Molecular pathogenesis of skin fibrosis: insight from animal models. Curr Rheumatol Rep 12(1):26–33
106. Michelotti GA et al (2013) Smoothened is a master regulator of adult liver repair. J Clin Invest 123(6):2380–2394
107. Swiderska-Syn M et al (2014) Myofibroblastic cells function as progenitors to regenerate murine livers after partial hepatectomy. Gut 63(8):1333–1344
108. Ubil E et al (2014) Mesenchymal-endothelial transition contributes to cardiac neovascularization. Nature 514(7524):585–590
109. Kitamura H et al (2011) Mouse and human lung fibroblasts regulate dendritic cell trafficking, airway inflammation, and fibrosis through integrin alphavbeta8-mediated activation of TGF-beta. J Clin Invest 121 (7):2863–2875
110. Lagares D et al (2017) ADAM10-mediated ephrin-B2 shedding promotes myofibroblast activation and organ fibrosis. Nat Med 23 (12):1405–1415
111. Chen YT et al (2014) Lineage tracing reveals distinctive fates for mesothelial cells and submesothelial fibroblasts during peritoneal injury. J Am Soc Nephrol 25(12):2847–2858
112. Ntokou A et al (2015) Characterization of the platelet-derived growth factor receptor-alpha-positive cell lineage during murine late lung development. Am J Physiol Lung Cell Mol Physiol 309(9):L942–L958
113. Miwa H, Era T (2015) Generation and characterization of PDGFRalpha-GFPCreERT2 knock-in mouse line. Genesis 53(5):329–336
114. Klotz L et al (2015) Cardiac lymphatics are heterogeneous in origin and respond to injury. Nature 522(7554):62–67
115. Kong P et al (2013) Lack of specificity of fibroblast-specific protein 1 in cardiac remodeling and fibrosis. Am J Physiol Heart Circ Physiol 305(9):H1363–H1372
116. Cheng N et al (2005) Loss of TGF-beta type II receptor in fibroblasts promotes mammary carcinoma growth and invasion through upregulation of TGF-alpha-, MSP- and HGF-mediated signaling networks. Oncogene 24(32):5053–5068
117. Sun L et al (2015) FSP1(+) fibroblast subpopulation is essential for the maintenance and regeneration of medullary thymic epithelial cells. Sci Rep 5:14,871
118. Trimboli AJ et al (2009) Pten in stromal fibroblasts suppresses mammary epithelial tumours. Nature 461(7267):1084–1091
119. Kobayashi A et al (2014) Identification of a multipotent self-renewing stromal progenitor population during mammalian kidney organogenesis. Stem Cell Reports 3(4):650–662
120. Zhou D et al (2014) Sonic hedgehog is a novel tubule-derived growth factor for interstitial fibroblasts after kidney injury. J Am Soc Nephrol 25(10):2187–2200
121. Li C et al (2015) Progenitors of secondary crest myofibroblasts are developmentally committed in early lung mesoderm. Stem Cells 33(3):999–1012

Chapter 21

Mouse Models of Lung Fibrosis

Olivia Mekhael, Safaa Naiel, Megan Vierhout, Aaron I. Hayat, Spencer D. Revill, Soumeya Abed, Mark D. Inman, Martin R. J. Kolb, and Kjetil Ask

Abstract

The drug discovery pipeline, from discovery of therapeutic targets through preclinical and clinical development phases, to an approved product by health authorities, is a time-consuming and costly process, where a lead candidates' success at reaching the final stage is rare. Although the time from discovery to final approval has been reduced over the last decade, there is still potential to further optimize and streamline the evaluation process of each candidate as it moves through the different development phases. In this book chapter, we describe our preclinical strategies and overall decision-making process designed to evaluate the tolerability and efficacy of therapeutic candidates suitable for patients diagnosed with fibrotic lung disease. We also describe the benefits of conducting preliminary discovery trials, to aid in the selection of suitable primary and secondary outcomes to be further evaluated and assessed in subsequent internal and external validation studies. We outline all relevant research methodologies and protocols routinely performed by our research group and hope that these strategies and protocols will be a useful guide for biomedical and translational researchers aiming to develop safe and beneficial therapies for patients with fibrotic lung disease.

Key words Pulmonary fibrosis, Experimental models of pulmonary fibrosis, Target engagement, Lung, Cytokines, Collagen deposition, Smooth muscle actin, Myofibroblast, Primary/secondary endpoints, Antifibrotic agents, Bleomycin

1 Introduction

The drug discovery and development pipelines are a crucial part of medicine and healthcare. The pharmaceutical industry has identified and implemented many steps and milestones that need to be successfully passed throughout the path toward the final approval of a novel medication [1]. The process usually starts with observational studies examining biological mechanisms of the disease, including its origin, development, resolution, or progression of

Olivia Mekhael and Safaa Naiel contributed equally to this work.

Boris Hinz and David Lagares (eds.), *Myofibroblasts: Methods and Protocols*, Methods in Molecular Biology, vol. 2299, https://doi.org/10.1007/978-1-0716-1382-5_21,

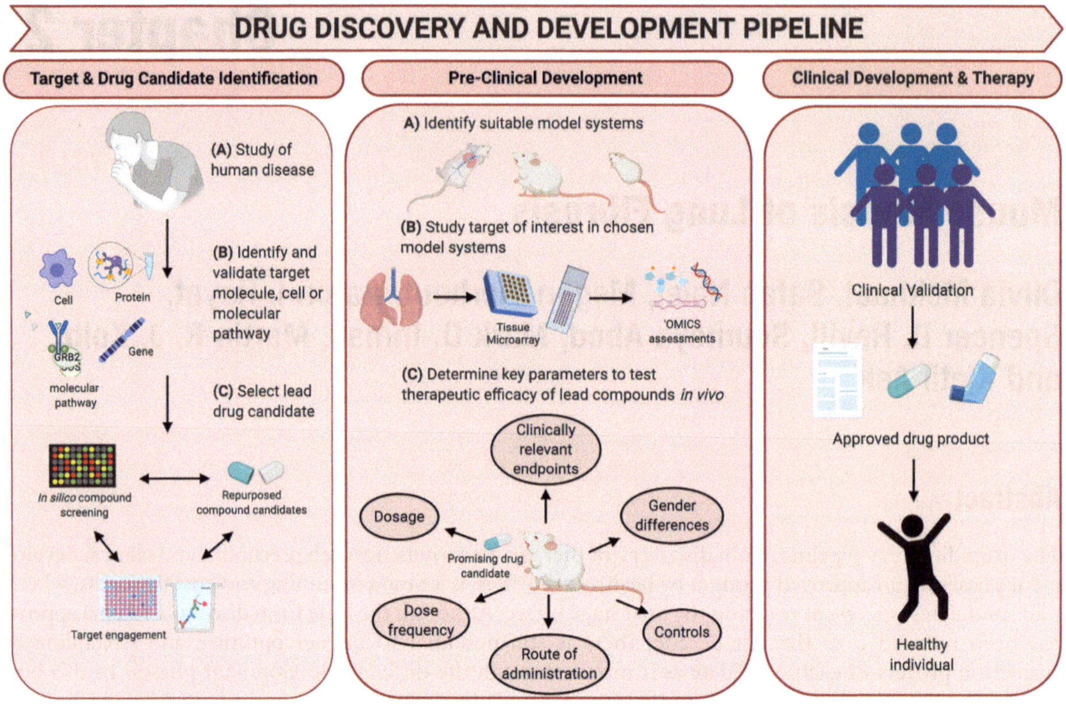

Fig. 1 *Drug discovery and development pipeline.* Promising drug candidates must pass through three key developmental phases of the pipeline before approval: (1) target and drug candidate identification: observational studies examining the molecular and cellular mechanisms of the disease of interest (A) will identify relevant targets (B) and drug candidates (C) that will initiate the pipeline; (2) preclinical development: in vivo studies will be conducted on appropriate experimental models that replicate human disease phenotype (A) and express the target of the interest (B), to ultimately test preclinical efficacy of the lead candidate by measuring clinically relevant outcomes (C); (3) clinical development and therapy: a lead candidate that has passed the preclinical stages and demonstrated preclinical efficacy must be clinically validated before being approved as a therapeutic product

the condition. Although the events that encompasses the cellular and molecular underpinnings of the disease and its resolution may be observed in this type of study, their exact nature may only be predicted, not tested. Hypothesized mechanisms may then be tested with specific interventional studies, and lead to improved understanding of the disease. Once a putative target, cell or molecular pathway is identified, that is believed to play a key role in the pathogenesis or progressive nature of the disease, the drug development pipeline starts [1–4] (Fig. 1). This process is often initiated by in silico or physical screening processes aimed at identifying lead compounds that have achieved a desired activity at the target site, and that can progress to the preclinical and clinical phases of the pipeline [2, 4]. Due to the significant time in development, high costs, and failure rates in the conventional drug development process, compound screens have evolved to also include health agency approved drugs with known toxicity and safety profiles [5, 6]. The preclinical stage involves proof-of-principle studies aimed to demonstrate safety, target engagement and therapeutic modulation that results in the amelioration of experimental disease-like outcomes [7–9]. Although both in vitro and in vivo models can be utilized to investigate target modulation and engagement, in vivo models are required for the validation of in situ target engagement within a live host organism and therapeutic efficacy in meaningful models of human disease [9]. In this book chapter, we describe current considerations and strategies to investigate the effect of therapeutic candidates in experimental models of pulmonary fibrosis. We will focus on the following: (1) selection of experimental rodent models that express the target(s) of interest; (2) selection of route of administration, dose and frequency; (3) selection of control groups and control treatments, statistical and species sex considerations; (4) analyzing a discovery cohort and selection of primary and secondary outcomes; and (5) design and execution of validation experiments. We have chosen to describe in detail the methodology of the well-known and widely used bleomycin model of pulmonary fibrosis. We believe that the overall principles and strategies outlined within this chapter, with slight modifications, can be relatively easily translated to investigate therapeutic efficacy in other experimental models of pulmonary fibrosis.

1.1 Selection of Experimental Rodent Models That Express The Target(s) of Interest

To successfully assess target validity and preclinical efficacy of any drug candidate, the first and most important consideration is the selection of the most appropriate experimental model system [10, 11]. To be labeled as a clinically relevant model for therapeutic efficacy studies, the experimental model would ideally need to replicate key elements of the human disease phenotype and express a clinically relevant target molecule or molecular pathway that a lead candidate intends to modulate [1, 12]. To date, many experimental models of pulmonary fibrosis have been reported, including bleomycin model, fluorescein isothiocyanate (FITC), irradiation, silica, overexpression of viral vectors, humanized models, asbestosis, age-dependent models, targeted type II alveolar epithelial cell (AEC) injury models, and various other lung injury models [10, 13–16]. Although no model on its own fully recapitulate progressive pulmonary fibrosis observed in the clinical setting, each of these models have been shown to share part of the fibrotic features observed in patients and could be useful candidate models, depending on the therapeutic target and pathway studied [10, 11, 15]. To achieve this, preliminary information about the temporal expression level of target molecules in each model needs to be available, as well as reasonable claims that the selected targets could be critically involved in the fibrogenic process [4, 17–19]. If the pathway or therapeutic target has not been formally studied and reported within the model, a formal investigation across model systems may be required to select the most suitable approach [15]. Ideally, standardized kinetic "multi-omic" datasets, including gene arrays (bulk and single-cell RNA sequencing—scRNAseq), proteomic, metabolomic, and lipidomic, would be generated and accessible from each model, so that subsequent bioinformatic assessments would help to select the appropriate model systems [4, 15, 18, 20, 21]. In-house repositories can also be developed, including the generation of tissue micro arrays (TMA) containing several model systems on a single block, with lungs extracted from multiple time-courses [15]. This is a relatively low-cost tool that can be developed and used to rapidly screen several model systems (depending on TMA design and available models) at the protein or mRNA level to select the most appropriate model [15]. In Subheading 3, we describe the methodology to develop a screening TMA containing four distinct experimental models of lung fibrosis, to allow for rapid evaluation of model suitability.

1.2 Selection of Route of Administration, Dose, and Frequency

The overall objective of therapeutic assessment trials in preclinical experimental models is to demonstrate a beneficial effect of the intervention. How do we ensure that our therapy reaches the right area, at the right time, with the right amount and adequate frequency? How do we make sure that the putative therapeutic target is engaged, and modulated appropriately? These are some of the pharmacodynamic questions that need to be carefully assessed before preclinical trials are designed. For experimental models of lung disease, most therapeutic options are formulated for oral or inhaled delivery [22–24]. Animal inhaled aerosols are a novel vehicle for the delivery of high concentrations of drugs into the airways[25]. Recent advancements in flexiVent technology allow for simultaneous lung function assessment and drug delivery. Regardless of the chosen method, when a new therapy is to be evaluated, we recommend that tolerability studies are conducted to address the safety levels of the drug and to determine whether the drug displays any short- or long-term toxic effects, followed by pharmacokinetic (PK) and pharmacodynamic (PD) studies, to address the kinetic profile of the drug [26, 27]. These experiments can be performed on a reduced number of animals, and both in naïve animals as well as in the appropriate model systems. When direct lung delivery is performed, we usually monitor animals for acute (same day) or latent (up to one week) distress signs post-administrations. Once a tolerable dose has been identified, PK and PD assessments need to be established. This is preferably performed in the model system where the drug will be evaluated as the injured and fibrotic lung might have a different drug metabolism and PK/PD capacities [24, 26, 28]. The timing of drug intervention also needs careful consideration and has been subject to extensive debates and review [16, 17, 29, 30]. Although many preclinical trials administer therapeutic options in a prophylactic regimen, where the therapy is given prior or at the same time as the agent inducing the disease, we have over the past decade seen a shift in the timing, where more therapeutic modalities have been introduced [11, 16, 30]. The latter is thought to better mimic the clinical setting, where the diagnosis of fibrotic lung disease and therapy are only given after diagnosis, and so far not to prevent the onset of the disease [11, 16, 17, 30].

1.3 Selection of Control Groups and Control Treatments, Statistical and Species Sex Considerations

Well-designed experimental trials benefit from the addition of treatment groups containing already approved therapies (nintedanib or pirfenidone) [10, 31, 32], and established benchmark candidates in various experimental models of lung fibrosis [28, 32, 33]. Other important considerations that need to be determined before the start of the trial are the number of treatment groups required and the appropriate group size, based on a priori power calculations [11, 34]. As an example, we consider a case where an investigational drug (X) is hypothesized to prevent the accumulation of myofibroblasts and subsequent pulmonary fibrosis (Fig. 2). In our hypothetical example, X will be delivered as an inhaled formulation and administered daily, with vehicle Y, at two concentrations to potentially establish a dose-dependent effect. The control drug to include (either nintedanib or pirfenidone) will be administered as per previously published studies, per os and with daily administrations using appropriate (and likely different) vehicle formulations. Taking into consideration the research question outlined above, the following additional questions should be addressed: How many control groups do we need in this case? How many mice should be included in each group? Which statistical tests should be used to examine outcomes? Should we perform the experiment in both male and females? Due to the significant costs of these trials, often times one sex is used for initial assessments. To account for sex influences, it is recommended to replicate experiments for the interventional aims in the opposite sex [17]. Based on our hypothetical example, a negative control group is required to establish the baseline level in the parenchyma of naïve mice (group 1). A positive control group will be required to establish the maximal amount of parenchymal fibrosis (group 2). Two potentially different vehicle control groups are required, one with the inhaled vehicle formulation (group 3) and one with oral vehicle formulation (nintedanib/pirfenidone arm, group 4), to account for potential effects of the formulations alone. Two treatment arms of the investigational, inhaled drug (X) may be required (group 5, 6) if a potential dose-dependency is to be examined. The result of these arms should be compared to the positive treatment arm (nintedanib/pirfenidone, oral administration, group 7). In all, based on this simple example, 7 investigational groups may be required. As described in depth in Subheadings 1.4 and 1.5, establishing the primary outcomes prior to the initiation of the study is essential as it

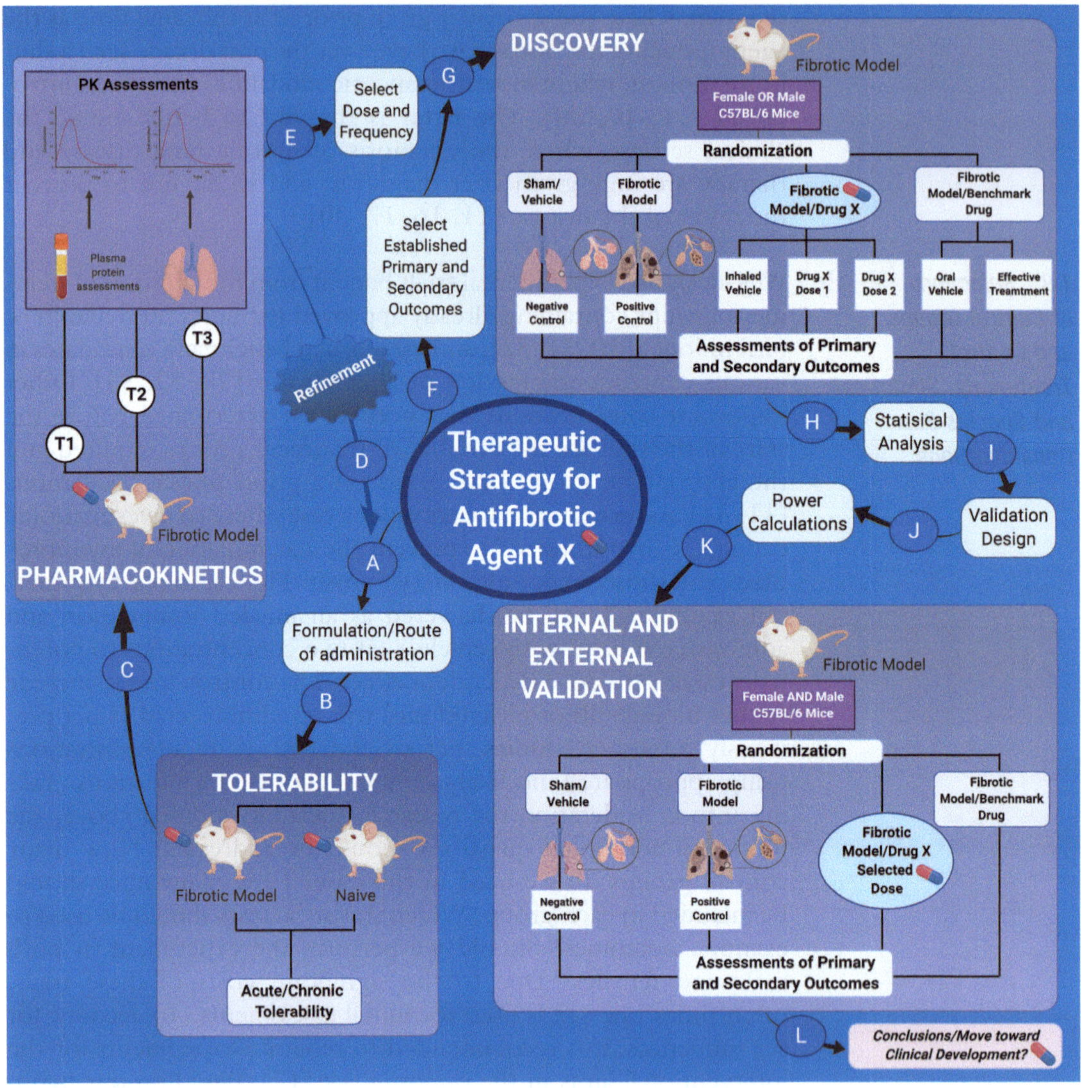

Fig. 2 *Therapeutic study design strategy for antifibrotic agent X.* To investigate whether the lead candidate, drug X, has antifibrotic properties, it must pass through several preclinical phases. To evaluate the efficacy of Drug X, the formulation (A) is initially tested in a tolerability study (B) to address the safety levels of the drug. When direct lung delivery is performed, animals are monitored for acute (same day) or latent (up to one week) distress signs post-administration to determine if the drug displays any short- or long-term effects, respectively. This is ideally tested in both naïve and in the fibrotic model system. Once a safe and tolerable dose is established, pharmacokinetic studies (PK) (C) are conducted to establish a kinetic profile of the drug. PK studies are preferably performed in the model system where the drug will be evaluated as the injured and fibrotic lung might have different drug metabolism and PK properties. PK studies for Drug X is usually performed by administering the drug at various time-points (T1, T2, T3) and collecting samples, such as plasma and lung, that are processed and analyzed for drug and relevant metabolites. If a favorable PK profile is not achieved, the formulation and administration route may require further optimization (D). A favorable PK profile will determine the selection of drug X dose(s) and frequency (E) and, after selection of established primary and secondary outcomes (F), a discovery cohort can be designed and evaluated (G). Based on this hypothetical example, a negative control group is required to establish the baseline level (group 1). A positive

prevents the researcher from mining through the measured study outcomes in search of a statistically significant effect, often referred to as "p-hacking" [34, 35]. When having many primary outcomes, it is difficult to perform power calculations for each readout as standard deviations may vary between the different assessments [36]. This will lead to the requirement of larger sample sizes to conclude that an observed effect is a true effect [34, 36]. In general, and as outlined in Fig. 2, we recommend that preclinical studies are conducted in several phases, and that after careful tolerability and PK evaluations, a smaller discovery cohort is designed and evaluated. This discovery cohort could be performed with a minimum sample size of five for all control and test groups, and that the data generated from this study is used to refine the design, consolidate the selected primary and secondary outcomes, and conduct appropriate power analysis calculations before a validation study is initiated. Other important considerations include animals to be randomly allocated to the treatment groups, where intervention initiation and assessments follow, and to practice operator and data analyst blindness before and during the study to deter experimental and observer biases [34, 37]. If at times operator blindless cannot occur, then analyst blindness with data uniquely coded is encouraged [34]. Sample and data elimination during the analysis process should be strongly discouraged and only in the case of a validated outlier test recommends the removal of select data points [38]. Furthermore, biases and outlier handling can be better managed by establishing solid inclusion and exclusion criteria during the planning of the study and prior to the initiation of the treatments [34, 38].

1.4 Analyzing a Discovery Cohort and Selection of Primary and Secondary Outcomes

For preclinical studies in lung fibrosis, it is not unusual to assess a multitude of outcomes. These outcomes can include the assessment of pulmonary physiology, various biochemical assessments of lung tissues, histological examinations, bronchoalveolar lavage fluid, or blood-borne components [17] (Fig. 3). With recent advances of "omics" technology, including, but not limited to, gene arrays,

Fig. 2 (continued) control group will be required to establish the level of fibrosis (group 2). Two potentially different vehicle control groups may be required, one with the inhaled vehicle formulation (group 3) and one with oral vehicle formulation (nintedanib/pirfenidone arm, group 4). Two treatment arms of the investigational, inhaled drug (X) may be required (group 5, 6) if a potential dose-dependency is to be examined. The result of these arms should be compared to the positive treatment arm (nintedanib/pirfenidone, oral administration, group 7). The data generated and analyzed (H) from this study is used to refine the design (I), primary and secondary outcomes and to conduct appropriate power analysis calculations (J) before an internal or external validation study is initiated (K). Due to the lack of reproducibility often observed in the preclinical field, and to improve translation to the clinical settings, we recommend sharing study designs and associated data with the scientific community to allow for external assessments and further validations. If Drug X displayed preclinical efficacy, it may facilitate the move toward clinical development (L)

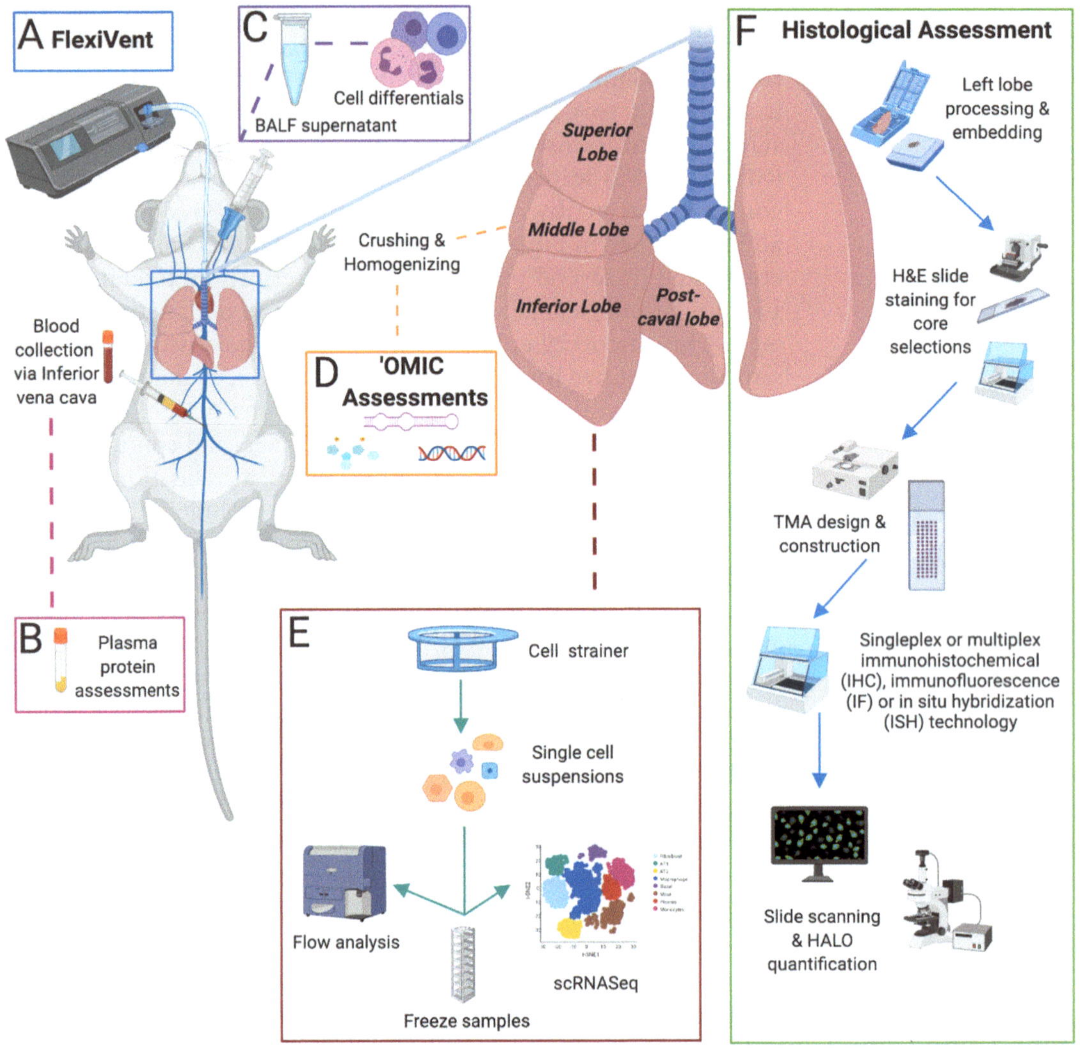

Fig. 3 *Processing and archiving of biological samples.* (A) Lung function assessments using a flexiVent mechanical respirator. (B) Blood is collected via the inferior vena cava and plasma isolated for potential protein analysis. (C) BALF is collected and processed for cell differential assessments. (D) Middle and post-caval lobes are snap-frozen for potential "omics" assessment. (E) Superior and inferior lobes are digested and single-cell suspensions are frozen for later flow cytometry and potential scRNAseq processing and assessments. (F) Left lung is fixed in formalin, embedded for TMA generation and downstream histological assessments (such as α-Smooth Muscle Actin, Masson's Trichrome, Picrosirius Red, and Hematoxylin and Eosin)

metabolomics, lipidomic and proteomic assessments on bulk or single-cell suspensions, plasma or in bronchoalveolar lavage (BAL) supernatant, single-cell RNAseq and multiplex strategies on histology specimens, gives the possibility to measure an endless number of variables [4, 19, 39]. Due to technological advances in the past decades, whole-body animal imaging, such as miniaturized X-ray and ultrasound, are being utilized more frequently to track the effects of the drug candidates in live animals, providing clinically relevant information and ultimately shortening the preclinical phase of the pipeline [17, 40–42]. Depending on the target engagement profile of the lead candidate and the research hypothesis being addressed, one quantifiable primary endpoint may not be sufficient to link preclinical to clinical readouts [8, 43]. For instance, when testing lead antifibrotic candidates in experimental models of lung fibrosis, certain studies have considered the quantification of collagen via hydroxyproline assessment as the primary readout for disease severity and success of antifibrotic therapy [17]. In other antifibrotic therapeutic trials, the combination of multiple primary endpoints such as respiratory mechanics assessment and histopathology has been used. A composite index has been developed in other fibrotic systems [44] and is strongly warranted for use in preclinical therapeutic interventions studies in lung fibrosis. Overall, limiting the number of primary outcomes will increase the robustness of the trial, allow for proper power-calculations, decrease the number of animals required, and decrease the chances of reporting false positive findings [36, 38, 45]. Secondary outcomes are additional variables that can be assessed and support primary outcomes [43]. Frequently, exciting findings are observed when analyzing secondary outcomes. For example, an anti-fibrotic intervention is seen to also reduce the extent of inflammatory cell infiltration. It is important to recognize that as a researcher increases the number of secondary outcome measurements there will be a proportional increase in the number of false positive observations. When this happens, there is obviously a temptation to switch the aim of the study, to the new, exciting data with strong statistical significance. While that is obviously an error, it is likely that this occurs frequently and contributes to a biased and false state of "knowledge." However, analyzing secondary outcome is likely the source of exciting new hypotheses. What is important, and this responsibility lies with the basic scientist, is that exciting observations made on secondary outcomes need to be confirmed with a brand new experiment or clearly declared as secondary findings, in need of validation, when presented in the final publication. To avoid this, we recommend here that discovery experiments are used to analyze multiple outcomes and to formally identify primary and secondary outcomes. Appropriate power calculations can then be conducted before internal or external validation experiments are conducted, as discussed below.

1.5 Design and Execution of Validation Experiments

Since 2004, based on recommendation made by the International Committee of Medical Journal Editors (ICMJE), all clinical trials must be registered online prior to enrolling the first trial participant. This achieves several aims, perhaps the most important of which is the declaration of the primary outcome variable before the data is analyzed. As mentioned above, this prevents authors from selecting the outcome that best supports the pre-trial hypothesis, and thereby markedly reduces bias and the extent to which knowledge is influenced by chance, or false positive results

[36, 46]. Currently, this same systematic oversight is not currently applied to basic or preclinical research. A basic researcher is still free to pick whichever outcome he/she chooses to include in the publication at the end of a research project. It is therefore the responsibility of the researcher to ensure that the data presented, and the conclusions drawn from that data, are the best attempt at representing the underlying truth, as opposed to the best attempt to support the pre-study hypothesis. Therefore, it is recommended that the primary outcomes (or pre-defined composite of multiple outcomes) should be selected and appropriately powered prior to the validation analysis, thereby imposing the same control that is systematically imposed in clinical research. In addition, we recommend validation studies to include both male and female groups and to potentially focus on specific groups identified in the discovery cohort and as shown in Fig. 2. Finally, validation studies executed by external research groups increase the confidence that "real" effects are seen, provided that shared protocols and methodologies are available[46]. To this effect, we have here assembled the different methodologies used in our research laboratory to assess the effect of anti-fibrotic therapies, focusing on the bleomycin-model of lung fibrosis.

2 Materials

2.1 Intratracheal Administration of Drugs

1. Biological safety cabinet (BSC).
2. Autoclave machine.
3. Anesthetic machine with an accompanying anesthetic chamber.
4. Gaseous isoflurane anesthetic.
5. Mouse scale.
6. Ear notcher.
7. 70% Ethanol.
8. Intratracheal intubation equipment: Intubation board, otoscope handle, operating otoscope that attaches to the handle, mouse intubation specula that fits on the end of the operating otoscope, 22G 1 inch Insyte IV catheter needle with the sharp end of the needle blunted, 200μL extended length gel loading pipette tips, blunt end forceps, and masking tape.
9. Bleomycin: Reconstitute bleomycin to a stock concentration of 1.6 units/mL by adding 9.38 mL of saline to a vial containing 15 units of bleomycin. Aliquot 500μL of bleomycin in 1.5 mL Eppendorf tubes and store in −80 °C. On the day of bleomycin instillation, dilute stock concentration to obtain the required doses of either 0.04 or 0.05 or 0.06 units (U)/mouse of bleomycin in a volume of 50μL sterile saline per mouse, and

gently vortex. Keep on ice throughout the procedure (*see* **Note 1**).

2.2 Endpoint Measures: Respiratory Mechanics

1. FlexiVent® mechanical respirator (flexiVent® SCIREQ, Montreal, PQ, Canada).
2. Ketamine hydrochloride/xylazine anesthesia in mice: To prepare a cocktail, in a sterile vial combine 0.75 mL (75 mg) of ketamine (100 mg/mL concentration), 0.25 mL (5 mg) of xylazine (20 mg/mL), and 4 mL of sterile water. For mouse, recommended dose for ketamine is 80–100 mg/kg and for xylazine is 10 mg/kg. Cocktail can be stored in a cool place protected from light for a maximum of 1 month.
3. Mouse cannula.
4. 25G Needle.
5. 1 mL Syringe.
6. Heating blanket.
7. Straight forceps.
8. 70% Ethanol.
9. Micro forceps.
10. Rocuronium bromide (paralytic): To prepare a total volume of 500μL, sufficient for 10 mice [40μL per mouse (1:5 dilution)], in a labeled 1.5 mL Eppendorf tube, add 400μL of sterile phosphate-buffered saline (PBS) to a 100μL of rocuronium (10 mg/mL) (extract using a 25-gauge (G) needle and a 1 mL syringe) and gently vortex. Prepare the working concentration on the day of experiment. Paralytic is sufficient for one day. Keep on ice throughout the procedure.

2.3 Endpoint Measures: Blood Collection for Plasma Analysis

1. 1 mL Syringe.
2. Heparin.
3. 23G Needle.
4. K2EDTA vacutainer blood collection tube.
5. Ice box.
6. 200μL pipette.

2.4 Endpoint Measures: BALF Collection for Immune Cell Differentials

1. Mouse cannula.
2. Phosphate-buffered saline (PBS): 137 mM NaCl, 2.7 mM KCl, 8 mM Na_2HPO_4, and 2 mM KH_2PO_4. Adjust pH to 7.4.
3. 1 mL Syringe.
4. Hemocytometer.
5. Trypan blue 0.4%
6. Centrifuge.
7. 200μL Pipette.

8. −80 °C.
9. Microscope slides.
10. Cytocentrifugation.
11. Wright-Giemsa stain kit.

2.5 Endpoint Measures: Lung Processing for Histopathology, Flow Cytometry, and Single-Cell RNA Sequencing

1. Surgical sutures.
2. Digestion Media: To a complete Roswell Park Memorial Institute (RPMI) media containing 10% fetal bovine serum (FBS), 1% L-glutamine, and 1% penicillin streptomycin (pen-strep), add the following: 50 U/mL deoxyribonuclease I from bovine pancreas (DNase I), and 300 U/mL collagenase type I.
3. Cell shaker.
4. Syringe plunger.
5. 40μm Cell strainer.
6. 50 mL Falcon tube.
7. RMPI media: 10% FBS, 1% L-glutamine, and 1% pen/strep.
8. Hemocytometer.
9. Centrifuge.
10. Fluorescence activated cell sorting (FACS) buffer: 0.5% bovine serum albumin (BSA) in PBS.
11. Freezing Media: Consists of 50% of complete RPMI media (10% FBS, 1% L-glutamine, and 1% pen-strep), 40% of FBS, and 10% of dimethyl sulfoxide (DMSO).
12. Cryovials.
13. Freezing materials: cold ice, dry ice, 1 large supply of liquid nitrogen, and freezing containers.
14. –80 °C Freezer.
15. 10% Formalin.
16. 70% Ethanol.
17. Histology cassettes.
18. Paraffin wax.
19. TMA.
20. Bond RX (Leica).
21. Lung crushing equipment for snap-frozen lung processing and assessment: Metal chamber (to localize the sample), piston (cylindrical metal piece to crush the sample, placed in metal chamber), wooden block, hammer, several metal spatulas, 1 Styrofoam container for freezing equipment and instruments, 1 Styrofoam container containing dry ice, 14 mL polypropylene tubes, tin foil, and a balance.
22. Lung homogenization equipment for snap-frozen lung processing and assessment: Mechanical homogenizer, a beaker, and

1 50 mL falcon tube of 70% ethanol, and 1 50 mL falcon tube of distilled water.

23. Radioimmunoprecipitation assay (RIPA) buffer containing protease inhibitor (*see* **Note 2**): To prepare 100 mL of RIPA buffer stock solution, in a sterile glass bottle combine 93 mL of PBS (1×), 1 mL of octylphenoxypolyethoxyethanol (IGEPAL CA-630) (100%), 5 mL of Na-deoxycholate (10%), and 1 mL of sodium dodecyl sulfate (SDS) (10%). To prepare inhibitor, per 1 mL of RIPA buffer, add: 5μL Na_3VO_4 (200 mM), 30μL aprotinin, 5μL phenylmethylsulfonyl fluoride (PMSF) (20 mg/mL), and 1μL of 1,4-Dithiothreitol (DTT) (1 M).
24. 1.5 mL Eppendorf tubes.
25. RNA assessments reagents/equipment: Nucleospin RNA Plus Isolation Kit, NanoDrop spectrophotometer, Bioanalyzer, NanoString, and RNA sequencing technologies.
26. Reagents and equipment for collagen quantification assessment: Sircol™ Soluble collagen assay, Hydroxyproline assay, and a microplate reader.

2.6 Analysis of Homogenized Tissue Samples

1. Histology cassettes.
2. Hot paraffin wax.
3. TMA coring tools.
4. Small oven (research only).
5. 4 °C Refrigerator.
6. Microscope slide digitizer and scanner.

2.7 Quantification Using Digitalized TMAs

1. Picrosirius red (PSR) stain kit.
2. Masson's trichrome stain kit.
3. Hematoxylin and Eosin (H&E) stain kit.
4. Alpha-smooth muscle actin (α-SMA) stain kit.

3 Methods

3.1 Intratracheal Administration of Drugs

Intratracheal intubation is a simple and reliable method of instilling drugs into mouse lungs [47]. This procedure involves putting mice under inhalable anesthesia, visualizing the trachea through the opening of vocal cords, inserting a catheter through the tracheal opening, and delivering drug in a syringe through the catheter [48]. Once a researcher has some experience with this method, it can be carried out with little damage to the soft tissues of the upper respiratory tract [47, 48]. Evaluating the presence of a fibrotic response brought about by intratracheal intubation of bleomycin shows all lobes to be affected, referring to even dispersal of drug into the lungs when procedure is done proficiently [47]. In this chapter, we describe how pulmonary fibrosis is induced in 8-12-week-old female C57BL/6J mice (Jackson Laboratories) by a single intratracheal instillation of bleomycin. Mice are anesthetized with gaseous isoflurane anesthesia prior to the procedure. The step-by-step procedure is outlined below:

1. After preparing the BSC (*see* **Note 3**), anesthetize mice with isoflurane using the anesthetic machine. Turn the oxygen to 1.5 L/min and isoflurane setting to 5%. Place the mouse (or mice) into the anesthetic chamber.
2. Ensure that the intubation board is set up at an angle of 60° relative to the working area (BSC counter), often referred to as a semi-suspended dorsal recumbency position [49].
3. When the mouse is deeply anesthetized (breathing rate is around one inhalation every two or 3 s), hang it on the intubation board by its teeth (upper incisors). Place a strip of masking tape over the mouse's ribcage to hold the chest in a position that ensures the airways are easily detected when looking inside with the otoscope.
4. Use the forceps to gently pull the mouse's tongue to the side of the mouth (*see* **Note 4**).
5. Turn on the otoscope and insert the assembled otoscope (*see* **Note 5**) into the mouse's mouth, looking for the vocal cords. They will appear as flaps of muscle that are opening and closing relative to the respiratory rate. If they cannot be seen, then the specula may have gone down too far, pull back and reinsert.
6. When the vocal cords can be seen, insert the syringe-catheter (*see* **Note 6**) ensemble through the vocal cord opening. Once through, hold the syringe steady and place the otoscope on the bench. Remove the needle and syringe from the catheter, leaving only the catheter in place.
7. To confirm the catheter is in the trachea, take the 200μL pipette containing water and air (*see* **Note** 7) and hold its pipette tip securely inside the catheter to form a tight seal. If the catheter is in the trachea, the water in the pipette tip will be drawn up and down by the mouse's breathing. Place this pipette back on the bench. If there is no movement of water, remove the catheter and start again. Proceed to **step 10** if you are confident the catheter is in the trachea.
8. While holding the catheter steady, take the 200μL pipette containing bleomycin (*see* **Note 8**) and hold its pipette tip securely inside the catheter to form a tight seal. The breathing of the mouse will draw the fluid into its airways. As this happens, gently push down on the pipette plunger to prevent fluid from coming back up the tip (*see* **Note 9**).

9. When all the fluid has been inhaled, hold the catheter in the trachea and place the blunt-ended needle into the catheter. This will expel any fluid remaining in the catheter.
10. Remove the catheter from the mouse. Remove the mouse from the intubation board. Place the mouse in the cage on its back in an upright position using the bedding to elevate its head; this will prevent suffocation. This also helps to prevent any fluid from draining from the lungs.
11. Monitor the mouse until it is conscious and has a normal respiration rate and activity level.

3.2 Endpoint Measures: Respiratory Mechanics

Lung function assessments are conducted using a flexiVent® mechanical respirator. FlexiVent® enables the measurement of lung compliance, or its inverse, lung elastance, which is a functional parameter derived from the pressure-driven pressure–volume loops. Instructions on flexiVent® procedures are previously described [50] and provided below:

1. Fully anesthetize mice via an intraperitoneal (i.p.) injection of a ketamine hydrochloride and xylazine cocktail at a dose of 0.06–0.07 mL/10 g of body weight using a 25G needle and 1 mL syringe (*see* **Notes 10** and **11**).
2. Place mouse on a heating blanket set on low to maintain body temperature.
3. Pinch the toes with straight forceps to verify that the subject has no reflexes.
4. Spray mouse with 70% ethanol solution.
5. Perform tracheal cannulation according to the following **steps 6–11**:
6. Using sterile surgical scissors and straight forceps, make a small incision of the skin in the throat area and cut laterally to expose a section of the anterior side of the neck.
7. Gently separate the submaxillary glands, separate muscle layer and any adhesive tissue, and remove the skin around the tracheal area to expose the trachea (*see* **Note 12**).
8. Gently slide sterile curved micro-forceps underneath the trachea.
9. Make a small incision at the top of the trachea, below the cricoid cartilage between two rings of cartilage.
10. Insert a 18G cannula into the incision and push the cannula into the trachea past 3–5 tracheal rings.
11. Place a surgical suture over the cannula and tie two knots over it.

12. Connect the mouse to the mechanical ventilator controlled by the SCIREQ software (FlexiWare 7.6.6) (*see* **Note 13**) and start ventilation.
13. To paralyze subjects and inhibit respiratory effort, administer rocuronium bromide (*see* **Note 14**) (10 mg/mL) at a dose of 4 mg/kg i.p. (40μL per mouse) using a 25G needle connected to a 1 mL syringe.
14. Ventilate subject for 10–30 s and then hyperinflate the lung to total lung capacity in order to attain standardized baseline conditions [51].
15. Start the flexiVent® script and the lung perturbations as previously described [50]. Mice are ventilated with 10 mL/kg of air at a rate of 150 breaths per min in between forced oscillation waveforms maneuvers. The changes in flow, volume, and pressure within the airways are recorded and the raw data is fit to the *single-compartment model* to assess lung elastance resistance, and the *pressure-volume loop Salazar Knowles equation* to assess quasi-static elastance [51].

3.3 Endpoint Measures: Blood Collection for Plasma Analysis

Following flexiVent® assessments, 0.5–1 mL of blood can be obtained from mouse inferior vena cava as follows:

1. Load heparin (or another appropriate anticoagulant) into 1 mL syringe to coat the surface.
2. Lift the abdomen skin and cut laterally from mouse pelvis/ prepuce to the xiphoid.
3. Gently move the intestines to the left side of the mouse to expose the inferior vena cava.
4. Gently insert a 23G needle attached to the 1 mL heparinized syringe into the inferior vena cava and slowly withdraw the blood.
5. Place extracted blood in a K2EDTA vacutainer blood collection tubes and immediately store on ice until later processing and plasma collection.
6. To collect plasma, spin blood at 1500 g for 10 mins at 4 °C.
7. Draw plasma from each tube using a 200μL pipette (*see* **Note 15**).
8. Store plasma at −80 °C in sterile cryovials for downstream assessments (*see* **Note 16**).

3.4 Endpoint Measures: BALF Collection for Immune Cell Differentials

Following blood collection, BALF can be harvested for immune cell differentials as follows:

1. Puncture and cut away diaphragm to stop breathing of mouse.
2. Open the thoracic cavity of the mouse.

3. As described in detail in Subheading 3.2, insert a cannula in the incision made on top of the trachea.
4. Inject 600μL of PBS into the lungs using a 1 mL syringe via the cannula for a first wash.
5. Briefly massage the lungs to wash out lung infiltrate cells and then extract PBS through negative pressure using the same 1 mL syringe.
6. Place the recovered lavage fluid (~400–500μL) in a 1.5 mL Eppendorf tube.
7. Inject 400μL of PBS into the lungs using a 1 mL syringe via the cannula for a second wash.
8. Repeat **steps 4** and **5** above.
9. Following BALF collection, place the tubes immediately on ice until later processing.
10. Perform total cell count using a hemocytometer from an aliquot of 20μL from each sample and add trypan blue stain 0.4% to these tubes to test for cellular viability.
11. Centrifuge each Eppendorf tube at 400 × *g* for 8 mins at 4 °C.
12. Draw supernatant from each tube using a 200μL pipette and store it at −80 °C in sterile cryovials for downstream assessments (*see* **Note 17**).
13. Suspend the cell pellet in 300–600μL (depending on cell density of each sample) of sterile cold PBS.
14. Use 120μL aliquot to cytospin onto a microscope slide.
15. Prepare smears by cytocentrifugation at a speed of 22 g for 3 min.
16. Stain the developed cytospin slides with Wright–Giemsa stain according to the manufacturer's protocol.
17. Conduct differential cell count in BALF under 40× magnification using a microscope. A sample size of 200 is chosen and morphological characteristics are assessed to identify the different cell types in the lung BALF (*see* **Note 18**).
18. Present each cell type as total numbers, based on the corresponding total cell count calculated from each sample.

3.5 Lung Processing for Histopathology, Flow Cytometry, and Single-Cell RNA Sequencing

1. Following collection of the BALF, heart and lungs are disconnected from mouse carcass and the four lobes of the right lung are separated from the left lung by tying them up with a surgical suture and excising them. The lung lobes are separated and processed as described below (*see* **Notes 19** and **20**).

3.5.1 Processing of the Right Superior and Inferior Lobes for Flow Cytometry or scRNAseq Assessment

2. Mince the superior and inferior lobes into small pieces, immediately place them into 5 mL of digestion media (refer to Subheading 2.5).
3. Keep lobes on a cell shaker at 37 °C for 1–2 h.
4. Subsequently, using a sterile syringe plunger, crush the digested lobes and push them through a 40μm cell strainer placed on a 50 mL falcon tube (*see* **Note 21**).
5. Centrifuge cells drained through filters at 400 × *g* for 5 min at 4 °C.
6. Decant supernatant and suspend pellets in 10 mL of complete RPMI medium (10% FBS, 1% L-glutamine, and 1% pen/strep) or PBS.
7. Perform cell count using a hemocytometer from an aliquot of 20μL from each sample and add trypan blue stain 0.4% to these tubes to test for cellular viability.
8. Centrifuge cells at 350 × *g* for 5 min at 4 °C.
9. Decant supernatant and resuspend pellets in cell staining FACS buffer (refer to Subheading 2.5) or PBS (*see* **Note 22**) at a concentration of 40×10^6 cells/mL to be stained for flow cytometry assessment of different cell markers.
10. Freeze remaining cells in freezing media (refer to Subheading 2.5) and place samples into cryovials (1 mL of freezing medium per each sample) (*see* **Note 23**).
11. Immediately, place cryovials into cold freezing containers and store containers at −80 °C for 24–48 h.
12. After 24–48 h, transfer frozen lung cells from −80 °C to liquid nitrogen for later flow or scRNAseq processing and analysis.

3.5.2 Processing of the Right Middle and Post-Caval Lung Lobes for RNA Analysis

1. Immediately snap-freeze the middle and post-caval lobes in liquid nitrogen.
2. Store the lobes at −80 °C for further protein and RNA analysis.

3.5.3 Processing of the Left Lung Lobe for Histology

1. Inflate the left lung with 10% formalin at a 30 cm H_2O of pressure for 3–5 min.
2. Submerge the left lung into 15 mL falcon tube filled with 10% formalin and fix them for 24 h before transferring them to 70% ethanol.

3. Later, cut each left lung (2 tissues per lung) and place them into histology cassettes.
4. Process and embed all tissue-containing cassettes in paraffin wax.
5. Generate a tissue microarray (TMA) as described in Subheading 3.7.
6. Following the development of a TMA containing mouse lung tissues, cut and stain with a Bond RX (Leica), as described previously [51, 52].

3.5.4 Crushing of Snap-Frozen Lung and Homogenization for RNA and Protein Assessment

1. Snap-frozen right middle and post-caval lungs can be retrieved from −80 °C and processed for further RNA and protein assessments. Processing and assessment protocols are as follows:
2. Prepare two sets of 14 mL polypropylene tubes labeled with animal identification numbers (*see* **Note 24**).
3. Add 1 mL each of RIPA buffer (refer to Subheading 2.4) into one set of tubes only.
4. Take and record weights of all tubes using a balance.
5. Place the snap-frozen lungs (removed from −80 °C freezer and stored in tin foil) in liquid nitrogen (*see* **Note 25**).
6. Freeze equipment (metal chamber, piston, and spatulas) by placing them in a Styrofoam container filled with liquid nitrogen (*see* **Note 26**).
7. Remove frozen metal chamber from Styrofoam container and place on wooden block. Add small amount of liquid nitrogen to frozen metal chamber.
8. Remove frozen tissue sample from tin foil and take the weight using a balance.
9. Place tissue in frozen metal chamber and slide piston into chamber on top of tissue (*see* **Note 27**).
10. Hammer the top of the piston approximately 10 times until lungs have been crushed into a fine powder.
11. Remove the piston and scrape any powder left on piston surface into the metal chamber using a frozen spatula (*see* **Note 28**).
12. Use spatula to stir and scrape tissue powder to ensure adequate mixing.
13. Quickly invert metal chamber onto a piece of clean tin foil to remove the tissue. Hammer the bottom of the metal chamber approximately 10–15 times to remove all powder (*see* **Note 29**).

14. Transfer the tin foil containing the powder to dry ice. Using frozen spatula, remove one scoop of powder and place into corresponding RIPA containing tube. Store tube on dry ice.
15. Place remaining powder in the respective empty 14 mL tube (for RNA analysis). Store the tube on dry ice (*see* **Note 30**).
16. Take weights of tubes now containing powder and subtract prior tube weights without tissue (taken in **step 3**) to determine the respective tissue weights (*see* **Note 31**).
17. Store all tubes at −80 °C.
18. For lung homogenization (*see* **Note 32**), place all tissue sample tubes on wet ice and let thaw.
19. After assembling the homogenizer, clean the metal cylindrical wand by dipping it into a 50 mL falcon tube of 70% ethanol and letting the homogenizer run for 10 s. Repeat this with distilled water and then wipe the wand dry.
20. Place the first tube containing tissue powder and RIPA buffer in a beaker containing ice and insert the metal wand of the homogenizer into the tube. Homogenize the sample for 10 s, followed by a 10 s pause, and another 10 s of homogenization. Remove the tube from the beaker and place on ice (*see* **Note 33**).
21. Clean the homogenizer with 70% ethanol and distilled water as described in **step 19**.
22. Centrifuge samples at 1500 rpm for 8 min.
23. Collect supernatant and store in 1.5 mL Eppendorf tubes.
24. Resuspend the pellet in PBS and store in 1.5 mL Eppendorf tubes.
25. Store the supernatant and pellet tubes at −80 °C.

3.6 Analysis of Homogenized Tissue Samples

If desired, powdered lung tissue stored at -80°C as described in Subheading 3.2.7 can be processed to extract RNA for downstream analysis.

3.6.1 RNA Extraction

1. Isolate RNA using a commercially available spin column RNA kit, according to the manufacturer's instructions.
2. With 1–2μL of isolated RNA use the Nanodrop spectrophotometer to determine the RNA concentration in each sample.
3. Examine RNA integrity and quality using Bioanalyzer.
4. Submit RNA samples to a core facility for gene expression analysis, NanoString Technologies, or bulk RNA sequencing.

3.6.2 Collagen Assessment

1. To quantify the levels of soluble collagen in the supernatant from the RIPA-homogenized lung tissues, the Sircol™ soluble collagen assay can be utilized. Perform the assay using

supernatants from lung homogenates, according to the manufacturer's protocol. Read absorbance using a microplate reader at a wavelength of 555 nm and determine the test sample collagen concentrations from the standard curve. Results are expressed as μg soluble collagen/mL solution.

2. To quantify the levels of insoluble collagen in the pellet from the RIPA-homogenized lung tissues, the colorimetric hydroxyproline assay can be utilized: Perform the assay as previously described [53] by reading absorbance using a microplate reader at a wavelength of 550 nm and determining the test sample collagen concentrations from standard curve. Results are expressed as μg hydroxyproline/ml solution.

3.7 Histology Tissue Microarray (TMA) Production and Digitization

3.7.1 Creation of a TMA

Assessing all formalin-fixed paraffin-embedded murine tissues on an individual basis is time consuming and can be quite costly. Extracting these tissues from each individual paraffin block and consolidating them into a single paraffin block provides the ability to analyze whole lung slices for all experimental mice in a single paraffin slice, as illustrated previously [15]. Creating a tissue microarray requires planning and knowledge of the precise location(s) the core(s) should be extracted. In this case, by using a puncher with a sufficiently large diameter size (1.5–2.0 mm), 1–2 cores are enough to extract the entirety of the mouse lung. Thus, deciding on a location for punching is not required as most of the tissue is included.

1. Organize all formalin-fixed paraffin-embedded tissue blocks from which cores will be extracted, ensuring each one has a meaningful block identifier (*see* **Note 34**).
2. Determine the limiting factor: what aspect of the TMA must or must not take place, such as the minimum core diameter not being less than 1.5 mm or ensuring that all animal tissues are present on a single TMA (as opposed to being spread out over 2 or 3 TMAs).
3. Design TMA layout(s) with limiting factor(s) in mind (*see* **Note 35**).
4. Prepare the paraffin receiving block as this is what will become the TMA: Fill a 37 mm × 24 mm block mold with hot paraffin wax to a depth of approximately 8 mm and place an empty tissue cassette face down and remove the fresh receiving block from mold once paraffin has solidified.
5. Drill out the designated number of holes into the receiving block with the desired diameter drill bit at a depth of approximately 7 mm according to the TMA layout design.
6. Once drilling is complete, remove excess wax filings by inverting drilled receiving block upside-down and tapping the

underside of the cassette lightly causing the filings to fall out of the drilled holes.

7. If specific areas of the specimen are required, ensure that these locations are chosen and identified ahead of time to allow for an accurate and smooth extraction of tissue from the donor to the recipient block (*see* **Note 36**).
8. With the receiving block drilled and all of the regions on the donor tissue blocks identified, begin punching.
9. Punch a tissue core in each designated donor tissue block and transfer these formalin-fixed paraffin-embedded cores into the receiving block's designated location, according to the TMA layout design. Continue until all desired tissues have been punched and all receiving block holes have been filled.
10. Once the receiving block (now a TMA) is full, place a new microscope slide on top such that it covers all the tissue cores, and invert the receiving block plus microscope slide such that the TMA is lying flat on the microscope slide (*see* **Note 37**).
11. Set the oven to 45 °C.
12. Place the new TMA inverted on a microscope slide directly onto one of the oven racks, and let it sit for 24 h. The gentle heat will cause a very slight melting of the paraffin filling in the gaps caused by the drilling and sealing the tissue cores into one cohesive unit. The weight of the cassette will also press the paraffin slightly, compacting the TMA.
13. Remove from the oven and place the TMA on the slide directly into a 4 °C refrigerator for a few hours. The paraffin should be hard to the touch.
14. Remove the TMA from the refrigerator. The microscope slide will be stuck to the top of the TMA; pull it off and discard.
15. The top of the TMA will be uneven from the drying process. Run the TMA through a microtome to smooth out the top, as this will ensure each successive cut will contain tissue.
16. The TMA is now ready for slicing onto a microscope slide and staining.

3.7.2 Tissue Array Digitization

Viewing tissue slides utilizing a microscope is limited by experience with a microscope, time of viewing, equipment availability and inability to utilize for assessment. Digitizing these slides enables easy viewing, without requiring experience with a microscope, on any computer. This allows for more people to view the slide simultaneously, even remotely, and side-by-side comparison of stains.

1. Load desired slides into a slide digitizer (*see* **Note 38**).
2. Once the digitizer has gone through a quick low magnification overview scan of the slide, select the region of the slide where

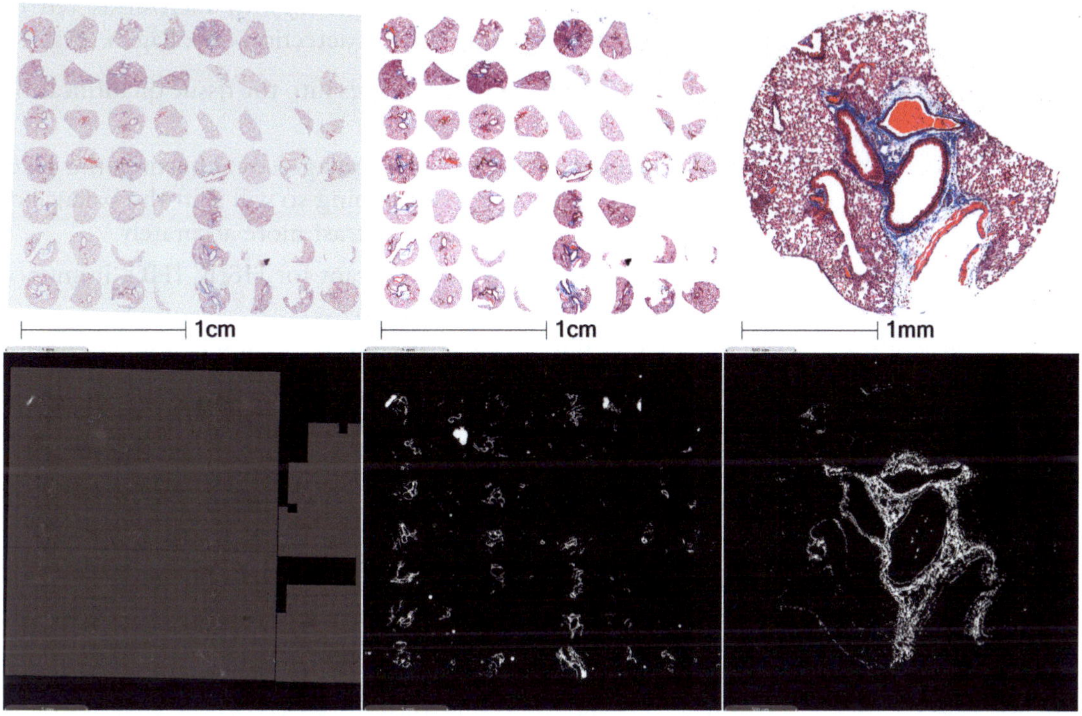

Fig. 4 *Importing and contrast-correcting images in HALO®*. (Top-Left) Tissue Microarray (TMA) stained with Masson's Trichrome, digitized and directly imported into HALO. (Top-Center) Pixel black input increased from 0 to 125 and white input decreased from 255 to 220. End result removes background and increases color vividness. (Top-Right) Zoom in on one specific core to see mouse lung tissue in reddish-purple and collagen in blue. (Bottom-Left) Same TMA serially sliced and stained with Picrosirius Red, digitized using polarized white light and directly imported into HALO. (Bottom-Center) Pixel black input increased from 0 to 13 and white input decreased from 255 to 15. End result removes background and increases whiteness. (Bottom-Right) Zoom in on the same specific core to see mouse lung collagen in white and everything else in black

the high magnification scan will take place, that being the tissue of interest.

3. Different stains require also different imaging techniques (*see* **Note 39**).
4. Once the scanning is complete, the slide has been digitized and is ready for viewing and/or quantification.

3.8 Quantification Using Digitized TMAs

With digitized tissues, image analyses can be performed to extract further quantifiable and qualitative information. The power of digital pathology made possible here is immense, however our foci here are the number of cells within the tissue as defined by nuclear staining, the proportion of cells that are positive for a protein or mRNA transcript as defined by Immunohistochemistry or ISH staining, respectively, and their proximity to previously defined cells. There is also the quantification of protein area with respect to the total tissue area, such as with Picrosirius red or Masson's trichrome staining for the detection of collagen (Fig. 4).

1. Import the digitized tissue images into the tissue quantification software.
2. Contrast correct the digitized images by adjusting the pixel input black and white levels. Doing so will shift the balance of color to increase the image contrast more accurately.
3. The colors should be more vibrant for H&E, IHC, immunofluorescence, C-ISH, and FISH.
4. The white fluorescence should be brighter for Picrosirius red-POL.
5. The background noise should be eliminated.
6. Recreate the design map within the quantification software, by adding the rows and columns according to the TMA layout.
7. Assign each core useful information for quantification, such as animal ID, group #, and potential drug treatment.

3.8.1 Protein and mRNA Quantification from Digitized TMAs

1. The algorithms used will require assigning a color for nuclei, which can be done by selecting a range of nuclear pixels on-screen and averaging those pixel colors, as well as colors for the protein(s)/mRNA(s) for which the image is stained (*see* **Note 40**).
2. Adjust the cellular parameters such that the nuclear size, shape, contrast threshold, and minimum optical density properly reflect the cells on-screen (*see* **Note 41**).
3. Adjust the protein/mRNA parameters such that the cellular compartment and minimum positive threshold(s) properly reflect the stain(s) on-screen.
4. The stain must present itself within the proximity of the cell for the algorithm to consider that cell to be positive for the selected protein/mRNA transcript (*see* **Note 42**).
5. The stain must also be of sufficient pixel intensity or minimum optical density for it to be considered true staining (*see* **Note 43**).
6. Run the protein/mRNA quantification software over the entire image for single tissue slides, or for TMAs, run it through the TMA layout so that each core is analyzed individually, and results are assigned on a core-by-core basis instead of treating the image as a single tissue.
7. Using the raw data (total cells, stain + cells, etc.), determine the proportion of cells that are positive for your stain of interest according to the experimental groups and perform statistical analyses.

3.8.2 Picrosirius Red Quantification

1. The fluorescence area quantification algorithm used will require assigning a fluorescence channel (POL-MONO) as the selected dye (*see* **Note 44**).
2. Adjust the dye minimum intensity such that only the white fluorescence of the collagen passes the positivity threshold (*see* **Note 45**).
3. Annotate the tissue region to determine the total tissue area.
4. Run the fluorescence area quantification algorithm to determine the total POL-MONO area (*see* **Note 46**).
5. If desired, excise the major airways and blood vessels from the lung by means of annotation and then run the algorithm again (*see* **Note 47**).
6. Using the raw data (POL-MONO area, total area, etc.), determine the proportion of the lung that is composed of collagen according to the experimental groups and perform statistical analyses.

4 Notes

1. It is recommended to test the aliquoted vial prior to each use, as the strength of drug and its response may vary depending on the provider, lot/batch number, the mouse (strain, gender, and age), and more. Subsequently, these tests will help determine the proper dose.
2. Protease inhibitor should be made fresh on the day of the experiment.
3. Ensure that the working area around the biological safety cabinet (BSC) is sterilized and approved for animal work and autoclave all materials listed in Subheading 2.1.
4. It is important to not pinch the tip of the animal's tongue to avoid injury.
5. Prior to use, attach the operating otoscope to the otoscope handle, and then attach the specula to the operating otoscope.
6. Prior to injection, blunt the ends of several 22G, 1-inch insyte IV catheter needles. Then, attach the 22G, 1-inch insyte IV catheter/blunt ended needle to a 1 mL syringe.
7. Prior to the start of the experiment, adjust a 200μL pipette to the 150μL setting and attach a standard-length pipette tip. Then, draw 150μL of double-distilled water into the pipette tip and leave a space at the end of the tip for a large volume of air. Set aside until use.
8. Prior to the start of the experiment, adjust an extended length gel loading pipette tip to a second 200μL and draw 50μL of bleomycin into the pipette tip. Set aside until use.

9. It is important the agent is inhaled and not forced into the lungs. Forced fluid can come out the nose, potentially being fatal.
10. If needed, top-up the anesthesia with 1/3 dose of ketamine only.
11. With this ketamine/xylazine cocktail, animals will remain surgically anesthetized for 20–30 min.
12. Carefully remove any fur and skin around the trachea, avoid cutting the jugular veins, and ensure that there is no excess blood in tracheal area as it is critical to not have blood entering the airways.
13. Software will differ depending on the flexiVent® provider.
14. Refer to Subheading 2.2 to prepare a working concentration of rocuronium bromide (1:5 dilution).
15. 0.5 mL—1.0 mL of blood results in a volume of approximately 200–500μL of plasma.
16. It is critical to aliquot the plasma (for example, in a volume of 50μL) and to avoid freeze-thaw cycles.
17. It is critical to aliquot the BALF supernatant and to avoid freeze-thaw cycles.
18. The following morphological characteristics are assessed to identify the different cell types in the lung BALF: Macrophages are the largest of the white blood cells with large purple kidney-shaped nucleus, abundant cytoplasm, and fine pink/purple granules. Lymphocytes are slightly larger than red blood cells (RBC), look like monocytes but with small spherical purple nucleus and not much cytoplasm. Neutrophils contain multi-lobed purple nuclei, cytoplasm contains purplish granules, and are the smallest of granulocytes but bigger than RBC. Eosinophils are larger than neutrophils, contain two lobed purple nuclei, and cytoplasm with pink granules. Bronchial epithelial cells are ciliated and columnar with purple nucleus.
19. To avoid red blood cells contamination with cells of interest in lungs, heart perfusion might be conducted. Perform heart perfusion prior to BALF collection as follows: After opening the thoracic cavity of the mouse, expose lungs and heart as much as possible by removing surrounding tissues and rip cage. Insert a 26G1/2 needle, connected to a 5 mL syringe that contains sterile PBS, into the right ventricle of mouse heart. Slowly inject 5 mL of PBS until lung color changes and turns white.
20. Caution must be taken while performing heart perfusion as lungs might be very fragile at specific timepoints. The fragility of the lungs depends on the time of investigation and the amount of injury. For instance, during inflammatory phase

(seven days following bleomycin administration), perfusing heart to remove blood might result in leakage in the lungs.

21. Immediately, place 50 mL falcon tube containing digested cells on ice.
22. Resuspending pellets in FACS buffer or PBS depends on the respective flow cytometry staining protocol.
23. To maximize total number of cells to be processed for flow cytometry, it is recommended to stain fresh viable cells before the freezing process. Following one freeze-thaw cycle, approximately 50% of total number of viable frozen cells are lost.
24. One set of tubes containing RIPA buffer will be for protein analysis, and the other empty set of tubes will be used for RNA analysis.
25. It is always important to keep tissues frozen throughout the entire procedure. Perform steps quickly when tissues are not in liquid nitrogen or dry ice.
26. Freeze the inside of the metal chamber by adding, swirling, and pouring out the liquid nitrogen.
27. Ensure that liquid nitrogen has evaporated from inside of chamber prior to adding the piston.
28. It is important to keep returning the piston back into Styrofoam freezing container.
29. Once sample is crushed and powder is transferred over onto a tin foil, quickly return the metal chamber back into the Styrofoam freezing container.
30. All tubes on dry ice should be transferred and stored in −80 °C as soon as possible.
31. To ensure that tissue is being properly conserved, store the tubes on dry ice when they are not being weighed.
32. This protocol should be performed in a fume hood.
33. All tissue powder must be dissolved in the RIPA buffer. Often, longer than 10 s homogenization cycles, or extra homogenization cycles, are required.
34. If multiple mice are present on a single block, include identifiers for each separate mouse whose tissues are present in the block.
35. In the case of an animal model TMA, it is likely more desirable to have tissues from each animal be present within a single TMA. Also, multiple TMAs can be made with this in mind [15]: One TMA could have tissues from all animals present in a single 2.0 mm core, allowing for larger field of views for visual phenotyping. Another could have tissues from all animals

present as 3 replicate 1.0 mm cores to allow for statistical analyses as well as analyses set over multiple regions in a single lung.

36. If regions must be determined by microscopic examination, have digitized images of full tissue slices annotated to making finding these locations during punching efficient.

37. This is done to ensure none of the cores accidentally slide out of the TMA during the inversion process, as the paraffin has not yet molded to the cores which will occur during the heating process.

38. Provide a descriptive, searchable file name that can be distinguished from other slides that may have similar tissues and/or stains.

39. For example, hematoxylin and eosin (H&E) slides should be scanned using the 40× objective lens, as this magnification is needed when pathologists need to determine cell types and tissue phenotypes; Immunohistochemistry (IHC) slides should be scanned using the 20× objective lens; less detail is required as compared to the H&E slide, since labeled proteins are very dark and spacious within or around cells, making them easily viewable and quantifiable; In situ hybridization (ISH) slides, both colorimetric and fluorescent, should be scanned using the 40× objective lens, as labeled mRNA transcripts are much smaller than labeled proteins.

40. For nuclear stains, the prevalent counterstains in the histology performed leaves the nuclei with a blue color. For IHC stains, such as α-smooth muscle actin, this is a deep brown color. Care must be made so as not to select a color that is too dark, as gray/black artifacts may be accidentally categorized as IHC stain, when they should be excluded.

41. For example, based on the average volume of alveolar macrophages [54] and the average diameter of lung cell nuclei [55], with the whole cell diameter being calculated from the volume and the nuclear diameter being subtracted from that value, the resultant diameter (and thus, radius) is found to be approximately 6.5μm.

42. Based the previous example and **Note 41**, we have defined the radius of the cytoplasm to be 6.5μm surrounding the nucleus.

43. This is often done with the assistance of the resident pathologists that have agreed to assist the laboratory with these situations.

44. POL-MONO is the specific term used by Olympus to refer to using the mono-color channel in conjunction with a polarizer filter lens.

45. Picrosirius Red staining enhances the natural birefringence of collagen making it shine white under polarized fluorescent light.
46. Under polarized fluorescent light, the image will only display black and white; white is the total amount of collagen present in the image, black is everything else.
47. By removing vasculature and airways, all the collagen present is considered as being part of the parenchyma only.

References

1. Hughes J, Rees S, Kalindjian S, Philpott K (2011) Principles of early drug discovery. Br J Pharmacol 162:1239–1249. https://doi.org/10.1111/j.1476-5381.2010.01127.x
2. Sioud M (2007) Main approaches to target discovery and validation. In: Sioud M (ed) Target discovery and validation reviews and protocols: volume 1, emerging strategies for targets and biomarker discovery. Humana Press, Totowa, NJ, pp 1–12
3. Naiel S, Tat V, Padwal M et al (2019) Protein misfolding and er stress in chronic lung disease: will cell-specific targeting be the key to the cure? Chest. https://doi.org/10.1016/j.chest.2019.11.009
4. Kaminski N, Rosas IO (2006) Gene expression profiling as a window into idiopathic pulmonary fibrosis pathogenesis: can we identify the right target genes? Proc Am Thorac Soc 3:339–344. https://doi.org/10.1513/pats.200601-011TK
5. Cha Y, Erez T, Reynolds IJ et al (2018) Drug repurposing from the perspective of pharmaceutical companies. Br J Pharmacol 175:168–180. https://doi.org/10.1111/bph.13798
6. Karatzas E, Bourdakou MM, Kolios G, Spyrou GM (2017) Drug repurposing in idiopathic pulmonary fibrosis filtered by a bioinformatics-derived composite score. Sci Rep 7. https://doi.org/10.1038/s41598-017-12849-8
7. Durham TB, Wiley MR (2017) Target engagement measures in preclinical drug discovery: theory, methods, and case studies. In: Bhattachar SN, Morrison JS, Mudra DR, Bender DM (eds) Translating molecules into medicines: cross-functional integration at the drug discovery-development interface. Springer International Publishing, Cham, pp 41–80
8. Durham TB, Blanco M-J (2015) Target engagement in lead generation. Bioorg Med Chem Lett 25:998–1008. https://doi.org/10.1016/j.bmcl.2014.12.076
9. Steinmetz KL, Spack EG (2009) The basics of preclinical drug development for neurodegenerative disease indications. BMC Neurol 9:S2. https://doi.org/10.1186/1471-2377-9-S1-S2
10. Tashiro J, Rubio GA, Limper AH et al (2017) Exploring Animal models that resemble idiopathic pulmonary fibrosis. Front Med (Lausanne) 4. https://doi.org/10.3389/fmed.2017.00118
11. Denayer T, Stöhr T, Van Roy M (2014) Animal models in translational medicine: validation and prediction. New Horizons Transl Med 2:5–11. https://doi.org/10.1016/j.nhtm.2014.08.001
12. Justice MJ, Dhillon P (2016) Using the mouse to model human disease: increasing validity and reproducibility. Dis Model Mech 9:101–103. https://doi.org/10.1242/dmm.024547
13. Moore BB, Hogaboam CM (2008) Murine models of pulmonary fibrosis. Am J Physiol Lung Cell Mol Physiol 294:L152–L160. https://doi.org/10.1152/ajplung.00313.2007
14. B Moore B, Lawson WE, Oury TD et al (2013) Animal models of fibrotic lung disease. Am J Respir Cell Mol Biol 49:167–179. https://doi.org/10.1165/rcmb.2013-0094TR
15. Yanagihara T, Chong SG, Vierhout M et al (2020) Current models of pulmonary fibrosis for future drug discovery efforts. Expert Opin Drug Discovery:1–11. https://doi.org/10.1080/17460441.2020.1755252
16. Kolb P, Upagupta C, Vierhout M et al (2020) The importance of interventional timing in the bleomycin model of pulmonary fibrosis. Eur Respir J. https://doi.org/10.1183/13993003.01105-2019
17. Jenkins RG, Moore BB, Chambers RC et al (2017) An official American thoracic society

workshop report: use of animal models for the preclinical assessment of potential therapies for pulmonary fibrosis. Am J Respir Cell Mol Biol 56:667–679. https://doi.org/10.1165/rcmb.2017-0096ST

18. Bauer Y, Tedrow J, de Bernard S et al (2015) A novel genomic signature with translational significance for human idiopathic pulmonary fibrosis. Am J Respir Cell Mol Biol 52:217–231. https://doi.org/10.1165/rcmb.2013-0310OC
19. Adams TS, Schupp JC, Poli S et al (2019) Single Cell RNA-seq reveals ectopic and aberrant lung resident cell populations in idiopathic pulmonary fibrosis. bioRxiv:759902. https://doi.org/10.1101/759902
20. Yu G, Ibarra GH, Kaminski N (2018) Fibrosis: lessons from OMICS analyses of the human lung. Matrix Biol 68–69:422–434. https://doi.org/10.1016/j.matbio.2018.03.014
21. Herazo-Maya JD, Kaminski N (2012) Personalized medicine: applying 'omics' to lung fibrosis. Biomark Med 6:529–540. https://doi.org/10.2217/bmm.12.38
22. Patil JS, Sarasija S (2012) Pulmonary drug delivery strategies: a concise, systematic review. Lung India 29:44–49. https://doi.org/10.4103/0970-2113.92361
23. Strong P, Ito K, Murray J, Rapeport G (2018) Current approaches to the discovery of novel inhaled medicines. Drug Discov Today 23:1705–1717. https://doi.org/10.1016/j.drudis.2018.05.017
24. Epstein-Shochet G, Pham S, Beck S et al (2020) Inhalation: a means to explore and optimize nintedanib's pharmacokinetic/pharmacodynamic relationship. Pulm Pharmacol Ther:101933. https://doi.org/10.1016/j.pupt.2020.101933
25. Darquenne C (2012) Aerosol deposition in health and disease. J Aerosol Med Pulm Drug Deliv 25:140–147. https://doi.org/10.1089/jamp.2011.0916
26. Tuntland T, Ethell B, Kosaka T et al (2014) Implementation of pharmacokinetic and pharmacodynamic strategies in early research phases of drug discovery and development at Novartis Institute of Biomedical Research. Front Pharmacol 5. https://doi.org/10.3389/fphar.2014.00174
27. Polson AG, Fuji RN (2012) The successes and limitations of preclinical studies in predicting the pharmacodynamics and safety of cell-surface-targeted biological agents in patients. Br J Pharmacol 166:1600–1602. https://doi.org/10.1111/j.1476-5381.2012.01916.x
28. Schaefer CJ, Ruhrmund DW, Pan L et al (2011) Antifibrotic activities of pirfenidone in animal models. Eur Respir Rev 20:85–97. https://doi.org/10.1183/09059180.00001111
29. Bonniaud P, Fabre A, Frossard N et al (2018) Optimising experimental research in respiratory diseases: an ERS statement. Eur Respir J 51:1702133. https://doi.org/10.1183/13993003.02133-2017
30. Moeller A, Ask K, Warburton D, et al (2008) The bleomycin animal model: a useful tool to investigate treatment options for idiopathic pulmonary fibrosis? Int J Biochem Cell Biol 40:362–382. doi: 10.1016/j.biocel.2007.08.011
31. King TE, Bradford WZ, Castro-Bernardini S et al (2014) A phase 3 trial of pirfenidone in patients with idiopathic pulmonary fibrosis. N Engl J Med 370:2083–2092. https://doi.org/10.1056/NEJMoa1402582
32. Richeldi L, du Bois RM, Raghu G et al (2014) Efficacy and safety of nintedanib in idiopathic pulmonary fibrosis. N Engl J Med 370:2071–2082. https://doi.org/10.1056/NEJMoa1402584
33. Richeldi L, Costabel U, Selman M et al (2011) Efficacy of a tyrosine kinase inhibitor in idiopathic pulmonary fibrosis. N Engl J Med 365:1079–1087. https://doi.org/10.1056/NEJMoa1103690
34. Curtis MJ, Bond RA, Spina D et al (2015) Experimental design and analysis and their reporting: new guidance for publication in BJP. Br J Pharmacol 172:3461–3471. https://doi.org/10.1111/bph.12856
35. Head ML, Holman L, Lanfear R et al (2015) The extent and consequences of p-hacking in science. PLoS Biol 13:e1002106. https://doi.org/10.1371/journal.pbio.1002106
36. Aban IB, George B (2015) Statistical considerations for preclinical studies. Exp Neurol 270:82–87. https://doi.org/10.1016/j.expneurol.2015.02.024
37. Huang W, Percie du Sert N, Vollert J, Rice ASC (2020) General principles of preclinical study design. In: Bespalov A, Michel MC, Steckler T (eds) Good research practice in non-clinical pharmacology and biomedicine. Springer International Publishing, Cham, pp 55–69
38. Festing MFW (2006) Design and statistical methods in studies using animal models of development. ILAR J 47:5–14. https://doi.org/10.1093/ilar.47.1.5
39. Kulkarni MM (2011) Digital multiplexed gene expression analysis using the NanoString nCounter system. Curr Protoc Mol Biol 25.

Unit25B.10. https://doi.org/10.1002/0471142727.mb25b10s94

40. Ask K, Labiris R, Farkas L et al (2008) Comparison between conventional and "clinical" assessment of experimental lung fibrosis. J Transl Med 6(16). https://doi.org/10.1186/1479-5876-6-16
41. Beckmann N, Kneuer R, Gremlich H-U et al (2007) In vivo mouse imaging and spectroscopy in drug discovery. NMR Biomed 20:154–185. https://doi.org/10.1002/nbm.1153
42. Schambach SJ, Bag S, Schilling L et al (2010) Application of micro-CT in small animal imaging. Methods 50:2–13. https://doi.org/10.1016/j.ymeth.2009.08.007
43. Vetter TR, Mascha EJ (2017) Defining the primary outcomes and justifying secondary outcomes of a study: usually, the fewer, the better. Anesth Anal 125:678–681. https://doi.org/10.1213/ANE.0000000000002224
44. Liang W, Menke AL, Driessen A et al (2014) Establishment of a general NAFLD scoring system for rodent models and comparison to human liver pathology. PLoS One 9. https://doi.org/10.1371/journal.pone.0115922
45. Noordzij M, Tripepi G, Dekker FW et al (2010) Sample size calculations: basic principles and common pitfalls. Nephrol Dial Transplant 25:1388–1393. https://doi.org/10.1093/ndt/gfp732
46. Bert B, Heinl C, Chmielewska J et al (2019) Refining animal research: the animal study registry. PLoS Biol 17. https://doi.org/10.1371/journal.pbio.3000463
47. Cai Y, Kimura S (2013) Noninvasive intratracheal intubation to study the pathology and physiology of mouse lung. J Vis Exp. https://doi.org/10.3791/50601
48. Vandivort TC, An D, Parks WC (2016) An improved method for rapid intubation of the trachea in mice. J Vis Exp. https://doi.org/10.3791/53771
49. Puri S, Li J, Xiong M et al (2015) Method for single intravenous anesthetic infusion in a rodent model. Open J Anesthesiol 05:96–104. https://doi.org/10.4236/ojanes.2015.55019
50. McGovern TK, Robichaud A, Fereydoonzad L et al (2013) Evaluation of respiratory system mechanics in mice using the forced oscillation technique. J Vis Exp e50172. https://doi.org/10.3791/50172
51. Ayaub EA, Kolb PS, Mohammed-Ali Z et al (2016) GRP78 and CHOP modulate macrophage apoptosis and the development of bleomycin-induced pulmonary fibrosis. J Pathol 239:411–425. https://doi.org/10.1002/path.4738
52. Ayaub EA, Dubey A, Imani J et al (2017) Overexpression of OSM and IL-6 impacts the polarization of pro-fibrotic macrophages and the development of bleomycin-induced lung fibrosis. Sci Rep 7:13,281. https://doi.org/10.1038/s41598-017-13511-z
53. Brown S, Worsfold M, Sharp C (2001) Microplate assay for the measurement of hydroxyproline in acid-hydrolyzed tissue samples. Biotechniques 30:38–40, 42. https://doi.org/10.2144/01301bm06
54. Krombach F, Münzing S, Allmeling AM et al (1997) Cell size of alveolar macrophages: an interspecies comparison. Environ Health Perspect 105:1261–1263. https://doi.org/10.1289/ehp.97105s51261
55. Crapo JD, Barry BE, Gehr P et al (1982) Cell number and cell characteristics of the normal human lung. Am Rev Respir Dis 126:332–337. https://doi.org/10.1164/arrd.1982.126.2.332

Chapter 22

Mouse Models of Kidney Fibrosis

Rafael Kramann and Sylvia Menzel

Abstract

Chronic kidney disease (CKD) affects over 10% of the worldwide population and kidney fibrosis is a main driver of CKD and considered a therapeutic target. The mechanisms leading to kidney fibrosis are highly complexed and can be best studied in rodent models. Here we describe the most commonly used kidney fibrosis models in mice, the unilateral ureteral obstruction (UUO) model and the ischemia reperfusion injury (IRI) model. Both models are easy to learn and can be applied in animals of different age, sex, and strain.

Key words Unilateral ischemia reperfusion, IRI, UUO, Hydronephrosis, Mouse model

1 Introduction

Chronic kidney disease (CKD) affects more than 10% of the worldwide population with an increasing prevalence. Kidney fibrosis is the common final pathway of virtually all chronic progressive injury to the kidney and a considered therapeutic target in CKD. Myofibroblasts are the effector cells that drive fibrosis but their cellular origin is still controversially discussed and their mechanisms of activation are complex and partly unclear [1, 2]. The mechanisms that drive kidney fibrosis by activation and expansion of myofibroblasts are very difficult to model ex vivo and thus rodent models are still the gold standard to study renal fibrosis in order to understand disease mechanisms and to test potential therapeutics.

Currently, there is no specific treatment for kidney fibrosis approved. Once CKD patients reach end-stage renal disease (ESRD) they need to undergo dialysis or kidney transplant; however, the waiting times for kidney transplant are very long and dialysis patients suffer from a very high morbidity and mortality. Thus, reliable and replicable in vivo models for kidney fibrosis are needed to study disease mechanisms, identify therapeutic targets and validate potential therapeutics. Table 1 shows commonly used models of kidney fibrosis in rodents.

Boris Hinz and David Lagares (eds.), *Myofibroblasts: Methods and Protocols*, Methods in Molecular Biology, vol. 2299, https://doi.org/10.1007/978-1-0716-1382-5_22, © Springer Science+Business Media, LLC, part of Springer Nature 2021

Table 1
Different rodent models of kidney fibrosis [3]

Model	Species	Comments
Unilateral ureteral Obstruction—UUO	Mouse, rat	Fibrosis model, hydronephrosis, tubular dilation, interstitial fibrosis, normal kidney function in contralateral kidney thus no CKD
Ischemia reperfusion injury—IRI (unilateral or bilateral)	Mouse, rat	Primarily an acute kidney injury (AKI) model, can also be utilized to study AKI to CKD development (bilateral), tubular injury/repair depends on severity of injury (time of artery clamping), strain and sex, normal kidney function in contralateral kidney (unilateral model)
Aristolochic acid	Mouse, rat	Tubule injury, resembles Balkan nephropathy or Chinese herb nephropathy in human, progressive tubulointerstitial nephritis, CKD
Folic acid	Mouse, rat	Acute tubular damage with cast formation, followed by tubulointerstitial fibrosis after 14 days, CKD
Adenine nephropathy	Mouse, rat	Crystal induced nephropathy Tubulointerstitial damage and fibrosis, CKD, often heavy weight loss as animals avoid eating
5/6 nephrectomy	Mouse, rat	Loss of functional tissue with hyperfiltration of the remnant kidney glomerulosclerosis with hyperfiltration and tubulointerstitial fibrosis, strain dependent elevation of BUN and serum creatinine. High complication rates
DOCA salt with unilateral nephrectomy	Mouse, rat	Tubulointerstitial fibrosis, strain dependent, V57BL/6 resistant to kidney damage

Here, we provide a guideline for the optimal use of two of the most commonly used kidney fibrosis models in mice: the unilateral ureteral obstruction (UUO) and the unilateral ischemia reperfusion injury (IRI).

1.1 UUO Model

In the UUO model, the ureter is ligated at one or two sites using a suture. This leads to an obstruction of the kidney outflow-tract on the ligated side so that the urine cannot drain anymore, causing hydronephrosis with tubular dilation and injury followed by immune cell infiltration and fibrosis. Mechanical stretching of tubular epithelial cells leads to injury and death of the tubular epithelium which is resulting in an inflammatory response with a tremendous influx of immune cells. Various mechanisms lead then to pericyte and fibroblast-to-myofibroblast differentiation and activation, driving the development and progression of kidney fibrosis [4, 14].

1.1.1 UUO Advantages

- Quick and reproducible model for renal fibrosis.
- Easy to learn.

- Method can be applied in animals of different age, sex, and strain.
- Internal control by contralateral kidney.

1.1.2 UUO Disadvantages

- No loss of kidney function/CKD.
- Clinically UUO like diseases are very rare, e.g., obstruction due to tumor or retroperitoneal fibrosis.

1.1.3 UUO Variations

- Reversed UUO.
- Bilateral UUO (different timepoints, reversed vs. non-reversed).

1.2 IRI Model

The IRI model is based on clamping the renal vascular pole for a certain period of time (usually 25–45 min dependent on injury severity and gender of mice), which ultimately results in diminished oxygen supply to the kidney with hypoxic injury. During ischemia, the removal of cellular waste products is also reduced. If blood supply returns (unclamping), the tissue is resupplied with oxygen and nutrients (reperfusion). Mismatch of oxygen supply and oxygen demand as well as the accumulation of waste products lead to inflammation and oxidative stress. In the kidney, tubular epithelial cells are injured first resulting in de-differentiation and cell death. Depending on the extent of tissue damage, tubular regeneration occurs after injury but it is widely accepted that there is a path forward from acute kidney injury (AKI) to CKD, as one episode of AKI already increases the risk of future CKD [5]. After severe or repetitive kidney injury, maladaptive tissue repair responses lead to the development of tubulo-interstitial fibrosis and CKD (bilateral model) [6, 7].

1.2.1 IRI Advantages

- Model for acute kidney injury and renal fibrosis in chronic kidney disease (bilateral model).
- Easy to learn.
- Method can be applied in animals of different age, sex, and strain.
- Internal control by contralateral non-treated kidney (only if unilateral model is applied).
- Mild model for long-term observation (one intact kidney).
- Severity of injury can be adjusted by changing ischemia duration.

1.2.2 IRI Disadvantages

- No loss of kidney function/CKD (unilateral model).
- High variability between males and females, so you will need different groups if you want to include both sexes.
- In terms of fibrosis IRI causes initially tubular injury with repair and fibrosis comes in at later stage only after severe injury. A

good timepoint to measure fibrosis is 14–28 days after severe IRI.

1.2.3 IRI Variations

- Bilateral IRI: both kidneys are clamped at the same time. Cave: extremely temperature/pressure sensitive (no injury vs. high lethality). CKD develops and can be estimated using urine and serum sampling [8].
- Unilateral IRI with contralateral nephrectomy: one kidney is clamped, the other kidney is removed. Healthy kidney for control before treatment, etc., leads to CKD Cave: higher mortality than in IRI without contralateral nephrectomy [9].
- Very rare: cold ischemia-warm reperfusion (body is cooled down to 32 °C).
- Mostly performed in mice and rats, sometimes also in pigs, rabbits, cats, and nonhuman primates.

2 Materials

All surgical instruments (Fig. 1) and devices should be checked for their functionality before beginning surgery. Instruments have to be cleaned thoroughly with a disinfectant (e.g., Sekusept plus); sterilization is not necessary.

2.1 Materials for UUO Model (See Notes 1–6)

- Heating pad with homeothermic monitoring system (for setup *see* Fig. 2).
- Hatching incubator (Fig. 3) or other device for warming after surgery (*see* **Note 8**).
- Small clipper (e.g., Aesculap Exacta).
- Tape for fixation of cables, etc.
- Small tissue forceps 1 × 2 teeth (approx. 12 cm).
- Surgical scissors (blunt + sharp, approx. 12 cm).
- Fine scissors (sharp + sharp, approx. 9 cm).
- 2 Iris forceps curved and serrated (approx. 7 cm).
- 2 Micro forceps, e.g., #4 and #7 jewelers.
- Suture tying forceps (optional).
- Needle holder (approx. 12 cm).
- Wound clip applier and wound clips 9 mm (optional).
- Nonabsorbable suture polyfil 5/0 (e.g., Mersilene).
- 10% povidone-iodine for skin disinfection.
- Sterile saline (0.9% NaCl).

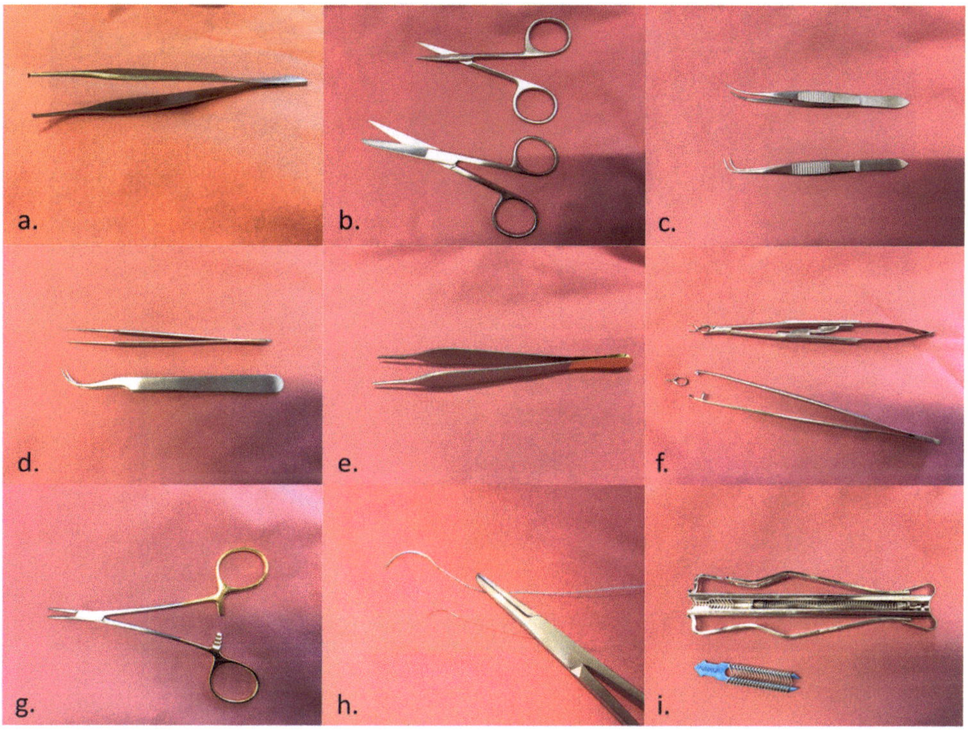

Fig. 1 *Instruments.* (**a**) Tissue forceps, (**b**) scissors; top: fine scissors, bottom: surgical scissors, (**c**) iris forceps; top: slightly curved, bottom: strongly curved, (**d**) micro forceps; top: #4 jewelers, bottom: #7 jewelers, (**e**) suture tying forceps, (**f**) micro-serrefine clamps, top: clamp applicator with lock, bottom: clamp forceps, (**g**) needle holder, (**h**) sutures, top: 5.0 suture, bottom: 7.0 suture, (**i**) wound clip applier with wound clips

- Sterile swabs.
- Q-tips or surgical spears.
- Eye ointment.
- Ketamine and xylazine for anesthesia (or any anesthetic of your choice, *see* **Note 9**).
- Buprenorphine for analgesia (or any analgesia of your choice).

2.2 Materials for IRI Model (See Notes 1–6)

- Heating pad with homeothermic monitoring system.
- Hatching incubator or other device for warming after surgery (*see* **Note 8**).
- Small clipper (e.g., Aesculap Exacta).
- Timer for ischemia time.
- Tape for fixation of cables, etc.
- Small tissue forceps 1 × 2 teeth (approx. 12 cm).
- Surgical scissors (blunt + sharp, approx. 12 cm).
- Fine scissors (sharp + sharp, approx. 9 cm).

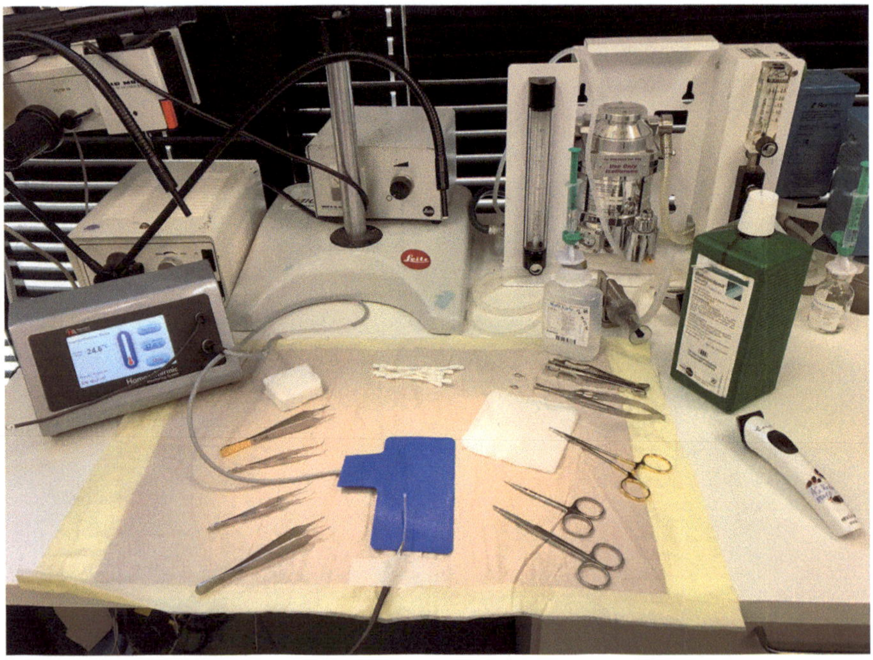

Fig. 2 Surgery setup

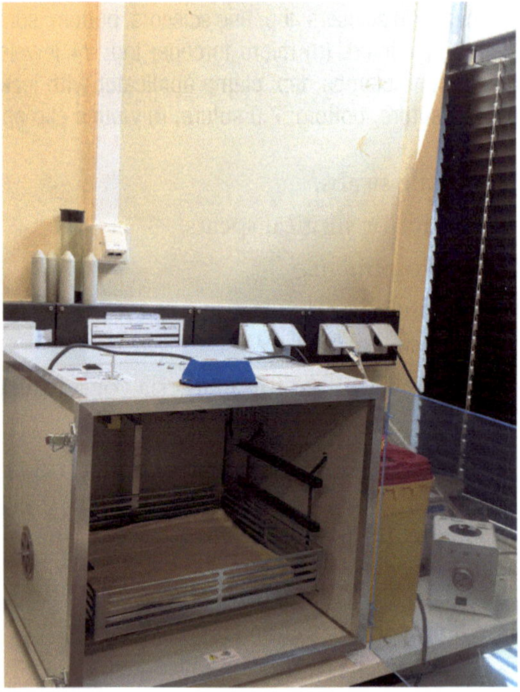

Fig. 3 *Hatching incubator.* Temperature and humidity can be regulated and feedback controlled. Due to constant ventilation, temperature is evenly distributed within the incubator

- 2 Iris forceps curved and serrated (approx. 7 cm).
- 2 Micro forceps, e.g., #4 and #7 jewelers.
- Suture tying forceps (optional).
- Needle holder (approx. 12 cm).
- 1 Micro-serrefine clamp with delicate, atraumatic serrations (jaw length 15–17 mm, gram press 85–125, *see* **Note 17**).
- Clamp applicator or clamp forceps.
- Wound clip applier and wound clips 9 mm (optional).
-
- Nonabsorbable suture polyfil 5/0 (e.g., Mersilene).
- 10% povidone-iodine for skin disinfection.
- Sterile saline (0.9% NaCl).
- Sterile swabs.
- Q-tips or surgical spears.
- Eye ointment.
- Ketamine and xylazine for anesthesia (or any anesthetic of your choice, *see* **Note 9**).
- Buprenorphine for analgesia (or any analgesia of your choice).

3 Methods

3.1 UUO Model

1. Make sure that the hatching incubator (Fig. 3) or any other device that is intended to be used for body temperature control after surgery is warmed up. Check the surgery setup so that everything is in place (Fig. 2).
2. 30 min before surgery analgesia is performed by a subcutaneous injection of 0.1 mg/kg BW buprenorphine. Anesthesia is induced by intraperitoneal injection of 120 mg/kg BW ketamine + 16 mg/kg BW xylazine. After loss of reflexes, the animal is placed on a heating pad (37 °C), eye ointment is applied and the left flank is shaved, followed by disinfection of the area with Betadine solution (10% povidone-iodine). Body temperature is constantly checked by use of a rectal temperature probe.
3. A lateral incision of approx. 2.5 cm length is performed in the left flank (*see* **Note 7**); the kidney is prepared with surgical spears or Q-tips (Fig. 4). The ureter is prepared with curved iris forceps or micro forceps #7 jewelers.
4. Lift the kidney slightly, and if necessary remove excessive adipose tissue carefully from the ureter using surgical swabs or Q-tips (*see* **Notes 11** and **12**). Place your 7.0 suture (*see* **Note**

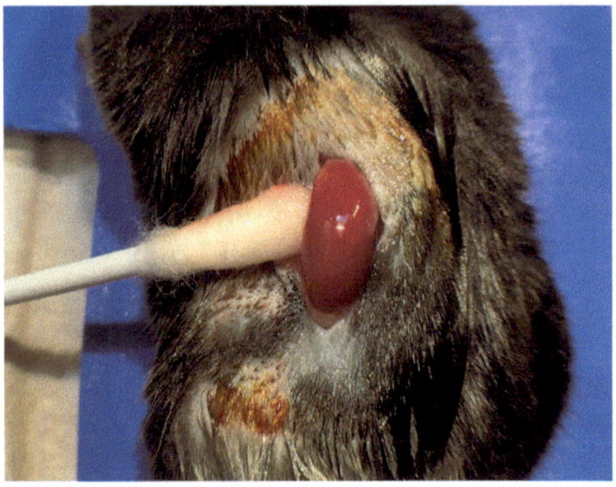

Fig. 4 Preparation of the left kidney

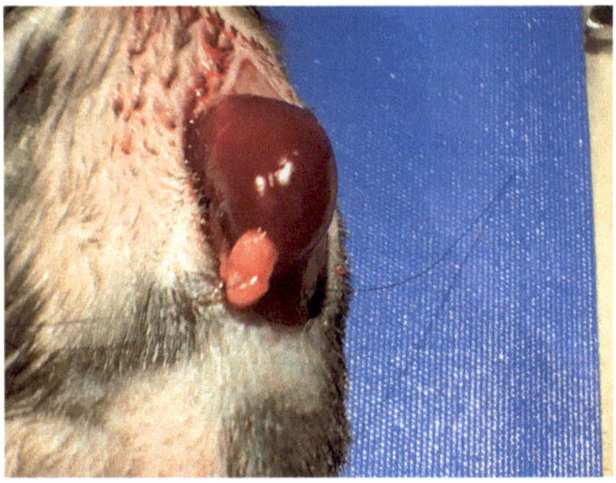

Fig. 5 Ligation close to the kidney's lower pole by surgeon's knot

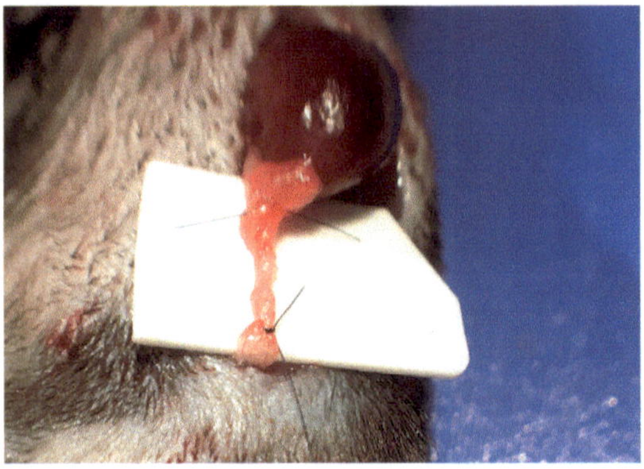

Fig. 6 Ligation and backup ligation

10) around the ureter close to the level of the lower pole and perform a surgeon's knot (Fig. 5, *see* **Note 13**). Place two further simple knots for additional security of the ligation. Cut the suture approx. 5 mm away from the knot. Repeat this procedure some millimeters below your first ligation for a backup ligation (Fig. 6, *see* **Note 14**). Place the kidney back into the body cavity.

5. Apply 500μL of warmed up sterile saline (37 °C) into the body cavity to prevent dehydration. Wound closure is performed by nonabsorbable suture (5.0) of the peritoneum and application of wound clips for the skin. The contralateral kidney is left undisturbed.
6. Place the animal in the hatching incubator until it is conscious again. Now the animal can be placed back into its cage. Analgesia is continued for three days after surgery with buprenorphine (subcutaneous injections of 0,1 mg/kg BW at 08:00 am 12:00 noon and 04:00 pm, and in drinking water over night with 0,009 mg/mL) [10].

3.2 IRI Model

1. Make sure that the hatching incubator (Fig. 3) or any other device that is intended to be used for warming after surgery is warming up. Check the surgery setup so that everything is in place (Fig. 2).
2. 30 min before surgery, analgesia is performed by a subcutaneous injection of 0,1 mg/kg BW buprenorphine. Anesthesia is induced by intraperitoneal injection of 120 mg/kg BW ketamine + 16 mg/kg BW xylazine. After loss of reflexes, the animal is placed on a heating pad (37 °C), eye ointment is applied and the left flank is shaved, followed by disinfection of the area with Betadine solution (10% povidone-iodine). Body temperature is constantly checked by use of a rectal temperature probe (*see* **Note 20**).
3. A lateral incision of approx. 2.5 cm length is performed at the left flank (*see* **Note 7**), the kidney is prepared with surgical spears or Q-tips (Fig. 4). The renal hilum is prepared with curved iris forceps or micro forceps #7 jewelers. The hilum is then clamped with an atraumatic micro serrefine clip (Fig. 7, *see* **Notes 16** and **21**). Ischemic kidneys are shown after 2 min (Fig. 8) and after 10 min (Fig. 9).
4. Switch the timer on (*see* **Note 22**). The kidney is carefully moved back, and the incision is covered with a small sterile pad to minimize evaporation and heat loss (Fig. 10, *see* **Note 18**). The animal is placed in the hatching incubator.
5. For male mice on a C57BL/6 genetic background, a suitable ischemia time would be 26 min but this varies a lot from lab to lab and thus this needs to be established. For severe bilateral IRI the time varies in males from 24 to 30 min and in female mice 35 to 50 min (*see* **Notes 15** and **19**). Make sure that you place the mouse back on the heating pad in time. Relocate the

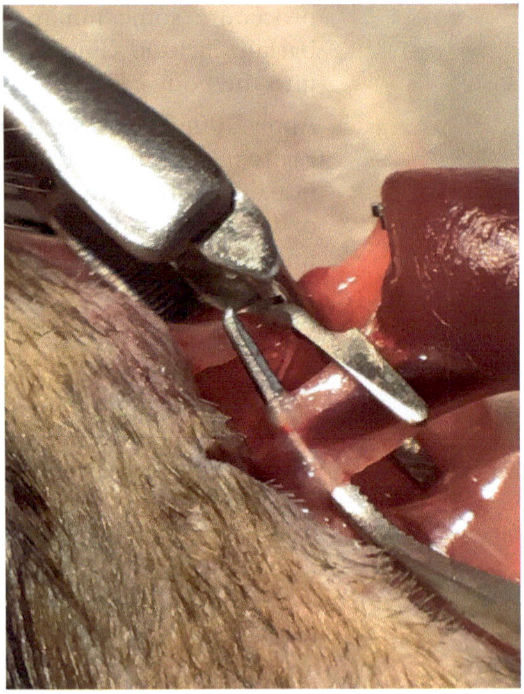

Fig. 7 Clamping the renal hilum

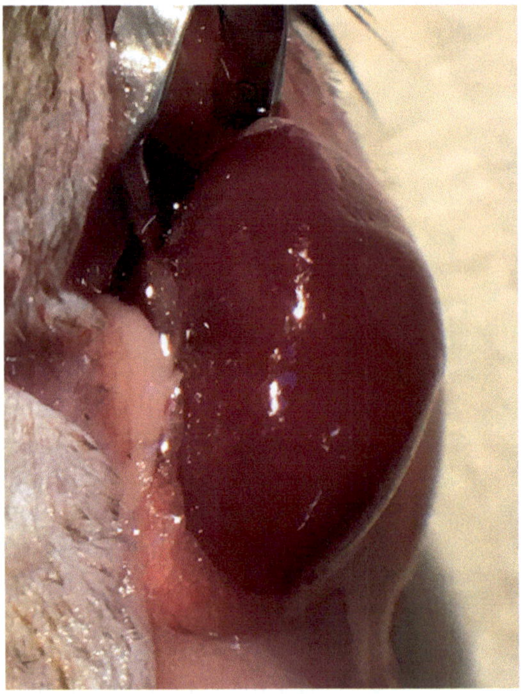

Fig. 8 Ischemic kidney 2 min after clamping

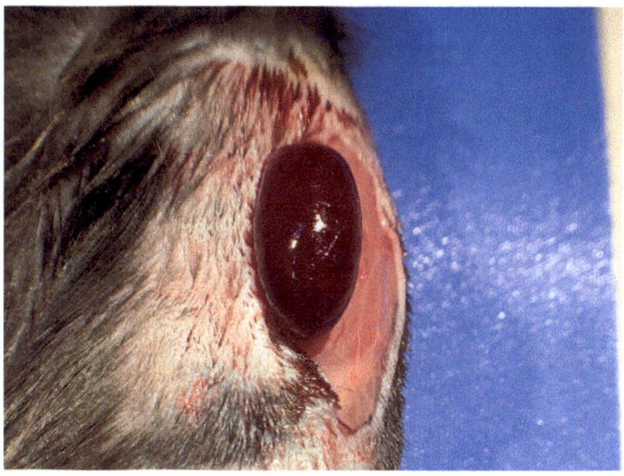

Fig. 9 Ischemic kidney 10 min after clamping

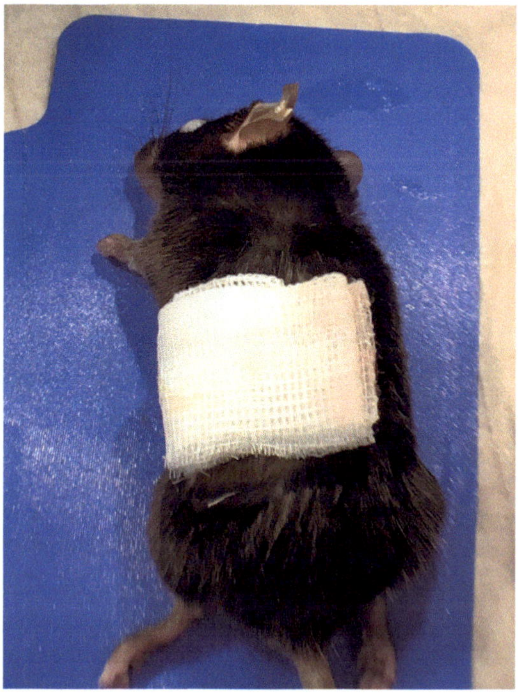

Fig. 10 Covering the incision with a sterile pad to minimize evaporation and heat loss

rectal probe. After the required period of time, pad and clip are removed carefully.

6. Control your work visually: shortly after successful reperfusion the kidney should regain a healthy pink color (Fig. 11).
7. Apply 500μL of warmed up sterile saline (37 °C) into the body cavity to prevent dehydration. Wound closure is performed by nonabsorbable suture (5.0) of the peritoneum and application

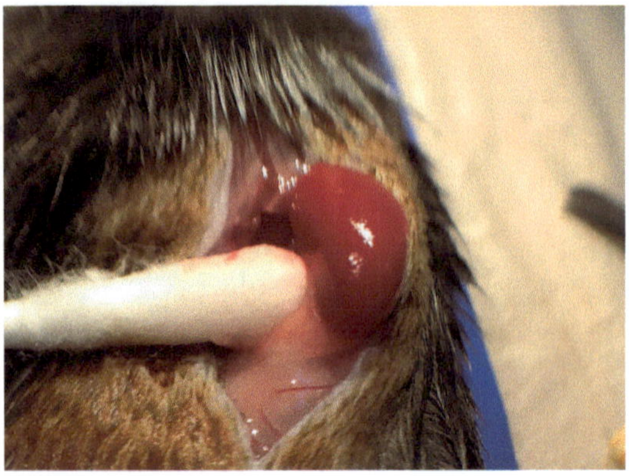

Fig. 11 Reperfused kidney after removal of the clamp

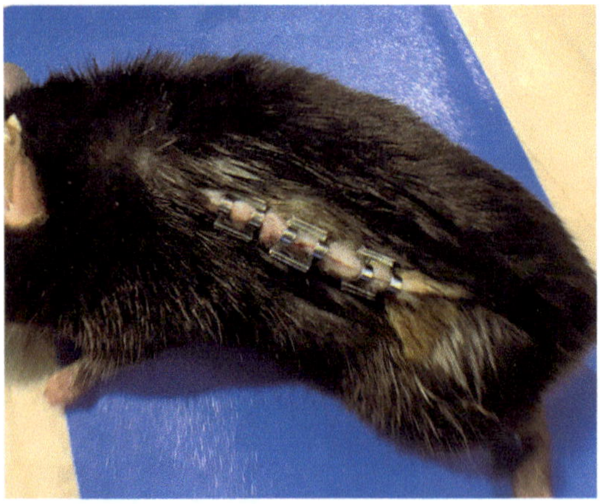

Fig. 12 Wound closure with clips

of wound clips for the skin (Fig. 12). The contralateral kidney is left undisturbed.

8. Place the animal in the hatching incubator until it is conscious again. Now the animal can be placed back into its cage. Analgesia is continued for three days after surgery with buprenorphine (subcutaneous injections of 0.1 mg/kg BW at 08:00 am 12:00 noon and 04:00 pm, and in drinking water over night with 0.009 mg/mL) [10].

3.3 Time Points

In the classic UUO model described here suitable time points for finalization are day 1, day 3, and day 10 [14]. On day 1 early changes in cell injury can be detected but this time point will give you only subtle histological changes such as enlarged tubular epithelial cells. On day 3 damage has increased and you will detect a more differentiated picture with first hits of interstitial fibrosis, atrophic tubular areas and immune cell infiltration. In our opinion day 10 is the most suitable time point if you want to see a full phenotype with severe tubule atrophy and interstitial fibrosis. Histological evidence for kidney damage in the form of hydronephrosis is actually visible to the naked eye. Additionally intermediate day 7 or a very late timepoint day 12 can be used but are less common.

In the unilateral IRI model day 1 and day 3 are good timepoints to study the acute injury, hypoxemia, dedifferentiation, and immune cell infiltration. To study regeneration after acute kidney injury day 5, 7, 12, and 14 might be of interest. If you are interested to study transition from AKI to CKD with fibrosis day 14, day 21 and day 28 might be of interest.

3.4 Downstream Techniques Suitable to Investigate These Processes

In the case of unilateral IRI combined with contralateral nephrectomy or a bilateral IRI you could analyze kidney function by collecting blood and urine samples from the living animal. Measuring protein, creatinine, and urea (blood) will give you valuable insights into the extent of kidney damage.

Kidney tissue from both UUO and IRI models can be investigated in a number of ways. On the histological level HE, PAS, Sirius Red, and Masson's Trichrome are very useful staining methods to visualize histological changes. Immunohistochemical or immunofluorescence stainings for fibrosis and or myofibroblast markers (e.g., collagen, fibronectin, aSMA) are commonly used as well as staining of infiltrating immune cells (CD45, F4/80, CD68, etc.) and markers of tubule injury such as KIM1 or staining of endothelium and capillary loss using, e.g., CD31 allow further investigation of the tissue in a spatial context. With suitable antibodies all antigens of interest could of course be subjected to flow cytometry, western blot, etc. If it comes to analysis on DNA or RNA level, techniques such as in situ hybridization (ISH) and quantitative real-time PCR are useful to look at the markers described above or individual marker genes. Obviously, there is a great suite of other tools that are not mentioned here and include novel genomic and proteomic technologies among others.

4 Notes

1. One thing that might be forgotten easily in the experimental setup is the scientist. Take care that your sitting posture is comfortable. A chair with adjustable height helps you to find the right position with the feet flat to the floor. Your neck and back should be straight and relaxed, your forearms and hands should find support at right angle to the desk.
2. Both methods can be performed with or without surgical microscope.

3. In performing animal experiments, every scientist should try to incorporate the 3 R's according to Russell & Burch in their work. The 3 Rs stand for: replacement, reduction, and refinement [11]. If you are new to mouse surgeries it might be a very good option to practice your eye–hand coordination with the following three ex vivo methods. Doing so will reduce the number of required mice considerably. Because your most valuable material are the mice.
4. Letter scraping: Place a piece of newspaper on the desk in front of you and fixate it with tape. Now choose a letter (e.g., all "c's") and erase them by scraping the newspaper with repeated soft strides of your forceps. Try not to pierce the paper. Aim for removing the print without damaging the paper. You can practice with both hands and several forceps types.
5. Suture at a glove: Span a glove on a petri dish so that you have an even surface formed by the palm of the glove. Make an incision with a scalpel and practice interrupted sutures and continuous sutures. Increase difficulty by tightening the glove.
6. Ligations on orange segments: Fixate an orange or tangerine segment on the desk in front of you, rounded side on the desk. Now perform ligations on the white skin as described for UUO. The more delicate the skin, the more difficult not to rupture it. Start with stronger sutures and at the central column, then switch to thinner sutures and more delicate parts.
7. When choosing your surgical method you should consider that laparotomy will decrease body temperature more than flank incision.
8. An important point regarding the experimental setup: The hatching incubator used as a device for maintaining homeothermy works very well for us. Nevertheless, if you want to use isoflurane (or any other inhalation anesthetics) for anesthesia, you have to consider reconstruction of the incubator in order to house an anesthesia manifold for ischemia time.
9. When choosing your anesthesia/analgesia scheme you should consider that there is some controversy about the possible protective effect that inhalation anesthetics such as isoflurane might have on ischemia [12]. As we use the combination of ketamine and xylazine for many years successfully in many different strains, we have not tried any isoflurane for that type of surgery until now.
10. When practicing the method it is important to prepare everything for comfortable working. Cut the suture for ligation in suitable length (for beginners approx. 10 cm). This might help you to grab ends but is also not too long to prevent a mess. With more experience, you can use shorter pieces of suture.

11. Try to clean the ureter from adipose tissue but be careful not to rupture the ureter. This works well by stroking the ureter with dry Q-tips.
12. In female mice you might find occasionally cyst-like structures close to the lower pole of the kidney. Check if the structures are connected to the ureter. If there is a connection, this animal should not be used for UUO as the results might become unpredictable. If not, clean your surgical field and proceed, taking care not to damage the cyst-like structures and to place the ligations in a suitable distance to the cysts.
13. Take care to add an extra twist in the first throw.
14. Avoid placing the second ligation directly onto your first as this might not provide a sufficient obstruction of the ureter.
15. Ischemia time in unilateral IRI can be performed even longer but is then resulting in very severe injury with reduced repair and a none-functional kidney.
16. Never apply micro serrefine clamps with hands or a normal forceps as they are prone to damage by shearing. Damage will be invisible but you might get incoherent results if the clamps do not close properly anymore. Always use an applicator or clamp forceps.
17. Different clamps have different closure pressures. This might influence your results.
18. If there is too much evaporation during surgery vessels may stick to the instruments and could rupture. Always keep the surgical field moist.
19. In males, kidney injury is more easily induced than in females (hence males need shorter ischemia period, [13]). This would require separate groups for each sex, otherwise effect size will decrease dramatically.
20. In the IRI model, body core temperature control is not only important for rapid recovery as in any rodent surgery but also essential for your model's success because lower temperatures will cause a decrease in ischemic injury and render your model ineffectively.
21. Watch the kidney carefully: the organ's color darkens quickly when perfusion is interrupted—ischemia time starts now.
22. If you do bilateral IRI, use one timer per kidney.

References

1. Kramann R, Machado F, Wu H, Kusaba T, Hoeft K, Schneider RK, Humphreys BD (2018) Parabiosis and single-cell RNA sequencing reveal a limited contribution of monocytes to myofibroblasts in kidney fibrosis. JCI Insight 3(9):e99561. https://doi.org/10.1172/jci.insight.99561

2. Kramann R, Wongboonsin J, Chang-Panesso-M, Machado FG, Humphreys BD (2017) Gli1[+] pericyte loss induces capillary rarefaction and proximal tubular injury. J Am Soc Nephrol 28(3):776–784
3. Rabe M, Schaefer F (2016) Non-transgenic mouse models of kidney disease. Nephron 133(1):53–61. https://doi.org/10.1159/000445171
4. Martínez-Klimova E, Aparicio-Trejo OE, Tapia E, Pedraza-Chaverri J (2019) Unilateral ureteral obstruction as a model to investigate fibrosis-attenuating treatments. Biomol Ther 9 (4):141. https://doi.org/10.3390/biom9040141
5. Heung M, Chawla LS (2014) Acute kidney injury: gateway to chronic kidney disease. Nephron Clin Pract 127(1–4):30–34. https://doi.org/10.1159/000363675
6. Bonventre JV, Yang L (2011) Cellular pathophysiology of ischemic acute kidney injury. JCI (121):4210–4221
7. Hesketh EE, Czopek A, Clay M, Borthwick G, Ferenbach D, Kluth D, Hughes J (2014) Renal ischaemia reperfusion injury: a mouse model of injury and regeneration. J Vis Exp:88. https://doi.org/10.3791/51816
8. Kramann R, Fleig SV, Schneider RK, Fabian SL, DiRocco DP, Maarouf O, Wongboonsin J, Ikeda Y, Heckl D, Chang SL, Rennke HG, Waikar SS, Humphreys BD (2015) Pharmacological GLI2 inhibition prevents myofibroblast cell-cycle progression and reduces kidney fibrosis. J Clin Invest 125 (8):2935–2951. https://doi.org/10.1172/JCI74929
9. Li L, Kang H, Zhang Q, D'Agati VD, Al-Awqati Q, Lin F (2019) FoxO3 activation in hypoxic tubules prevents chronic kidney disease. J Clin Invest 129(6):2374–2389. https://doi.org/10.1172/JCI122256
10. Sauer M, Fleischmann T, Lipiski M, Arras M, Jirkof P (2016) Buprenorphine via drinking water and combined oral-injection protocols for pain relief in mice. Appl Anim Behav Sci 185:103–112. https://doi.org/10.1016/j.applanim.2016.09.009
11. Russell WMS, Burch RL (1959) The principles of humane experimental technique. Methuen, London, pp 69–154
12. Ham A, Kim M, Kim JY, Brown KM, Yeh J, D'Agati VD, Lee HT (2013) Critical role of interleukin-11 in isoflurane-mediated protection against ischemic acute kidney injury in mice. Anesthesiology 119(6):1389–1401. https://doi.org/10.1097/ALN.0b013e3182a950da
13. Aufhauser DD Jr, Wang Z, Murken DR, Bhatti TR, Wang Y, Ge G, Redfield RR 3rd, Abt PL, Wang L, Svoronos N, Thomasson A, Reese PP, Hancock WW, Levine MH (2016) Improved renal ischemia tolerance in females influences kidney transplantation outcomes. J Clin Invest 126(5):1968–1977. https://doi.org/10.1172/JCI84712
14. Kuppe C, Ibrahim MM, Kranz J, Zhang X, Ziegler S, Perales-Patón J, Jansen J, Reimer KC, Smith JR, Dobie R, Wilson-Kanamori JR, Halder M, Xu Y, Kabgani N, Kaesler N, Klaus M, Gernhold L, Puelles VG, Huber TB, Boor P, Menzel S, Hoogenboezem RM, Bindels EMJ, Steffens J, Floege J, Schneider RK, Saez-Rodriguez J, Henderson NC, Kramann R. Decoding myofibroblast origins in human kidney fibrosis. Nature. 2021 Jan;589 (7841):281–286. https://doi.org/10.1038/s41586-020-2941-1

Chapter 23

Mouse Models of Liver Fibrosis

Aashreya Ravichandra and Robert F. Schwabe

Abstract

Liver fibrosis is defined as excessive accumulation of extracellular matrix, and results from maladaptive wound healing processes that occur in response to chronic liver injury and inflammation. The main etiologies of liver fibrosis include nonalcoholic fatty liver disease (NAFLD), chronic viral hepatitis, as well as alcoholic and cholestatic liver disease. In patients, liver fibrosis typically develops over several decades and can progress to cirrhosis, and liver failure due to replacement of functional liver tissue with scar tissue. Additionally, advanced fibrosis and cirrhosis are associated with an increased risk for the development of hepatocellular carcinoma. On a cellular level, hepatic fibrosis is mediated by activated hepatic stellate cells, the primary fibrogenic cell type of the liver. Murine models are employed to recapitulate, understand, and therapeutically target mechanisms of fibrosis and hepatic stellate cell activation. Here, we summarize different mouse models of liver fibrosis focusing on the most commonly used models of toxic, biliary, and metabolically induced liver fibrosis, triggered by treatment with carbon tetrachloride (CCl_4), thioacetamide (TAA), bile duct ligation (BDL), 3,5-diethoxycarbonyl-1,4-dihydrocollidine (DDC), and high-fat diets.

Key words Liver fibrosis, Hepatic stellate cell, Fatty liver, Injury, Fibroblast

1 Introduction

Liver fibrosis contributes to many of the deadly complications of chronic liver disease such as portal hypertension and loss of liver function (via replacement of functional parenchyma) and is associated with the development of liver cancer. Accordingly, liver fibrosis has been shown to be a main determinant of outcomes for patients with chronic liver disease [1, 2]. Currently, there are no approved anti-fibrotic drugs for the prevention or treatment of liver fibrosis. Fibrosis can be prevented or even reverted by successful treatment of the underlying disease, best demonstrated for chronic viral hepatitis, where viral eradication has been associated with fibrosis regression in a substantial percentage of patients [3]. However, successful treatment of the underlying disease cannot always be achieved, particularly in patients with nonalcoholic fatty liver disease, alcoholic liver disease, and several cholestatic liver diseases.

Boris Hinz and David Lagares (eds.), *Myofibroblasts: Methods and Protocols*, Methods in Molecular Biology, vol. 2299,
https://doi.org/10.1007/978-1-0716-1382-5_23,

Animal models have been useful to understand the underlying pathophysiology and determine the cellular networks and signaling pathways that drive fibrosis. Hepatic stellate cells (HSC) are in the center of this profibrogenic network [4, 5], and there are considerable efforts to develop therapeutic strategies to inhibit their activation or promote cell death of activated HSC to reduce their numbers. While HSC activation can also be achieved in vitro, cell culture models lack the complex hepatic environment of the liver. Accordingly, gene expression patterns of HSC activating in vitro differ significantly from HSC activating within a fibrotic liver [6, 7]. Hence, mouse models represent an important approach to understand mechanisms of HSC activation and fibrosis. Ideally, mouse models of liver fibrosis would mimic key features of human liver fibrosis including the underlying cause, transcriptomic changes, and pathways as well as key cell-cell interactions. At the same time, these models should be reproducible; result in robust fibrosis in a reasonable time frame; be amenable to study fibrosis reversal as main therapeutic goal for translational studies; and be standardized among the scientific community.

Unfortunately, there is currently not a single model that fulfills all these criteria. As such, the most commonly used model is CCl_4-induced liver fibrosis, which induces robust and rapid fibrosis, is standardized and easy to use, but considerably faster than human liver fibrosis, developing in days to weeks as opposed to typically years to decades in patients. Moreover, this model induces hepatocyte necrosis rather than apoptosis that is commonly seen in patients and lacks a human equivalent. Other models such as specific high-fat diets are more closely reproducing human pathophysiology but are considerably slower in their onset, often only induce mild histological fibrosis and HSC activation, and are far less standardized with considerable variations in terms of the specific diet composition and in some cases require specific transgenic or inbred mice. Biliary liver fibrosis can be mimicked by bile duct ligation, 3,5-diethoxycarbonyl-1,4-dihydrocollidine (DDC) diet or genetic knockout of Mdr2 and reproduces many features of human cholestatic liver disease. Finally, alcoholic liver disease-associated fibrosis has been difficult to model in rodents with only minimal HSC activation and fibrosis. Here, we review the most common models and provide detailed step-by-step guidelines, focusing on toxic, biliary, and nonalcoholic steatohepatitis (NASH)-associated liver fibrosis (Fig. 1).

1.1 General Considerations

Different murine genetic backgrounds display profound differences regarding the development of liver fibrosis. While the majority of studies are done in C57BL/6 mice due to fact that most knockout strains are in this genetic background, BALB/c mice tend to have more profound liver fibrosis in the majority of models [8, 9]. The genetic background is particularly important for studying NASH-

	Toxic fibrosis	Biliary Fibrosis	NASH fibrosis
Common models	CCl_4 TAA	BDL DDC Mdr2-/-	*Dietary*: CD-HFD/CDAA HFHCD/FPC *Genetic+dietary:* Mc4R-/- DIAMOND
Duration to achieve histological fibrosis	2-24 weeks (CCl_4) 8-24 weeks (TAA)	2-3 weeks (BDL, DDC) > 6 weeks age (Mdr2-/-)	>16-24 weeks (HFHCD/FPC) > 12-15 (CD-HFD/CDAA) > 20 weeks (Mc4R-/-) > 16 weeks (DIAMOND)
Histological pattern	Septal fibrosis	Periportal fibrosis, Onion-skinning like	Chicken-wire

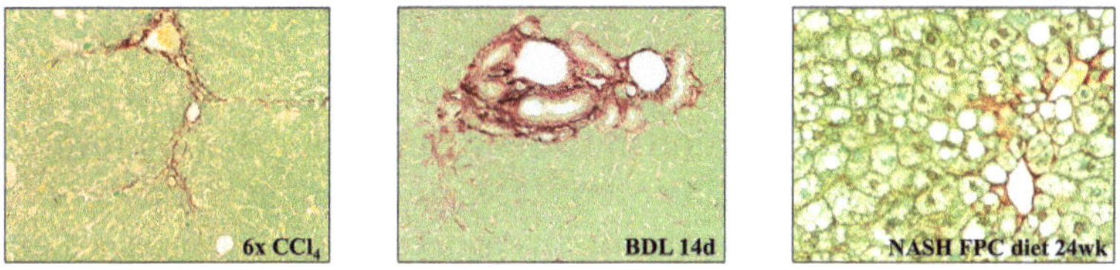

Fig. 1 *Summary of toxic, biliary, and NASH-associated fibrosis models.* Common rodent liver fibrosis models with duration of treatment and histopathological pattern as stained by picrosirius red. (Abbreviations: CCl_4 Carbon tetrachloride, BDL Bile duct ligation, FPC Fructose, palmitate and cholesterol-containing diet.)

associated liver fibrosis which should be performed in the C57Bl/6 background owing to more robust development of obesity, insulin resistance, and fatty liver. Most studies are performed in male mice for reasons of standardization. However, female mice can also be used, and may in some models, such as Mdr2−/− biliary fibrosis model, even display more robust fibrosis [10]. While most of the field employs mice aged 8–10 weeks to study fibrosis, using older mice may be a better reflection of human disease due to the later onset of fibrosis in most patients. Finally, the use of older and male mice may be warranted when fibrosis is induced with the primary purpose of isolating HSC, as yields will be considerable higher [11].

1.2 Hepatotoxin-Induced Fibrosis Models

Toxins such as CCl_4 and thioacetamide (TAA) are widely used to model liver fibrosis. CCl_4 and TAA are metabolically activated by centrilobular hepatocytes resulting in liver injury, inflammation, and subsequently fibrosis. Fibrosis caused by these toxins is first seen around the pericentral veins and over time leads to the formation of fibrotic septa, typically in a central-central pattern. The pericentral fibrotic patterns seen in rodent toxin-induced fibrosis models may differ from the periportal and lobular fibrotic patterns seen in humans [12]. Likewise, CCl_4 and TAA predominantly induce necrotic cell death whereas many human liver diseases leading to liver fibrosis have been associated with apoptotic cell death. Despite this, CCl_4 and TAA represent the most used murine fibrosis models because of their ease of use, high robustness and reproducibility. Additionally, chronic liver injury followed by cessation of

CCl_4 or TAA may be used as a method to study regression of fibrosis [13]. It is important to note, however, that there is no common human equivalent disease to these fibrosis models.

1.2.1 Carbon Tetrachloride (CCl_4)

CCl_4 is one of the most used hepatotoxins to induce liver fibrosis. Repeated administration of CCl_4 induces liver injury that over time leads to fibrosis. Once injected, CCl_4 is metabolically activated in hepatocytes by P450 cytochromes, mainly CYP2E1, leading to the formation of trichloromethyl (CCl_3) radicals, which trigger hepatocyte death via lipid peroxidation. In response to hepatocyte injury and death, resident and bone marrow-derived macrophages are activated and/or recruited, generating the release of pro-fibrotic and pro-inflammatory molecules that contribute to the activation and expansion of HSC and the development of liver fibrosis. Crucial pathways in this process include the TGFß and PDGF pathways [4]. CCl_4-induced liver fibrosis is well characterized with numerous publications showing the biochemical changes, inflammation, injury, and HSC activation associated with this model.

1.2.2 Thioacetamide (TAA)

Like CCl_4, TAA requires bioactivation in hepatocytes via cytochromes P450 such as CYP2E1 to become hepatotoxic. This process, results in the oxidation of TAA to TAA-*S*-oxide (TAA-SO) and then to TAA-*S,S*-oxide (TAA-SO_2), resulting in increased oxidative stress, centrilobular hepatocyte necrosis, and, subsequently, liver fibrosis. While the pattern of fibrosis and injury induced by CCl_4 and TAA are similar, TAA also induces some degree of periportal hepatocyte and cholangiocyte injury and ductal proliferation [14]. Accordingly, chronic TAA administration can lead to cholangiocarcinoma development [14, 15].

1.3 Biliary Fibrosis Models

Cholestatic liver disease is caused by impairing bile secretion at various levels. The histopathological changes associated with this disease are periportal inflammation, bile duct proliferation (also termed ductular reaction), and biliary liver fibrosis. Many well-established models of cholestatic liver disease have been developed to study the progression of the disease. The most used surgical technique of inducing biliary fibrosis is bile duct ligation (BDL). Diet-induced biliary fibrosis and ductular reaction can be observed using DDC diet, choline-deficient ethionine-supplemented diet (CDE), and α-naphthyl-isothiocyanate (ANIT) diet. Genetic modifications of mice, such as the deletion of the multidrug resistance-associated protein 2 (Mdr2) gene, the mouse homolog of human MDR3, results in spontaneous biliary fibrosis and models various human cholestatic diseases associated with mutations of MDR3 [16].

1.3.1 Bile Duct Ligation (BDL)

Bile duct ligation is a commonly used model to induce cholestasis and periportal fibrosis in the liver. First established in rats and later in mice, this highly reproducible model consists of the double ligation of the common bile duct resulting in the gradual development of fibrosis. Upon BDL, cholestasis is first observed by increase of hepatocyte death due to increased biliary pressure, often seen as peribiliary areas of necrosis termed "bile infarcts." Cholestasis and hepatocyte injury subsequently trigger inflammation, recruitment of immune cells, cholangiocyte proliferation, HSC activation, and significant periportal fibrosis within 2–3 weeks after BDL. This model is nonreversible unless a biliodigestive anastomosis operation is performed.

1.3.2 3,5-Diethoxycarbonyl-1,4-Dihydrocollidine (DDC) Diet

The DDC liver fibrosis model involves the administration of a diet rich in 3,5-diethoxycarbonyl-1,4-dihydrocollidine, a porphyrinogen. DDC diet promotes the hepatic accumulation of protoporphyrin, a dark brown pigment, and the appearance of intraductal porphyrin plugs, which obstruct small bile ducts, damaging biliary epithelia and promoting liver injury as well as ductular reaction. As a consequence, periportal liver fibrosis develops, typically within 2–3 weeks [17, 18]. The ease of this model is that it can be purchased as a specialized diet directly from approved vendors where DDC is added to a regular rodent chow diet to achieve a final concentration of 0.1%. As a diet, this model is reversible with sufficient time on a regular chow diet.

1.3.3 Mdr2−/− Mice

The multidrug resistance associated protein 2 (Mdr2, encoded by *Abcb4* in mice and MDR3 in humans) is a phospholipid flippase part of the ABC canalicular transporters that plays a role in phospholipid excretion. In humans, mutations of the MDR3 gene have been linked to a range of hepatobiliary disorders including progressive familial intrahepatic cholestasis (PFIC3) and primary sclerosing cholangitis (PSC) [16]. As such, the rodent Mdr2 −/− model, which displays an absence of phospholipid secretion into the bile and spontaneous liver injury and periportal fibrosis, is ideal to study these cholangiopathies [19]. Mdr2 −/− mice start developing spontaneous biliary fibrosis at 4 weeks of age. In contrast to patients with MDR3 mutations, who predominantly develop cholangiocarcinoma, almost all liver tumors seen in Mdr2−/− mice are hepatocellular carcinoma (HCC), developing at the age of 12–18 months. Liver fibrosis is more profound in female than in male Mdr2−/− mice.

1.4 NASH-Associated Fibrosis

Obesity and obesity-associated conditions such as diabetes and fatty liver disease are increasing in prevalence worldwide. Nonalcoholic fatty liver disease (NAFLD), defined as hepatic steatosis in the absence of excessive alcohol intake or viral hepatitis, may progress to nonalcoholic steatohepatitis (NASH), a more severe form of

Table 1
Models of NASH

Diet	Description	Weight gain	Hepatic steatosis	Liver injury	Insulin resistance	Fibrosis	HCC development	Reference
CDAA	Choline-deficient, 60% fat, 21% sucrose, 0.1% methionine	–	++	+	–	+	+	[29]
CD-HFD	Choline-deficient, 45% fat, 24% sucrose	+	++	++	+	+	+	[30]
MCD diet	10% fat, 40% sucrose, deficient of choline and methionine	–	++	+	+	+	–	[28]
FPC diet	55% fat, 2% cholesterol, 35% sucrose + fructose/glucose drinking water	++	++	+	++	++	–	[27]
ALIOS diet	45% fat, 25% sucrose + fructose/glucose drinking water	++	++	+	+	++	+	[26, 34]
Western/fast food diet	45% fat, 0.2% cholesterol, 35% sucrose + fructose/glucose drinking water	++	++	+	+	+	–	[24, 25]
Mc4R-deficient hyperphagic mice	KO of the melanocortin-4-receptor gene	++	+++	–	++	++	+	[20]
ob/ob mice	Mutant mice homozygous for a spontaneous point mutation in the gene encoding leptin	+++	+++	–	+	–	–	[21]
DIAMOND	Diet-induced animal model of non-alcoholic fatty liver disease mice are an inbred isogenic strain (C57/Bl6J, 40% S129S1/svlm) that become obese on a HFD diet (42% fat, 0.1% cholesterol + fructose/glucose drinking water)	+++	++	+	+	+	+	[22]
STAM model	Administration of streptozotocin to neonatal mice	–	++	++	–	+++	+	[23]

NAFLD characterized by hepatocellular injury and death, hepatic ballooning, and inflammation. Over time, these events may lead to varying degrees of fibrosis, and can progress to cirrhosis, liver failure, and development of HCC. An ideal mouse model would combine key metabolic features of human NAFLD such as obesity and insulin resistance with strong hepatic steatosis, inflammation, and development of fibrosis, initially as "chicken-wire" fibrosis that progresses to bridging fibrosis. While there is no perfect model that incorporates all aspects of human fibrosis, several rodent models can replicate many of the metabolic aspects as well as key histopathological features of the disease including hepatocyte ballooning, steatosis, and fibrosis. The most used models of NASH are diet-induced and are typically rich in fat, glucose, fructose, or sucrose, and combined with either high levels of cholesterol or choline deficiency. These diets can be readily purchased from approved vendors and can be stored at 4 °C or − 20 °C. Diets in mouse cages must be replaced twice a week to prevent them from going rancid. Several genetic models of NASH are also available and involve dietary modifications to trigger disease progression, such as melanocortin-4 receptor-deficient mice which are hyperphagic and replicate the most common genetic cause of obesity in humans [20, 21]; or inbred strains such as the DIAMOND mice which develop liver fibrosis when fed high-fat diet [22]. The "STAM" model combines ablation of pancreatic beta cells by streptozotocin with high-fat diet. While this replicates type I rather than type II diabetes (the latter being associated strongly with NASH development), the STAM model induces NASH with injury and fibrosis [23]. Key features of these NAFLD/NASH models are summarized in Table 1.

1.4.1 High-Fat High-Cholesterol Diet (HFHCD) Models

Several diets in this category such as the Western/Fast food diet [24, 25], the *A*merican *L*ifestyle-*I*nduced *O*besity *S*yndrome (ALIOS) Diet [26], and the *F*ructose, *P*almitate, *C*holesterol-containing "FPC" diet [27] can be used as models of NASH fibrosis. These models replicate to various degrees, key features of NAFLD such as obesity and metabolic syndrome, and induce the development of hepatic steatosis, injury, inflammation, and fibrosis. The timeframe of the experiment for the development of profound histological fibrosis often exceeds 6 months, but mild histological fibrosis and upregulation of fibrogenic and HSC activation genes can be observed earlier. These dietary NASH models are well established and have a high degree of reproducibility but note that there may be some variability due to the amount of diet ingested by individual mice. Key injury- and fibrosis-promoting factors in these diets are high amounts of cholesterol (typically 1–2%), high levels of fructose (often supplemented in drinking water and thus replicating high fructose corn syrup-based soft drinks), and disease-promoting fatty acids such as palmitate.

1.4.2 Choline-Deficient High-Fat Diet Models

While classic choline-deficient diets like the methionine-choline-deficient (MCD) diet induce hepatic steatosis, liver injury, and fibrosis, they do not reproduce typical metabolic alterations seen in most NASH patients such as obesity and insulin resistance. In fact, MCD diet induces loss of body weight and peripheral adipose tissues as well as insulin hypersensitivity [28]. To achieve fibrosis seen in choline deficiency and typical metabolic alterations found in NASH, high-fat diets have been combined with choline deficiency. Two well-established models include the choline-deficient amino acid-supplemented (CDAA) diet and choline-deficient high-fat diet (CD-HFD), which differ in their composition but follow the same principle. While the CDAA diet produces more profound fibrosis but no or little weight gain [29], CD-HFD results in milder fibrosis but higher weight gain [30].

2 Materials

2.1 CCl_4 Injection

1. Carbon tetrachloride (CCl_4).
2. Corn oil.
3. 30 mL Glass vial.
4. 0.2μm Filter.
5. 27G 1 mL Syringe (or 1 mL syringe and multiple 27G needle heads).
6. Weighing scale.

2.2 BDL Model

1. Circulating water warming pad.
2. Heating pad with temperature regulator.
3. Sterile surgical gloves.
4. Hair clippers or hair removal cream.
5. 70% Ethanol for disinfection.
6. 10% Povidone-iodine (betadine) for disinfection.
7. Ophthalmic ointment.
8. Isoflurane.
9. Oxygen/Isoflurane vaporizer or other anesthetic of choice.
10. Anesthesia induction chamber.
11. Buprenorphine.
12. Carprofen.
13. Sterile cotton swabs.
14. Tape to fix the nose cone and other cables.
15. 11.5 cm Surgical scissors.
16. Serrated forceps and curved serrated forceps.

17. Colibri retractor.
18. 0.9% Sterile sodium chloride (NaCl) solution or sterile Phosphate-Buffered Saline (PBS).
19. 6–0 Sterile silk sutures.
20. 5–0 Sterile Vicryl sutures.
21. Sterile wound clips with applicator.
22. Clips remover.

2.3 HFHCD Diet

1. ALIOS diet (TD.06303, Envigo), Western Diet (TD.88137, Envigo), or FPC diet (TD. 160785, Envigo).
2. D-(+)-Glucose monohydrate.
3. D-(−)-Fructose.
4. Acidified water.
5. Glass bottles (autoclaved).
6. 0.2μM Filter.

2.4 Choline-Deficient High-Fat Diet Models

1. CDAA diet (A06071302, Research Diets), CD-HFD (D05010402, Research Diets).

2.5 Harvesting Tissue and Analysis

1. Isoflurane.
2. Anesthesia induction chamber.
3. Surgical scissors.
4. Serrated forceps/curved serrated forceps.
5. 70% Alcohol.
6. Weighing scale.
7. Petri dish.
8. Formalin solution, neutral buffered, 10%.
9. Tissue embedding cassette.
10. Eppendorf 2.0 mL tubes.
11. Paraformaldehyde.
12. 30% Sucrose solution.
13. Picrosirius red solution/Masson's trichrome.
14. Antibodies for staining (α-SMA, F4/80, etc.).
15. RNA isolation kit.
16. qPCR probes.

3 Methods

3.1 CCl_4 Model

While most often injected intraperitoneally, CCl_4 can also be administered in rodents subcutaneously, by gavage or by inhalation. Liver fibrosis increases with the dosage and the duration of toxin exposure. The dosage of injected CCl_4 ranges from 0.2–2μL/g of body weight diluted in corn oil depending on the study. For example, CCl_4 can be used to induce acute liver injury and HSC activation with a single dose, administered twice or thrice a week continuously for 4–6 weeks to study early fibrosis, administered twice a week for 8 weeks in rats to study early reversible cirrhosis; or administered twice a week for 12 weeks in rats to study partially reversible cirrhosis [31]. Advanced fibrosis may also be studied by escalating the dose of CCl_4 over the course of the study. It is important to note that dosage, duration, and interval may vary depending on the strain of mice being used in the study. The most common strains susceptible to CCl_4-induced fibrosis are C57BL/6 and BALB/C mice, with the latter developing more pronounced fibrosis but C57BL/6 mice being more commonly used for genetic knockout studies. In terms of endpoint, if immediate pro-inflammatory alterations and HSC activation are of interest, tissue should be harvested 24–48 h after a single acute dose. For chronic fibrosis and/or cirrhosis studies, tissue should be harvested 2 days after the last injection, when the peak of hepatocyte injury is still high.

Like CCl_4, TAA (protocol not provided here) can be dissolved in sterile NaCl or 1% sterile phosphate buffer saline (PBS) and administered intraperitoneally, two times a week in the following dose escalating manner for a minimum of 6 weeks: first dose at 100 mg/kg; subsequent doses week 1–2 at 200 mg/kg; week 3–4, 300 mg/kg; week 4–6, 400 mg/kg [32] TAA can also be added to the drinking water at a dose of 300 or 600 mg/L for 20–30 continuous weeks to induce liver fibrosis [33], with longer treatments also inducing cholangiocarcinoma.

1. Under the fume hood, dilute CCl_4 in corn oil at a ratio of 1:4 (a higher proportion of CCl_4 with result in peritonitis) and filter through a 0.2μm filter into a 30 mL glass vial. In our laboratory, to model chronic injury, mice are injected with of 2μL/g body weight of the 1:4 CCl_4:corn oil solution (*see* **Notes 1–3**).
2. Gently remove the mouse from the cage and place on the weighing scale and note the weight to calculate the volume to be injected.
3. Carefully remove the needle cap and draw up into the syringe the desired quantity of the prepared CCl_4 solution. Remember

to gently flick the syringe to remove any air bubbles (*see* **Notes 4–5**).

4. Place the mouse on a textured surface (such as the wire cage) and restrain. With the mouse in the correct position, ideally lowering the head of the mouse by about 30 degrees, identify the lower right/left quadrant of the mouse (*see* **Notes 6–8**).
5. Holding the syringe at a 45° angle, insert the syringe (bevel side up) through the abdominal wall (about 4–5 mm) deep and inject the calculated volume of CCl_4.
6. Place the mouse back into the cage and observe for any discomfort.
7. Discard the needle and syringe into a sharp container without recapping. Use a new syringe or needle head for every mouse (*see* **Note 9**).

3.2 BDL Model

For BDL models, mice may be sacrificed 5–7 days post-surgery to investigate early HSC activation, 12–16 days post-surgery to investigate early fibrosis, and 21–28 d post-surgery to investigate profound fibrosis.

1. 2–8 h before the BDL procedure, an analgesic is administered, e.g., by the subcutaneous injection of buprenorphine (0.15 mg/kg).
2. Place the heating pad under a new clean cage for post-surgery monitoring and warm the circulating warm water pad to ensure that the animal body temperature is regulated during and post-surgery. Set up all surgical instruments on a clean surface (*see* **Note 10**).
3. Remove the mouse from the cage and place in the anesthesia induction chamber with a mix of 3–4% isoflurane and 1 L/min oxygen. Measure depth of anesthesia by hind limb toe pinch reflex. Once no response to pinch is observed, the appropriate depth of anesthesia is reached.
4. Inject the mouse with a pre-surgical dose of carprofen (5 mg/kg body weight) subcutaneously.
5. Using hair clippers or hair removal cream remove abdominal fur to minimize risk of infection.
6. Apply ophthalmic ointment over the eyes to prevent drying.
7. Move the mouse to the circulating warm water pad and fix the limbs with tape. Maintain general anesthesia via inhalation at 1–2% isoflurane and 0.8 L/min oxygen. Ensure that the snout of the mouse is firmly fixed into the cone (*see* **Note 11**).
8. Clean the shaved area. Scrub can be alternated between 70% alcohol and 10% povidone-iodine (betadine) three times in a circular fashion away from the center to perimeter of the area.

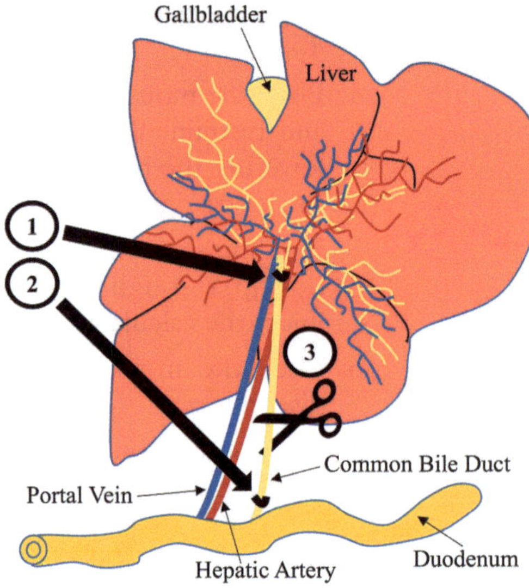

Fig. 2 *Schematic representation of the BDL procedure.* The common bile duct is located in close proximity to the portal vein and hepatic artery. After careful separation, the common bile duct should be ligated at sites indicated with black arrows (Step 1 and 2). After ligation, the common bile duct should be cut in between the ligatures (Step 3)

9. Using a pair of surgical scissors, perform a midline laparotomy of approximately 1–2 cm in length below the sternum. Using the scissors or a 0.9% sterile NaCl soaked cotton swab as a spreader, spread the connective tissue to expose the peritoneum.
10. Cut along the linea alba to expose the peritoneal cavity. Insert the Colibri retractor into the incision to expose the internal organs.
11. Carefully, using a wet cotton swab, move the liver and the intestines away to expose the bile duct (*see* **Note 12**).
12. Carefully separate the bile duct from the portal vein and using the curved scissors, place 6–0 sterile silk sutures around the common bile duct just above the duodenum and tie swiftly with a double surgical knot. Perform a second ligature just a few millimeters above the first ligature near the hilum of the liver (*see* **Note 13**) (Fig. 2).
13. Cut the bile duct between the two ligatures to ensure complete ligation (Fig. 2). Remove excess sutures by cutting off the superfluous ends. Rinse the peritoneal cavity with 0.9% sterile NaCl solution.
14. Gently move the organs back into place using a wet cotton swab and remove the Colibri retractor.
15. Using sterile 5–0 Vicryl sutures, close the inner abdominal incision. Pull the skin together and close the incision by applying wound clips.

16. With the surgery complete, move the mouse to a warmed cage (set up pre-surgery) and observe until it is fully awake and active. Move the mouse back to its regular cage and monitor every 2 h for the first 4 h.
17. 24 h post-surgery, administer a second dose of anesthetic such as carprofen (5 mg/kg body weight, subcutaneously). Mice are observed daily post-surgery till experimental endpoint. Follow all postoperative protocols and monitor the mice in accordance to the guidelines set by your institution (*see* **Note 14**).
18. Mice should be sacrificed, and liver tissue harvested 7–21 days post-surgery.

3.3 HFHCD Diet

The Western or fast food diet can induce obesity, metabolic syndrome, and steatosis after 2 months of diet; increased inflammation and immune cell infiltration after 3 months of diet; NASH and fibrosis within 6–9 months of diet; hepatic ballooning and severe fibrosis within 14–20 months [25, 26]. Rodents fed the ALIOS diet become obese and have increased alanine aminotransferase (ALT) levels and steatosis 4 months after diet. Fibrosis and inflammation are evident after about 12 months [26, 34]. The FPC diet combines the formula of the Western and ALIOS diets. Mice fed this diet develop insulin resistance, inflammation, NAFLD, hepatocyte death, HSC activation, and mild fibrosis within 4 months of feeding and more profound fibrosis after 8 months of feeding [27]. For the composition of all the diets, refer to Table 1.

1. Remove regular chow diet from the cage and add HFHCD of your choice (*see* **Notes 15** and **16**).
2. To prepare a 42g/L glucose and fructose (55%/45%, w/w) drinking water solution, add 19 g D-(+)-glucose monohydrate and 23 g D-(−)-fructose to 1 L of acidified drinking water. Stir till completely dissolved (*see* Notes **17** and **18**).
3. Under a sterile hood, filter the water through a 0.2μM filter into an autoclaved glass bottle and store at 4 °C.
4. Replace the diet and water solution at least twice a week. Weigh mice weekly. To monitor liver injury, serum ALT can be measured at various time points using retro-orbital bleeds.

3.4 Choline-Deficient High-Fat Diet Models

Mice fed the CDAA diet develop hepatomegaly, steatosis, and inflammation 3 weeks after diet, with visible fibrosis observed 6–9 weeks after diet depending on the strain of the mouse. Moderate-to-severe fibrosis occurs at 20 weeks and HCC has been reported with high incidence in mice 24 weeks post diet [29]. When fed the CD-HFD diet, mice develop steatosis and increased ALT levels within 20–24 weeks, ballooned hepatocytes at around 45 weeks and HCC after one year on CD-HFD diet [30].

1. Remove regular chow diet from the cage and replace with CDAA or CD-HFD (*see* **Note 19**).
2. Monitor the cages daily to ensure that enough food is in the cage (*see* **Note 20**). Weigh mice weekly. To monitor liver injury, serum ALT can be measured at various time points using retro-orbital bleeds.

3.5 Harvesting Tissue and Analysis

At the endpoint of the experiment, all mice should be humanely euthanized according to the animal guideline of your institution.

1. Remove the mouse from the cage and place in the anesthesia induction chamber with a mix of 3–4% isoflurane and 1 L/min oxygen. Measure depth of anesthesia by hind limb toe pinch reflex. Once no response to pinch is observed, the appropriate depth of anesthesia is reached.
2. Draw a small amount of blood retro-orbitally, from the sub-mandibular vein or from the portal vein. Centrifuge at 3000 × *g* for 15 min and analyze ALT and AST (aspartate aminotransferase) levels.
3. Weigh the mouse, dissect to harvest the liver, and weigh the liver to calculate liver/body weight ratio.
4. One piece of each lobe of the harvested liver should be snap frozen for RNA extractions in 2.0 mL sterile tubes. Quantitative PCR may be used to measure fibrogenic and HSC activation markers such as *Acta2*, *Col1a1*, *Col1a2*, *Lox*, or *Timp1* and to determine inflammatory mediators such as cytokines and interleukins in liver tissue.
5. The rest of the liver is formalin-fixed and paraffin embedded. For fibrosis evaluation, blocks can be cut into 5μM slides to be stained with Sirius Red or Masson's trichrome staining (Fig. 1). To evaluate HSC activation, slides can also be stained for α-SMA. Immune cell infiltration and inflammation can be viewed by staining of immune cell markers such as CD45, F4/80, or CD3.
6. One piece of each lobe may be placed in 4% paraformaldehyde (PFA) for 24–48 h and then transferred to a 30% sucrose solution for 24–48 h. This tissue can now be used to make frozen bocks/tissue and fibrosis can be quantified by hydroxyproline measurements.

4 Notes

1. Prior to any injection or addition of diet, please check the expiry date. In long models, reagents and diets will need to be purchased multiple times. Surgical instruments should be checked for functionality and autoclaved before each usage.

2. While CCl_4 may be stored at 4 °C, we recommend storing it at room temperature. After the CCl_4-corn oil solution is prepared, it should be stored in a glass bottle and protected from light.
3. CCl_4 is an extremely toxic compound. Be extremely cautious and careful when preparing the solution, handling the syringe, and disposing the syringe to prevent accidental piercing of the skin or splashing your eyes. Wearing of personal protective equipment including safety glasses is always recommended.
4. The CCl_4/oil solution is viscous. The easiest way to fill the syringe is to remove the needle cap, aspirate the solution directly into the syringe, and then place the needle cap back on.
5. A helpful tip is to turn the needle so the bevel points upward and the numbers are aligned to be read easily.
6. To make the process more efficient and less stressful for the animal, it is best to create a table listing the injection volumes for different mouse weights.
7. The easiest way to restrain a mouse for intraperitoneal injection is to employ the scruff restraint. To do so, grab the mouse by the tail and place onto a textured surface. Tug slightly on the tail allowing the mouse to grab onto the surface with its fore limbs. Using your other hand (specifically the thumb and index finger) grab the mouse firmly by the scruff. Only perform the injection when the mouse is firmly grasped.
8. Once restrained, tilt the mouse slightly downward so that the head is lower than the hind. This allows for organs in the abdomen to shift and reduces the chance of accidentally puncturing any organs.
9. Disposal of CCl_4 needles, syringes, glass vials and pipettes must be done in clearly labeled toxic waste containers. Please be aware and follow waste disposal requirements set by your university.
10. All instruments must be autoclaved before surgery. A helpful tip is to arrange all surgical instruments, sutures, and clips in the order that they are being used. If multiple surgeries are being performed, all instruments should be sterilized in a bead sterilizer and cleaned with ethanol after each surgery to minimize the risk of infection and cross-contamination.
11. Throughout the surgery, the depth of anesthesia should be measured and adjusted based on hind limb toe pinch reflex. Mouse breathing and heartbeat can be monitored visually by looking at chest elevations.
12. The bile duct may sometimes be adhered to the portal vein. Make sure to carefully separate the two. This can be done by using a moistened cotton swab and micro-forceps. Do not use

surgical scissors as accidentally cutting/nicking of the portal vein will result in extensive bleeding.

13. To ensure complete ligation, the surgical knots should be tight and firm at each ligation junction. It is extremely important that no bile leakage occurs which may prevent cholestasis and instead lead to peritonitis.

14. Postoperative follow-up is key. Any signs of distress, lack of mobility and/or lethargy may indicate complications such as internal bleeding or peritonitis. At this point, the mouse should be humanely euthanized in compliance with protocols set by your institution.

15. These diets can be readily purchased from approved vendors, stored at 4 °C for the at least 6 months, and must be replaced twice a week. Some investigators prefer storing diets at −20 °C.

16. The FPC diet our laboratory uses is currently is available in a powdered form and must be formed into soft pellets before being added to the cages. Please consult with the diet manufacturer and follow all instructions before using any diet.

17. Like the diet, the drinking water solution must also be replaced twice a week. To simplify things, we recommend making a 10× stock of the sucrose/fructose solution and storing at 4 °C or − 20 °C and diluting with acidified water when necessary.

18. In our experience, it is best to use acidified drinking water directly from the animal facility. Using deionized or ultrapure water may deprive animals of key electrolytes. Other protocols may be appropriate in mouse facilities that use autoclaved instead of acidified water.

19. These diets can be readily purchased from approved vendors and can be stored at 4 °C or −20 °C.

20. This diet may turn rancid quickly and must be replaced frequently. In our experience, to monitor food intake and ensure uniformity, it is best to weigh out about 10–15 g of diet/mouse in the cage at least twice a week.

References

1. Vilar-Gomez E, Calzadilla-Bertot L, Wai-Sun Wong V, Castellanos M, Aller-de la Fuente R, Metwally M, Eslam M, Gonzalez-Fabian L, Alvarez-Quinones Sanz M, Conde-Martin AF, De Boer B, McLeod D, Hung Chan AW, Chalasani N, George J, Adams LA, Romero-Gomez M (2018) Fibrosis severity as a determinant of cause-specific mortality in patients with advanced nonalcoholic fatty liver disease: a multi-National Cohort Study. Gastroenterology 155(443–457):e417. https://doi.org/10.1053/j.gastro.2018.04.034

2. Angulo P, Kleiner DE, Dam-Larsen S, Adams LA, Bjornsson ES, Charatcharoenwitthaya P, Mills PR, Keach JC, Lafferty HD, Stahler A, Haflidadottir S, Bendtsen F (2015) Liver fibrosis, but no other histologic features, is associated with long-term outcomes of patients with nonalcoholic fatty liver disease.

Gastroenterology 149(389–397):e310. https://doi.org/10.1053/j.gastro.2015.04.043

3. Ellis EL, Mann DA (2012) Clinical evidence for the regression of liver fibrosis. J Hepatol 56:1171–1180. https://doi.org/10.1016/j.jhep.2011.09.024
4. Tsuchida T, Friedman SL (2017) Mechanisms of hepatic stellate cell activation. Nat Rev Gastroenterol Hepatol 14:397–411. https://doi.org/10.1038/nrgastro.2017.38
5. Mederacke I, Hsu CC, Troeger JS, Huebener P, Mu X, Dapito DH, Pradere JP, Schwabe RF (2013) Fate tracing reveals hepatic stellate cells as dominant contributors to liver fibrosis independent of its aetiology. Nat Commun 4:2823. https://doi.org/10.1038/ncomms3823
6. De Minicis S, Seki E, Uchinami H, Kluwe J, Zhang Y, Brenner DA, Schwabe RF (2007) Gene expression profiles during hepatic stellate cell activation in culture and in vivo. Gastroenterology 132:1937–1946. https://doi.org/10.1053/j.gastro.2007.02.033
7. Sancho-Bru P, Bataller R, Gasull X, Colmenero J, Khurdayan V, Gual A, Nicolas JM, Arroyo V, Gines P (2005) Genomic and functional characterization of stellate cells isolated from human cirrhotic livers. J Hepatol 43:272–282. https://doi.org/10.1016/j.jhep.2005.02.035
8. Hillebrandt S, Goos C, Matern S, Lammert F (2002) Genome-wide analysis of hepatic fibrosis in inbred mice identifies the susceptibility locus Hfibl on chromosome 15. Gastroenterology 123:2041–2051. https://doi.org/10.1053/gast.2002.37069
9. Shi Z, Wakil AE, Rockey DC (1997) Strain-specific differences in mouse hepatic wound healing are mediated by divergent T helper cytokine responses. Proc Natl Acad Sci U S A 94:10663–10668. https://doi.org/10.1073/pnas.94.20.10663
10. van Nieuwerk CM, Groen AK, Ottenhoff R, van Wijland M, van den Bergh Weerman MA, Tytgat GN, Offerhaus JJ, Oude Elferink RP (1997) The role of bile salt composition in liver pathology of mdr2 (−/−) mice: differences between males and females. J Hepatol 26:138–145. https://doi.org/10.1016/s0168-8278(97)80020-7
11. Mederacke I, Dapito DH, Affo S, Uchinami H, Schwabe RF (2015) High-yield and high-purity isolation of hepatic stellate cells from normal and fibrotic mouse livers. Nat Protoc 10:305–315. https://doi.org/10.1038/nprot.2015.017
12. Delire B, Starkel P, Leclercq I (2015) Animal models for fibrotic liver diseases: what we have, what we need, and what is under development. J Clin Transl Hepatol 3:53–66. https://doi.org/10.14218/JCTH.2014.00035
13. Kim YO, Popov Y, Schuppan D (2017) Optimized mouse models for liver fibrosis. Methods Mol Biol 1559:279–296. https://doi.org/10.1007/978-1-4939-6786-5_19
14. Wallace MC, Hamesch K, Lunova M, Kim Y, Weiskirchen R, Strnad P, Friedman SL (2015) Standard operating procedures in experimental liver research: thioacetamide model in mice and rats. Lab Anim 49:21–29. https://doi.org/10.1177/0023677215573040
15. Yeh CN, Maitra A, Lee KF, Jan YY, Chen MF (2004) Thioacetamide-induced intestinal-type cholangiocarcinoma in rat: an animal model recapitulating the multi-stage progression of human cholangiocarcinoma. Carcinogenesis 25:631–636. https://doi.org/10.1093/carcin/bgh037
16. Trauner M, Fickert P, Wagner M (2007) MDR3 (ABCB4) defects: a paradigm for the genetics of adult cholestatic syndromes. Semin Liver Dis 27:77–98. https://doi.org/10.1055/s-2006-960172
17. Meerman L, Koopen NR, Bloks V, Van Goor H, Havinga R, Wolthers BG, Kramer W, Stengelin S, Muller M, Kuipers F, Jansen PL (1999) Biliary fibrosis associated with altered bile composition in a mouse model of erythropoietic protoporphyria. Gastroenterology 117:696–705. https://doi.org/10.1016/s0016-5085(99)70464-6
18. Fickert P, Stoger U, Fuchsbichler A, Moustafa T, Marschall HU, Weiglein AH, Tsybrovskyy O, Jaeschke H, Zatloukal K, Denk H, Trauner M (2007) A new xenobiotic-induced mouse model of sclerosing cholangitis and biliary fibrosis. Am J Pathol 171:525–536. https://doi.org/10.2353/ajpath.2007.061133
19. Smit JJ, Schinkel AH, Oude Elferink RP, Groen AK, Wagenaar E, van Deemter L, Mol CA, Ottenhoff R, van der Lugt NM, van Roon MA et al (1993) Homozygous disruption of the murine mdr2 P-glycoprotein gene leads to a complete absence of phospholipid from bile and to liver disease. Cell 75:451–462. https://doi.org/10.1016/0092-8674(93)90380-9
20. Itoh M, Suganami T, Nakagawa N, Tanaka M, Yamamoto Y, Kamei Y, Terai S, Sakaida I, Ogawa Y (2011) Melanocortin 4 receptor-deficient mice as a novel mouse model of nonalcoholic steatohepatitis. Am J Pathol 179:2454–2463. https://doi.org/10.1016/j.ajpath.2011.07.014

21. Trak-Smayra V, Paradis V, Massart J, Nasser S, Jebara V, Fromenty B (2011) Pathology of the liver in obese and diabetic Ob/Ob and db/db mice fed a standard or high-calorie diet. Int J Exp Pathol 92:413–421. https://doi.org/10.1111/j.1365-2613.2011.00793.x
22. Asgharpour A, Cazanave SC, Pacana T, Seneshaw M, Vincent R, Banini BA, Kumar DP, Daita K, Min HK, Mirshahi F, Bedossa P, Sun X, Hoshida Y, Koduru SV, Contaifer D Jr, Warncke UO, Wijesinghe DS, Sanyal AJ (2016) A diet-induced animal model of non-alcoholic fatty liver disease and hepatocellular cancer. J Hepatol 65:579–588. https://doi.org/10.1016/j.jhep.2016.05.005
23. Fujii M, Shibazaki Y, Wakamatsu K, Honda Y, Kawauchi Y, Suzuki K, Arumugam S, Watanabe K, Ichida T, Asakura H, Yoneyama H (2013) A murine model for non-alcoholic steatohepatitis showing evidence of association between diabetes and hepatocellular carcinoma. Med Mol Morphol 46:141–152. https://doi.org/10.1007/s00795-013-0016-1
24. Kohli R, Kirby M, Xanthakos SA, Softic S, Feldstein AE, Saxena V, Tang PH, Miles L, Miles MV, Balistreri WF, Woods SC, Seeley RJ (2010) High-fructose, medium chain trans fat diet induces liver fibrosis and elevates plasma coenzyme Q9 in a novel murine model of obesity and nonalcoholic steatohepatitis. Hepatology 52:934–944. https://doi.org/10.1002/hep.23797
25. Charlton M, Krishnan A, Viker K, Sanderson S, Cazanave S, McConico A, Masuoko H, Gores G (2011) Fast food diet mouse: novel small animal model of NASH with ballooning, progressive fibrosis, and high physiological fidelity to the human condition. Am J Physiol Gastrointest Liver Physiol 301:G825–G834. https://doi.org/10.1152/ajpgi.00145.2011
26. Tetri LH, Basaranoglu M, Brunt EM, Yerian LM, Neuschwander-Tetri BA (2008) Severe NAFLD with hepatic necroinflammatory changes in mice fed trans fats and a high-fructose corn syrup equivalent. Am J Physiol Gastrointest Liver Physiol 295:G987–G995. https://doi.org/10.1152/ajpgi.90272.2008
27. Wang X, Zheng Z, Caviglia JM, Corey KE, Herfel TM, Cai B, Masia R, Chung RT, Lefkowitch JH, Schwabe RF, Tabas I (2016) Hepatocyte TAZ/WWTR1 promotes inflammation and fibrosis in nonalcoholic steatohepatitis. Cell Metab 24:848–862. https://doi.org/10.1016/j.cmet.2016.09.016
28. Machado MV, Michelotti GA, Xie G, Almeida Pereira T, Boursier J, Bohnic B, Guy CD, Diehl AM (2015) Mouse models of diet-induced nonalcoholic steatohepatitis reproduce the heterogeneity of the human disease. PLoS One 10:e0127991. https://doi.org/10.1371/journal.pone.0127991
29. Wei G, An P, Vaid KA, Nasser I, Huang P, Tan L, Zhao S, Schuppan D, Popov YV (2020) Comparison of murine steatohepatitis models identifies a dietary intervention with robust fibrosis, ductular reaction, and rapid progression to cirrhosis and cancer. Am J Physiol Gastrointest Liver Physiol 318: G174–G188. https://doi.org/10.1152/ajpgi.00041.2019
30. Wolf MJ, Adili A, Piotrowitz K, Abdullah Z, Boege Y, Stemmer K, Ringelhan M, Simonavicius N, Egger M, Wohlleber D, Lorentzen A, Einer C, Schulz S, Clavel T, Protzer U, Thiele C, Zischka H, Moch H, Tschop M, Tumanov AV, Haller D, Unger K, Karin M, Kopf M, Knolle P, Weber A, Heikenwalder M (2014) Metabolic activation of intrahepatic CD8+ T cells and NKT cells causes nonalcoholic steatohepatitis and liver cancer via cross-talk with hepatocytes. Cancer Cell 26:549–564. https://doi.org/10.1016/j.ccell.2014.09.003
31. Constandinou C, Henderson N, Iredale JP (2005) Modeling liver fibrosis in rodents. Methods Mol Med 117:237–250. https://doi.org/10.1385/1-59259-940-0:237
32. Popov Y, Sverdlov DY, Sharma AK, Bhaskar KR, Li S, Freitag TL, Lee J, Dieterich W, Melino G, Schuppan D (2011) Tissue transglutaminase does not affect fibrotic matrix stability or regression of liver fibrosis in mice. Gastroenterology 140:1642–1652. https://doi.org/10.1053/j.gastro.2011.01.040
33. Seki E, De Minicis S, Osterreicher CH, Kluwe J, Osawa Y, Brenner DA, Schwabe RF (2007) TLR4 enhances TGF-beta signaling and hepatic fibrosis. Nat Med 13:1324–1332. https://doi.org/10.1038/nm1663
34. Mells JE, Fu PP, Kumar P, Smith T, Karpen SJ, Anania FA (2015) Saturated fat and cholesterol are critical to inducing murine metabolic syndrome with robust nonalcoholic steatohepatitis. J Nutr Biochem 26:285–292. https://doi.org/10.1016/j.jnutbio.2014.11.002

Chapter 24

Mouse Models of Muscle Fibrosis

Antonio L. Serrano and Pura Muñoz-Cánoves

Abstract

Fibrosis in skeletal muscle is the natural tissue response to persistent damage and chronic inflammatory states, cursing with altered muscle stem cell regenerative functions and increased activation of fibrogenic mesenchymal stromal cells. Exacerbated deposition of extracellular matrix components is a characteristic feature of human muscular dystrophies, neurodegenerative diseases affecting muscle and aging. The presence of fibrotic tissue not only impedes normal muscle contractile functions but also hampers effective gene and cell therapies. There is a lack of appropriate experimental models to study fibrosis. In this chapter, we highlight recent developments on skeletal muscle fibrosis in mice and expand previously described methods by our group to exacerbate and accelerate fibrosis development in murine muscular dystrophy models and to study the presence of fibrosis in muscle samples. These methods will help understand the molecular and biological mechanisms involved in muscle fibrosis and to identify novel therapeutic strategies to limit the progression of fibrosis in muscular dystrophy.

Key words Fibrosis, Skeletal muscle regeneration, Duchenne muscular dystrophy, mdx mice, Exercise, Chronic injury

1 Introduction

Fibrosis is defined as the pathological accumulation of excessive components of the extracellular matrix (ECM) in a tissue, typically resulting from chronic inflammation [1]. It is the distinguishing mark of a failure of the natural tissue repair mechanism in different organs of the body, and is a major causative factor of clinical illness. Fibrosis development can be considered the adaptive response of the tissue to persistent damage, leading to cellular dysfunction and parenchymal scar and ultimately to organ failure and death. Highlighting its importance, it has been estimated that chronic fibroproliferative disease is involved in about 45% of all deaths in the developed world [2, 3].

Different stages have been proposed to be required for fibrosis development after the primary tissue injury, beginning with the recruitment of inflammatory cells, including lymphocytes, polymorphonuclear leukocytes, eosinophils, basophils, mast cells, and

Boris Hinz and David Lagares (eds.), *Myofibroblasts: Methods and Protocols*, Methods in Molecular Biology, vol. 2299,
https://doi.org/10.1007/978-1-0716-1382-5_24,

several types of macrophages [2, 4, 5]. As a general mechanism leading to organ fibrosis, cytokines, inflammatory mediators, and other released factors induce the activation of different mesenchymal-derived effector cells (comprising fibroblasts, fibrocytes, tissue-specific pericytes, and myofibroblasts), which proliferate and start to produce extracellular matrix components. The positive balance between deposition and resorption of these ECM molecules promotes fibrosis progression, which is often amplified by autocrine loops of signaling [1, 5].

1.1 Cellular Mechanisms in Skeletal Muscle Fibrosis

The common pathway leading to fibrosis in different organs is equally applicable to skeletal muscle [2, 4, 6–10]. Because of its remarkable ability to regenerate after severe injury, this tissue has been used as a model system of the mammalian wound-healing response to study the interactions between tissue-specific stem cells (satellite cells), different inflammatory cell types, effector cells, and the components of the ECM [6].

The strong regenerative ability of skeletal muscle relies primarily on its stem cells, which are in a quiescent state in resting muscle. A few hours after injury, they become activated and expand exponentially during the next days to generate enough numbers of myogenic progenitors needed to reconstitute the damaged areas [11]. In this period, other niche cell components, including both tissue-resident cells and different infiltrating immune cell types, act coordinately to support the function of satellite cells and their progeny to repair successfully the injured tissue [12, 13]. Of particular importance in this process are the fibrogenic cells, a population of tissue-resident mesenchymal cells expressing PDGFRα, identified as progenitors of fibroblasts and adipocytes, and termed fibro/adipogenic progenitors (FAPs) [14, 15].

FAPs are located in the interstitial space between myofibers and, through not yet defined mechanisms, were recently found to be necessary for effective muscle regeneration and the homeostatic maintenance of muscle mass [16]. After an injury of skeletal muscle, FAPs expand briefly and provide the temporary ECM required to reconstitute tissue architecture [14, 17]. The spatiotemporal dynamics of critical soluble factors present in the regenerative microenvironment regulate the fate of FAPs in skeletal muscle after injury. Whereas tumor necrosis factor α (TNFα) has a pro-apoptotic effect on these cells, transforming growth factor β1 (TGF-β1) promotes FAPs survival and their differentiation into matrix-producing cells [17], acting at least in part, via a mechanism involving αv integrins [18] and being modulated by PDGFRα signaling [19].

During normal regeneration, FAPs expansion is limited to the first stages after injury, due to the increased TNFα levels released by inflammatory macrophages. In case of extensive, chronic damage and in pathological situations cursing with elevated and continuous

TGF-β1 levels, FAPs persistence leads to a regenerative failure and drive fibrosis and scar formation by exacerbated production of ECM components [4]. In the last years, several molecules secreted by various cell types present in the damaged muscles have been identified to directly influence FAPs fate acting in a paracrine fashion, whereas FAPs themselves also influence other cellular populations (reviewed in [20]).

In summary, skeletal muscle fibrosis involves tissue damage, the presence of inflammatory cells that favors activation of TGF-β1-dependent pathways and increased FAPs fibrogenic activity. Primary damage of skeletal muscle can be caused experimentally by severe trauma or be the consequence of genetic diseases coursing with muscle degeneration. The persistence of the injury over time and the inhibition of the normal regenerative activities of muscle stem cells are also required for the accumulation of excessive amount of ECM components [7, 21–23].

1.2 Muscle Fibrosis in Mice: Muscular Dystrophy Models

Skeletal muscle fibrosis is a prominent feature of muscular dystrophies, a group of genetic diseases of different severity characterized by mutations of the protein complex that connects the actin cytoskeleton of the myofibers to the ECM [10]. This link is established trough different subunits of the dystrophin glycoprotein complex (DGC). The DGC spans the muscle plasma membrane and provides stability maintaining the structural integrity of the myofibers [24]. Among the different types of muscular dystrophies, Duchenne muscular dystrophy (DMD) is the most common inherited type of muscle disease and is caused by mutations in the dystrophin gene that encodes a subsarcolemmal protein that plays an essential role in the DGC. In DMD, muscle fibers undergo continuous cycles of necrosis and repair. Moreover, satellite cells also express dystrophin, which has a function in regulating asymmetric cell division to yield stem cells and committed myogenic progenitors [25]. The decreased number of asymmetric divisions leads to a strong reduction in the number of myogenic progenitors needed for proper muscle regeneration [25, 26]. Therefore, in DMD, besides the existing defect in the structural stability of the myofibers, there is an intrinsic regenerative failure which results in the replacement of the myofibers by fibrous and fatty tissue at late stages of the disease. Recent reviews on the mechanisms regulating muscle fibrosis can be found in references (27, 28).

The most extensively used mouse model of an inherited muscle degenerative disease is the mdx mice which, similarly to DMD patients, harbors a mutation in the dystrophin gene and is widely used in research. However, in spite of having elevated serum creatine kinase (CK) and developing a myopathy with mild fibrosis, which is more evident in male mice [29], the natural disease history of the mdx mouse model differs substantially from that of DMD patients, with less severe symptoms and only a small reduction of life span. Diaphragm muscle is the only skeletal muscle of mdx mice in which pathological features of DMD (including overt fibrosis) are recapitulated, with ten times higher collagen content than limb muscles at 16 months of age [30]. A possible explanation for these discrepancies between DMD and mdx may include species-specific compensatory mechanisms, differences in the self-renewal ability of satellite cells, muscle loading and body size, as well as in the duration of the growth phase [31, 32]. In any case, mdx mice remain instrumental since it is a cost-efficient model, they are easy to manipulate and have a short generation time, being used routinely by many laboratories to gain understanding of the disease and in preclinical tests of DMD therapies. Since diaphragm muscle only represents a small portion of the total muscle mass, is not easily accessible, and appreciable ECM deposition takes months to develop, procedures aiming at accelerating and increasing fibrosis development should be designed for research purposes.

1.3 Increasing Naturally Occurring Fibrosis in mdx Mice by Genetic Strategies

Several strategies have been used to develop mouse models that recapitulate the DMD phenotype (including the presence of muscle fibrosis) more faithfully. These include producing a series of double knockout mouse models, combining the mdx dystrophin mutation with mutations in other genes to exacerbate the disease phenotypes, and generating a number of alternative versions of mdx in different genetic backgrounds. See references (33–35) for detailed information on different mdx models.

Some examples of the first strategy are the increased fibrosis observed in mdx mice with haploinsufficiency of utrophin (an autosomally encoded homolog of dystrophin (mdx/utrn+/−)) [36, 37] (available at https://www.jax.org/strain/019014), with differences from mdx mice in limb muscles already evident at 2 months of age and increased collagen deposition in diaphragm and quadriceps muscles at 6 months of age; mdx mice with an inactivating deletion of the CMAH (cytidine monophosphate–sialic acid hydroxylase) gene (Cmah−/−mdx), which eliminates the biosynthesis of the N-glycolylneuraminic sialic acid and have appreciable fibrosis in the quadriceps muscle at 2 months of age [38] (available at https://www.jax.org/strain/017929), or mdx mice lacking the RNA component of telomerase (mdx/mTR), which display limb fibrosis at 76 weeks of age [39] (available at https://www.jax.org/strain/023535).

Introduction of the point mutation in exon 23 of the Dmd gene into the DBA/2J (D2) genetic background (by intercrossing mdx mice in the C57BL/10 background, C57BL/10-mdx with DBA/2J mice) to generate D2-mdx mice (available at https://www.jax.org/strain/013141) resulted in a more severe pathology accompanied by a myogenic deficit and prominent fibrosis in limb muscles, already evident at 6 months of age [31]. A complete phenotypic characterization of the D2-mdx strain has been recently published [32]. Indeed, in D2-mdx mice, the failure of regenerative myogenesis and increased FAPs accumulation leading to prominent fibrosis has been shown to depend on enhanced TGF-β1 signaling with respect to other mdx strains which show a milder dystrophic phenotype [40].

1.4 Accelerating and Expanding Fibrosis Development in mdx Mice

As described previously, even if muscle degeneration cycles, chronic inflammation, and aberrant regeneration are features of mdx mice, excessive ECM deposition is not prominent until late stages of the disease. This hampers fibrosis research in dystrophic models and we have proposed various strategies to accelerate and extend the appearance of fibrosis, allowing using younger mice, easily accessible limb muscles, and increasing the amount of available muscles developing fibrosis to examine [23]. Early studies showed that lack of dystrophin in muscles of mdx mice make myofibers particularly sensitive to damage after exercise, due to the induction of local micro-damages, which allow the transient leakage of proteins from the fibers [41, 42]. More recent data confirm that eccentric type of exercise causes more damage to mdx muscles than predominately concentric contractions [43]. Implementing exercise protocols is a noninvasive physiological method to periodically induce transient muscle damage and inflammation, fostering the process of accumulation of ECM components [23].

2 Materials

2.1 Materials for Traumatic Skeletal Muscle Fibrosis in mdx Mice

- Dystrophic mice of the desired strain and age. We have successfully used mdx mice of 2, 3, and 5 months of age to produce limb fibrosis.
- Commercially available treadmill apparatus for rodents. For instance, Exer-3/6 Open Treadmill (Columbus Instruments, USA) or 76-0896 LE8710MTS Touchscreen Treadmill (Harvard Apparatus, USA) models (*see* **Note 1** and Fig. 1).
- Grip strength meter (Grip Strength Meter, Bioseb Harvard Apparatus, USA).

2.2 Morphological Assessment of Fibrosis by Picro-Sirius Red Stain

- Isopentane.
- Liquid nitrogen.
- Cryostat.
- Standard glass adhesion slides.
- Bouin's solution fixative: Prepared with saturated picric acid (0.9% w/v), formaldehyde (9% w/v), and acetic acid (5% v/v) (Sigma-Aldrich HT10132).

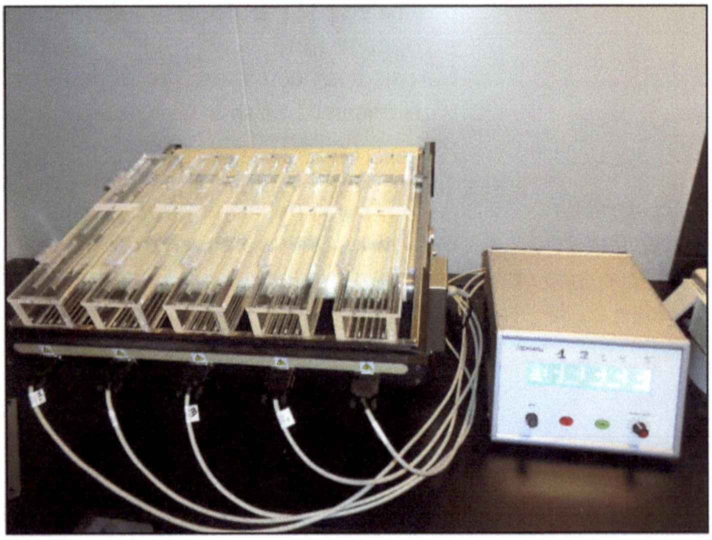

Fig. 1 Picture of the mouse treadmill apparatus. Five mice can be exercised simultaneously

- Picro-sirius Red solution: 0.5 g of Direct Red 80, also known as Sirius red F3BA or Sirius red F3B (Sigma-Aldrich 365548) in 500 mL of saturated picric acid solution 1.3% in H_2O (P6744 Sigma-Aldrich). This solution is stable for at least 3 years.
- Acetic acid solution 2% (v/v): Acetic acid (glacial) (A6283 Sigma-Aldrich) 2 mL in 100 mL of H_2O.
- Standard light microscope.

2.3 Biochemical Determination of Collagen on Muscle Sections

- Direct Red 80, also known as Sirius red F3BA or Sirius red F3B (Sigma-Aldrich 365548).
- Fast green FCF (F7252 Sigma-Aldrich).
- Saturated picric acid solution 1.3% in H_2O (P6744 Sigma-Aldrich).
- 0.1 M NaOH in absolute methanol solution (1:1 v/v).

3 Methods

3.1 Traumatic Induction of Skeletal Muscle Fibrosis in mdx Mice

Skeletal muscles of mdx mice are more sensitive to repeated contractions than non-dystrophic muscles. Forced treadmill exercise increases the severity of dystrophic pathology and is a useful method to exacerbate muscle damage and accelerate the development of fibrosis in limb muscles.

1. Experimental and housing conditions should be maintained as stable as possible to increase the reproducibility of the method and diminish the stress of the animals.

2. The time of the day in which the training is performed during the repeated sessions must be kept constant.
3. Experimental groups should be randomized and every single mouse belonging to an experimental group must be unambiguously identified.
4. Mice should follow an acclimation period of at least 3 days prior initiation of the experiment. During these days, allow the animals to stay on the treadmill without the motor turned on for about 5 min. Then, subject mice to a daily session of 20 min at 6 m per minute speed to get used to run on the treadmill.
5. Following the acclimation period, mdx mice are exercised three times per week on a horizontal treadmill for 30 min at a constant speed of 12 m per minute, with a rest of 5 min every 10 min of exercise. We have used this protocol for up to 3 months (*see* **Notes 2–4**). Compared to age-matched non--exercised control mice, a month of training is enough to aggravate the dystrophic phenotype. A longer duration of the training program further increases the presence of muscle fibrosis [23].
6. To monitor the effectiveness of the traumatic exercise protocol at inducing muscle tissue damage, blood samples obtained the day before starting, every week during the exercise period and at the end of the training regime (3 months) can be used to analyze the release of muscle proteins such as creatine kinase, a marker of body-wide muscle damage, into the bloodstream (*see* **Note 5**).
7. Performing periodical grip strength tests as a quick, simple, and noninvasive method to evaluate mouse muscle force in vivo is advisable. After positioning horizontally the grip strength meter, mice held by their tail are allowed to grab the metal grid of the instrument and are then pulled manually backward in the horizontal plane. The force applied to the grid just before it loses grip is recorded as the peak tension by as sensor in the apparatus (*see* **Note 6** and Fig. 2).

3.2 Morphological Assessment of Fibrosis by Picro-Sirius Red Stain

This protocol allows the histological visualization of collagen fibers in muscle tissue sections. With a standard light microscope, collagen appears red and the cytoplasm of muscle fibers has a yellow color. Under polarized light microscope, resulting collagen fibers are birefringent (*see* **Note** 7 and Fig. 3).

1. Skeletal muscle samples, previously frozen in isopentane cooled with liquid nitrogen and stored at −80 °C, are sectioned on a cryostat and placed on glass adhesion slides to minimize tissue loss during staining (*see* **Note 8**).

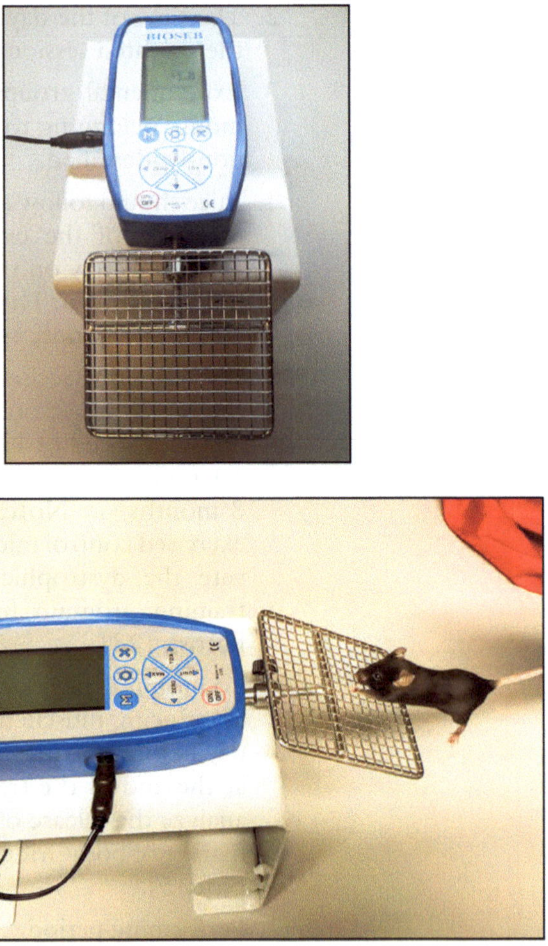

Fig. 2 Measurement of muscle force in mdx mice by the Grip Strength Test

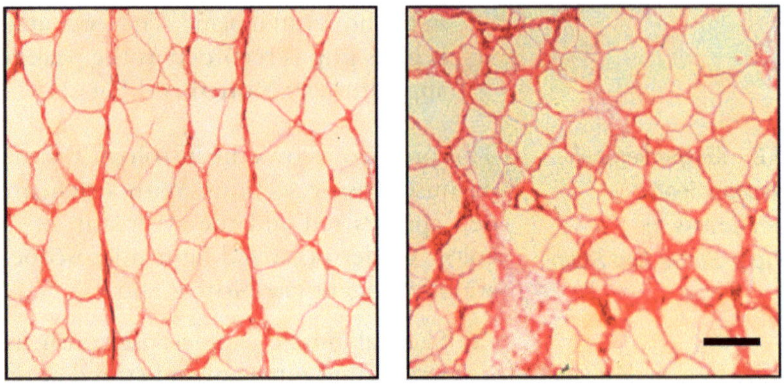

Fig. 3 Illustrative examples of Sirius Red stains of muscle cryosections from non-exercised mdx mice (left) and exercised mdx mice (right). Scale bar: 50 microns

2. After drying the slides at room temperature (RT), they are incubated overnight in Bouin's fixative.
3. Wash with tap water 5 min and incuba te in Picro-sirius Red solution for at least 1 h at RT.
4. Wash rapidly with 2% acetic acid solution.
5. Dehydrate with EtOH 70%/96%/100% 5 min each.
6. Clear with Xylol 5 min and mount in a solvent-based mounting medium.

3.3 Biochemical Determination of Collagen on Muscle Sections

This simple procedure allows sensitive and quantitative determination of collagen and protein content in muscle cryosections. It is based on the selective binding of collagen by the Direct Red 80 dye, and the binding of Fast green FCF to non-collagenous proteins [44]. Once bound to the sample, both dyes are eluted simultaneously with a solution of NaOH-methanol readily and the amount of collagen and protein is determined based on the absorbances obtained at 540 and 605 nm. This protocol also works on formalin-fixed paraffin-embedded samples and the results on collagen content match those obtained by hydroxyproline analyses. It has the advantage of using very small amount of sample, which sometimes is a limiting factor in fibrosis research in skeletal muscle.

1. Prepare the staining solution containing 0.1% Direct Red 80 and 0.1% Fast Green FCF in saturated picric acid solution (100 mg Direct Red 80 and 100 mg of Fast Green in 100 mL of saturated picric acid solution).
2. Place at least ten cryosections (10μm thickness each) in a 1.5 mL microcentrifuge tube (sections are collected with a needle inside the cryostat) and cover the tube with aluminum foil.
3. Preincubate the tubes with a solution of 0.1% Fast Green FCF in saturated picric acid solution for 15 min at RT on a rotary shaker (this step prevents unspecific binding of Direct Red80 to non-collagenous proteins). A volume of 200μL per tube is enough.
4. Wash with distillated water three times.
5. Incubate tubes with the staining solution for 30 min at RT on a rotary shaker.
6. Wash with distillated water three times until water is colorless.
7. Wash the samples with a solution containing 0.1 M NaOH in absolute methanol (1:1, v/v) for a few seconds, until all the color is eluted.
8. Measure the absorbance of the eluate in a spectrophotometer at 540 and 605 nm corresponding to the maximal absorbances of Direct Red 80 and Fast green FCF, respectively.

9. Calculations:
   - 9.1 Correct the absorbance at 540 nm by the following formula: absorbance 540 nm − (absorbance at 605 nm × 0.291). This step subtracts the interference of the absorbance of Fast green FCF on the absorbance of Direct Red 80.
   - 9.2 Divide the absorbance at 605 nm by 2.08. This indicates the net amount of non-collagenous proteins in the sections.
   - 9.3 Divide the corrected absorbance at 540 nm by 38.4. This indicates the net amount of collagen proteins.
   - 9.4 Calculate the relative amount of collagen per mg of protein by dividing the value obtained in **step 9.3** by the sum of values of **steps 9.2** + **9.3** (total protein). The results are expressed as mg of collagen/mg of total protein.

3.4 Other Downstream Techniques Suitable to Investigate Muscle Fibrosis

A variety of techniques can be used to assess fibrosis in skeletal muscle. Biochemical methods quantifying collagen content or digital image analysis of histochemical or immunostaining of ECM components are the gold standard. Noteworthy, complementary analyses such as analyzing FAPs numbers on muscle samples or determination of the levels of fibrosis-related genes do not correlate with amount of collagen in tissue samples, particularly when fibrosis has already developed. For instance, in D2-mdx mice, expression levels of genes related with fibrosis (Tgf-β, connective tissue growth factor (Ctgf) and collagen) diminished with aging (from 10 to 34 weeks) and did not parallel increased amount of fibrosis based on histological observations [45]. Therefore, care should be taken when interpreting isolated data on fibrosis based exclusively on gene expression levels.

4 Notes

1. This type of instruments consists of motor-driven rolling belts in which speed and slope can be easily adjusted and reproduced on different animals to allow the desired forced exercise training protocol. The suggested models, with 5 or 6 different lanes, allow training several mice simultaneously, saving a significant amount of time.
2. Eventually some mice will stop on the belt and will need an external motivating stimulation to continue running. Treadmill apparatuses are normally equipped with an electrified grid that can apply light electric foot shocks to induce the mice return to running on the belt. Alternative stimulating methods are the use of air-puff accessories of the apparatus or gentle encouragement by the operator using objects like a stick and touching the tail and/or applying a sound stimulus.

3. Eccentric type of exercise has been shown to increase muscle damage. This effect can be induced by placing the treadmill with a downstairs inclination. Reports in the literature on this type of protocols range from 10° to 16° degrees inclination, average running speeds of 10 m/min and duration of the sessions from 5 to 90 min (see [46] for review). Repeated downhill treadmill sessions in rat showed that, to maintain damage ongoing, sessions should be repeated every 4 days, and that a beneficial muscle adaptation and reduced damage was observed if the frequency of the sessions was reduced to one every 7 days [47]. If a downhill running protocol wants to be adopted, the intensity and durations of the sessions should be adapted to the age and particular conditions of the animals, since a proportion of mice may not be able to complete the protocol.
4. Lower exercise intensities of the above indicated may not reach the necessary threshold to increase fibrosis in mdx muscles. For example, a training protocol in horizontal treadmill at speeds of 4 or 8 m/min, for 30 min, three times a week during 24 weeks did not cause damage nor increased muscle fibrosis in 4–5-month-old D2-mdx mice [48].
5. This type of protocol requires continuous supervision of the mice during the training sessions. All parameters of the session should be carefully monitored. Some experimenters perform video recording of the sessions to identify possible behavioral alterations.
6. Practical details on this protocol can be found in [49].
7. On a light microscope, collagen fibers will appear red. Digital pictures can easily be analyzed following these steps with the public domain software ImageJ (available at https://imagej.nih.gov/ij/download.html):
   (a) Open the image file and transform the color image into a 8-bit grayscale image (command sequence File, Open, Image, Type, 8-bit).
   (b) Convert the image into a black and white binary image by adjusting the threshold (command sequence Image, Adjust, Threshold, Apply).
   (c) Measure the total area of the image (command: Analyze, Measure) and the area occupied by collagen (by selecting the option Limit to threshold in the menu Analyze, Set Measurements and then Analyze, Measure).
   (d) Calculate the relative area occupied by the collagen staining in the image versus the total area using the data of the Results window.

(e) Once the images are open, specific areas of interest can be selected on the pictures with the Freehand selection tool and the commands Edit, Clear outside.

8. This protocol is also suitable for paraffin-embedded samples. In this case, deparaffinization and rehydration of the samples are required as initial steps of the protocol. We prefer to use frozen samples since the formalin-fixation and paraffin-embedding procedures tend to shrink muscle fibers and alter interstitial spaces.

Acknowledgments

The authors acknowledge funding from the Spanish Ministry of Science, Innovation and Universities, Spain (grants SAF2015-67369-R, RTI2018-096068-B-I00, and SAF 2015-70270-REDT, a María de Maeztu Unit of Excellence award to UPF [MDM-2014-0370], and the UPF-CNIC collaboration agreement, ERC-2016-AdG-741966, La Caixa-HEALTH (HR17-00040), MDA, UPGRADE-H2020-825825, AFM, DPP-E and Fundació La Marató de TV3 (grant no. 202033 to A.L.S. and no. 202021 to P.M.C.). The CNIC is supported by the Instituto de Salud Carlos III (ISCIII), the Ministerio de Ciencia, Innovación y Universidades (MCNU) and the Pro CNIC Foundation, and is a Severo Ochoa Center of Excellence (SEV-2015-0505).

References

1. Wynn TA, Ramalingam TR (2012) Mechanisms of fibrosis: therapeutic translation for fibrotic disease. Nat Med 18(7):1028–1040
2. Gieseck RL 3rd, Wilson MS, Wynn TA (2018) Type 2 immunity in tissue repair and fibrosis. Nat Rev Immunol 18(1):62–76
3. Wynn TA (2008) Cellular and molecular mechanisms of fibrosis. J Pathol 214 (2):199–210
4. Munoz-Canoves P, Serrano AL (2015) Macrophages decide between regeneration and fibrosis in muscle. Trends Endocrinol Metab 26 (9):449–450
5. Rockey DC, Bell PD, Hill JA (2015) Fibrosis--a common pathway to organ injury and failure. N Engl J Med 373(1):96
6. Mann CJ, Perdiguero E, Kharraz Y, Aguilar S, Pessina P, Serrano AL et al (2011) Aberrant repair and fibrosis development in skeletal muscle. Skelet Muscle 1(1):21
7. Pessina P, Kharraz Y, Jardi M, Fukada S, Serrano AL, Perdiguero E et al (2015) Fibrogenic cell plasticity blunts tissue regeneration and aggravates muscular dystrophy. Stem Cell Reports 4(6):1046–1060
8. Serrano AL, Mann CJ, Vidal B, Ardite E, Perdiguero E, Munoz-Canoves P (2011) Cellular and molecular mechanisms regulating fibrosis in skeletal muscle repair and disease. Curr Top Dev Biol 96:167–201
9. Serrano AL, Munoz-Canoves P (2010) Regulation and dysregulation of fibrosis in skeletal muscle. Exp Cell Res 316(18):3050–3058
10. Serrano AL, Munoz-Canoves P (2017) Fibrosis development in early-onset muscular dystrophies: mechanisms and translational implications. Semin Cell Dev Biol 64:181–190
11. Sousa-Victor P, Gutarra S, Garcia-Prat L, Rodriguez-Ubreva J, Ortet L, Ruiz-Bonilla V et al (2014) Geriatric muscle stem cells switch reversible quiescence into senescence. Nature 506(7488):316–321
12. Bentzinger CF, Wang YX, Dumont NA, Rudnicki MA (2013) Cellular dynamics in the muscle satellite cell niche. EMBO Rep 14 (12):1062–1072

13. Wosczyna MN, Rando TA (2018) A muscle stem cell support group: coordinated cellular responses in muscle regeneration. Dev Cell 46 (2):135–143
14. Joe AW, Yi L, Natarajan A, Le Grand F, So L, Wang J et al (2010) Muscle injury activates resident fibro/adipogenic progenitors that facilitate myogenesis. Nat Cell Biol 12 (2):153–163
15. Uezumi A, Fukada S, Yamamoto N, Takeda S, Tsuchida K (2010) Mesenchymal progenitors distinct from satellite cells contribute to ectopic fat cell formation in skeletal muscle. Nat Cell Biol 12(2):143–152
16. Wosczyna MN, Konishi CT, Perez Carbajal EE, Wang TT, Walsh RA, Gan Q et al (2019) Mesenchymal stromal cells are required for regeneration and homeostatic maintenance of skeletal muscle. Cell Rep 27(7):2029–2035. e5
17. Lemos DR, Babaeijandaghi F, Low M, Chang CK, Lee ST, Fiore D et al (2015) Nilotinib reduces muscle fibrosis in chronic muscle injury by promoting TNF-mediated apoptosis of fibro/adipogenic progenitors. Nat Med 21 (7):786–794
18. Murray IR, Gonzalez ZN, Baily J, Dobie R, Wallace RJ, Mackinnon AC et al (2017) Alphav integrins on mesenchymal cells regulate skeletal and cardiac muscle fibrosis. Nat Commun 8 (1):1118
19. Contreras O, Cruz-Soca M, Theret M, Soliman H, Tung LW, Groppa E et al (2019) Cross-talk between TGF-beta and PDGFRalpha signaling pathways regulates the fate of stromal fibro-adipogenic progenitors. J Cell Sci 132(19)
20. Biferali B, Proietti D, Mozzetta C, Madaro L (2019) Fibro-adipogenic progenitors cross-talk in skeletal muscle: the social network. Front Physiol 10:1074
21. Contreras O, Rebolledo DL, Oyarzun JE, Olguin HC, Brandan E (2016) Connective tissue cells expressing fibro/adipogenic progenitor markers increase under chronic damage: relevance in fibroblast-myofibroblast differentiation and skeletal muscle fibrosis. Cell Tissue Res 364(3):647–660
22. Gonzalez D, Contreras O, Rebolledo DL, Espinoza JP, van Zundert B, Brandan E (2017) ALS skeletal muscle shows enhanced TGF-beta signaling, fibrosis and induction of fibro/adipogenic progenitor markers. PLoS One 12(5):e0177649
23. Pessina P, Cabrera D, Morales MG, Riquelme CA, Gutierrez J, Serrano AL et al (2014) Novel and optimized strategies for inducing fibrosis in vivo: focus on Duchenne muscular dystrophy. Skelet Muscle 4:7
24. Belhasan DC, Akaaboune M (2020) The role of the dystrophin glycoprotein complex on the neuromuscular system. Neurosci Lett 722:134833
25. Dumont NA, Wang YX, von Maltzahn J, Pasut A, Bentzinger CF, Brun CE et al (2015) Dystrophin expression in muscle stem cells regulates their polarity and asymmetric division. Nat Med 21(12):1455–1463
26. Chang NC, Sincennes MC, Chevalier FP, Brun CE, Lacaria M, Segales J et al (2018) The dystrophin glycoprotein complex regulates the epigenetic activation of muscle stem cell commitment. Cell Stem Cell 22(5):755–768. e6
27. Mahdy MAA (2019) Skeletal muscle fibrosis: an overview. Cell Tissue Res 375(3):575–588
28. Smith LR, Barton ER (2018) Regulation of fibrosis in muscular dystrophy. Matrix Biol 68–69:602–615
29. Salimena MC, Lagrota-Candido J, Quirico-Santos T (2004) Gender dimorphism influences extracellular matrix expression and regeneration of muscular tissue in mdx dystrophic mice. Histochem Cell Biol 122 (5):435–444
30. Stedman HH, Sweeney HL, Shrager JB, Maguire HC, Panettieri RA, Petrof B et al (1991) The mdx mouse diaphragm reproduces the degenerative changes of Duchenne muscular dystrophy. Nature 352(6335):536–539
31. Fukada S, Morikawa D, Yamamoto Y, Yoshida T, Sumie N, Yamaguchi M et al (2010) Genetic background affects properties of satellite cells and mdx phenotypes. Am J Pathol 176(5):2414–2424
32. van Putten M, Putker K, Overzier M, Adamzek WA, Pasteuning-Vuhman S, Plomp JJ et al (2019) Natural disease history of the D2-mdx mouse model for Duchenne muscular dystrophy. FASEB J 33(7):8110–8124
33. McGreevy JW, Hakim CH, McIntosh MA, Duan D (2015) Animal models of Duchenne muscular dystrophy: from basic mechanisms to gene therapy. Dis Model Mech 8(3):195–213
34. Rodrigues M, Echigoya Y, Fukada SI, Yokota T (2016) Current translational research and murine models for Duchenne muscular dystrophy. J Neuromuscul Dis. 3(1):29–48
35. Yucel N, Chang AC, Day JW, Rosenthal N, Blau HM (2018) Humanizing the mdx mouse model of DMD: the long and the short of it. NPJ Regen Med 3:4
36. Zhou L, Rafael-Fortney JA, Huang P, Zhao XS, Cheng G, Zhou X et al (2008) Haploinsufficiency of utrophin gene worsens skeletal muscle inflammation and fibrosis in mdx mice. J Neurol Sci 264(1–2):106–111

37. Gutpell KM, Hrinivich WT, Hoffman LM (2015) Skeletal muscle fibrosis in the mdx/utrn+/− mouse validates its suitability as a murine model of Duchenne muscular dystrophy. PLoS One 10(1):e0117306
38. Chandrasekharan K, Yoon JH, Xu Y, de Vries S, Camboni M, Janssen PM et al (2010) A human-specific deletion in mouse Cmah increases disease severity in the mdx model of Duchenne muscular dystrophy. Sci Transl Med 2(42):42ra54
39. Sacco A, Mourkioti F, Tran R, Choi J, Llewellyn M, Kraft P et al (2010) Short telomeres and stem cell exhaustion model Duchenne muscular dystrophy in mdx/mTR mice. Cell 143(7):1059–1071
40. Mazala DA, Novak JS, Hogarth MW, Nearing M, Adusumalli P, Tully CB et al (2020) TGF-beta-driven muscle degeneration and failed regeneration underlie disease onset in a DMD mouse model. JCI Insight 5(6)
41. Brussee V, Tardif F, Tremblay JP (1997) Muscle fibers of mdx mice are more vulnerable to exercise than those of normal mice. Neuromuscul Disord 7(8):487–492
42. Vilquin JT, Brussee V, Asselin I, Kinoshita I, Gingras M, Tremblay JP (1998) Evidence of mdx mouse skeletal muscle fragility in vivo by eccentric running exercise. Muscle Nerve 21 (5):567–576
43. Mathur S, Vohra RS, Germain SA, Forbes S, Bryant ND, Vandenborne K et al (2011) Changes in muscle T2 and tissue damage after downhill running in mdx mice. Muscle Nerve 43(6):878–886
44. Lopez-De Leon A, Rojkind M (1985) A simple micromethod for collagen and total protein determination in formalin-fixed paraffin-embedded sections. J Histochem Cytochem 33(8):737–743
45. Gordish-Dressman H, Willmann R, Dalle Pazze L, Kreibich A, van Putten M, Heydemann A et al (2018) "Of mice and measures": a project to improve how we advance Duchenne muscular dystrophy therapies to the clinic. J Neuromuscul Dis 5(4):407–417
46. Hyzewicz J, Ruegg UT, Takeda S (2015) Comparison of experimental protocols of physical exercise for mdx mice and Duchenne muscular dystrophy patients. J Neuromuscul Dis 2 (4):325–342
47. Marqueste T, Giannesini B, Fur YL, Cozzone PJ, Bendahan D (2008) Comparative MRI analysis of T2 changes associated with single and repeated bouts of downhill running leading to eccentric-induced muscle damage. J Appl Physiol (1985) 105(1):299–307
48. Zelikovich AS, Quattrocelli M, Salamone IM, Kuntz NL, McNally EM (2019) Moderate exercise improves function and increases adiponectin in the mdx mouse model of muscular dystrophy. Sci Rep 9(1):5770
49. Aartsma-Rus A, van Putten M (2014) Assessing functional performance in the mdx mouse model. J Vis Exp 85

Chapter 25

Mouse Models of Skin Fibrosis

Aleix Rius Rigau, Markus Luber, and Jörg H. W. Distler

Abstract

Systemic sclerosis (SSc) is a rare systemic autoimmune disease associated with a high mortality. The first histopathological hallmarks are vasculopathy and inflammation, followed by fibrosis of the skin and internal organs. The molecular and cellular mechanisms are incompletely understood. Rodent models provide important insights into the pathogenesis of SSc and are a mainstay for the development of novel targeted therapies. Here we describe the mechanistic insights of inducible and genetic models, and also discuss in detail the limitations and pitfalls of the most frequently used SSc mouse models. We also describe protocols for running the established bleomycin-induced scleroderma skin fibrosis model.

Key words Bleomycin, Skin fibrosis, Systemic sclerosis, Chronic graft-versus-host disease, Mouse model

1 Introduction

Systemic sclerosis (SSc), also known as systemic scleroderma, is a complex and heterogeneous disease. The limited-cutaneous form affects mainly the skin, whereas the diffuse-cutaneous form often involves internal organs. The main hallmarks of this disease are vasculopathy, activation of immune system, autoimmunity, and fibrosis [1]. The molecular pathogenesis is complex and still incompletely understood. However, cytokines and growth factors such as TGF-β, PDGF, WNT [2], and Hedgehog [3, 4] as well as transcriptions factors (SMADs, AP1 family members such as FRA2 [5], ATF3 [6], and ETS family members such as PU.1 and FLI1 [7]) have been identified as key players in the pathogenesis of fibrotic tissue remodeling.

The earliest events in the disease are vascular injury and endothelial cell apoptosis. The damaged vessels are surrounded by perivascular infiltrates, consisting of T and B lymphocytes, macrophages, and mast cells. All these perivascular leucocytes release profibrotic mediators. This promotes fibroblast-to-

Boris Hinz and David Lagares (eds.), *Myofibroblasts: Methods and Protocols*, Methods in Molecular Biology, vol. 2299, https://doi.org/10.1007/978-1-0716-1382-5_25,

myofibroblast differentiation and accumulation of extracellular matrix (ECM), resulting in tissue fibrosis [8].

In this context, animal models are necessary to better understand the cellular and molecular mechanisms involved in SSc. Many murine models are available to study this disease; however, none of them mimics the whole spectrum of histopathological features. A single model is also not able to recapitulate the heterogeneity of SSc with its broad spectrum of different clinical manifestations.

Here, we discuss the pathogenic drivers of the most commonly used mouse models to study SSc, their advantages, disadvantages, and limitations. We are thereby aiming to provide a guideline for the optimal use of mouse models in SSc research.

1.1 Bleomycin-Induced Skin Fibrosis

Bleomycin was originally isolated from *Streptomyces verticillus* and used in cancer treatment [9]. Higher doses of bleomycin were frequently associated with lung injury and subsequent fibrosis, and these side effects limit its therapeutic application [10]. Cases of skin fibrosis as an adverse event of bleomycin treatment were also reported. These tissues express the lowest levels of bleomycin-hydrolase compared to other organs, being thus particularly sensitive to bleomycin-induced injury. For this reason, bleomycin is used in murine models to induce skin and lung fibrosis.

The bleomycin model of skin fibrosis was established in 1999 by Yamamoto and coworkers, and nowadays is one of the most frequently used animal models in SSc research. Fibrosis is induced by repeated subcutaneous injections of bleomycin in a defined skin area for 3-4 weeks, leading to fibrosis that persists for appr. 6 weeks after the last injection [11]. Similarly, intratracheal administration of bleomycin leads to lung fibrosis, showing the strongest phenotype around 4 weeks after one single injection [12].

Interestingly, there is variation in the sensitivity to bleomycin among different mouse strains. For example, C3H/He, DBA/2, B10.D2, and B10.A strains develop more intense fibrosis compared to C57BL/6 [13]. A study showed a correlation between the bleomycin-hydrolase activity and the bleomycin-induced lung fibrosis phenotype in two different mouse strains [14].

Bleomycin-induced skin fibrosis mimics the early inflammatory changes that occur in SSc patients (*see* **Note 1**). Bleomycin injections in the skin induce the release of reactive oxygen species (ROS) and promote the expression of adhesion molecules and influx of immune cells as T and B lymphocytes, macrophages, eosinophils, and mast cells [8]. The inflammatory infiltrates release pro-fibrotic and pro-inflammatory cytokines, growth factors, and chemokines like TGF-β, IL-4, IL-6, IL-13, PDGF, and MCP-1 with subsequent accumulation of myofibroblasts and excessive ECM deposition [8, 15]. This model is also characterized by the occurrence of antinuclear antibodies (ANA), anti-topoisomerase-I, anti-U1 RNP, and anti-histone antibodies [16]. The levels of these autoantibodies are, however, low and they occur only infrequently with most protocols. Injections of higher doses of bleomycin into the skin may lead to plasma concentrations that are high enough to induce pulmonary fibrosis. The histological features of lung fibrosis in this case yet differ from that induced by intratracheal injection of bleomycin, with more perivascular fibrosis as compared to peribroncheolar damage [11, 17].

1.2 Sclerodermatous Chronic Graft-Versus-Host Disease (SclGvHD) Model

Chronic graft-versus-host disease (cGvHD) is the major complication after allogenic hematopoietic cell transplantation, occurring in 60–80% of long-term survival patients [18], and it decreases the success of transplantation by increasing the risk of death and disability [19]. Sclerodermatous GvHD (SclGvHD) is a form of cGvHD that manifests as severe fibrosis of skin and fascia, and accounts for 20% of all cases with cGvHD [20]. SclGvHD shows many commonalities with the inflammatory phase of diffuse SSc [21].

Jaffe and Claman [22] developed 35 years ago a murine cGvHD model to study human SSc. This model is based on the transplantation of hematopoietic cells with a mismatch in the minor histocompatibility antigens from donor into recipient mice. Although many different models have been described, the most common model is the B10.D2 $^{(H\text{-}2d)}$→BALB/c $^{(H\text{-}2d)}$ model of SclGvHD. In this model, splenocytes and bone marrow cells of B10.D2 are isolated and injected into the tail veins of sublethally irradiated Balb/c mice. Other models use, e.g., splenocytes from LP/J mice and C57BL/6 as recipient mice (LP/J $^{(H\text{-}2b)}$→C57/Bl6 $^{(H\text{-}2b)}$) [22, 23].

This model demonstrates severe inflammatory infiltrates composed mainly of T cells, but also monocytes/macrophages, B cells, and mast cells [8, 24]. The leukocytes stimulate resident fibroblast to release large amounts of ECM leading to tissue fibrosis. The fibrotic injury is most severe in the skin, but most models also develop mild-to-moderate pulmonary fibrosis and may develop intestinal and kidney fibrosis [25].

Alloreactive T cells play a key role in the pathogenesis of this model, but the infiltrates are heterogeneous, and different T cell population may synergize to drive disease in these models [26]. Th1 differentiation was long considered to represent the paradigm of T cell response in cGvHD, but recent studies have also shown the implication of Th17 cells [27]. Moreover, although donor-derived B cell antibodies are not essential to initiate the tissue injury, they promote cutaneous Th17 infiltration and perpetuate the SclGvHD [28].

In the B10.D2 $^{(H\text{-}2d)}$→BALB/c $^{(H\text{-}2d)}$ model, first manifestations of SclGvHD are detectable at around 3 weeks after transplantation, and they are characterized by loss of dermal fat, hair follicle destruction, mononuclear cell infiltration, and increased collagen deposition [29]. Systemic manifestations such as weight loss also occur [29, 30].

1.3 Fibrosis Induced by Recombinant DNA Topoisomerase I and Complete Freund's Adjuvant

Autoimmunity is a key feature of SSc. Moreover, the presence of autoantibodies against multiple intracellular antigens is a serological hallmark of the disease. These antibodies are helpful biomarkers and can be found in more than 95% of the patients [31]. One of these autoantibodies is the anti-topoisomerase I, which is associated with diffuse-cutaneous SSc and a poorer prognosis. SSc patients with anti-topoisomerase I antibodies have higher risk of pulmonary fibrosis and cardiac involvement [32].

Yoshizaki et al. have generated a model, in which subcutaneous administration of DNA topoisomerase I in Complete Freund's adjuvant (CFA) induces dermal and lung fibrosis within 8 weeks [33].

The pathogenesis of this model is not completely understood, but apart from the autoimmunity, increased levels of IL-6, IL-17, and TGF-β and decreased IL-10 production were observed along with Th2 and Th17 T cell differentiation [33].

1.4 Tight Skin Mice 1 (Tsk1) Mice

The Tight skin 1 (Tsk1) phenotype was first described in the mouse strain B10.D2(58N)/Sn in 1976 [34]. However, the underlying genetic mutation was not identified until 1996 [35]. The Tsk1 phenotype is caused by an autosomal dominant mutation on chromosome 2 that consists an in-frame tandem partial replication of the fibrillin 1 gene (*Fbn1*). Homozygous embryos degenerate *in utero* at 8–10 days of gestation. Heterozygous mice develop SSc-like features with skin fibrosis and autoimmunity [36].

The mechanisms causing the skin fibrosis in the Tsk1 mice are not clear, but abnormal fibrillin might activate the TGF-β signaling and thereby promote myofibroblast differentiation [37]. These mice demonstrate thickening of the hypodermal layer and fascial layers without striking thickening of the dermis. The lung manifestations also differ from those of human SSc patients: Tsk1 mice develop lung emphysema rather than lung fibrosis [38].

Tsk1 mice also develop autoantibodies, in particular against fibrillin, which is also observed in a subset of SSc patients [31, 39, 40]. Of particular interest, fibroblasts explanted from Tsk1 mice demonstrate a persistently activated phenotype in vitro, even after several passages. Tsk1 fibroblasts thus resemble the activated phenotype of SSc fibroblasts.

The Tsk1 mouse strain is used as a model for the later stages of human SSc, when inflammation has largely subsided in the majority of the patients.

1.5 Fra2-Transgenic Mice

Fra2 (Fos related antigen 2) is a transcription factor that belongs to the Activator Protein 1 (AP1) family. Fra2 is implicated in cell proliferation, inflammation, wound healing [5], osteoblast differentiation, and collagen production [41]. Epigenetics alterations at the Fra2 promoter have been described in SSc patients [42].

The group of Erwin Wagner developed a transgenic mouse strain where the *Fra2* expression is under the control of the ubiquitous MHC-I antigen H-2kb promotor. This strain develops systemic inflammation and fibrosis in multiple organs, including the skin, the lungs, and the heart [43, 44]. One important characteristic of the *Fra2*-transgenic mice is the vascular phenotype: apoptosis of microvascular endothelial cells is observed at an age of 9 weeks, but is not present at birth. Fibrosis of the lungs and skin follows at an age of 12 weeks [45]. Mice also demonstrate proliferation of the vascular smooth muscle cells with subsequent development of pulmonary arterial hypertension.

This model is particularly useful to study the complex interplay between different disease features in SSc, e.g., between vasculopathy and fibrosis.

1.6 Other Genetic Models

In 1986, Peters et al. described the tight skin 2 (Tsk2) mouse model. In this strain, the mutation was induced by a chemical compound (ethylnitrosourea) and consists of an autosomal dominant mutation on chromosome 1 [46], specifically a deletion in the Col3a1 gene that results in a gain-of-function mutation [47]. Similar to Tsk1 mice, homozygous embryos degenerate *in utero*. The cutaneous changes in the heterozygous mice become apparent at 3–4 weeks of age with thickening of the dermis and excessive deposition of thick collagen fibers, which extended deeply into the subdermal adipose tissue [46]. Cultured fibroblasts isolated from Tsk2 express increased levels of Col1a1 and Col3a1 mRNA [48].

TGF-β is a key player of fibroblast activation and ECM production in SSc. For this reason, Sonnylal and coworkers inserted a STOP cassette flanked by LoxP sites /in front of the $TBRI^{CA}$ mouse gene. When cross-breed with mice expressing a proα2 (I) collagen promoter-dependent Cre-ER, these mice express active TGF-β receptor type I (TGFβRI) selectively in fibroblasts upon tamoxifen-challenge [49].

Hydroxyltamoxifen injection leads to selective excision of the stop cassette and TGFβRI expression in fibroblasts. These mice then develop progressive and generalized dermal fibrosis. This model does not develop pulmonary fibrosis; however thickening of the vascular wall in arteries of the kidneys, lungs, and adrenal glands have been reported [49].

This mouse model is very useful to study TGF-β signaling *in vivo*, but do not recapitulate the complex interplay of different profibrotic pathways that are simultaneously activated in SSc patients [8]. Other features of this disease such as autoimmunity and inflammation are absent in this model.

Alternative to fibroblast-specific Cre-mediated overexpression of $TBRI^{CA}$, TGFβRI overexpression can also be achieved using vectors such as adenoviruses. This approach has been extensively used to study the TGF-β signaling in skin fibrosis in vivo [50–54].

Denton et al. also developed a mouse strain with aberrant TGFβRII [55]. Using the same transcriptional enhancer of the proα2(I) collagen gene as in the TBR^{CA};CreER model, they generated a mouse strain, in which the fibroblasts express mutant type II TGF-β receptor that lacks the intracellular kinase domain (TβRIIΔk). While overexpression of TβRIIΔk inhibits TGFβ signaling in fibroblasts in vitro [56], TβRIIΔk mice demonstrate enhanced TGFβ signaling [55]. The mechanisms of this paradoxical TGF-β activation are not fully understood.

TβRIIΔk mice develop fibrosis of the lungs, the skin, the heart, and the intestine. Vascular fibrosis is also observed. Newborn TβRIIΔk transgenic mice appear normal, but from 6 weeks after birth approximately 25% of the animals failed to thrive and lose weight. Contrasting with the sporadic lung abnormalities, all transgenic mice develop increased dermal thickness at 12 weeks of age [55].

Development of skin fibrosis is the only common characteristic present in all SSc models outlined above. Indeed, no mouse model mimics all features of SSc. Moreover, a single model alone is not sufficient to capture the heterogeneity of SSc with a wide range of different clinical manifestations in different patient subgroups.

In summary, a perfect moue model for SSc does not exist. However, a careful combination of different mouse models enables modeling of different aspects, distinct stages, and various subgroups of SSc patients. The pipeline for preclinical development of novel therapeutic compounds should not be restricted to a single mouse model, but should rather include different models in conjunction with studies on human cells and tissues.

In the next lines, we will describe in detail the material and reagents needed as well as the protocols of the most used SSc animal models: bleomycin-induced skin fibrosis.

2 Materials

2.1 Bleomycin-Induced Skin Fibrosis

1. DBA mice (*see* **Notes 2** and **3**), 6 week old (*see* **Note 4**), same sex.
2. Bleomycin, 0.5 mg/mL in 0.9% NaCl.
3. 0.9% NaCl sterile.
4. 27G Needles and 1 mL syringes.
5. Shaver.
6. Skin marker pen.
7. Ketamine chloride, 100 mg/mL and xylazine chloride 2% (or any anesthetic of your choice).
8. Infrared lamp.

3 Methods

3.1 Bleomycin-Induced Skin Fibrosis

1. The animal welfare needs to be respectful of specifically and strict ethical rules (*see* **Note 5**).
2. Prepare the anesthesia by mixing 0.9% NaCl, ketamine chloride, and xylazine chloride in 2:1:1 proportion.
3. Anesthetize the mice injecting intraperitoneal 30μL of the anesthetic mixture prepared.
4. Wait until the mouse falls asleep. Apply 0.9% NaCl to the eyes to prevent blindness.
5. Shave a large area of 2 cm × 2 cm between the shoulders at the back of the mouse.
6. With the skin marker pen, mark a square of 1 cm × 1 cm between the shoulders at the back of the mouse.
7. Inject 100μL of Bleomycin intradermally (use 100μL of 0.9% NaCl for the control mice) (*see* **Note 6**).
8. Prevent excessive cooling down of the mouse during anesthesia by using an infrared lamp.
9. During the next 6 weeks, every other day bleomycin is again injected intradermally at varying injection sites (*see* **Note 7**). For this, an assistant should hold the mouse to prevent moving, so no further anesthesia is required (*see* **Note 8**).
10. If necessary, anesthetize the mice again for reshaving or refreshing the marks on the back (*see* **Note 9**).
11. Sacrifice the mice 4 weeks after the first injection (*see* **Notes 10** and **11**).

3.2 Assessment of Skin Fibrosis

Skin fibrosis can be assessed by histological analysis via H&E, Picrosirius Red, and Masson's Trichrome stainings. Immunohistochemical stainings for fibrosis (collagen type I, fibronectin) or myofibroblast (α SMA) markers are commonly used. The thickness of the dermal and the hypodermal layers are also good surrogate markers of skin fibrosis. The dermal thickness is typically quantified by measuring the distance from the epidermal/dermal border to the dermal/hypodermal border at four random sites per mouse. Biochemical measurement of collagen type I via determination of hydroxyproline levels in skin tissue samples is also a standard method to determine skin fibrosis.

4 Notes

1. Although the bleomycin-induced skin fibrosis model is widely used as a model for scleroderma, it does not capture the

heterogeneity of SSc. The advantages and disadvantages of this model are discussed in the following summary:

(a) Advantages:
- Very well documented and characterized in detail
- Mimics the inflammatory features of SSc, thus, useful to evaluate anti-inflammatory and anti-fibrotic therapies
- Can be applied to numerous mouse strains
- Easy to handle

(b) Disadvantages:
- No systemic effect, since fibrosis is limited to the site of injection
- The vascular phenomena that precedes fibrosis in human scleroderma is usually absent in the bleomycin model
- The anti-fibrotic effects of anti-inflammatory therapies can be overestimated.

2. Depending on the experimental setup and purpose, different mouse strains can be used. It is important to take into account that the sensivity to bleomycin can vary among different mouse strains, so the protocol should be adapt to each strain.
3. The method described above is also well established in C57BL/6 mice, a strain easier to handle than DBA/2.
4. Don't use mice younger than 6 weeks old since they may suffer from increased toxicity.
5. For ethical reasons, during in vivo experiments all scientists must take care about animal welfare. One of the most important rules to follow is the well known 3 R's. This principle consists of: replacement (methods to replace the use of animals), reduction (methods to minimize the number of animals used) and refinement (methods to minimize animals suffering and improve welfare) [57]. Moreover, to avoid animal stress, the experiment must be started at least one week after the animals arrive in the facility.
6. Be sure that the bleomycin is administrated intradermal and not subcutaneous. A wrong administration can affect dramatically the experiment results.
7. To ensure that all the marked area has a homogeneous fibrosis, we recommend dividing it in five small areas: 1) top left; 1) top right; 3) bottom right; 4) bottom left; 5) center. After the first

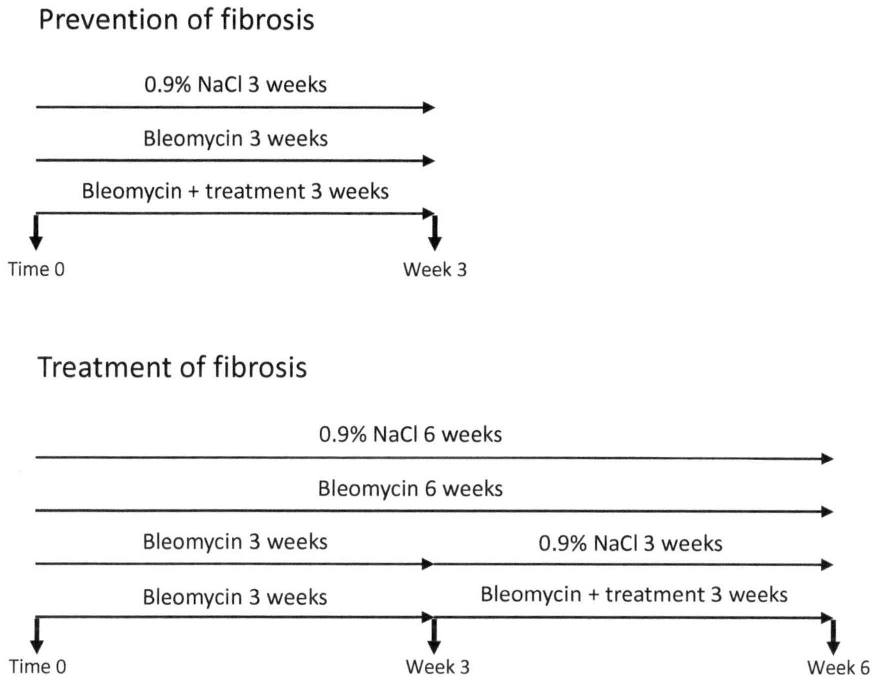

Fig. 1 Protocols for testing of drug candidates in bleomycin-induced skin fibrosis

injection in position 1, the mice will be injected every other day at the next position, until the circle restarts every 10 days.

8. To reduce the amount of anesthesia for the mice, the bleomycin injections can be performed with the help of an assistant. He should hold the tail and the ears of the animal to leave the back region, allowing a proper injection procedure.
9. Reshave the mouse back (under anesthesia) and remark the injection field as often as needed to inject the bleomycin always in the same skin area.
10. To test therapeutic compounds for the treatment of skin fibrosis, different protocols can be used depending if the effects of the drug should be tested in a preventive or therapeutic approach (see Fig. 1).

(a) Preventive model: Run the animal modelfor 3 weeks with the following three groups:

- Healthy control (injection with 0.9% NaCl).
- Fibrotic control (injection with bleomycin).
- Test group (injection with bleomycin + treatment)

(b)Therapeutic model: Run the animal model6 weeks, with the following four groups:

- Healthy control (injection with 0.9% NaCl).

- Second control (injection with Bleomycinin the first 3 weeks and with 0.9% NaCl for the last 3 weeks).
- Fibrotic control (injection with bleomycinfor 6 weeks).
- Test group (injection with bleomycinfor 6 weeks, and add the treatment the last 3 weeks).

11. Many techniques can be applied to study the fibrotic phenotype. Those include studies on transcriptional changes using qPCR or RNA sequencing...), protein analysis, e.g., with Western blot, histology (from both, paraffin embedded tissue and cryosections), and cell isolation for flow cytometry or cell culture, as well as peripheral blood tests.

    The mice can be sacrificed by cervical dislocation under narcosis (CO_2 or isoflurane). Once the mice are dead, the hair of the fibrotic skin area can be removed with a depilatory cream and the treated 1 cm × 1 cm skin area can be removed and put in a petri dish (with the epidermis to the top). Depending on the readouts that you plan to perform, a specific processing of the tissue may be needed.

    (a) Genetic analysis: A 3 mm punch is usually enough to extract enough mRNA for qPCR or RNAseq.

    (b) Histology: Ideally, a larger area, e.g., samples of 1 cm × 0.3 cm should be collected.

    (c) Flow cytometry: It is important to obtain a high-quality single-cell suspension from the enzymatic digestion. Depending on the localization of the cells of interest you should remove the fat layer or do a dispase epidermal digestion. It is also necessary to filter the samples (70μm restrainer minimum) and add some debris removal step to avoid blocking of the FACS device. If you want to study small cell, it may be necessary to pool samples from several mice. The same digesting protocol can be used to culture the cell population of interest.

    (d) Protein: Homogenize at least a 3 mm punch with the appropriate protein lysis buffer.

References

1. Gabrielli A, Avvedimento EV, Krieg T (2009) Scleroderma. N Engl J Med 360 (19):1989–2003
2. Distler JHW, Gyorfi AH, Ramanujam M, Whitfield ML, Konigshoff M, Lafyatis R (2019) Shared and distinct mechanisms of fibrosis. Nat Rev Rheumatol 15(12):705–730
3. Liang R, Sumova B, Cordazzo C, Mallano T, Zhang Y, Wohlfahrt T et al (2017) The transcription factor GLI2 as a downstream mediator of transforming growth factor-beta-

induced fibroblast activation in SSc. Ann Rheum Dis 76(4):756–764

4. Liang R, Kagwiria R, Zehender A, Dees C, Bergmann C, Ramming A et al (2019) Acyl-transferase skinny hedgehog regulates TGFbeta-dependent fibroblast activation in SSc. Ann Rheum Dis 78(9):1269–1273
5. Wagner EF, Eferl R (2005) Fos/AP-1 proteins in bone and the immune system. Immunol Rev 208:126–140
6. Mallano T, Palumbo-Zerr K, Zerr P, Ramming A, Zeller B, Beyer C et al (2016) Activating transcription factor 3 regulates canonical TGFbeta signalling in systemic sclerosis. Ann Rheum Dis 75(3):586–592
7. Wohlfahrt T, Rauber S, Uebe S, Luber M, Soare A, Ekici A et al (2019) PU.1 controls fibroblast polarization and tissue fibrosis. Nature 566(7744):344–349
8. Beyer C, Schett G, Distler O, Distler JH (2010) Animal models of systemic sclerosis: prospects and limitations. Arthritis Rheum 62 (10):2831–2844
9. Umezawa H, Ishizuka M, Maeda K, Takeuchi T (1967) Studies on bleomycin. Cancer 20 (5):891–895
10. Muggia FM, Louie AC, Sikic BI (1983) Pulmonary toxicity of antitumor agents. Cancer Treat Rev 10(4):221–243
11. Yamamoto T, Takagawa S, Katayama I, Yamazaki K, Hamazaki Y, Shinkai H et al (1999) Animal model of sclerotic skin. I: local injections of bleomycin induce sclerotic skin mimicking scleroderma. J Invest Dermatol 112(4):456–462
12. Moore BB, Hogaboam CM (2008) Murine models of pulmonary fibrosis. Am J Physiol Lung Cell Mol Physiol 294(2):L152–L160
13. Yamamoto T, Kuroda M, Nishioka K (2000) Animal model of sclerotic skin. III: histopathological comparison of bleomycin-induced scleroderma in various mice strains. Arch Dermatol Res 292(11):535–541
14. Haston CK, Wang M, Dejournett RE, Zhou X, Ni D, Gu X et al (2002) Bleomycin hydrolase and a genetic locus within the MHC affect risk for pulmonary fibrosis in mice. Hum Mol Genet 11(16):1855–1863
15. Yamamoto T, Nishioka K (2005) Cellular and molecular mechanisms of bleomycin-induced murine scleroderma: current update and future perspective. Exp Dermatol 14(2):81–95
16. Yoshizaki A, Iwata Y, Komura K, Ogawa F, Hara T, Muroi E et al (2008) CD19 regulates skin and lung fibrosis via Toll-like receptor signaling in a model of bleomycin-induced scleroderma. Am J Pathol 172(6):1650–1663
17. Harrison JH Jr, Lazo JS (1987) High dose continuous infusion of bleomycin in mice: a new model for drug-induced pulmonary fibrosis. J Pharmacol Exp Ther 243(3):1185–1194
18. Ratanatharathorn V, Ayash L, Lazarus HM, Fu J, Uberti JP (2001) Chronic graft-versus-host disease: clinical manifestation and therapy. Bone Marrow Transplant 28(2):121–129
19. Lee SJ (2017) Classification systems for chronic graft-versus-host disease. Blood 129 (1):30–37
20. Arai S, Pidala J, Pusic I, Chai X, Jaglowski S, Khera N et al (2016) A randomized phase II crossover study of imatinib or rituximab for cutaneous sclerosis after hematopoietic cell transplantation. Clin Cancer Res 22 (2):319–327
21. Strong Rodrigues K, Oliveira-Ribeiro C, de Abreu Fiuza Gomes S, Knobler R (2018) Cutaneous graft-versus-host disease: diagnosis and treatment. Am J Clin Dermatol 19(1):33–50
22. Jaffee BD, Claman HN (1983) Chronic graft-versus-host disease (GVHD) as a model for scleroderma, I. Description of model systems. Cell Immunol 77(1):1–12
23. Rogai V, Lories RJ, Guiducci S, Luyten FP, Matucci Cerinic M (2008) Animal models in systemic sclerosis. Clin Exp Rheumatol 26 (5):941–946
24. Artlett CM (2014) Animal models of systemic sclerosis: their utility and limitations. Open Access Rheumatol 6:65–81
25. Zhang Y, McCormick LL, Desai SR, Wu C, Gilliam AC (2002) Murine sclerodermatous graft-versus-host disease, a model for human scleroderma: cutaneous cytokines, chemokines, and immune cell activation. J Immunol 168(6):3088–3098
26. Santos ESP, Bennett CL, Chakraverty R (2018) Unraveling the mechanisms of cutaneous graft-versus-host disease. Front Immunol 9:963
27. Forcade E, Paz K, Flynn R, Griesenauer B, Amet T, Li W et al (2017) An activated Th17-prone T cell subset involved in chronic graft-versus-host disease sensitive to pharmacological inhibition. JCI Insight 2(12)

28. Jin H, Ni X, Deng R, Song Q, Young J, Cassady K et al (2016) Antibodies from donor B cells perpetuate cutaneous chronic graft-versus-host disease in mice. Blood 127 (18):2249–2260
29. Chu YW, Gress RE (2008) Murine models of chronic graft-versus-host disease: insights and unresolved issues. Biol Blood Marrow Transplant 14(4):365–378
30. Levy S, Nagler A, Okon S, Marmary Y (2000) Parotid salivary gland dysfunction in chronic graft-versus-host disease (cGVHD): a longitudinal study in a mouse model. Bone Marrow Transplant 25(10):1073–1078
31. Kayser C, Fritzler MJ (2015) Autoantibodies in systemic sclerosis: unanswered questions. Front Immunol 6:167
32. Mehra S, Walker J, Patterson K, Fritzler MJ (2013) Autoantibodies in systemic sclerosis. Autoimmun Rev 12(3):340–354
33. Yoshizaki A, Yanaba K, Ogawa A, Asano Y, Kadono T, Sato S (2011) Immunization with DNA topoisomerase I and Freund's complete adjuvant induces skin and lung fibrosis and autoimmunity via interleukin-6 signaling. Arthritis Rheum 63(11):3575–3585
34. Green MC, Sweet HO, Bunker LE (1976) Tight-skin, a new mutation of the mouse causing excessive growth of connective tissue and skeleton. Am J Pathol 82(3):493–512
35. Siracusa LD, McGrath R, Ma Q, Moskow JJ, Manne J, Christner PJ et al (1996) A tandem duplication within the fibrillin 1 gene is associated with the mouse tight skin mutation. Genome Res 6(4):300–313
36. Jimenez SA, Millan A, Bashey RI (1984) Scleroderma-like alterations in collagen metabolism occurring in the TSK (tight skin) mouse. Arthritis Rheum 27(2):180–185
37. Isogai Z, Ono RN, Ushiro S, Keene DR, Chen Y, Mazzieri R et al (2003) Latent transforming growth factor beta-binding protein 1 interacts with fibrillin and is a microfibril-associated protein. J Biol Chem 278 (4):2750–2757
38. Baxter RM, Crowell TP, McCrann ME, Frew EM, Gardner H (2005) Analysis of the tight skin (Tskl/+) mouse as a model for testing antifibrotic agents. Lab Invest 85 (10):1199–1209
39. Bocchieri MH, Henriksen PD, Kasturi KN, Muryoi T, Bona CA, Jimenez SA (1991) Evidence for autoimmunity in the tight skin mouse model of systemic sclerosis. Arthritis Rheum 34(5):599–605
40. Shibata S, Muryoi T, Saitoh Y, Brumeanu TD, Bona CA, Kasturi KN (1993) Immunochemical and molecular characterization of anti-RNA polymerase I autoantibodies produced by tight skin mouse. J Clin Invest 92(2):984–992
41. Bozec A, Bakiri L, Jimenez M, Schinke T, Amling M, Wagner EF (2010) Fra-2/AP-1 controls bone formation by regulating osteoblast differentiation and collagen production. J Cell Biol 190(6):1093–1106
42. Bergmann C, Brandt A, Merlevede B, Hallenberger L, Dees C, Wohlfahrt T et al (2018) The histone demethylase Jumonji domain-containing protein 3 (JMJD3) regulates fibroblast activation in systemic sclerosis. Ann Rheum Dis 77(1):150–158
43. Eferl R, Hasselblatt P, Rath M, Popper H, Zenz R, Komnenovic V et al (2008) Development of pulmonary fibrosis through a pathway involving the transcription factor Fra-2/AP-1. Proc Natl Acad Sci U S A 105 (30):10,525–10,530
44. Venalis P, Kumanovics G, Schulze-Koops H, Distler A, Dees C, Zerr P et al (2015) Cardiomyopathy in murine models of systemic sclerosis. Arthritis Rheumatol 67(2):508–516
45. Maurer B, Busch N, Jungel A, Pileckyte M, Gay RE, Michel BA et al (2009) Transcription factor fos-related antigen-2 induces progressive peripheral vasculopathy in mice closely resembling human systemic sclerosis. Circulation 120(23):2367–2376
46. Christner PJ, Peters J, Hawkins D, Siracusa LD, Jimenez SA (1995) The tight skin 2 mouse. An animal model of scleroderma displaying cutaneous fibrosis and mononuclear cell infiltration. Arthritis Rheum 38 (12):1791–1798
47. Long KB, Li Z, Burgwin CM, Choe SG, Martyanov V, Sassi-Gaha S et al (2015) The Tsk2/+ mouse fibrotic phenotype is due to a gain-of-function mutation in the PIIINP segment of the Col3a1 gene. J Invest Dermatol 135(3):718–727
48. Christner PJ, Hitraya EG, Peters J, McGrath R, Jimenez SA (1998) Transcriptional activation of the alpha1(I) procollagen gene and up-regulation of alpha1(I) and alpha1(III) procollagen messenger RNA in dermal fibroblasts from tight skin 2 mice. Arthritis Rheum 41 (12):2132–2142

49. Sonnylal S, Denton CP, Zheng B, Keene DR, He R, Adams HP et al (2007) Postnatal induction of transforming growth factor beta signaling in fibroblasts of mice recapitulates clinical, histologic, and biochemical features of scleroderma. Arthritis Rheum 56(1):334–344
50. Beyer C, Zenzmaier C, Palumbo-Zerr K, Mancuso R, Distler A, Dees C et al (2015) Stimulation of the soluble guanylate cyclase (sGC) inhibits fibrosis by blocking non-canonical TGFbeta signalling. Ann Rheum Dis 74(7):1408–1416
51. Palumbo-Zerr K, Zerr P, Distler A, Fliehr J, Mancuso R, Huang J et al (2015) Orphan nuclear receptor NR4A1 regulates transforming growth factor-beta signaling and fibrosis. Nat Med 21(2):150–158
52. Zhang Y, Dees C, Beyer C, Lin NY, Distler A, Zerr P et al (2015) Inhibition of casein kinase II reduces TGFbeta induced fibroblast activation and ameliorates experimental fibrosis. Ann Rheum Dis 74(5):936–943
53. Akhmetshina A, Palumbo K, Dees C, Bergmann C, Venalis P, Zerr P et al (2012) Activation of canonical Wnt signalling is required for TGF-beta-mediated fibrosis. Nat Commun 3:735
54. Chakraborty D, Sumova B, Mallano T, Chen CW, Distler A, Bergmann C et al (2017) Activation of STAT3 integrates common profibrotic pathways to promote fibroblast activation and tissue fibrosis. Nat Commun 8(1):1130
55. Denton CP, Zheng B, Evans LA, Shi-wen X, Ong VH, Fisher I et al (2003) Fibroblast-specific expression of a kinase-deficient type II transforming growth factor beta (TGFbeta) receptor leads to paradoxical activation of TGFbeta signaling pathways with fibrosis in transgenic mice. J Biol Chem 278 (27):25109–25119
56. Brand T, MacLellan WR (1993) Schneider MD. A dominant-negative receptor for type beta transforming growth factors created by deletion of the kinase domain. J Biol Chem 268(16):11500–11503
57. MacArthur Clark J (2018) The 3Rs in research: a contemporary approach to replacement, reduction and refinement. Br J Nutr 120(s1): S1–S7

Chapter 26

Mouse Models of Intestinal Fibrosis

Jiannan Li, Dina Dejanovic, Megan T. Zangara, Jyotsna Chandra, Christine McDonald, and Florian Rieder

Abstract

Mouse models are essential for investigation of underlying disease mechanisms that drive intestinal fibrosis, as well as assessment of potential therapeutic approaches to either prevent or resolve fibrosis. Here we describe several common mouse models of intestinal inflammation and fibrosis, including chemically driven colitis models, a bacterially triggered colitis model, and spontaneous intestinal inflammation in genetically susceptible mouse strains. Detailed protocols are provided for dextran sodium sulfate (DSS) colitis, 2,4,6-trinitro-benzene sulfonic acid (TNBS) colitis, adherent-invasive *Escherichia coli* (AIEC)-triggered colitis, the interleukin-10 knockout (IL-10KO) mouse model of spontaneous colitis, and the SAMP/YitFc model of spontaneous ileocolitis.

Key words Mouse models, Dextran sodium sulfate (DSS), 2,4,6-Trinitro-benzene sulfonic acid (TNBS), Adherent-invasive *Escherichia coli* (AIEC)-triggered colitis, IL-10 knockout, SAMP/YitFc

1 Introduction

Over half of Crohn's disease (CD) patients develop disease complications during their disease course that can include fibrosis-induced intestinal obstruction and result in debilitating symptoms [1] and ultimately require surgery [2]. Although less appreciated, intestinal fibrosis also occurs in ulcerative colitis (UC) patients [3]. Short gut syndrome, need for hospitalizations, as well as the tremendous financial burden of fibrosis-associated complications pose significant concerns to these patients and their healthcare providers. No specific anti-fibrotic therapies are currently available, which emphasizes the critical need to obtain a mechanistic understanding of the drivers of intestinal fibrosis. The goal remains to develop novel preventive and effective therapeutic approaches.

Human studies have been limited in this field due to lack of available tissue for in vitro investigations and clinical trials endpoints. Hence, well-characterized animal models of intestinal fibrosis are filling this void to better define and understand the

Boris Hinz and David Lagares (eds.), *Myofibroblasts: Methods and Protocols*, Methods in Molecular Biology, vol. 2299, https://doi.org/10.1007/978-1-0716-1382-5_26,

pathophysiology of fibrosis. Animal models enable us to perform mechanistic studies, and facilitate the testing of novel, pathophysiology-based therapeutic approaches at the same time.

This chapter provides detailed protocols for several frequently used animal models of intestinal fibrosis. These models include chemically induced intestinal inflammation, such as dextran sodium sulfate (DSS) [4–6] and 2,4,6-trinitro-benzene sulfonic acid (TNBS) [7–9], that when applied repetitively, result in the development of intestinal fibrosis. Another model described is chronic inflammation and fibrosis induced in response to colonization with a specific adherent-invasive *Escherichia coli* (AIEC) strain [10, 11]. Finally, genetically driven spontaneous interleukin-10 (IL-10) knockout [12–14] colitis, as well as the SAMP1/YitFc [15] spontaneous ileocolitis model are discussed (Table 1).

Unless otherwise noted, these models are performed with mice of the ages of 6–8 weeks and between 18 and 20 g of weight at the start of the experiments. For all these models, disease severity is strongly influenced by genetic background and animal housing facility. Therefore, if performing these models using mice with a mixed strain background, littermate controls are required for robust interpretation of the results. Additionally, each model will require optimization for each animal facility. Generally, a sample size of 10 mice per group repeated in two independent experiments provides robust statistical power analysis. However, it is recommended that a formal power analysis is performed using the degree of effect and experimental variation expected for the hypothesis being tested. All models require the review and approval of the institutional animal care and use committee (IACUC) and must comply with the local regulations of the animal care facility.

2 Materials

2.1 DSS Colitis Model

1. DSS (30,000–50,000 colitis grade) solution: Prepare 1–10% (w/v) DSS solutions in sterile deionized water in 1% steps and store at room temperature. Confirm that the DSS powder is completely dissolved in water; otherwise it may clog the water bottle sipper tube.
2. Animals: 6–8-week-old male mice of preferred strains. This model can be performed in both male and female mice; however, the estrus cycle influences disease activity in this model and increases the variability of response. DSS induces colitis in a broad range of standard inbred mouse strains (e.g., C57BL/6 [16], BALB/c [17], as well as CD1 mice [18]), but severity of colitis is influenced by genetic background.
3. Hemoccult Sensa Single Slides(Beckman Coulter, Indianapolis, IN, USA).

Table 1
Summary of mouse models

	Summary	Advantages	Disadvantages
DSS colitis model	Multiple cycles of DSS induces chronic colitis and fibrosis	Easy to perform, good reproducibility	Limited human IBD relevance; limited cell types involved
TNBS colitis model	Intrarectal application of hapten TNBS: results in chronic colitis leading to fibrosis	Widely used for studying intestinal fibrosis	Requires susceptible mice strain to induce colitis; great variability
AIEC colitis model	Oral gavaging of AIEC induces gut inflammation	Mimics the inflammation pattern of human CD; Great tool for microbiota research in IBD	Require ABSL2 facility
IL-10 knockout colitis model	Spontaneous chronic intestinal fibrosis model	Spontaneous colitis model, while chemicals and infectious components could be added to accelerate the onset of disease	Disease development is slow
SAMP1/YitFc ileocolitis model	Spontaneous colitis models resemble all stages of human CD	Strain closely mimics all stages of stricture formation in CD patients; symptoms start as early as 4 weeks	Low breeding rate; not commercialized currently

2.2 TNBS Colitis Model

1. Pre-sensitization solution: Prepare pre-sensitization solution freshly on the day of pre-sensitization. Mix acetone and olive oil in a 4:1 volume ratio by vortexing rigorously. Prepare 1% (w/v) TNBS solution: Mix 4 volumes of acetone/olive oil with 1 volume of 5% TNBS solution to obtain 1% (w/v) TNBS. Mix the solution by rigorous vortexing (*see* **Note 1**).
2. Control pre-sensitization solution: prepare acetone and olive oil in a 4:1 volume ratio.
3. TNBS solution for intrarectal administration: Combine ethanol, double-distilled water (ddH_2O), and TNBS to create 0, 0.5, 1, 2, 3, and 4% (w/v) TNBS in 70% ethanol. Ensure mixing through vigorous vortexing. Prepare enough solution to administer 100 μL/mouse with an additional 300 μL volume to account for loss during administration.
4. Vehicle solution for intrarectal administration: 70% ethanol diluted in ddH2O. Prepare enough solution to administer 100 μL/mouse with an additional 300 μL volume to account for loss during administration.
5. French (F) catheter at least 6 cm in length.
6. Lubricant to help inserting catheter.
7. Electric razor.
8. 6–8-week-old mice with TNBS-sensitive genetic background [19].
9. Isoflurane for anesthesia.

2.3 AIEC Colitis Model

1. Bacterial strain: AIEC strain NRG857c (O83:H1); isolated from an ileal tissue biopsy from a CD patient in the Charité Hospital (Berlin, Germany) [10] (*see* **Note 2**).
2. Bacterial culture reagents: NRG857c is cultured in LB broth and is resistant to chloramphenicol (34 μg/mL) and ampicillin (100 μg/mL). Selective LB agar plates should be prepared with one or both antibiotics for monitoring colonization of mice by fecal AIEC levels and determination of tissue colonization and dissemination.
3. Animals: C57BL/6, CD1, DBA/2, C3H, or 129e mice housed in an animal biosafety level 2 (ABSL2) facility. Due to variation in colonization levels, we recommend performing the experiment with $n = 4$–5/group, independently repeating the experiment 3–4 times.
4. Antibiotics: Streptomycin (100 mg/mL) or vancomycin (17.5 mg/mL) diluted in sterile water for gavage into mice to disrupt the intestinal microbiota and provide a niche for AIEC colonization [10].

5. Commercially available phosphate-buffered saline.
6. Enzyme-linked immunosorbent assay (ELISA) reader.

2.4 IL-10-Deficiency Colitis Model

1. Mice: The strain background and housing conditions of the IL-10KO mouse are major contributing factors for the severity of disease and the time to onset. Another factor to consider is the age of the mouse at the time of intervention or study. To examine factors that accelerate disease onset (e.g., diet), mice should be challenged between 3 and 5 weeks of age. For experiments involving chemical or infectious agents to result in earlier or more severe colitis, animals should be between 8 and 10 weeks of age at the initiation of the accelerant.
2. ELISA reader.

2.5 SAMP1/YitFc Ileocolitis Model

1. Mice: Specific pathogen-free (SPF) environment does not affect the time of onset of disease; however, the severity of disease could be less, characterized by less inflammation [20]. For prevention of fibrosis, therapy can be started at 6–8 weeks of age. This stage is characterized by intense immune response without severe lesions histologically [21]. The long duration of this model to 30 weeks of age, however, is making daily administration of drugs bothersome or in case of intraperitoneal (IP) injections may even lead to risk of peritonitis over time. Alternative approaches could be implantation of drug pumps [22]. For therapeutic approaches to already established fibrosis, mice should be treated starting week 30 onward for approximately 8–10 weeks [15, 21].
2. ELISA reader.

3 Methods

3.1 DSS Colitis Model

Mice drinking the sugar polymer DSS (36,000–50,000 molecular weight fragments) for several days develop a highly reproducible colitis with bloody diarrhea, ulcerations, and weight loss [23, 24]. DSS is believed to have a toxic effect on intestinal epithelial cells leading to disruption of the epithelial barrier. Subsequently, activation of intestinal macrophages might contribute to intestinal inflammation. DSS-induced colitis is characterized by weight loss, bloody diarrhea, ulcer formation, loss of epithelial cells, and infiltration of neutrophils, which resembles human UC [24]. In certain strains, chronic, intermittent administration of low concentrations DSS for several cycles results in chronic colitis with the development of a substantial amount of fibrosis [6]. From a technical perspective, this model is easy and yields reproducible results, explaining its popularity and widespread use [25]. The main disadvantage of this model is the questionable relevance to human IBD, as IBD is not caused by a chemical and more cell types than epithelial cells and macrophages are involved [26]. Mouse genetic background and microbiota [27] affects the degree of induced colitis as well, which makes it essential to perform a DSS dose titration in a specific animal facility.

3.1.1 DSS Titration (5–8 d)

DSS dose titration is recommended for each mouse strain and housing facility prior to deciding on the DSS dose for the chronic experiments. Disease severity in this model is strongly influenced by strain background and housing conditions (e.g., environmental microbes and/or intestinal microbiota).

1. On day 7, start training the mice to use water bottles as their only source of water. Acclimatization to the water bottles will reduce stress and prevent weight loss due to the change of water source.
2. On day 0, mark the mice for identification and determine their weight and baseline water volume (e.g., bottle weight). Collect stool to check occult blood and stool consistency.
3. Administer DSS solution (0% as control, then 1%, 2%, 3%, etc.) to the cage water. Estimated water consumption is ~5 mL of DSS solution per mouse per day. Control mice should receive sterile deionized water without DSS.
4. Determine the mouse weight, water consumed, assess stool consistency and stool occult blood every day when the mice are on DSS as well as control mice (*see* **Note 3**). The above indicators determine mouse clinical disease severity, which is also referred to as clinical score. Briefly, weight loss of 1–5%, 5–10%, 10–20%, and more than 20% can be scored as 1, 2, 3, and 4, respectively. Stool consistency is scored as 0 for well-formed pellets, 2 for pasty and semi-formed stools, which does not stick to the anus, 3 for pasty unformed stool, and 4 for liquid stools that remain adhesive to the anus. Bleeding is scored 0 for no blood in hemoccult, 2 for positive hemoccult, 3 for obvious blood stick to the surface of stool, and 4 for gross bleeding from the rectum. Occult blood is tested by Hemoccult Sensa Single Slides(Beckman Coulter, Indianapolis, IN, USA). Weight, stool consistency, and bleeding sub-scores will be added and divided by 3, resulting in a total clinical score ranging from 0 (healthy) to 4 (maximal activity of colitis).
5. At the time points of choice (generally on days 5–8), euthanize the mice and proceed to harvest mouse colons. We measure the colon length, and collect terminal ileum, proximal colon, transverse colon, and rectum segments. For each segment, we collect one piece immersed in HistoChoice (VWR, PA, USA) overnight for paraffin-embedded tissue sections, freeze one piece in OCT for frozen sections, one piece for mRNA extraction, and one piece for protein extraction.

6. After harvesting the tissue, use the paraffin-embedded tissue sections for hematoxylin and eosin (H&E) staining, as well as Masson Trichrome staining, to evaluate the level of inflammation and fibrosis. The degree of inflammation and fibrosis should be evaluated ideally by a trained histopathologist blinded to treatment conditions using a standardized scoring systems [28, 29]. The most preferable DSS concentration is determined by weight loss between 15% and 20% within 5–8 d of DSS administration with 100% viability of the animals. The preferable DSS dosage will be used in the chronic colitis model.

3.1.2 DSS Induction of Chronic DSS Colitis (30–42 d)

1. On day 7, start training the mice to use water bottles as their only source of water.
2. On day 0, mark the mice for identification and determine their weight and water bottle weight. Collect stool to check occult blood and stool consistency.
3. Administer the preferred DSS dose in cage water bottles. Estimate water consumption as ~5 mL of DSS solution per mouse per day. Control mice should receive sterile deionized water without DSS.
4. Determine the mice weight and water bottle weight and check the stool consistency and stool occult blood every day when the mice are on DSS as well as control mice (*see* **Note 4**). Mouse disease severity is also determined by the clinical score, combining the score of weight loss, stool consistency, and occult blood as above mentioned. Briefly, weight loss of 1–5%, 5–10%, 10–20%, and more than 20% can be scored as 1, 2, 3, and 4, respectively. Stool consistency is scored as 0 for well-formed pellets, 2 for pasty and semiformed stools, which does not stick to the anus, 3 for pasty unformed stool, and 4 for liquid stools that remain adhesive to the anus. Bleeding is scored 0 for no blood in hemoccult, 2 for positive hemoccult, 3 for obvious blood stick to the surface of stool, and 4 for gross bleeding from the rectum. Occult blood is tested by Hemoccult Sensa Single Slides (Beckman Coulter, Indianapolis, IN, USA). Weight, stool consistency, and bleeding sub-scores will be added and divided by 3, resulting in a total clinical score ranging from 0 (healthy) to 4 (maximal activity of colitis).
5. The preferred DSS dose is administered in the drinking water for a cycle of 5–7 d, followed by 10 d of recovery on water. As mentioned above there are differences in susceptibility of different strains to DSS (e.g., C57B6 background mice are more sensitive to DSS, while BALB/c are more resistant) and disease activity is also influenced by housing conditions. Change the water bottle to sterile deionized water in all experimental animals for recovery. Continue monitoring the mice for disease activity as described in **step 4** throughout the experiment.

6. A second cycle of DSS is performed, administering the same concentration of DSS for the same length of time as cycle one followed by 10 d of recovery. Harvest according to Subheading 3.1.1, **steps 5** and **6**, after two cycles of DSS and two cycles of recovery.

3.2 TNBS Colitis Model

Colitis can be induced in susceptible murine strains by intrarectal instillation of the hapten TNBS diluted in ethanol [30]. It is critical to check in the literature the susceptibility of the mouse strain to TNBS [19]. For instance, BALB/c is considered to be susceptible to TNBS, while C57BL/6 is considered resistant [19]. Comparable to the DSS models, a dose titration is strongly recommended, as disease severity is influenced by genetic background and housing facility (e.g., environmental microbes and/or microbiota). Mechanistically, ethanol breaks the epithelial barrier and TNBS is believed to modify colonic proteins and activate a delayed-type hypersensitivity reaction. This depends on T cells and leads to transmural inflammation [9]. When given weekly for 6 weeks of escalating doses of TNBS (chronic model), chronic colitis leads to fibrosis. This model is one of the most widely used to study intestinal fibrosis in vivo. Its limitations are the high variability [31] and again the nature of a chemical inducing colitis.

3.2.1 TNBS Titration (5–8 d)

1. Weigh and mark the mice to allow identification.
2. Lightly anesthetize the mouse via a nose cone with 1.5% isoflurane-mixed gas while it is resting on an isothermal warming pad.
3. Pre-sensitization: On day 1, carefully shave a 1.5 × 1.5 cm field of the skin of the mouse using an electric razor. To avoid the mouse from grooming TNBS off the skin (may potentially induce oral tolerance), we preferentially select an area on the back between the shoulders. While holding the lightly anesthetized mouse with one hand, we apply 150 μL pre-sensitization solution using a 200 μL pipette to the shaved back skin. The solution is absorbed by the skin quickly. Control mice are treated with pre-sensitization solution without TNBS.
4. On day 8, weigh the mouse.
5. Anesthetize the mouse by using 1.5% isoflurane-mixed gas. Attach a 1-mL syringe to a 3.5 F catheter, and load it with 0, 0.5, 1, 2, 3, and 4% (w/v) TNBS solution in 70% EtOH. Lubricate the catheter tip and insert the catheter through the anus into the colon ~4 cm (*see* **Note 5**).
6. Slowly inject 100 μL of TNBS solution into the lumen of the colon, and remove the catheter from the colon, and position the mouse head down for 1 min (*see* **Note 6**).

7. Closely monitor the animals for signs of distress, perform daily measurement of weight, stool consistency, and blood in stool [32].
8. At the timepoint of choice, euthanize the animal and proceed to harvest mouse colons. We measure the colon length, collect transverse colon and rectum. For each segment, we collect one piece immersed in HistoChoice overnight for paraffin-embedded tissue sections, one piece for optimal cutting temperature (OCT) compound embedded frozen sectioning, one piece for mRNA extraction, and one piece for protein extraction.
9. After harvesting the tissue, use the paraffin-embedded tissue sections for H&E staining, as well as Masson Trichrome staining, to evaluate the level of inflammation and fibrosis. The degree of inflammation and fibrosis should be evaluated ideally by a trained histopathologist blinded to treatment conditions using a standardized scoring system [28, 29]. The most preferable TNBS concentration is determined by weight loss between 15% and 20% within 5–8 d of TNBS administration with 100% viability of the animals.

3.2.2 Chronic TNBS Colitis (56 d)

Mice tolerize to TNBS over the duration of the experiment. For the chronic model we hence use escalating doses of TNBS with starting with the optimal dose from the titration experiments and increase by 0.5 mg increments every 2 weeks for up to 6 weeks.

1. Weigh and mark the mice to allow identification.
2. Lightly anesthetize the mouse via a nose cone with 1.5% isoflurane-mixed gas while it is resting on an isothermal warming pad.
3. Pre-sensitization: On day 1, we carefully shave a 1.5 × 1.5 cm field of the skin of the mouse using an electric razor. To avoid the mouse from grooming TNBS off the skin (may potentially induce oral tolerance), we preferentially select an area on the back between the shoulders. While holding the lightly anesthetized mouse with one hand, we apply 150 μL pre-sensitization solution using a 200 μL pipette to the shaved back skin. The solution is absorbed by the skin quickly. Control mice are treated with pre-sensitization solution without TNBS.
4. On day 8, weigh the mouse.
5. Anesthetize the mouse by using 1.5% isoflurane-mixed gas. Attach a 1-mL syringe to a 3.5 F catheter and load it with either (1) the optimal TNBS dose, (2) vehicle control solution for administration rectally. Lubricate the catheter tip and insert the catheter through the anus into the colon ~4 cm (*see* **Note** 7).

6. Closely monitor the animals for signs of distress, perform daily measurement of weight, stool consistency, and blood in stool [32].
7. Repeat **steps 5** and **6** weekly, escalating the TNBS dose by 0.5 mg every 2 weeks, up to 6 weeks. Harvest at 8 weeks according to Subheading 3.2.1, **steps 8** and **9**.
8. At week 8, euthanize the animal and proceed to harvest mouse colons. We measure the colon length, collect transverse colon and rectum. For each segment, we collect one piece immersed in HistoChoice overnight for paraffin-embedded tissue sections, one piece for OCT embedded frozen sectioning, one piece for mRNA extraction, and one piece for protein extraction.
9. After harvesting the tissue, use the paraffin-embedded tissue sections for H&E staining, as well as Masson Trichrome staining, to evaluate the level of inflammation and fibrosis. The degree of inflammation and fibrosis should be evaluated ideally by a trained histopathologist blinded to treatment conditions using a standardized scoring system [28, 29].

3.3 AIEC Colitis Model

The microbiota plays a key role in the initiation and perpetuation of chronic inflammation in IBD [33]. A singular pathogenic microbial species has not be identified; rather, it is thought that the combination of host genetic susceptibility and the opportunistic expansion of commensal microbes initiates and perpetuates IBD [33]. *Escherichia coli,* a Gram-negative commensal member of the *Proteobacteria* phyla, is a normal component of the intestinal microbiota and contributes to intestinal homeostasis [10]. However, CD patients have been found to harbor higher than normal levels of *E. coli* that have acquired virulence factors [34, 35]. AIEC is one such mucosa-associated bacterium often found in ileal biopsies of CD patients [34, 36]. Several strains of AIEC have been shown to exacerbate or even initiate experimental colitis in murine models [10, 37–41]. However, several of these models require specific genetically altered mice [38–41] and/or mono-colonization [40, 41] to induce chronic intestinal inflammation. The exception to this is a model that utilizes the AIEC strain NRG857c that can colonize a broad variety of standard inbred mouse strains after antibiotic treatment [10]. AIEC strain NRG857c induces a strong Th17 response in the mouse intestine, which drives chronic intestinal inflammation and transmural fibrosis, mainly in the cecum, with limited mortality [10]. Extracellular matrix deposition can be detected by Masson trichrome stain or picrosirius red stain as early as 7 d post-infection and persists up to 63 d post-infection. These features have been observed in multiple standard inbred mouse strains, such as CD1, DBA/2, C3H, 129e, and C57BL/6, adding to its utility as a model of intestinal fibrosis [10, 11].

3.3.1 Induction of AIEC Colitis (7–63 d)

1. Day-4: Weigh mice and identify individual mice by ear punch or ear tag. If not already housed in an ABSL2 facility, transfer mice to the ABSL2 facility.
2. Day-2: Prepare a fresh AIEC NRG857c stock plate on selective LB agar.
3. Day-1: Antibiotic pretreatment: Remove food and water from the cage 4 h prior to gavage and weigh the mice. Orally gavage mice with 20 mg streptomycin or 3.5 mg of vancomycin in 200 μL sterile water. Sedation or anesthesia is not recommended before performing oral gavage, as it may increase the risk of aspiration pneumonia. Return food and water to cage after gavage.
4. Day-1: Inoculate duplicate aliquots of 4 mL sterile LB containing chloramphenicol/ampicillin with a single colony of AIEC NRG857c from the stock plate. Incubate overnight at 37 °C while shaking at 220 rpm.
5. Day 0: Mouse preparation: Remove food and water from the cage 4 h prior to gavage and weigh the mice.
6. Day 0: AIEC inoculation preparation: Measure the density of the overnight AIEC bacterial culture by reading the absorbance at 600 nm in a spectrophotometer. Calculate the volume of culture required to deliver 2×10^9 colony forming units (cfu) to twice the number of mice intended for colonization. Concentrate overnight culture by centrifuging for 10 min at $4000 \times g$. Aspirate supernatant and gently, but thoroughly, resuspend bacterial pellet in sterile PBS at a volume of 200 μL per mouse. Inoculum should look milky white, but easily pass through the gavage needle. Use a small aliquot of this final inoculum to dilute and plate on duplicate selective LB agar plates to confirm the actual concentration of the inoculum. Incubate plate overnight at 37 °C and count colonies the next morning to calculate the actual amount of AIEC delivered to each mouse.
7. Day 0: Bacterial inoculation: Orally gavage mice with 2×10^9 cfu in 200 μL volume inside a biosafety cabinet. Return food and water after gavage. As AIEC NRG857c is a human pathobiont, mouse handling must take place as specified in an approved biosafety protocol (e.g., within a biosafety cabinet during inoculation, and all subsequent work with these mice must also take place in a biosafety cabinet using BSL2 precautions).
8. Monitoring: Mouse weight and activity is evaluated daily as a gross measure of disease activity. The productivity of AIEC colonization can be monitored by AIEC cfu in stool homogenates as early as day 1 post-inoculation and every 3–4 d after inoculation. Stool pellets are collected fresh from mice and

placed in a pre-weighed screw-cap tube containing 500 μL of sterile PBS. Tubes are then weighed to determine the amount of stool/mL. Stool samples are then homogenized in a biosafety cabinet by vortexing vigorously. Stool homogenates are then serially diluted in a sterile 96-well plate containing sterile PBS and 5 μL of the dilutions spot plated on duplicate selective LB agar plates to quantify the number of AIEC cfu per weight of the stool sample. Approximately 10^7–10^9 AIEC cfu/g of stool can be expected on day 1 post-infection [11]. Fecal loads generally peak at day 3, then steadily decline over the next 1–3 weeks post-inoculation depending upon the mouse strain background [10].

9. Endpoint collections (Day 0, 7, 14, 21, 63 post-inoculation): After euthanasia, collect blood for the analysis of systemic inflammation markers by ELISA (e.g., serum amyloid A). Collect intestinal tissue, including jejunum, terminal ileum, cecum, proximal colon, transverse colon as well as distal colon segments for fixation and histologic analysis of tissue damage, fibrosis, and inflammation. Stool, and segments of the colon, cecum, small intestine, liver, spleen, and mesenteric lymph nodes can be collected into pre-weighed screw-cap tubes containing sterile PBS for analysis of AIEC tissue colonization and dissemination. These tubes are weighed before processing to determine tissue weight and then homogenized in a biosafety cabinet using a tissue grinder. The volume of PBS for homogenization ranges between 0.5 and 5 mL depending on the day of collection and tissue type (use a lower volume for non-intestinal tissues or later in the experimental time course). Serially dilute and plate supernatants in duplicate from tissue homogenates on selective LB agar plates as performed with stool samples. Final AIEC tissue or stool levels are reported as cfu/g.
10. Analysis of inflammation and fibrosis: Sections from fixed tissues embedded in paraffin blocks are stained with H&E for evaluation of tissue histology and inflammation or Masson Trichrome stain to evaluate the deposition of extracellular matrix/fibrosis. The degree of inflammation and fibrosis should be evaluated ideally by a trained histopathologist blinded to treatment conditions using a standardized scoring system [10, 42, 43].

3.4 IL-10-Deficiency Colitis Model

IL-10 is an important immunoregulatory cytokine with anti-inflammatory properties that are normally employed to ramp down inflammation caused by the host immune response to pathogens [44]. Therefore, IL-10 is a major regulator of mucosal homeostasis and IL-10-deficient mice (IL-10KO) have been shown to develop chronic enterocolitis in response to even commensal microbiota stimuli [44, 45]. The severity and scope of colitis developed by IL-10KO mice is dependent on the milieu of microbes they are exposed to; germ-free IL-10KO mice will not develop inflammation, SPF mice will develop mild colitis generally restricted to the large intestine, and conventional or experimentally infected mice will develop severe enterocolitis. The other major contributor to disease activity and time to disease onset in this model is genetic background. The strains most susceptible to disease development are C3H/HeJBir and 129/SvEv, while the C57BL/6 J and C57BL/10 strains are the most resistant [14, 25, 44]. Taken together, depending on the strain background and the housing conditions, development of robust colitis in the IL-10KO model can vary between 12 and 20 weeks of age and must be experimentally determined in each animal facility.

The gradual onset of disease in the IL-10KO model is ideal for testing the contribution of environmental factors (such as food products or lifestyle choices like smoking tobacco) or specific immune system components to the acceleration or amelioration of disease onset [45–47]. However, since the development of spontaneous colitis in IL-10KO mice is slow, alternate methods are being investigated to accelerate the onset of disease. One approach relies on the administration of an anti-mouse IL-10R neutralizing antibody, rather than genetic deletion, to initiate disease more effectively [48–50]. Other approaches include the addition of intestinal irritants (e.g., DSS [51] or piroxicam [13]) or infectious agents (e.g., *Helicobacter* spp. [48, 49, 52] or AIEC [41]) to accelerate disease.

3.4.1 Spontaneous Colitis in IL-10 Knockout Mice (~20 Weeks)

1. Mice between 3 and 5 weeks of age should be assigned to groups and identified (e.g., ear punch or ear tag) for monitoring.
2. Measure the weight of the mice weekly and monitor stool consistency, occult blood in stool, and any sign of stress of the mice. Biomarkers of inflammation, such as fecal lipocalin-2 (Lcn-2) or circulating serum amyloid A can monitored weekly by ELISA to assess the levels of intestinal and systemic inflammation, respectively [53].
3. At weeks 12, 16, and 20, euthanize animals, collect blood for the analysis of systemic inflammation markers by ELISA (e.g., serum amyloid A), and proceed to excise the intestine. Observe any gross pathology of the intestine (e.g., intestinal thickening, strictures, intestinal shortening, poorly formed stool/diarrhea) and measure the colon length. Collect sections from the terminal ileum, proximal colon, transverse colon, and rectum for fixation and paraffin-embedding for histologic analysis of tissue damage, fibrosis, and inflammation. A key component of the histologic analysis is to select tissue from the same position

(e.g., x cm from the cecum) in every mouse and every experiment, due to the heterogeneity of disease severity along the length of the intestine [54]. Collect additional pieces of ileal and colon segments for RNA extraction and/or for protein extraction, if desired. Collect stool in pre-weighed screw-cap tubes containing 500 μL PBS for homogenization and measurement of fecal Lcn2 levels (pg/mg stool) by ELISA.

4. Analysis of inflammation and fibrosis: Sections from fixed tissues embedded in paraffin blocks are stained with either H&E for evaluation of tissue histology and inflammation or Massons' Trichrome stain to evaluate the deposition of extracellular matrix/fibrosis. The degree of inflammation and fibrosis should be evaluated by a trained IBD histopathologist blinded to treatment conditions using a standardized scoring system [42, 43, 54, 55].

3.5 SAMP1/YitFc Ileocolitis Model

While the abovementioned models develop mainly colitis, the SAMP1/Yit mouse exhibits spontaneous inflammation at 10–20 week of age with almost 100% penetrance after 30 weeks [20]. The strain was created by breeding a senescence-accelerated mouse line with inflammation being most severe in the terminal ileum [56]. What makes this model attractive is the spontaneous development of ileocolitis, the location in the terminal ileum, the segmental and transmural nature of the inflammation, and granulomas on histopathology. Foremost from a fibrosis perspective, this model develops accumulation of extracellular matrix in the small and large bowel and a thickening of the muscularis mucosa, which resemble features of human CD-associated intestinal strictures [15, 56]. Drawbacks of the SAMP/Yit mouse are the poor breeding ability, which makes very large colonies necessary, as well as the inconsistent location and severity of inflammation in younger mice in this model.

3.5.1 Spontaneous Colitis in SAMP1/YitFc Mice (~40 Weeks)

1. Mice between 3 and 7 weeks of age should be assigned to groups and identified (e.g., ear punch or ear tag) for monitoring (*see* **Note 8**).
2. Measure the weight of the mice weekly and monitor stool consistency, occult blood in stool, and any sign of stress of the mice [32].
3. At the timepoints delineated above, euthanize animals, collect blood for the analysis of systemic inflammation markers by ELISA, and proceed to harvest the intestine. Observe any gross pathology of the intestine (e.g., intestinal thickening, intestinal shortening, poorly formed stool/diarrhea) and measure the colon length. Collect sections from the terminal ileum, cecum, proximal colon, transverse colon, and rectum [56] for fixation and paraffin-embedding for histologic analysis of tissue

damage, fibrosis, and inflammation. A key component of the histologic analysis is to select tissue from the same position (e.g., x cm from the cecum) in every mouse and every experiment, due to the heterogeneity of disease severity along the length of the intestine [54]. Collect additional pieces of ileal and colon segments for RNA extraction and/or for protein extraction, if desired. Collect stool in pre-weighed screw-cap tubes for further experiments.

4. Analysis of inflammation and fibrosis: Sections from fixed tissues embedded in paraffin blocks are stained with either H&E for evaluation of tissue histology and inflammation or Masson Trichrome stain to evaluate the deposition of extracellular matrix/fibrosis. The degree of inflammation and fibrosis should be evaluated by a trained IBD histopathologist blinded to treatment conditions using a standardized scoring system [20].

4 Notes

1. Both acetone and TNBS are harmful. Handle them using appropriate safety equipment and measures, including eye protection, gloves, and lab coat. Always prepare the solutions freshly on the day of the intrarectal challenge. Protect TNBS from light.
2. This is a biosafety level 2 (BSL2) microorganism; therefore, use and handling procedures for this bacterial strain must be approved by the local institutional biosafety committee. Animals colonized with AIEC NRG857c require housing in an ABSL2 facility.
3. If the mice lose 20% weight as compared to Day 0 weight, the mice may to be euthanized (dependent on local regulations). Subcutaneous injection of saline solution, as well as giving hydrogels, may be provided as supportive care to alleviate dehydration of the mice.
4. If the mice lose 20% weight as compared to Day 0 weight, the mice may to be euthanized (dependent on local regulations). Subcutaneous injection of saline solution, as well as giving hydrogels may be provided as supportive care to alleviate dehydration of the mice.
5. Place the tube in the colon gently since it is easy to cause damage and perforation of the colon wall. If any resistance is sensed during catheter insertion, remove the catheter, and reinsert gently.

6. The mouse might wake up during the recovery period and cause TNBS to leak out of the colon. It is essential to make sure the TNBS solution remains in the colon, which can be achieved by keeping the mouse upside down and attempt to calm the animal. Then use a heating pad to make sure that the mouse stays warm.
7. The mouse might wake up during the recovery period and cause TNBS to leak out of the colon. It is essential to make sure the TNBS solution remains in the colon, which can be achieved by keeping the mouse upside down and attempt to calm the animal. Then use a heating pad to make sure that the mouse stays warm.
8. If any treatments need to be applied to prevent the early onset of the inflammation, 3-week-old mice are preferable [57].

Acknowledgments

This work was supported by grants from the National Institutes of Health K08 (DK110415 to F.R.), the Department of Defense Peer Reviewed Medical Research Program Expansion Award (PR181846 to C.M.), the Crohn's and Colitis Foundation of America, CURE 4 IBD, and the Helmsley Charitable Trust.

References

1. Cosnes J, Nion-Larmurier I, Beaugerie L, Afchain P, Tiret E, Gendre JP (2005) Impact of the increasing use of immunosuppressants in Crohn's disease on the need for intestinal surgery. Gut 54:237–241
2. Farmer RG, Whelan G, Fazio VW (1985) Long-term follow-up of patients with Crohn's disease. Relationship between the clinical pattern and prognosis. Gastroenterology 88:1818–1825
3. Gordon IO, Agrawal N, Goldblum JR, Fiocchi C, Rieder F (2014) Fibrosis in ulcerative colitis: mechanisms, features, and consequences of a neglected problem. Inflamm Bowel Dis 20:2198–2206. https://doi.org/10.1097/MIB.0000000000000080
4. Dieleman LA, Ridwan BU, Tennyson GS, Beagley KW, Bucy RP, Elson CO (1994) Dextran sulfate sodium-induced colitis occurs in severe combined immunodeficient mice. Gastroenterology 107:1643–1652
5. Kitajima S, Morimoto M, Sagara E, Shimizu C, Ikeda Y (2001) Dextran sodium sulfate-induced colitis in germ-free IQI/Jic mice. Exp Anim 50:387–395. https://doi.org/10.1538/expanim.50.387
6. Lund PK, Zuniga CC (2001) Intestinal fibrosis in human and experimental inflammatory bowel disease. Curr Opin Gastroenterol 17:318–323
7. Fichtner-Feigl S, Fuss IJ, Young CA, Watanabe T, Geissler EK, Schlitt HJ, Kitani A, Strober W (2007) Induction of IL-13 triggers TGF-beta1-dependent tissue fibrosis in chronic 2,4,6-trinitrobenzene sulfonic acid colitis. J Immunol 178:5859–5870
8. Lawrance IC, Wu F, Leite AZ, Willis J, West GA, Fiocchi C, Chakravarti S (2003) A murine model of chronic inflammation-induced intestinal fibrosis down-regulated by antisense NF-kappa B. Gastroenterology 125:1750–1761
9. Neurath MF, Fuss I, Kelsall BL, Presky DH, Waegell W, Strober W (1996) Experimental granulomatous colitis in mice is abrogated by induction of TGF-beta-mediated oral tolerance. J Exp Med 183:2605–2616

10. Small CL, Reid-Yu SA, McPhee JB, Coombes BK (2013) Persistent infection with Crohn's disease-associated adherent-invasive Escherichia coli leads to chronic inflammation and intestinal fibrosis. Nat Commun 4:1957. https://doi.org/10.1038/ncomms2957
11. Oberc AM, Fiebig-Comyn AA, Tsai CN, Elhenawy W, Coombes BK (2019) Antibiotics Potentiate Adherent-Invasive E. coli Infection and Expansion. Inflamm Bowel Dis 25:711–721. https://doi.org/10.1093/ibd/izy361
12. Berg DJ, Davidson N, Kuhn R, Muller W, Menon S, Holland G, Thompson-Snipes L, Leach MW, Rennick D (1996) Enterocolitis and colon cancer in interleukin-10-deficient mice are associated with aberrant cytokine production and CD4(+) TH1-like responses. J Clin Invest 98:1010–1020. https://doi.org/10.1172/JCI118861
13. Berg DJ, Zhang J, Weinstock JV, Ismail HF, Earle KA, Alila H, Pamukcu R, Moore S, Lynch RG (2002) Rapid development of colitis in NSAID-treated IL-10-deficient mice. Gastroenterology 123:1527–1542. https://doi.org/10.1053/gast.2002.1231527
14. Kuhn R, Lohler J, Rennick D, Rajewsky K, Muller W (1993) Interleukin-10-deficient mice develop chronic enterocolitis. Cell 75:263–274
15. Rivera-Nieves J, Bamias G, Vidrich A, Marini M, Pizarro TT, McDuffie MJ, Moskaluk CA, Cohn SM, Cominelli F (2003) Emergence of perianal fistulizing disease in the SAMP1/YitFc mouse, a spontaneous model of chronic ileitis. Gastroenterology 124:972–982
16. Chassaing B, Aitken JD, Malleshappa M, Vijay-Kumar M (2014) Dextran sulfate sodium (DSS)-induced colitis in mice. Curr Protoc Immunol 104:15–25. https://doi.org/10.1002/0471142735.im1525s104
17. Hoffmann M, Schwertassek U, Seydel A, Weber K, Falk W, Hauschildt S, Lehmann J (2018) A refined and translationally relevant model of chronic DSS colitis in BALB/c mice. Lab Anim 52:240–252. https://doi.org/10.1177/0023677217742681
18. Marcon R, Bento AF, Dutra RC, Bicca MA, Leite DF, Calixto JB (2013) Maresin 1, a pro-resolving lipid mediator derived from omega-3 polyunsaturated fatty acids, exerts protective actions in murine models of colitis. J Immunol 191:4288–4298. https://doi.org/10.4049/jimmunol.1202743
19. Scheiffele F, Fuss IJ (2002) Induction of TNBS colitis in mice. Curr Protoc Immunol 15:19. https://doi.org/10.1002/0471142735.im1519s49. Chapter 15: Unit
20. Pizarro TT, Pastorelli L, Bamias G, Garg RR, Reuter BK, Mercado JR, Chieppa M, Arseneau KO, Ley K, Cominelli F (2011) SAMP1/YitFc mouse strain: a spontaneous model of Crohn's disease-like ileitis. Inflamm Bowel Dis 17:2566–2584. https://doi.org/10.1002/ibd.21638
21. Vidrich A, Buzan JM, Barnes S, Reuter BK, Skaar K, Ilo C, Cominelli F, Pizarro T, Cohn SM (2005) Altered epithelial cell lineage allocation and global expansion of the crypt epithelial stem cell population are associated with ileitis in SAMP1/YitFc mice. Am J Pathol 166:1055–1067. https://doi.org/10.1016/S0002-9440(10)62326-7
22. Odashima M, Bamias G, Rivera-Nieves J, Linden J, Nast CC, Moskaluk CA, Marini M, Sugawara K, Kozaiwa K, Otaka M, Watanabe S, Cominelli F (2005) Activation of A2A adenosine receptor attenuates intestinal inflammation in animal models of inflammatory bowel disease. Gastroenterology 129:26–33. https://doi.org/10.1053/j.gastro.2005.05.032
23. Mahler M, Bristol IJ, Leiter EH, Workman AE, Birkenmeier EH, Elson CO, Sundberg JP (1998) Differential susceptibility of inbred mouse strains to dextran sulfate sodium-induced colitis. Am J Phys 274:G544–G551
24. Okayasu I, Hatakeyama S, Yamada M, Ohkusa T, Inagaki Y, Nakaya R (1990) A novel method in the induction of reliable experimental acute and chronic ulcerative colitis in mice. Gastroenterology 98:694–702
25. Rieder F, Kessler S, Sans M, Fiocchi C (2012) Animal models of intestinal fibrosis: new tools for the understanding of pathogenesis and therapy of human disease. Am J Physiol Gastrointest Liver Physiol 303:G786–G801. https://doi.org/10.1152/ajpgi.00059.2012
26. Lavelle A, Sokol H (2020) Gut microbiota-derived metabolites as key actors in inflammatory bowel disease. Nat Rev Gastroenterol Hepatol. https://doi.org/10.1038/s41575-019-0258-z
27. Sovran B, Planchais J, Jegou S, Straube M, Lamas B, Natividad JM, Agus A, Dupraz L, Glodt J, Da Costa G, Michel ML, Langella P, Richard ML, Sokol H (2018) Enterobacteriaceae are essential for the modulation of colitis severity by fungi. Microbiome 6:152. https://doi.org/10.1186/s40168-018-0538-9
28. Suzuki K, Arumugam S, Yokoyama J, Kawauchi Y, Honda Y, Sato H, Aoyagi Y, Terai S, Okazaki K, Suzuki Y, Mizumoto S, Sugahara K, Atreya R, Neurath MF,

Watanabe K, Hashiguchi T, Yoneyama H, Asakura H (2016) Pivotal Role of Carbohydrate Sulfotransferase 15 in Fibrosis and Mucosal Healing in Mouse Colitis. PLoS One 11: e0158967. https://doi.org/10.1371/journal.pone.0158967

29. Theiss AL, Fuller CR, Simmons JG, Liu B, Sartor RB, Lund PK (2005) Growth hormone reduces the severity of fibrosis associated with chronic intestinal inflammation. Gastroenterology 129:204–219. https://doi.org/10.1053/j.gastro.2005.05.019
30. Morris GP, Beck PL, Herridge MS, Depew WT, Szewczuk MR, Wallace JL (1989) Hapten-induced model of chronic inflammation and ulceration in the rat colon. Gastroenterology 96:795–803
31. Wirtz S, Popp V, Kindermann M, Gerlach K, Weigmann B, Fichtner-Feigl S, Neurath MF (2017) Chemically induced mouse models of acute and chronic intestinal inflammation. Nat Protoc 12:1295–1309. https://doi.org/10.1038/nprot.2017.044
32. Zhao S, Dejanovic D, Yao P, Bhilocha S, Sadler T, Schirbel A, West G, Doyon G, Lopez R, Mao R, Kurada S, El Ouali S, Grassl G, Fox PL, Cruise M, Worthley DL, de la Motte C, Fiocchi C, Rieder F (2020) Selective deletion of MyD88 signaling in alpha-SMA positive cells ameliorates experimental intestinal fibrosis via post-transcriptional regulation. Mucosal Immunol. https://doi.org/10.1038/s41385-020-0259-9
33. Sartor RB (2008) Microbial influences in inflammatory bowel diseases. Gastroenterology 134:577–594. https://doi.org/10.1053/j.gastro.2007.11.059
34. Darfeuille-Michaud A, Boudeau J, Bulois P, Neut C, Glasser AL, Barnich N, Bringer MA, Swidsinski A, Beaugerie L, Colombel JF (2004) High prevalence of adherent-invasive Escherichia coli associated with ileal mucosa in Crohn's disease. Gastroenterology 127:412–421
35. Martinez-Medina M, Aldeguer X, Lopez-Siles-M, Gonzalez-Huix F, Lopez-Oliu C, Dahbi G, Blanco JE, Blanco J, Garcia-Gil LJ, Darfeuille-Michaud A (2009) Molecular diversity of Escherichia coli in the human gut: new ecological evidence supporting the role of adherent-invasive E. coli (AIEC) in Crohn's disease. Inflamm Bowel Dis 15:872–882. https://doi.org/10.1002/ibd.20860
36. Barnich N, Darfeuille-Michaud A (2007) Adherent-invasive Escherichia coli and Crohn's disease. Curr Opin Gastroenterol 23:16–20. https://doi.org/10.1097/MOG.0b013e3280105a38
37. Carvalho FA, Barnich N, Sauvanet P, Darcha C, Gelot A, Darfeuille-Michaud A (2008) Crohn's disease-associated Escherichia coli LF82 aggravates colitis in injured mouse colon via signaling by flagellin. Inflamm Bowel Dis 14:1051–1060. https://doi.org/10.1002/ibd.20423
38. Carvalho FA, Barnich N, Sivignon A, Darcha C, Chan CH, Stanners CP, Darfeuille-Michaud A (2009) Crohn's disease adherent-invasive Escherichia coli colonize and induce strong gut inflammation in transgenic mice expressing human CEACAM. J Exp Med 206:2179–2189. https://doi.org/10.1084/jem.20090741
39. Martinez-Medina M, Denizot J, Dreux N, Robin F, Billard E, Bonnet R, Darfeuille-Michaud A, Barnich N (2014) Western diet induces dysbiosis with increased E coli in CEABAC10 mice, alters host barrier function favouring AIEC colonisation. Gut 63:116–124. https://doi.org/10.1136/gutjnl-2012-304119
40. Chassaing B, Koren O, Carvalho FA, Ley RE, Gewirtz AT (2014) AIEC pathobiont instigates chronic colitis in susceptible hosts by altering microbiota composition. Gut 63:1069–1080. https://doi.org/10.1136/gutjnl-2013-304909
41. Schmitz JM, Tonkonogy SL, Dogan B, Leblond A, Whitehead KJ, Kim SC, Simpson KW, Sartor RB (2019) Murine adherent and invasive E. coli induces chronic inflammation and immune responses in the small and large intestines of monoassociated IL-10−/− mice independent of long polar fimbriae adhesin A. Inflamm Bowel Dis 25:875–885. https://doi.org/10.1093/ibd/izy386
42. Machtaler S, Knieling F, Luong R, Tian L, Willmann JK (2015) Assessment of inflammation in an acute on chronic model of inflammatory bowel disease with ultrasound molecular imaging. Theranostics 5:1175–1186. https://doi.org/10.7150/thno.13048
43. Ding S, Walton KL, Blue RE, McNaughton K, Magness ST, Lund PK (2012) Mucosal healing and fibrosis after acute or chronic inflammation in wild type FVB-N mice and C57BL6 procollagen alpha1(I)-promoter-GFP reporter mice. PLoS One 7:e42568. https://doi.org/10.1371/journal.pone.0042568
44. Iyer SS, Cheng G (2012) Role of interleukin 10 transcriptional regulation in inflammation and autoimmune disease. Crit Rev Immunol 32:23–63. https://doi.org/10.1615/critrevimmunol.v32.i1.30
45. Rennick DM, Fort MM (2000) Lessons from genetically engineered animal models. XII.

IL-10-deficient (IL-10(−/−)) mice and intestinal inflammation. Am J Physiol Gastrointest Liver Physiol 278:G829–G833. https://doi.org/10.1152/ajpgi.2000.278.6.G829

46. Moore KW, O'Garra A, de Waal Malefyt R, Vieira P, Mosmann TR (1993) Interleukin-10. Annu Rev Immunol 11:165–190. https://doi.org/10.1146/annurev.iy.11.040193.001121
47. Chassaing B, Koren O, Goodrich JK, Poole AC, Srinivasan S, Ley RE, Gewirtz AT (2015) Dietary emulsifiers impact the mouse gut microbiota promoting colitis and metabolic syndrome. Nature 519:92–96. https://doi.org/10.1038/nature14232
48. Ahern PP, Schiering C, Buonocore S, McGeachy MJ, Cua DJ, Maloy KJ, Powrie F (2010) Interleukin-23 drives intestinal inflammation through direct activity on T cells. Immunity 33:279–288. https://doi.org/10.1016/j.immuni.2010.08.010
49. Izcue A, Hue S, Buonocore S, Arancibia-Carcamo CV, Ahern PP, Iwakura Y, Maloy KJ, Powrie F (2008) Interleukin-23 restrains regulatory T cell activity to drive T cell-dependent colitis. Immunity 28:559–570. https://doi.org/10.1016/j.immuni.2008.02.019
50. O'Farrell AM, Liu Y, Moore KW, Mui AL (1998) IL-10 inhibits macrophage activation and proliferation by distinct signaling mechanisms: evidence for Stat3-dependent and -independent pathways. EMBO J 17:1006–1018. https://doi.org/10.1093/emboj/17.4.1006
51. Li B, Alli R, Vogel P, Geiger TL (2014) IL-10 modulates DSS-induced colitis through a macrophage-ROS-NO axis. Mucosal Immunol 7:869–878. https://doi.org/10.1038/mi.2013.103
52. Kullberg MC, Jankovic D, Feng CG, Hue S, Gorelick PL, McKenzie BS, Cua DJ, Powrie F, Cheever AW, Maloy KJ, Sher A (2006) IL-23 plays a key role in Helicobacter hepaticus-induced T cell-dependent colitis. J Exp Med 203:2485–2494. https://doi.org/10.1084/jem.20061082
53. Chassaing B, Srinivasan G, Delgado MA, Young AN, Gewirtz AT, Vijay-Kumar M (2012) Fecal lipocalin 2, a sensitive and broadly dynamic non-invasive biomarker for intestinal inflammation. PLoS One 7:e44328. https://doi.org/10.1371/journal.pone.0044328
54. Delday M, Mulder I, Logan ET, Grant G (2019) Bacteroides thetaiotaomicron ameliorates colon inflammation in preclinical models of Crohn's disease. Inflamm Bowel Dis 25:85–96. https://doi.org/10.1093/ibd/izy281
55. Scheinin T, Butler DM, Salway F, Scallon B, Feldmann M (2003) Validation of the interleukin-10 knockout mouse model of colitis: antitumour necrosis factor-antibodies suppress the progression of colitis. Clin Exp Immunol 133:38–43. https://doi.org/10.1046/j.1365-2249.2003.02193.x
56. Matsumoto S, Okabe Y, Setoyama H, Takayama K, Ohtsuka J, Funahashi H, Imaoka A, Okada Y, Umesaki Y (1998) Inflammatory bowel disease-like enteritis and caecitis in a senescence accelerated mouse P1/Yit strain. Gut 43:71–78. https://doi.org/10.1136/gut.43.1.71
57. Bamias G, Marini M, Moskaluk CA, Odashima M, Ross WG, Rivera-Nieves J, Cominelli F (2002) Down-regulation of intestinal lymphocyte activation and Th1 cytokine production by antibiotic therapy in a murine model of Crohn's disease. J Immunol 169:5308–5314. https://doi.org/10.4049/jimmunol.169.9.5308

Chapter 27

A Rodent Model of Hypertrophic Scarring: Splinting of Rat Wounds

Dong Ok Son and Boris Hinz

Abstract

Human hypertrophic scars are the result of imperfect healing of skin, which is particularly evident from the scars developing after severe burns. In contrast, mouse and rat full-thickness skin wounds heal normally without forming visible scar tissue, which reduces the suitability of rodent models for the study of skin scarring. We here provide a simple procedure to splint the edges of full-thickness rodent skin with a sutured plastic frame that prevents wound closure by granulation tissue contraction. The resulting mechanical tension in the wound bed and the lack of neo-epithelium amplify myofibroblast formation and generate hypertrophic features, not unlike those of human skin. In addition to producing scar tissue, the splint provides a reservoir that can be used for the delivery of cellular and acellular wound treatment regimen. Despite being simple and almost historical, wound splinting is a robust and reliable model to study myofibroblast biology.

Key words α-Smooth muscle actin, Stress fibers, Mechanical stress, Hypertrophic scarring, Wound healing, Splint, Myofibroblast

1 Introduction

Myofibroblast biology is most widely studied in the context of fibrosis where myofibroblast activities result in connective tissue deformations that impede organ function, sometimes to the point of failure [1]. However, the discovery of the myofibroblast in 1971 by Gabbiani and coworkers was originally motivated by the desire to understand the fundamental principles of dermal wound closure—or more generally, how mammalian tissues repair after injury and damage [2, 3]. It has been known since the work of Carrel in the early twentieth century that the forces of wound contraction are generated within the healing tissue [4], which is called granulation tissue because of the granular appearance of neo-vessels in histology sections. Until five decades ago, two potential mechanisms were still highly debated to be at the origin of these forces: either active cellular contraction or rearrangement of

Boris Hinz and David Lagares (eds.), *Myofibroblasts: Methods and Protocols*, Methods in Molecular Biology, vol. 2299, https://doi.org/10.1007/978-1-0716-1382-5_27,

collagen extracellular matrix by a "shrinking"-like process. The discovery of non-muscle cells (fibroblasts) with muscle features, i.e., contractile actin-myosin bundles, within granulation tissue [2] has largely ended this debate.

Full-thickness skin wounds produced in rabbit skin and ears and in rodent dorsal skin are the prototypical small animal models to study wound healing processes [5–7]. However, in contrast to humans, such loose skin animals possess a layer of striated muscle located within the subcutaneous layer of the skin, called panniculus *carnosus* [8] (Fig. 1). The presence of "true" muscle as part of rodent skin has led to the occasional misconception that wound contraction is promoted by the panniculus *carnosus*. However, this muscle layer is excised in all full-thickness wound models and it has been shown in numerous studies that excised strips of wound granulation tissue, devoid of this skeletal muscle structure, contract in response to smooth muscle agonists [9, 10]. Nevertheless, there are important differences between rodent and human skin wound healing, one of which is the degree of scarring after wound closure. Healing of larger adult human skin wounds is almost always imperfect and stiff hypertrophic scar tissue remains at the original wound site even decades after the insult. Hypertrophic scarring is particularly evident in burn patients that suffer from the severe scarring of large area skin wounds, often for the rest of their lives [11]. Thus, achieving scarless healing is an important surgical aim—far beyond mere esthetic considerations [12, 13].

One crucial element driving human skin scarring—as opposed to rodent "scarless" healing—is mechanical stress, partly exerted by the higher mechanical resistance of tight human versus loose rodent skin [14–17]. To enhance mechanical stress in the granulation tissue of rodents various models of hypertrophic scarring have been developed [18–21]. In rodent skin, splints are used to restrain contraction of full-thickness wounds passively or actively using actuated expander devices sutured to the edges of incisional wounds [21–24]. The possibly simplest approach can be traced back at least to the ground-breaking works of Abercrombie and coworkers, who splinted full-thickness rabbit skin wounds with a plastic frame, superglued to the wound edges [23]. The procedure that we describe herein for rat wounds is essentially derived from this seminal study, which is highly recommended to readers who are interested in the younger history of wound healing research [23]. Mechanical fixation of the wound edges prevents re-epithelialization which is also a characteristic of human large area wounds and accelerates the buildup of disorganized granulation tissue and formation of myofibroblasts [21]. In our own studies, we have used this approach to demonstrate enhanced formation of myofibroblast features in granulation tissue by stress [21, 25, 26] and to treat hypertrophic scarring in vivo with anti-myofibroblast compounds [27]. Another advantage of the splint

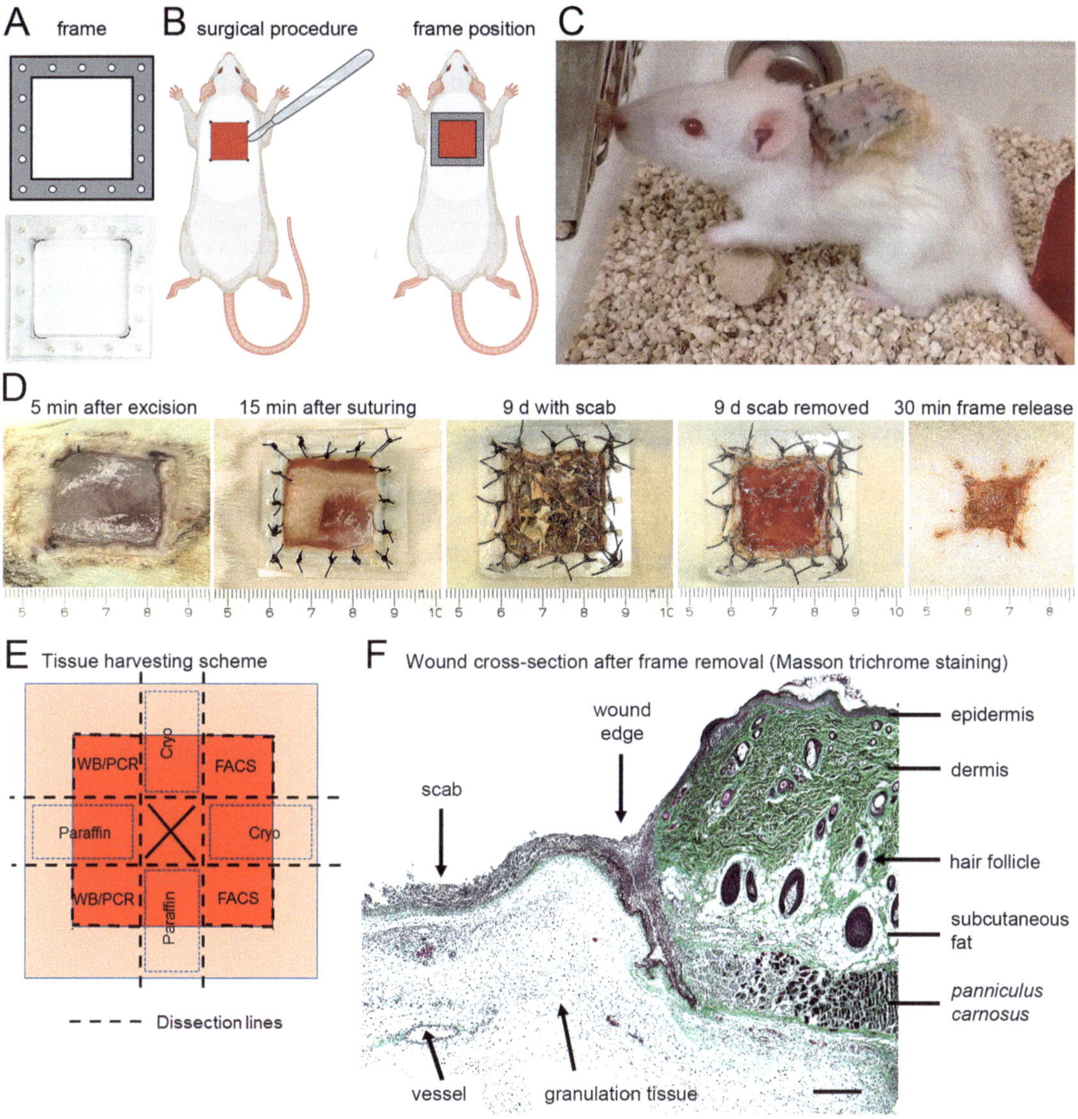

Fig. 1 Splinted wounds. (**a**) Acrylic plastic frame design and picture. (**b**) Schematics to illustrate positioning of the frame on the rat dorsum and side view of wounded rat skin layers with superposed frame. (**c**) Rats recover quickly after the surgery and are astonishingly little affected by the presence of a frame on their back. (**d**) Few minutes after excision of a 2.5 × 2.5 cm skin area, the wound edges expand beyond the initial size due to release of skin resting tension. The frame is sutured to maintain a 2.5 × 2.5 cm wound area over several days post-wounding. In open wounds, a scab develops within one day and needs to be carefully removed with forceps just before removing the splint by cutting the sutures. Following removal of the frame, the wound contracts within 1 h to ~50% of its initial surface area due to the myofibroblast-derived tension release. (**e**) Scheme provides an example how wound tissue can be collected for further analysis. For histological analysis such as after Masson trichome staining in (**f**), it is advised to include parts of the adjacent normal skin as "internal" control tissue. Scale bar: 5 mm

model is that the frame provides a reservoir that can be used to apply therapeutic cells, such as mesenchymal stromal cells [28], which will be briefly indicated in the following protocol.

2 Materials

2.1 Experimental Animals: Rats

1. Animal protocols approved by animal ethics committees.
2. Female Wister strain rats obtained from a commercial breeder, 10–12 weeks old (225–250 g).
3. Animal housing room with automated light-dark cycle control and free access to food and water.
4. Clean cages with bedding, autoclaved water bottle, food, and animal toys.
5. Animal scale.

2.2 Anesthesia, Operation Preparation, and Personal Protective Equipment (See Note 1)

1. Hair caps.
2. Surgical face mask.
3. Gown.
4. Surgical gloves.
5. Anesthesia setup with isoflurane and oxygen tank.
6. Nail clipper.
7. Hair clipper.
8. Gauze.
9. 70% Alcohol.
10. 10% Povidone-iodine (betadine).
11. Analgesics (e.g., slow release (SR) buprenorphine).
12. Needles (23G).
13. Syringes (1 mL).
14. Saline solution (0.9% NaCl).
15. Eye cream to prevent animal eye drying during surgery.
16. Alcohol resistant marker.
17. Thermally regulated water blanket to keep animal warm during surgery.

2.3 Surgical Instruments

1. Sterilized surgical packs: scalpel handle, tweezers, scissors, needle holder, gauzes (*see* **Note 2**).
2. Scalpel blades #10.
3. Nonabsorbable surgical silk threads (2-0 and 3-0).
4. Sterilized or disposable drape.

5. Dry glass beads sterilizer.
6. Sterile acrylic plastic frames for wound splinting (*see* **Note 3**) (Fig. 1a).

2.4 Post-Operation Treatment

1. Optional for cell delivery to the wounds: thrombin (500 μL of 50 mg/mL per rat) and fibrinogen (500 μL of 4 IU/mL per rat).
2. Tegaderm (optional-only needed for cell transplant studies; *see* **Note 4**).
3. Local anesthetic cream (e.g., Emla from AstraZeneca, 2.5% Lidocaine and 2.5% prilocaine) for pain control.
4. Anti-inflammatory ointment (e.g., Polysporin).
5. Heat lamp.

2.5 Sampling

1. One-side razor blades.
2. Forceps.
3. Dissection board (e.g., Styrofoam).
4. Needles to pin skin tissue.
5. 50 mL Tubes.
6. Cryo mold.
7. Embedding compound, e.g., optimal cutting temperature (OCT) compound.
8. 1.5 mL Cryovials.
9. Liquid nitrogen.
10. 3% Paraformaldehyde (PFA).
11. Scissors (e.g., Iris scissors) to chop tissues.
12. Freshly prepared collagenase/DNase (7.5 mg collagenase + 2 mg DNase + 5 mL HBSS).
13. Phosphate-buffered saline (PBS) and 70% ethanol to wash sampling instruments.

3 Methods

3.1 Animal Order and Preparation

1. Order animals in advance and acclimatize animals in the animal facility for 7 d or according to local regulations.
2. Book operation room and surgical preparation equipment in advance.
3. Autoclave surgical instruments with indicator strip.
4. Check animal health condition and weigh animals 24–48 h before surgery to prepare appropriate dose of analgesics.

3.2 Anesthesia and Operative Preparation

1. Set the preparation table with anesthesia machine, hair clipper, nail clipper, eye cream, 70% ethanol, and betadine.
2. Set the operation table with anesthesia machine, water blanket with cover, surgical instruments, analgesic, and saline solution.
3. Prepare the recovery stage with clean cages lined with paper towel under a safe heat source to pre-warm.
4. Calculate analgesics volumes based on concentration of the drug, animals' weight, and drug dosages. Prepare each volume of analgesics and 1 mL of saline in an appropriately sized syringe with needle (e.g., 1 mL syringes with 23G needles for rats).
5. Transfer animals from the housing room to the operation room.
6. Anesthetize rats with 5% isoflurane (usually less than 5 min) (*see* **Note 5**) and continuously monitor breathing pattern, color of extremities, temperature, and toe pinch response.
7. Maintain rats under 2.0–2.5% isoflurane with an oxygen flow of 1 L/min.
8. Apply eye cream to rat to prevent drying of the conjunctiva.
9. Inject 1 mL of warm saline solution subcutaneously to avoid dehydration.
10. Inject slow release (SR) buprenorphine (1.2 mg/kg) subcutaneously once.
11. Record the concentration and volume of buprenorphine, time of administration, animal registration number, and weight on the anesthesia log sheet.
12. Shave the animal back fur thoroughly to remove hairs in an area that is 2 cm wider (16 cm^2) than the wound excision area (4 cm^2) to reduce contamination of the wound bed with fur (Fig. 1b).
13. Clip animal nails to minimize scratching of the wound by the rat post-operation.
14. Place position marks with alcohol-resistant pen on the animal dorsal skin, using the four corners of the frame (Fig. 1b). The shoulder level in the middle of the dorsum will be good position to prevent removal of splint by the animal.
15. Disinfect rat skin with 70% ethanol and 10% betadine three times (last betadine on operation table).
16. Change to new gloves, clean gown, mask, and hair net to avoid contamination of the surgical wounds with rat hair from the shaving procedure.

17. Transfer animal to a thermally regulated water blanket and keep under 2.0–2.5% isoflurane via nose cone to maintain general anesthesia on operation table.
18. Disinfect plastic frames with 70% ethanol and 10% betadine three times on the operation table (*see* **Note 3**).

3.3 Generation of Splinted Wounds

1. Open all surgical packs on the operation table.
2. Wash your hands and put on surgical gloves.
3. Cover the animal with drape except the surgical area.
4. Generate a full-thickness dorsal wound of 2.5 × 2.5 cm at the shoulder level centered to the animal's axis on the using sterile surgical scissors (*see* **Note 6**) (Fig. 1b–d): Lift skin with tissue forceps and make a sharp cut including the cutaneous muscle with scalp. Create the full wound with Metzenbaum scissors (curved blunt surgical scissors) not to damage any other tissues.
5. Collect the excised skin patch for control histology if required.
6. Suture the edges of the wound onto a plastic frame with suture holes, using nonabsorbable surgical silk (thickness 3-0). Double suture four corners with 2-0 thinker threads (*see* **Note 7**) (Fig. 1d). For studies comparing normally healing to mechanically restrained wounds [21, 25, 26], it is good practice but not required to also perform a mock suture without the frame being present.
7. Check animal breathing pattern, color of extremities, temperature, and toe pinch response regularly during operation (*see* **Note 8**).
8. Optional—only needed for therapeutic cell delivery: Using a 1 mL syringe without needle, mix thrombin and fibrinogen with suspended therapeutic cells (e.g., mesenchymal stromal cells [28]) to a final volume of 1 mL without forming bubbles and immediately apply onto the splinted wound.
9. Required when applying therapeutic cells; otherwise optional: Cover the wound by wrapping Tegaderm around the frame edges (*see* **Note 4**).
10. Apply local anesthetic cream (e.g., Emla) for rat pain control.
11. Turn the isoflurane concentration to 0% but keep oxygen flow running.

3.4 Post-Operative Management

1. Transfer animals individually to a clean cage with a heat source and ensure that the animal cannot overheat or burn. Monitor animals continuously until they are fully awake (*see* **Note 9**) (Fig. 1c).

2. Place cage cards on each cage, including information which procedure has been performed, and all drugs that have been administrated, in addition to all information required by local regulations. Complete monitoring sheets with operation time, drug administration time, and animal information.
3. Clean surgical instruments with soap water and sterilize in a dry glass beads sterilizer to be ready for the next animal surgery. Ensure that instruments cool down before using them; you will notice.
4. Return animals to regular housing room and check daily for body temperature, food intake, locomotion, behavior, and any complications (*see* **Note 10**) (Fig. 1c).

3.5 Sample Collection and Analysis

1. After defined time points (e.g., 2, 4, 6, 9, and 12 d post-wounding), euthanize animals by exposing to CO_2 followed by cervical dislocation.
2. Carefully remove Tegaderm from covered wounds. In open wounds, a scab will develop within one day and needs to be carefully taken off using general-purpose broad-tipped dressing forceps just before detaching the splint.

   Detach the frame by cutting all sutures with a scalpel or Iris scissors. Observe wound contraction for 15–30 min (*see* **Note 11**) (Fig. 1d). Wound area reduction following splint release can be quantified from images as a measure for wound tension (initial wound area of 4 cm^2/wound area × min after release).
3. Cut a large flap of skin with the wound area using straight-bladed Mayo scissors (at least 2 cm wider than the wound area) and place it and pin four covers with needles on a dissection board. Dissect the skin with one-side razor blade.
4. Harvest samples for protein analysis (e.g., flow cytometry or immunoblotting), RNA analysis (e.g., real-time qPCR), and immunohistochemical analysis (e.g., immunofluorescence staining of cryo-sections or paraffin sections) according to the pattern shown in Fig. 1e (*see* **Notes 12** and **13**).
5. Other parameters can be assessed to quantify wound quality: Collagen density and alignment can be determined by image analysis of second-harmonic generation images and histological sections stained with Sirius red [25] or Masson's trichrome (Fig. 1f). Wound-breaking strength can be measured using a mechanical strain tester [29] and wound stiffness using atomic force microscopy [30]. Splinted wounds will not re-epithelialize due to the presence of the frame; however epidermal thickness at the wound edge can be determined using morphometric analysis of histological sections.

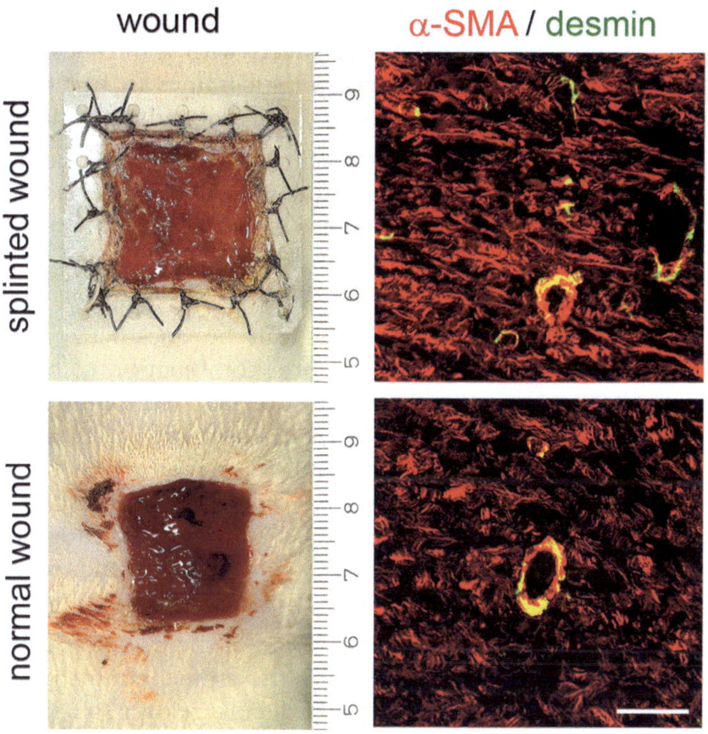

Fig. 2 Wound harvest and analysis. Splinted rat wounds are compared to normally healing wounds after 9 d of healing. Cross-sections of cryo-embedded wound granulation tissue has been stained for alpha smooth muscle actin (α-SMA) and vascular smooth muscle marker desmin. Occurrence of desmin-negative and α-SMA positive myofibroblasts is strongly enhanced by wound splinting. Scale bar: 40 μm

6. Myofibroblasts can be detected in immunofluorescence as α-SMA-positive, desmin-negative cells (Fig. 2). Additional myofibroblast markers in the ECM of granulation tissue are the extradomain-A (ED-A) splice variant of fibronectin [21] or periostin [31].

4 Notes

1. Use additional protective garments such as shoe covers, goggles, double gloves, and disposable gowns when handling contamination level 2 (CL2) materials including virus and primary human cells for transplant studies.
2. Surgical tools must be cleaned and disinfected with a dry glass beads sterilizer between every animal surgery. Surgical packs should be changed every four animals to avoid contamination and instruments becoming dull over time.

3. For our experiments, plastic frames were cut out of acrylic plastic sheets with a thickness of 2 mm using a laser cutter. Acrylic plastic frames are not autoclavable but can be sterilized with 70% ethanol and betadine preceding the operation.
4. Tegaderm will keep the wound moist and change healing characteristics. Dry wounds develop scar features better compared to moist wounds. There is also a risk that rats will remove overhanging Tegaderm cover, leading to uneven healing of the partly exposed wounds.
5. Rodents can be anesthetized with inhalant gas, injectable drugs, or a combination of both. Inhalants (e.g., isoflurane) are easy to control but the surgeon may be exposed if there is a leak in the system; symptoms are nausea, dizziness, fatigue, and headache. Injectable drugs (ketamine/xylazine at 90 mg/kg for rats) only affect the experimental animal but are more difficult to adjust. In overdose animals will die, in underdose animals will wake up and repeated drug injection will be required during surgery.
6. The position of the wound is chosen to prevent removal of the splint by the animal. Immediately after excision, the wound surface will enlarge over the 2.5×2.5 cm excision due to skin strain over the rodent body axis. Because of higher strain along the caudal-dorsal axis, non-splinted wounds will become somewhat asymmetric.
7. High granulation tissue tension will build up especially in the frame corners which can lead to detachment of the skin when one suture is cutting through.
8. Continuous animal condition check is important during operation. If extremities are of light blue color, reduce isoflurane and continue to supply oxygen via nose cone. If the rat stops breathing, massage its heart. If animals respond to a toe pinch test, increase isoflurane concentration.
9. Usually, there is no bedding in the recovery cage. Because it is delicate to lift the rat with a frame on their back, line paper towels on top of beddings and remove the paper towels once animal is fully awake. Provide food, water, and toys when animals are recovered and start moving around.
10. If rat wounds have been provided with biohazardous agents such as human-derived cells, virus, or toxic chemicals, animals will have to be maintained in biosafety level 2 facilities for at least 48 h and then returned to regular housing room. This period and regulations will change according to local authorities.

11. Following removal of the frame, the wound can spontaneously contract within 1 h by 25–75% of its initial surface area, depending on wound age and amount of myofibroblast-derived stored tension [28]. Be aware that this tension gradually increases over time and can eventually exceed the strength of the frame plastic, and/or sutures will cut through the skin. The resulting damage will result in uneven tension distribution and results will not be interpretable. We have not been able to keep an intact splint beyond day 12 post-wounding using the rat model.
12. For morphological assessment by histology, immunohistochemistry, or immunofluorescence, dissect wound tissue with adjacent normal skin which will serve as internal control in each sample. For any analysis that includes loss of positional information (e.g., immunoblotting, PCR, flow cytometry), make sure to excise comparable regions of the wound as indicated (Fig. 1e). Rodent healing of full-thickness wounds starts from the edges rather than from the central deep wound bed that has been excised in full-thickness wounds [32]. Consequently, there will be spatial differences in cell and ECM composition between the wound edges and the center (Fig. 1f).
13. For protein and RNA extraction, chop tissues into small pieces with Iris scissors on cryovial and snap freeze in liquid nitrogen. For flow cytometry assay, cut into small pieces with Iris scissors on 50 mL tubes and add 5 mL collagenase/DNase to digest and isolate singe cells. For Cryo sections, place the tissue on the mold and add embedding compound (e.g., optimal cutting temperature compound) and snap freeze in liquid nitrogen. You can keep your samples in −80 °C until you process samples. For paraffin sections, put tissues in 50 mL tube and add 3% paraformaldehyde (PFA) for 24–48 h to fix tissues. *See* our chapter on "Myofibroblast markers and microscopy detection methods in cell culture and histology" in this issue and follow the detailed procedures described therein.

Acknowledgments

The research of BH is supported by a foundation grant from the Canadian Institutes of Health Research (#375597) and support from the John Evans Leadership funds (#36050 and #38861) and innovation funds ("Fibrosis Network, #36349") from the Canada Foundation for Innovation (CFI) and the Ontario Research Fund (ORF).

References

1. Hinz B, Lagares D (2020) Evasion of apoptosis by myofibroblasts: a hallmark of fibrotic diseases. Nat Rev Rheumatol 16:11–31. https://doi.org/10.1038/s41584-019-0324-5
2. Gabbiani G, Ryan GB, Majno G (1971) Presence of modified fibroblasts in granulation tissue and their possible role in wound contraction. Experientia 27:549–550
3. Ribatti D, Tamma R (2019) Giulio Gabbiani and the discovery of myofibroblasts. Inflamm Res 68:241–245. https://doi.org/10.1007/s00011-018-01211-x
4. Carrel A (1910) The treatment of wounds. A first article. J Am Med Assoc 55:2148–2150
5. Dorsett-Martin WA (2004) Rat models of skin wound healing: a review. Wound Repair Regen 12:591–599. https://doi.org/10.1111/j.1067-1927.2004.12601.x
6. Nabai L, Ghahary A (2017) Hypertrophic scarring in the rabbit ear: a practical model for studying dermal fibrosis. Methods Mol Biol 1627:81–89. https://doi.org/10.1007/978-1-4939-7113-8_6
7. Li J, Wang J, Wang Z, Xia Y, Zhou M, Zhong A, Sun J (2020) Experimental models for cutaneous hypertrophic scar research. Wound Repair Regen 28:126–144. https://doi.org/10.1111/wrr.12760
8. Naldaiz-Gastesi N, Bahri OA, Lopez de Munain A, McCullagh KJA, Izeta A (2018) The panniculus carnosus muscle: an evolutionary enigma at the intersection of distinct research fields. J Anat 233(3):275–288. https://doi.org/10.1111/joa.12840
9. Gabbiani G, Hirschel BJ, Ryan GB, Statkov PR, Majno G (1972) Granulation tissue as a contractile organ. A study of structure and function. J Exp Med 135:719–734
10. Tomasek JJ, Gabbiani G, Hinz B, Chaponnier C, Brown RA (2002) Myofibroblasts and mechano-regulation of connective tissue remodelling. Nat Rev Mol Cell Biol 3:349–363
11. Amini-Nik S, Yousuf Y, Jeschke MG (2018) Scar management in burn injuries using drug delivery and molecular signaling: current treatments and future directions. Adv Drug Deliv Rev 123:135–154. https://doi.org/10.1016/j.addr.2017.07.017
12. Barnes LA, Marshall CD, Leavitt T, Hu MS, Moore AL, Gonzalez JG, Longaker MT, Gurtner GC (2018) Mechanical forces in cutaneous wound healing: emerging therapies to minimize scar formation. Adv Wound Care (New Rochelle) 7:47–56. https://doi.org/10.1089/wound.2016.0709
13. Gurtner GC, Werner S, Barrandon Y, Longaker MT (2008) Wound repair and regeneration. Nature 453:314–321
14. Kuehlmann B, Bonham CA, Zucal I, Prantl L, Gurtner GC (2020) Mechanotransduction in wound healing and fibrosis. J Clin Med 9(5):1423. https://doi.org/10.3390/jcm9051423
15. Hinz B (2010) The myofibroblast: paradigm for a mechanically active cell. J Biomech 43:146–155. https://doi.org/10.1016/j.jbiomech.2009.09.020
16. Hinz B, McCulloch CA, Coelho NM (2019) Mechanical regulation of myofibroblast phenoconversion and collagen contraction. Exp Cell Res 379:119–128. https://doi.org/10.1016/j.yexcr.2019.03.027
17. Padmanabhan J, Maan ZN, Kwon SH, Kosaraju R, Bonham CA, Gurtner GC (2019) In vivo models for the study of fibrosis. Adv Wound Care (New Rochelle) 8:645–654. https://doi.org/10.1089/wound.2018.0909
18. Davidson JM, Yu F, Opalenik SR (2013) Splinting strategies to overcome confounding wound contraction in experimental animal models. Adv Wound Care 2(4):142–148
19. Kwan P, Hori K, Ding J, Tredget EE (2009) Scar and contracture: biological principles. Hand Clin 25:511–528. https://doi.org/10.1016/j.hcl.2009.06.007
20. Carlson MA, Longaker MT, Thompson JS (2003) Wound splinting regulates granulation tissue survival. J Surg Res 110:304–309
21. Hinz B, Mastrangelo D, Iselin CE, Chaponnier C, Gabbiani G (2001) Mechanical tension controls granulation tissue contractile activity and myofibroblast differentiation. Am J Pathol 159:1009–1020
22. Aarabi S, Bhatt KA, Shi Y, Paterno J, Chang EI, Loh SA, Holmes JW, Longaker MT, Yee H, Gurtner GC (2007) Mechanical load initiates hypertrophic scar formation through decreased cellular apoptosis. FASEB J 21:3250–3261
23. Abercrombie M, James DW, Newcombe JF (1960) Wound contraction in rabbit skin, studied by splinting the wound margins. J Anat 94:170–182
24. Davidson JM, Yu F, Opalenik SR (2013) Splinting strategies to overcome confounding wound contraction in experimental animal models. Adv Wound Care (New Rochelle) 2:142–148. https://doi.org/10.1089/wound.2012.0424
25. Coelho NM, Arora PD, van Putten S, Boo S, Petrovic P, Lin AX, Hinz B, McCulloch CA (2017) Discoidin domain receptor 1 mediates

myosin-dependent collagen contraction. Cell Rep 18:1774–1790. https://doi.org/10.1016/j.celrep.2017.01.061

26. Klingberg F, Chow ML, Koehler A, Boo S, Buscemi L, Quinn TM, Costell M, Alman BA, Genot E, Hinz B (2014) Prestress in the extracellular matrix sensitizes latent TGF-beta1 for activation. J Cell Biol 207:283–297. https://doi.org/10.1083/jcb.201402006

27. Hinz B, Gabbiani G, Chaponnier C (2002) The NH2-terminal peptide of alpha-smooth muscle actin inhibits force generation by the myofibroblast in vitro and in vivo. J Cell Biol 157:657–663

28. Li CX, Talele NP, Boo S, Koehler A, Knee-Walden E, Balestrini JL, Speight P, Kapus A, Hinz B (2017) MicroRNA-21 preserves the fibrotic mechanical memory of mesenchymal stem cells. Nat Mater 16:379–389. https://doi.org/10.1038/nmat4780

29. Zebardast N, Lickorish D, Davies JE (2010) Human umbilical cord perivascular cells (HUCPVC): a mesenchymal cell source for dermal wound healing. Organogenesis 6:197–203

30. Goffin JM, Pittet P, Csucs G, Lussi JW, Meister JJ, Hinz B (2006) Focal adhesion size controls tension-dependent recruitment of alpha-smooth muscle actin to stress fibers. J Cell Biol 172:259–268

31. Elliott CG, Wang J, Guo X, Xu SW, Eastwood M, Guan J, Leask A, Conway SJ, Hamilton DW (2012) Periostin modulates myofibroblast differentiation during full-thickness cutaneous wound repair. J Cell Sci 125:121–132. https://doi.org/10.1242/jcs.087841

32. Correa-Gallegos D, Jiang D, Christ S, Ramesh P, Ye H, Wannemacher J, Kalgudde Gopal S, Yu Q, Aichler M, Walch A, Mirastschijski U, Volz T, Rinkevich Y (2019) Patch repair of deep wounds by mobilized fascia. Nature 576:287–292

Chapter 28

Three-Dimensional Model of Hypertrophic Scar Using a Tissue-Engineering Approach

Veronique J. Moulin

Abstract

Following wound healing, skin is replaced by a specialized tissue called scar. Sometime, this scar can become pathologic, called hypertrophic scar, with a high amount of extracellular matrix, capillaries, and myofibroblast persistence. To understand the mechanisms at the origin of the fibrosis is paramount to treat patients, but despite few animal models and in vitro studies using mainly human pathological cells cultured on plastic on monolayer, the treatment of these fibrotic scars remains unsatisfactory. As in tissue, cells are most often imbedded in extracellular matrix, we have developed, using a tissue engineering method, new in vitro models to study human fibrotic skin pathologies as hypertrophic scars. Human cells isolated from hypertrophic scars are used to reconstitute a three-dimensional fibrotic skin comprising both dermal and epidermal parts. This method called the self-assembly approach is based on the cell capacity to reconstitute their own environment as in vivo. In this chapter, the described methods include extraction and culture of human scar keratinocytes and fibroblasts from cutaneous biopsies as well as the protocols to produce fibrotic skin that can be used to study pathological process.

Key words Fibrosis, Hypertrophic scar, Skin, Skin substitutes, Reconstructed tissues, Tissue engineering

1 Introduction

Hypertrophic scars are fibrotic cutaneous pathologies that result from a perturbed wound healing process. They are often observed following large burn wound area and are thought to be one of the main long-term problems for these patients. The treatment of fibrotic scars is limited because of the limited knowledge concerning the mechanisms at the origin of these problems. In addition to a few animal models [1–3], the pathology can be studied using human cells cultured on plastic on monolayer. These studies have been useful to determine that fibrotic fibroblasts produce a higher quantity of extracellular matrix elements and/or

Boris Hinz and David Lagares (eds.), *Myofibroblasts: Methods and Protocols*, Methods in Molecular Biology, vol. 2299, https://doi.org/10.1007/978-1-0716-1382-5_28,

lower quantities of degradative enzymes than fibroblasts isolated from normal skin [4, 5]. However, cell behavior is known to be different according to the environment. Moreover, in tissue cells are most often imbedded in extracellular matrix, whose composition can also modulate cell phenotype [6–8]. Using a mix of extracellular matrix and cells, the aim of tissue engineering method is to recreate a tissue in vitro that mimics the architecture and the functionality of the original tissue.

The research group LOEX from Université Laval in Quebec city, Canada, has used the natural properties of cells to secrete and organize matrix as well as to spontaneously organize themselves into biological structures as capillary network [9, 10] or multilayer epithelium [11, 12]. Using specific culture conditions, cells can reproduce a tissue that is very similar than those observed in vivo. We call this the self-assembly method. According to the used cells, we have developed various tissues like skin [11, 12], blood vessel [13], adipose tissues [14], or cornea [15] that can be used to replace defective tissues in patients [16, 17]. However, these in vitro reconstituted tissues can also be used to understand physiological mechanisms as well as pathological ones [18–20].

My team has utilized the self-assembly method to study pathological fibrotic mechanisms, using keratinocytes and fibroblasts that were isolated from hypertrophic scars or normal skin obtained after reconstructive surgery [20–22]. Three-dimensional (3D) fibrotic scar is then reconstructed using the self-assembly approach and compared with 3D skin, reconstructed with cells isolated from normal skin. A mix of cells can also be used to better understand cell-cell interactions. We have demonstrated that the origin of the keratinocytes is crucial on the final thickness of the dermis: keratinocytes originating from fibrotic scars can stimulate fibrotic dermis formation, even when normal skin fibroblasts are used to reconstitute dermis [20, 22]. This result allows to hypothesize that epidermis can play a role into fibrotic development of the skin.

In this chapter, the methods for extraction and culture of fibroblasts and keratinocytes from the same skin biopsy are described in detail, followed by protocols for the reconstruction of the fibrotic scars using the self-assembly approach. This is an update of the previously published protocols in [23]. Our protocols allow to obtain a thick dermis constituted of fibroblasts imbedded into a complex extracellular matrix as well as a differentiated epidermis (Fig. 1). This model is useful to further understand pathophysiological process of the fibrosis.

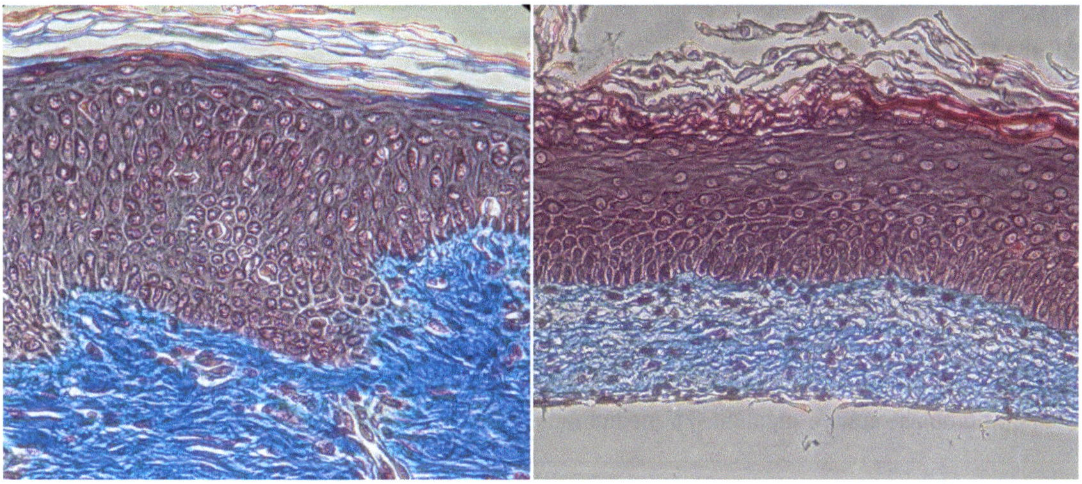

Fig. 1 Histological cross-section of human skin (left) and human reconstructed skin after 2 weeks of culture at the air/liquid interface (right) stained by Masson's trichrome

2 Materials

2.1 Culture Media

2.1.1 Tissue Transport Medium

1. Dulbecco's modified Eagle's medium (DME): 13.5 g/L DME, 3.7 g/L $NaHCO_3$ (44 mM). Dissolve in apyrogenic ultrapure water and adjust pH to 7.1. Sterilize by filtration through a 0.22 μm low-binding disposable filter. Distribute in 450 mL aliquots and store protected from light at 4 °C.
2. Fetal calf serum: Thaw fetal calf serum (e.g., cat. no. 088-150, Wisent, Saint-Bruno, QC, Canada) in cold water. Inactivate complement proteins in 56 °C hot water for 30 min. Distribute in 50 mL aliquots and store at −20 °C until use (*see* **Note 1**).
3. Antibiotics solution: Dissolve 50,000 IU/mL of penicillin G and 12.5 mg/mL of gentamicin sulfate in apyrogenic ultrapure water to make a 500× stock solution. Sterilize by filtration through a 0.22 μm low-binding disposable filter, distribute in 1 mL aliquots, and store at −80 °C.
4. Fungizone: Dissolve 0.25 mg/mL of amphotericin B (0.27 mM) in apyrogenic ultrapure water to make a 500× stock solution. Sterilize by filtration through a 0.22 μm low-binding disposable filter, distribute in 500 μL aliquots, and store at −80 °C.
5. Thaw all components at 4 °C and mix according to Table 1. Aliquoted transport medium in sterile container can be stored at 4 °C (1 month) or −20 °C (6 months).

Table 1
Tissue transport medium

Component	Quantity	Final concentration
DME	900 mL	90% (v/v)
Fetal calf serum	100 mL	10% (v/v)
Penicillin G/Gentamicin 500×	2 mL	Penicillin G 100 IU/mL, Gentamicin 25 μg/mL
Fungizone 500×	2 mL	0.5 μg/mL

Table 2
Human fibroblast culture medium (Fb medium)

Component	Quantity	Final concentration
DME	900 mL	90% (v/v)
Fetal calf serum	100 mL	10% (v/v)
Penicillin G/Gentamicin 500×	2 mL	Penicillin G 100 IU/mL, Gentamicin 25 μg/mL

2.1.2 Human Fibroblast Culture Medium (Fb Medium)

1. To prepare Fb medium, thaw and mix selected components described in Subheading 2.1.1. at 4 °C according to Table 2: 90% DME, 10% Fetal Calf Serum, Penicillin G/Gentamicin.
2. Fb medium can be stored protected from light at 4 °C for 10 d.

2.1.3 Human Keratinocyte Culture Medium (K Medium)

1. DME-Ham: Mix DME and Ham's F12 medium in 3:1 ratio. Add 3.07 g/L $NaHCO_3$ (36.54 mM), 24.3 mg/L adenine (0.18 mM). Dissolve in apyrogenic ultrapure water and adjust pH to 7.1. Sterilize by filtration through a 0.22 μm low-binding disposable filter. Aliquot and store protected from light at 4 °C.
2. FetalClone II serum: Thaw FetalClone II serum (cat. no. SH30066.03, HyClone, GEHealthcare Life Sciences, Mississauga, Ont, Canada) in cold water. Inactivate in hot water (56 °C) for 30 min. Distribute in single use aliquots and store at −20 °C.
3. Insulin stock solution: Dissolve 250 mg insulin in 50 mL 0.005 N HCl to make a 1000× stock solution (0.87 mM). Sterilize by filtration through a 0.22 μm low-binding disposable filter, distribute in single use aliquots, and store at −80 °C.
4. Hydrocortisone stock solution: Dissolve 25 mg hydrocortisone in 5 mL of 99% ethanol and complete to 125 mL with DME-Ham to make a 500× stock solution (0.55 mM). Sterilize by filtration through a 0.22 μm low-binding disposable filter, distribute in single use aliquots, and store at −80 °C.

5. Isoproterenol hydrochloride stock solution: Dissolve 21.2 mg isoproterenol hydrochloride in 100 mL of HCl 0.001 N to make a 1000× stock solution (8.55×10^{-4} M). Sterilize by filtration through a 0.22 μm low-binding disposable filter, distribute in single use aliquots, and store at −80 °C.
6. Epidermal growth factor stock solution: Dissolve 500 μg epidermal growth factor in 2.5 mL of 10 mM HCl. Complete to 50 mL with DME-Ham (*see* Subheading 2.1.3, **step 1**) supplemented with 10% (v/v) fetal clone II serum to make a 1000× stock solution. Sterilize by filtration through a 0.22 μm low-binding disposable filter, distribute in single use aliquots, and store at −80 °C.
7. Antibiotics solution as described in Subheading 2.1.1.
8. To prepare K medium, thaw all components at 4 °C and mix according to Table 3 (*see* **Note 2**). K medium can be stored at 4 °C for 10 d.

2.1.4 Air Liquid Epidermis Culture Medium (A/L Medium)

1. Thaw all components described in Subheading 2.1.3, at 4 °C, except for epidermal growth factor according to Table 4 (*see* **Note 2**). A/L medium can be stored at 4 °C for 10 d.

2.1.5 Freezing Medium

1. Fetal calf serum as described in Subheading 2.1.1.
2. Dimethyl sulfoxide (DMSO): Distribute the stock solution (99.7%) in single-use aliquots and store at −20 °C.
3. To prepare freezing medium, thaw all components at 4 °C and mix according to Table 5. Keep on ice at 4 °C (*see* **Note 3**).

Table 3
Human keratinocyte culture medium (K medium)

Component	Quantity	Final concentration
DME-Ham	950 mL	95% (v/v)
Fetal Clone II	50 mL	5% (v/v)
Insulin 1000×	1 mL	5 μg/mL
Hydrocortisone 500×	2 mL	0.4 μg/mL
Isoproterenol 1000×	1 mL	0.212 μg/mL
Epidermal Growth Factor 1000×	1 mL	10 ng/mL
Penicillin G/Gentamicin 500×	2 mL	Penicillin G 100 IU/mL, Gentamicin 25 μg/mL

Table 4
Air/liquid culture medium (A/L medium)

Component	Quantity	Final concentration
DME-Ham	950 mL	95% (v/v)
Fetal Clone II	50 mL	5% (v/v)
Insulin 1000×	1 mL	5 μg/mL
Hydrocortisone 500×	2 mL	0.4 μg/mL
Isoproterenol 1000×	1 mL	0.212 μg/mL
Penicillin G/Gentamicin 500×	2 mL	Penicillin G 100 IU/mL, Gentamicin 25 μg/mL

Table 5
Freezing medium

Component	Quantity	Final concentration
Fetal calf serum	0.8 mL	80% (v/v)
DMSO	0.2 mL	20% (v/v)

2.2 Monolayer Culture of Human Fibroblasts

2.2.1 Extraction and Culture of Human Fibroblasts

1. The source of cells are small pieces of fibrotic skin (hypertrophic scar) removed by surgery and following procedures approved by the institution's committee for the protection of human subjects.
2. Transport medium (*see* Subheading 2.1.1).
3. 50 mL Reaction tubes.
4. Phosphate-buffered saline (PBS) with -Penicillin G/Gentamicin/Fungizone: 137 mM NaCl, 2.7 mM KCl, 6.5 mM Na_2HPO_4, 1.5 mM KH_2PO_4 in apyrogenic ultrapure water and verify that pH is 7.4. Sterilize by filtration through a 0.22 μm low-binding disposable filter. Store at room temperature. Before use, add Penicillin G-Gentamicin 500× stock solution and Fungizone 500× stock solution (*see* Subheading 2.1.1) by diluting these additives to 1 × .
5. Culture dish 100 × 15 mm (not tissue-culture treated).
6. HEPES solution: First, make a 10× HEPES stock solution in apyrogenic ultrapure water, using 0.1 M HEPES, 67 mM KCl, 1.42 M NaCl, and adjust pH to 7.3. Protect from light and store at 4 °C. To prepare HEPES 1× solution, dilute the 10× stock solution to 1× with apyrogenic ultrapure water and add 1 mM $CaCl_2$ and adjust pH to 7.45. Protect from light (*see* **Note 4**) and store at 4 °C.

7. Thermolysin stock solution: Dissolve 500 μg/mL thermolysin in HEPES 1× and sterilize by filtration through a 0.22 μm low-binding disposable filter and store at 4 °C (*see* **Note 3**).
8. Dissecting curved forceps.
9. Scalpel #4 and blade size 22.
10. Parafilm.
11. Collagenase H: Dissolve 0.125 U/mL collagenase H in Fb medium (*see* Subheading 2.1.2) and sterilize by filtration through a 0.22 μm low-binding disposable filter (*see* **Note 3**).
12. Trypsinization unit (e.g., Celstir suspension culture flask, Wheaton Sciences Products).
13. Tissue culture flask, 75 cm^2.
14. Fb medium (*see* Subheading 2.1.2).

2.2.2 Subculture of Human Fibroblasts (Passage)

1. Fb medium (*see* Subheading 2.1.2).
2. Trypsin/EDTA stock solution: 2.8 mM D-glucose, 0.05% (w/v) trypsin 1–500, 0.00075% (v/v) phenol red, 100,000 IU/L penicillin G, 25 mg/L gentamicin, 0.01% (w/v) ethylenediaminetetraacetic acid (EDTA, disodium salt). Dissolve in 1× PBS, adjust to pH 7.45, and sterilize by filtration through a 0.22 μm low-binding disposable filter. Distribute in single use aliquots, and store at −20 °C.
3. 15 mL reaction tubes.
4. Tissue culture flask, 75 cm^2.

2.2.3 Cryopreservation of Human Fibroblasts

1. Freezing medium (*see* Subheading 2.1.5).
2. Sterile cryogenic vials.
3. Freezing container, precooled at 4 °C.

2.2.4 Thawing of Human Fibroblasts

1. Fb medium (*see* Subheading 2.1.2).
2. 15 mL Reaction tubes.
3. Tissue culture flask, 75 cm^2.

2.3 Monolayer Culture of Human Keratinocytes

2.3.1 Extraction and Culture of Human Keratinocytes

1. Use as source of cells the same small piece of fibrotic human skin (hypertrophic scar) used to isolate fibroblasts.
2. Trypsin/EDTA (*see* Subheading 2.2.2).
3. Trypsinization unit (e.g., Celstir suspension culture flask, Wheaton Sciences Products).
4. 50 mL Reaction tubes.
5. Tissue culture flask, 75 cm^2.
6. K medium (*see* Subheading 2.1.3).

7. Fibroblast feeder layer: Irradiated human newborn skin fibroblasts obtained as described in this protocol are irradiated with 6000 rads with a selenium source (e.g., Gammacell® Elite irradiator) and are kept frozen in nitrogen liquid tank until used. 600,000 cells are then seeded in 75cm^2 flasks with 20 mL of K medium and kept in 8% CO2, 100% humidity atmosphere at 37 °C at least 7 d before to be used. Change medium once per week and seed keratinocytes at least 24 h after last medium change (*see* **Note 5**).

2.3.2 Subculture of Human Keratinocytes (Passaging)

1. K medium (*see* Subheading 2.1.3).
2. Feeder layer (*see* Subheading 2.3.1).
3. Trypsin/EDTA (*see* Subheading 2.2.2).
4. 50 mL Reaction tubes.
5. Tissue culture flask, 75 cm^2.

2.3.3 Cryopreservation of Human Keratinocytes

1. As described in Subheading 2.2.3 for fibroblasts.

2.3.4 Thawing of Human Keratinocytes

1. K medium (*see* Subheading 2.1.3).
2. Tissue culture flask, 75 cm^2.
3. Feeder layer (*see* Subheading 2.3.1).
4. 15 mL Reaction tube.

2.4 Formation of Dermal Sheets

1. Confluent fibroblasts between passages 2 and 6 (*see* Subheading 2.2.2).
2. Ascorbic acid stock solution: Dissolve 10 mg/mL of ascorbic acid in DME (*see* Subheading 2.1.2) to make a 200× stock solution, and sterilize by filtration through a 0.22 μm low-binding disposable filter (*see* **Note 6**).
3. Fb medium (*see* Subheading 2.1.2), containing 50 μg/mL ascorbic acid.
4. Ingots, stainless steel grade # 316.
5. Curved forceps.
6. Six-well polystyrene plate (*see* **Note** 7).
7. Anchoring paper (*see* **Note** 7): Cut Whatman sheet paper in a circle with a diameter that allows to place it in the wells of a 6-well plate. Remove the concentric inside disk of 25 mm diameter after cutting.
8. PBS (*see* Subheading 2.2.1).

2.5 Keratinocyte Addition on the Dermis

1. Keratinocytes at 80% confluency between their third and fifth passages (co-cultured with feeder layer) (Subheading 2.3.1).
2. Trypsin\EDTA (*see* Subheading 2.2.2).
3. K medium (*see* Subheading 2.1.3) containing 50 μg/mL ascorbic acid.

2.6 Skin Assembly

1. Woven Membrane Filters (e.g., Spectra Mesh®, Spectrum® Laboratories).
2. Air–liquid stand, stainless steel grade #316.
3. Curved forceps.
4. Cell culture petri dish, 60 mm diameter (not tissue-culture treated).
5. A/L medium (*see* Subheading 2.1.4) containing 50 μg/mL ascorbic acid.
6. Ligaclips and Ligaclip applicator.

3 Methods

3.1 Monolayer Culture of Human Fibroblasts

3.1.1 Extraction and Culture of Human Fibroblasts

1. Transport and conservation: In the surgery room, put the skin specimen (~1 cm^2) into a sterile container filled with cold (4 °C) transport medium (at least 30 mL). Skin specimen can be stored at 4 °C for a maximum of 24 h before to be treated. All further manipulations are performed under a sterile laminar flow hood cabinet. All solutions added on cells must be previously warmed at 37 °C.
2. Remove and discard the subcutaneous fat. Wash the skin specimen in a 50 mL tube containing 30 mL PBS with Penicillin G/Gentamicin/Fungizone. Agitate vigorously. With sterile forceps, transfer the skin specimen in another tube filled with PBS. Repeat this step three times.
3. Spread out the skin specimen, epidermis on the top, into a 100 mm cell culture dish.
4. Cut the skin into 3 mm slices width with a scalpel blade 22.
5. Add 10 mL of cold (4 °C) thermolysin. Seal the culture dish with parafilm.
6. Incubate overnight at 4 °C.
7. With two curved forceps, separate the epidermis from the dermis. Put dermal pieces within a culture dish and cut the dermis in as small as possible pieces using the scalpel.
8. Place dermal pieces within a trypsinization unit containing 20 mL of collagenase H. Incubate under agitation for 4 h at 37 °C.

9. Collect the cellular suspension. Put into a 50 mL tube and add 20 mL of Fb medium. Wash the trypsinization unit with 10 mL of Fb medium and add it to the tube (total 40 mL) (*see* **Note 8**).
10. Count cells and measure the viability by trypan blue staining.
11. Centrifuge cell suspension at 300 × *g* for 10 min at room temperature.
12. Remove supernatant and add 10 mL of Fb medium on the pellet. Resuspend cell pellet by gentle pipetting.
13. Seed 3–4 × 10^6 fibroblasts in a 75 cm^2 culture flask with 15 mL of Fb medium. Incubate in 8% CO_2, 100% humidity atmosphere at 37 °C. Change culture medium three times a week.
14. When the fibroblasts reach confluency, subculture or freeze cells.

3.1.2 Subculture of Human Fibroblasts (Passaging)

1. All following manipulations are described for one 75 cm^2 culture flask of fibroblasts and are performed under a sterile laminar flow hood cabinet. All solutions added to cells have been previously warmed to 37 °C.
2. Remove medium.
3. Wash the culture flask with 2 mL of trypsin/EDTA and discard it.
4. Add 3 mL of trypsin/EDTA. Incubate at 37 °C until the cells are detached from the flask. Gently tap the flask after than 4 min to stimulate detachment.
5. Add 3 mL of Fb medium. Collect the cellular suspension. Put into a 15 mL conical tube. Wash the flask with 4 mL of Fb medium, collect the cellular suspension, and add it to the 15 mL tube (total 10 mL).
6. Count the cells and measure the viability by trypan blue staining. The cell viability should be higher than 95%. Usually, a confluent 75 cm^2 culture flask contains 4 × 10^6 cells.
7. Centrifuge cell suspension at 300 × *g* for 10 min at room temperature.
8. Resuspend cell pellet in Fb medium.
9. Seed 3 × 10^5 fibroblasts in a 75 cm^2 culture flask with 15 mL of Fb medium. Incubate in 8% CO_2, 100% humidity atmosphere at 37 °C. Change culture medium three times a week.

3.1.3 Cryopreservation of Human Fibroblasts

1. All following manipulations are performed under a sterile laminar flow hood cabinet.
2. Following Subheading 3.1.2, resuspend fibroblasts in a given volume of 4 °C cold fetal calf serum in order to obtain 2 × 10^6 fibroblasts/mL.

3. Drop by drop with gently mixing, dilute the cell suspension to 1×10^6 fibroblasts/mL with the freezing medium (*see* **Note 9**). Aliquot 1000 μL of cells in sterile cryogenic vials.
4. Put vials in a freezing container that has previously been cooled at 4 °C and freeze overnight at −80 °C, in the freezing container.
5. Store in liquid nitrogen tank.

3.1.4 Thawing of Human Fibroblasts

All following manipulations are performed under a sterile laminar flow hood cabinet. All solutions must be previously warmed at 37 °C before being added to cells.

1. Put the cryogenic vial in the water bath at 37 °C until complete thawed.
2. Using a pipette, transfer the content of the cryogenic vial into a new 50 mL tube and add drop by drop with manual shaking 9 mL of warm Fb medium. Do not vortex.
3. Count the cells and measure viability by trypan blue staining.
4. Centrifuge cell suspension at 300 × *g* for 10 min at room temperature.
5. Resuspend cell pellet in Fb medium.
6. Seed 5×10^5 fibroblasts in a 75 cm^2 culture flask with 15 mL of Fb medium. Incubate in 8% CO_2, 100% humidity atmosphere at 37 °C. Change culture medium three times a week.

3.2 Monolayer Culture of Human Keratinocytes

3.2.1 Extraction and Culture of Human Keratinocytes

1. All following manipulations are performed under a sterile laminar flow hood cabinet. All solutions added to cells must be previously warmed at 37 °C.
2. Surgically removed epidermis is placed into a trypsinization unit containing 20 mL of 37 °C warm trypsin/EDTA. Incubate under agitation during 15–30 min at 37 °C.
3. Collect the cell suspension. Put into a 50 mL tube and add 20 mL of warm (37 °C) K medium. Wash the trypsinization unit with 10 mL of same medium, add it to the tube, and complete the volume to a total of 40 mL with K medium.
4. Count the cells and measure the viability by trypan blue staining. Cell viability should be superior to 80%.
5. Centrifuge cell suspension at 300 × *g* for 10 min at room temperature.
6. Resuspend cell pellet in K medium.
7. Seed 1×10^6 keratinocytes in a 75 cm^2 culture flask containing feeder layer and 24 h-conditioned medium. Incubate in 8% CO_2, 100% humidity atmosphere at 37 °C. Change culture medium three times a week.
8. When the keratinocytes reach 80% confluency, subculture or freeze cells.

3.2.2 Subculture of Human Keratinocytes (Passage)

All following manipulations are performed under a sterile laminar flow hood cabinet. All solutions added on cells must be previously warmed at 37 °C. Procedures are described for a 75 cm^2 culture flask of keratinocytes.

1. Remove medium.
2. Wash the culture flask with 2 mL of trypsin/ EDTA and remove after last rinse.
3. Add 8 mL of trypsin/EDTA. Incubate at 37 °C until the cells are detached from the flask (5–6 min) following a gentle tap on the flask.
4. Add 8 mL of K medium. Collect the cell suspension and transfer to a 50 mL tube. Wash the flask with 2 mL of K medium, collect the cell suspension, and add it to the 50 mL tube to a total of 18 mL.
5. Count the cells and measure the viability by trypan blue staining. Cell viability should be superior to 80%.
6. Centrifuge cell suspension at 300 × *g* for 10 min at room temperature.
7. Resuspend cell pellet in K medium.
8. Seed 3×10^5 to 7×10^5 keratinocytes in a 75 cm^2 culture flask, containing fibroblast feeder layer and 24 h conditioned medium. Incubate in 8% CO_2, 100% humidity atmosphere at 37 °C. Change culture medium three times a week.

3.2.3 Cryopreservation of Human Keratinocytes

Performed as described for fibroblasts in Subheading 3.1.3.

3.2.4 Thawing of Human Keratinocytes

All following manipulations are performed under a sterile laminar flow hood cabinet. All solutions must be pre-warmed to 37 °C before being added to cells.

1. Put the cryogenic vial in 37 °C water bath until complete thawed.
2. Using a pipette, transfer the content of the cryogenic vial into a new 50 mL tube and add drop by drop with manual shaking 9 mL of warm K medium. Do not vortex.
3. Count the cells and measure cell viability by trypan blue staining.
4. Centrifuge cell suspension at 300 × *g* for 10 min at room temperature.
5. Resuspend cell pellet in 37 °C warm K medium.
6. Seed 5×10^5 to 7×10^5 keratinocytes in a 75 cm^2 culture flask containing feeder layer and 24 h conditioned medium. Incubate in 8% CO_2, 100% humidity atmosphere at 37 °C. Change culture medium three times a week.

3.3 Human Skin Reconstruction by the Self-Assembly Approach

All following manipulations are performed under a sterile laminar flow hood cabinet. All solutions must be pre-warmed to 37 °C before being added to cells.

3.3.1 Formation of Dermal Sheets

1. Put one anchoring paper in each well into the 6-well plate (*see* **Note 7**). Rinse anchoring papers with PBS, three times to eliminate paper fragments and then remove PBS. Add two ingots on the anchoring papers to prevent moving of the anchorage.
2. Seed a cell suspension corresponding to 50,000 fibroblasts/well in 2 mL of Fb medium.
3. Incubate in an 8% CO_2, 100% humidity atmosphere at 37 °C.
4. Two days after cell seeding, replace medium by 2 mL of Fb medium containing 50 μg/mL ascorbic acid into each well. From then on, change culture media (2 mL) three times a week.

3.3.2 Keratinocyte Addition

1. Typically, after 28 d of fibroblast culture, trypsinize keratinocytes as described in Subheading 3.2.2. Centrifugate and make a suspension of 10^6 keratinocytes/mL of K medium.
2. Remove the culture medium and the ingots from half of wells containing the dermal sheets and seed 2×10^6 keratinocytes (2 mL) with the ring. Incubate in 8% CO_2, 100% humidity atmosphere at 37 °C.
3. Two hours later, add 2 mL of K medium containing 50 μg/mL ascorbic acid. Incubate in 8% CO_2, 100% humidity atmosphere at 37 °C. Change culture medium (4 mL) every day over 4 d.

3.3.3 Skin Assembly

1. Place an air-liquid stand in an empty culture dish and add a membrane filter on the top of the stand.
2. Carefully detach a fibroblast sheet from the bottom of the well by holding the paper anchorage using fine forceps and transfer into a culture dish.
3. Repeat **step 1** and transfer a sheet containing keratinocytes on top of the previous fibroblast sheet. Typically, two cell sheets are superimposed to produce thicker tissues for routine experiments, but higher number of fibroblast sheets can be used. Cell sheets are then clipped together through the anchoring device using 2–3 Ligaclips.
4. Using curved forceps, place the skin on the membrane on the air–liquid stand.
5. Add 10 mL of A/L medium containing 50 μg/mL ascorbic acid (*see* **Note 10**). Incubate in 8% CO_2, 100% humidity atmosphere at 37 °C. Change culture medium three times a week (*see* **Note 11**).

6. After 2 weeks, reconstructed skin is then ready for various types of analysis after removal of the anchoring device.
7. The reconstructed skin can be cultivated at the air–liquid interface more than 28 d. A well-organized basement membrane is obtained after 14 d of culture at the air–liquid interface [11].

4 Notes

1. Serum must be screened to select the serum lot ensuring maximal cell performance according to the used cells (growth but also differentiation status and matrix production). Avoid changing serum lot between experiments.
2. Serum must be added first followed by addition of insulin with a new sterile plastic pipette.
3. This solution must be prepared the day of its use.
4. When exposed to light, HEPES buffer may undergo degradation and become toxic.
5. Feeder layer can be kept 1 month in 8% CO_2, 100% humidity atmosphere at 37 °C with changing medium every week. The presence of at least 24 h-feeder layer conditioned medium is crucial for the keratinocyte growth.
6. Ascorbic acid stock solution must be prepared immediately before use and protected from light.
7. We usually perform this model using 6-well plates but also successfully used 12-well plates or 25 cm^2 culture flask according to the need of the experiment. The size of the cell culture dish to be used is thus not limited and is to be chosen according to the desired type of reconstruction and experiment to be performed. Therefore, anchoring devices, membrane, and air-liquid stand should be customized to the culture dish format.
8. If pieces of dermis remain visible, allow dermis decantation, collect the supernatant, and further add new collagenase solution before to return unit at 37 °C for 1–2 h until complete dissociation.
9. DMSO is an oxidative agent toxic for the cells, especially at temperature above 10 °C. Thus, it must be used at 4 °C when added on cells.
10. The lower surface of the reconstructed skin must be in direct contact with culture medium. Avoid bubble formation between culture medium and dermis.
11. If medium become yellow after 2 d, change it every day.

Acknowledgments

The authors would like to thank current and former members of the LOEX laboratory. A special thank you to the members of the Moulin's team, who have contributed to adapt the protocols to obtain this fibrotic skin model.

References

1. Aksoy MH, Vargel I, Canter IH, Erk Y, Sargon M, Pinar A, Tezel GG (2002) A new experimental hypertrophic scar model in guinea pigs. Aesthet Plast Surg 26:388–396
2. Zhu KQ, Engrav LH, Tamura RN, Cole JA, Muangman P, Carrougher GJ, Gibran NS (2004) Further similarities between cutaneous scarring in the female, red Duroc pig and human hypertrophic scarring. Burns 30:518–530
3. Reid RR, Mogford JE, Butt R, deGiorgio-Miller A, Mustoe TA (2006) Inhibition of procollagen C-proteinase reduces scar hypertrophy in a rabbit model of cutaneous scarring. Wound Repair Regen 14:138–141
4. Dasu MR, Hawkins HK, Barrow RE, Xue H, Herndon DN (2004) Gene expression profiles from hypertrophic scar fibroblasts before and after IL-6 stimulation. J Pathol 202:476–485
5. Varedi M, Tredget EE, Scott PG, Shen YJ, Ghahary A (1995) Alteration in cell morphology triggers transforming growth factor-b1, collagenase, and tissue inhibitor of metalloproteinases-I expression in normal and hypertrophic scar fibroblasts. J Invest Dermatol 104:118–123
6. Nakagawa S, Pawelek P, Grinnell F (1989) Extracellular matrix organization modulates fibroblast growth and growth factor responsiveness. Exp Cell Res 182:572–582
7. Shook BA, Wasko RR, Rivera-Gonzalez GC, Salazar-Gatzimas E, Lopez-Giraldez F, Dash BC, Munoz-Rojas AR, Aultman KD, Zwick RK, Lei V, Arbiser JL, Miller-Jensen K, Clark DA, Hsia HC, Horsley V (2018) Myofibroblast proliferation and heterogeneity are supported by macrophages during skin repair. Science 362. https://doi.org/10.1126/science.aar2971
8. Stunova A, Vistejnova L (2018) Dermal fibroblasts—a heterogeneous population with regulatory function in wound healing. Cytokine Growth Factor Rev 39:137–150. https://doi.org/10.1016/j.cytogfr.2018.01.003
9. Black AF, Berthod F, L'Heureux N, Germain L, Auger FA (1998) In vitro reconstruction of a human capillary-like network in a tissue-engineered skin equivalent. FASEB J 12:1331–1340
10. Mayrand D, Laforce-Lavoie A, Larochelle S, Langlois A, Genest H, Roy M, Moulin V (2012) Angiogenic properties of myofibroblasts isolated from normal human skin wounds. Angiogenesis 15:199–212
11. Michel M, L'Heureux N, Pouliot R, Xu W, Auger FA, Germain L (1999) Characterization of a new tissue-engineered human skin equivalent with hair. In Vitro Cell Dev Biol Anim 35:318–326
12. Dubé J, Rochette-Drouin O, Lévesque P, Gauvin R, Roberge C, Auger FA, Goulet D, Bourdages M, Plante M, Germain L, Moulin V (2010) Restoration of the transepithelial potential within tissue-engineered human skin in vitro and during the wound healing process in vivo. Tissue Eng Part A 16:3055–3063
13. L'Heureux N, Paquet S, Labbé R, Germain L, Auger FA (1998) A completely biological tissue-engineered human blood vessel. FASEB J 12:47–56
14. Vermette M, Trottier V, Menard V, Saint-Pierre L, Roy A, Fradette J (2007) Production of a new tissue-engineered adipose substitute from human adipose-derived stromal cells. Biomaterials 28:2850–2860. https://doi.org/10.1016/j.biomaterials.2007.02.030
15. Carrier P, Deschambeault A, Audet C, Talbot M, Gauvin R, Giasson CJ, Auger FA, Guerin SL, Germain L (2009) Impact of cell source on human cornea reconstructed by tissue engineering. Invest Ophthalmol Vis Sci 50:2645–2652. https://doi.org/10.1167/iovs.08-2001
16. Boa O, Cloutier CB, Genest H, Labbe R, Rodrigue B, Soucy J, Roy M, Arsenault F, Ospina CE, Dube N, Rochon MH, Larouche D, Moulin VJ, Germain L, Auger FA (2013) Prospective study on the treatment of lower-extremity chronic venous and mixed ulcers using tissue-engineered skin substitute made by the self-assembly approach. Adv Skin Wound Care 26:400–409. https://doi.org/10.1097/01.ASW.0000433102.48268.2a

17. Germain L, Larouche D, Nedelec B, Perreault I, Duranceau L, Bortoluzzi P, Beaudoin Cloutier C, Genest H, Caouette-Laberge-L, Dumas A, Bussière A, Boghossian E, Kanevsky J, Leclerc Y, Lee J, Moulin VJ, Auger AA (2018) Severely burned patients treated with autologous bilayered self-assembled skin substitutes (SASS) as permanent grafts: a case series of 14 patients. Eur Cell Mater 6:128–141
18. Jean J, Soucy J, Pouliot R (2011) Effects of retinoic acid on keratinocyte proliferation and differentiation in a psoriatic skin model. Tissue Eng Part A 17(13–14):1859–1868. https://doi.org/10.1089/ten.TEA.2010.0463
19. Corriveau M, Boufaied I, Lessard J, Chabaud S, Senécal J, Grodzicky T, Chartier S, Raymond Y, Moulin V (2009) Fibrotic phenotype of Systemic Sclerosis fibroblasts varies with disease duration and severity of skin involvment. J Pathol 217:534–542
20. Bellemare J, Roberge CJ, Bergeron D, Lopez-Valle CA, Roy M, Moulin VJ (2005) Epidermis promotes dermal fibrosis: role in the pathogenesis of hypertrophic scar. J Pathol 206:1–8
21. Larochelle S, Langlois C, Thibault I, Lopez-Valle CA, Roy M, Moulin V (2004) Sensitivity of myofibroblasts to H_2O_2-mediated apoptosis and their antioxidant cell network. J Cell Physiol 200:263–271
22. Simon F, Bergeron D, Lopez Vallé C, Genest H, Armour A, Moulin VJ (2012) Enhanced secretion of TIMP-1 by hypertrophic scar keratinocytes could contribute to the fibrosis. Burns 38:421–427
23. Moulin VJ (2013) Reconstitution of skin fibrosis development using a tissue engineering approach. Methods Mol Biol 961:287–303. https://doi.org/10.1007/978-1-62703-227-8_19

Chapter 29

Methods for the Study of Renal Fibrosis in Human Pluripotent Stem Cell-Derived Kidney Organoids

Alejandro Moran-Horowich and Dario R. Lemos

Abstract

The mechanisms of kidney injury and fibrosis can now be studied using kidney organoids derived from human pluripotent stem cells (hPSCs). Mature kidney organoids contain nephrons and stromal cells with fibrogenic potential, spatially organized in a manner that resembles the anatomy of the kidney. Organoid nephron damage and interstitial fibrosis can be induced under well-controlled experimental conditions in vitro, making this an ideal system for the study of tissue-intrinsic cell signaling and intercellular crosstalk mechanisms in the absence of systemic signals and immune cells that are present in vivo. Here we describe methods for the generation of kidney organoids from a widely used hPSC line, and for the induction and analysis of nephron damage and interstitial fibrosis.

Key words Organoid, Nephron, Injury, Fibrosis

1 Introduction

Recently developed protocols allow the specification of hPSCs into metanephric mesenchymal cell cultures with the developmental potential to form renal vesicles that self-organize into nephron structures and interstitial stroma [1, 2]. These protocols involve directed differentiation of hPSCs in a stepwise fashion, by sequential addition of the GSK-3β inhibitor CHIR99021 (CHIR), Activin A and FGF9 into the media, to mimic the process of nephrogenesis that occurs during embryonic development [3]. When the differentiation is performed on ultralow-attachment culture surfaces, three-dimensional kidney organoids are formed, which contain major nephron structures including glomeruli, proximal tubules, and distal tubules. Kidney organoids also contain non-epithelial stromal cells with fibroblastic phenotype expressing the cell surface marker platelet-derived growth factor receptor beta (PDGFRβ). Incubation of kidney organoids with nephrotoxic molecules such as cisplatin or with acute inflammatory cytokines, such as interleukin 1β(IL1β), results in nephron epithelial cell damage and

Boris Hinz and David Lagares (eds.), *Myofibroblasts: Methods and Protocols*, Methods in Molecular Biology, vol. 2299,
https://doi.org/10.1007/978-1-0716-1382-5_29,

interstitial fibrosis [2, 4–6]. Organoid fibrosis recapitulates key cellular events observed in kidney fibrosis, including early metabolic activation and proliferation of PDGFRβ$^+$ stromal cells, and their accumulation around damaged tubules within a period of 48 h [4]. As the proliferative response progresses, PDGFRβ$^+$ stromal cells adopt a fibrogenic phenotype that is characterized by expression of the myofibroblast marker ACTA2 [4]. This process is concomitant with increased synthesis and deposition of extracellular matrix proteins such as collagen 1, fibronectin 1, and fibrillin [4]. The resulting anatomical pattern of fibrosis accumulation in the organoid interstitium largely resembles the process of tubulointerstitial scarring observed in the kidneys of patients with acute or chronic kidney disease [7, 8].

Fibrosis development in organoids can be studied using fluorescence microscopy and quantitative real-time polymerase chain reaction (qPCR). While fluorescence microscopy provides anatomical and semiquantitative information on the molecules analyzed, qPCR permits detecting quantitative changes in the expression of injury- and fibrosis-associated mRNAs. Protein quantification remains a challenge due to the low amount of tissue obtained from organoids. As described below, the combination of fluorescence imaging and qPCR permits remarkably precise analysis of the mechanisms of renal fibrosis.

2 Materials

2.1 Plating hPSCs for Differentiation

1. 6-Well plates.
2. 24-Well plates.
3. 96-Well ultralow-attachment round-bottom plates.
4. H9 human embryonic cell line (passage 27–65). Permission to use this cell line should be granted by institutional embryonic stem cell research oversight committee and by institutional regulatory board, prior to its use.
5. Complete mTeSR™1 medium: prepare by adding 5× mTeSR™1 Supplement to mTeSR™1 Basal Medium, to get 1× final concentration. Keep aliquots at −20 °C and thaw at 4 °C overnight. Use aliquot within a week.
6. Complete mTeSR™1 medium with ROCK inhibitor ($ROCK_i$, Y27632): Prepare aliquot of complete mTeSR™1 medium containing 10 nM $ROCK_i$ (working concentration). Store aliquots at 4 °C.
7. Cell dissociation reagent Accutase®: Store at 4 °C.
8. Matrigel: Store frozen aliquots at −20 °C. Thaw an aliquot overnight in refrigerator at 4 °C for plate coating the following day.

9. Dulbecco's modified Eagle medium and Ham's F-12 (DMEM/F12) media: Store at 4 °C.
10. Trypan Blue.
11. Luna II Automated Cell Counter.
12. Luna Cell Counting Slides.
13. Micropipettes.
14. Centrifuge.

2.2 Differentiation of hPSCs into Kidney Organoids

1. Basic differentiation medium: Add 5 mL of l-GlutaMAX to 500 mL of Advanced Roswell Park Memorial Institute (RPMI) 1640. Store at 4 °C and use it within a month.
2. CHIR (glycogen synthase kinase 3 inhibitor): Prepare 10 mM stock aliquots by reconstituting 5 mg of CHIR in 1.074 mL of DMSO. Make 10 μL aliquots and store them at −80 °C. Thaw aliquots at room temperature 10 min prior to preparing the Phase I differentiation medium. Once it has been thawed, the aliquot should be stored in a refrigerator at 4 °C. Use it within 1 week.
3. Recombinant Activin A: Prepare 50 μg/mL stock aliquots by reconstituting 50 μg of Activin A in 1 mL of PBS. Make 10 μL aliquots and store them at −80 °C. Thaw aliquots on ice, 15 min prior to preparing the Phase II differentiation medium (see below). After thawing, the aliquot should be stored in the refrigerator at 4 °C and use it within 1 week.
4. Fibroblast Growth Factor 9 (FGF9): Prepare 50 μg/mL stock aliquots by reconstituting 25 μg of FGF9 in 500 μL of PBS with 0.1% (wt/vol) bovine serum albumin (BSA). Make 20 μL aliquots and store them at −80 °C. Thaw aliquots on ice, 15 min prior to preparing the Phase III solution. After thawing, the aliquot should be stored in the refrigerator at 4 °C and use it within 1 week.
5. Phase I differentiation medium: Prepare by adding CHIR from 10 mM stock aliquot to basic differentiation medium to a final concentration of 8 μM.
6. Phase II differentiation medium: Prepare by adding Activin A from 50 μg/mL stock aliquot to basic differentiation medium to a final concentration of 10 ng/mL.
7. Phase III differentiation medium: Prepare by adding FGF9 from 50 μg/mL stock aliquot to basic differentiation medium to a final concentration of 10 ng/mL.

2.3 Injury of Kidney Organoids

1. Cisplatin: Prepare 5 mM stock aliquots by reconstituting 15 mg of cisplatin in 10 mL of sterile water. Make 20 μL aliquots and store them at −80 °C. Thaw aliquots at room temperature 10 min prior to addition to the medium. Once an aliquot has been thawed, store it at 4 °C. Use it within 1 week.

2. Interleukin 1 beta (IL1β Prepare 50 μg/mL aliquots of recombinant human IL1β by reconstituting 50 μg in 1 mL of PBS. Make 10 μL aliquots and store them at −80 °C. Thaw aliquots on ice, 15 min prior to addition to the medium. After thawing, the aliquot should be stored at 4 °C and use it within 1 week.
3. Basic differentiation medium with cisplatin: Add cisplatin from 5 mM stock aliquot to basic differentiation medium to a final concentration of 20 μM.
4. Basic differentiation medium with cisplatin IL1β: Add IL1β from 50 μg/mL stock aliquot to basic differentiation medium to a final concentration of 50 ng/mL.

2.4 Organoid Fixation and Cryopreservation

1. Paraformaldehyde (PFA) 4% (wt/vol): Prepare 20 mL by adding 5 mL of 16% (wt/vol) PFA to 15 mL PBS.
2. Sucrose: Prepare 100 mL of 20% (wt/vol) sucrose solution for cryopreservation by mixing 20 g of sucrose in 100 mL PBS.

2.5 Organoid Analysis by Confocal Microscopy

1. Antibodies and lectin: hPDGFRβ (Novus, AF385) 1:200; human Actin Alpha 2 (hACTA2; Sigma-Aldrich, F3777) 1:500 dilution; human Collagen 1 (hCOL1; Rockland) 1:300 dilution; human Fibronectin 1 (hFN 1; Abcam, ab154211Ab) 1:200 dilution; E-cadherin (CDH1; Abcam, cat. no. ab11512) 1:500 dilution; hKi67 (Abcam, ab15580) 1:200; hKIM1 1:500 dilution; Lotus Tetragonolobus Lectin (LTL; Vector lab, cat.no. B-1325) at 1:200 dilution.
2. Blocking buffer containing 0.3% Triton X-100 and 2.5% donkey serum: Prepare by adding 30 μL of Triton X-100 and 250 μL donkey serum to 10 mL of PBS. Shake the solution for 10 min to dissolve the Triton X-100 and the serum thoroughly. Store at 4 °C and use it within 2 weeks.
3. Antibody dilution buffer: Prepare by adding 400 mg of BSA to 40 mL of PBS with Triton X-100 (PBST). Shake the solution for 10 min to dissolve BSA thoroughly. Store at 4 °C and use it within 1 month.
4. Vectashield mounting media with DAPI.
5. Confocal microscope.

2.6 Organoid Analysis by Real-Time PCR (qPCR)

1. RNeasy Plus Mini Kit.
2. iScript Reverse Transcription Supermix, oligo(dT) and random primers.
3. Real-Time PCR System.

Taqman assays: GAPDH Hs02758991_g1; ACTA2 Hs00426835_g1; COL1A1 Hs00164004_m1; FN1 Hs01549976_m1; KIM1 (HAVCR1) Hs00930379_g1.

3 Methods

The following procedures must be performed in a sterile laminar flow hood unless otherwise specified.

3.1 Differentiation of hPSCs into Metanephric Mesenchyme (Fig. 1)

Cell culture incubator settings must be 37 °C and 5% CO_2.

1. Coat 24-well plates with Matrigel by diluting 15 μL of Matrigel in 1 mL of DMEM/F12, use 250 μL per well. Incubate plates at room temperature for 1 h. While coating takes place, prepare single cell suspensions.
2. Wash 6-well plate wells containing hPSC cultures with PBS.
3. Add 500 μL Accutase® into each well containing hPSC cultures to be used in differentiation assays, and incubate plate at 37 °C for 7 min.
4. Stop Accutase® activity by adding 500 μL of complete mTeSR™1 medium.
5. Lift cells using a 1000 μL micropipette and transfer to a 15 mL tube.
6. Rinse cells by adding another 1 mL of mTeSR™1 medium.
7. Add mTeSR™1 medium to the 15 mL tube to a final wash volume of 6 mL.
8. Centrifuge the tube at 900 rpm for 5 min at room temperature.

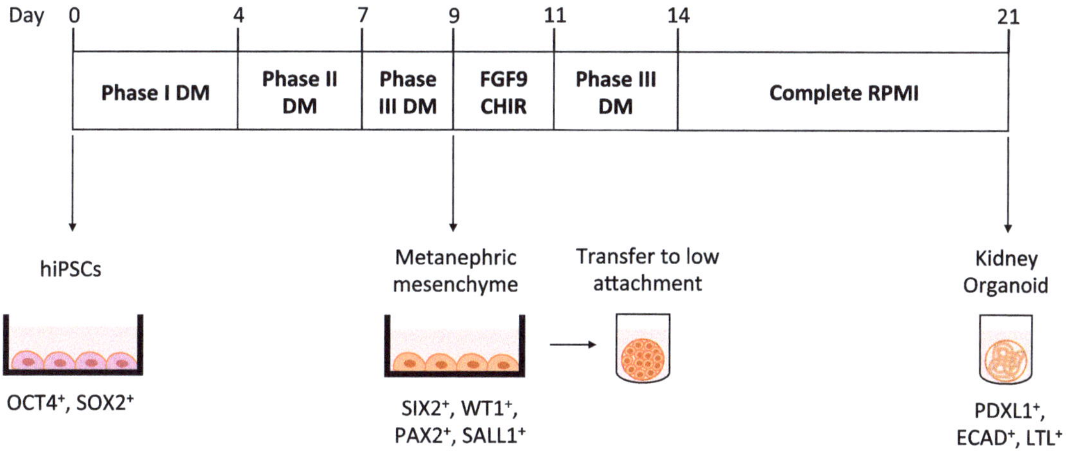

Fig. 1 Protocol for generation of kidney organoids from hPSCs. (**a**) Sequential exposure to well-defined factors drives hPSC differentiation into metanephric mesenchymal cells capable of generating nephrons in vitro. Incubation times (Days) in Phase I, Phase II, and Phase III differentiation medium (DM) are indicated. Transfer of single cell suspensions of metanephric mesenchymal cells to ultralow-attachment wells is indicated. OCT4, POU class 5 homeobox 1; SOX2, SRY-box 2; WT1, Wilms tumor 1; PAX2, paired box 2; SALL1, spalt-like transcription factor 1; Podocalyxin 1 (PDXL1); Cadherin 1 (CDH1); Lotus Tetranoglobus Lectin (LTL)

9. Discard supernatant without disturbing the cell pellet.
10. Resuspend the pellet in 1 mL of mTeSR™1 medium with $ROCK_i$.
11. Take 50 μL of the cell suspension and mix with 50 μL of Trypan Blue. Load the mix into a Luna cell counting slide and count the number of viable cells. *See* **Note 1**.
12. Plate 100,000 cells *per* well in a 24-well plate pre-coated with Matrigel, adjust the final volume to 300 μL. Allow the cultures to settle down overnight.
13. Next day confirm that cells are at approximately 40–50% confluency and wash the cultures with PBS. Add Phase I differentiation medium and place in incubator. Change half the volume of medium in each well every day for 4 days.
14. At the beginning of day 5, remove Phase I medium, wash the cultures with PBS and add 300 μL of Phase II differentiation medium and place in incubator. Change half the volume of Phase II medium in each well every day, for a total of 3 days. *See* **Note 2**.
15. On day 7 remove Phase II differentiation medium, wash the cultures with PBS and add 300 μL of Phase III differentiation medium. Place cells in the incubator. Change half the volume of medium in each well every day, for 2 days.
16. On day 9 remove Phase III differentiation medium, wash the cultures with PBS.
17. Prepare Phase III differentiation medium supplemented with 3 μM CHIR. To that end, add CHIR from 10 mM stock aliquot to the Phase III differentiation medium to a final concentration of 3 μM.
18. Treat cell cultures with 300 μL of Accutase per well of the 24-well plate. Incubate the cells at 37 °C in 5% CO_2 incubator for 10 min.
19. Dissociate cells with a 1000 μL micropipette, then transfer them to a 15 mL tube containing 300 μL of Phase III medium with 3 μM CHIR.
20. Stop Accutase® activity by adding 300 μL of complete RPMI medium.
21. Dissociate cells with a 1000 μL micropipette and transfer cells to a 15 mL tube.
22. Rinse cells by adding another 300 μL of complete RPMI medium.
23. Complete the volume of the 15 mL tube to 1 mL.

24. Take 20 μL of the cell suspension and mix with 20 μL Trypan Blue. Add 15 μL to a Luna cell counting slide and count the number of viable cells.
25. Complete the volume of the 15 mL tube to 6 mL.
26. Centrifuge at 900 rpm for 5 min at room temperature.
27. Resuspend cell pellet with 1 ml of Phase III medium supplemented with CHIR 3 μM.
28. Plate 200,000 cells/well in 200 μL of Phase III medium supplemented with CHIR 3 μM onto 96-well ultralow-attachment round-bottom plates. Centrifuge the plates at 850 rpm for 15 s, and incubate for 2 days.
29. On day 11, gently aspirate as much Phase III medium supplemented with CHIR 3 μM as possible and rinse with PBS. Add 200 μL of Phase III differentiation medium (without CHIR) and incubate. Change half the volume of Phase III medium in each well every day for 3 days. *See* **Notes 3** and **4**.
30. On day 14, carefully remove Phase III differentiation medium, and add complete RPMI medium. Incubate the organoids for 7 days changing medium every other day. *See* **Notes 3** and **4**.
31. Kidney organoids are maintained in complete RPMI medium in the incubator. *See* **Note 5**.

3.2 Induction of Organoid Nephron Damage and Interstitial Fibrosis (Fig. 2)

1. Incubation with IL1β. Incubation with this inflammatory cytokine recapitulates essential cellular and molecular events of acute kidney injury in vivo.
   (a) Remove organoid culture medium and add 200 μL of medium with IL1β. Place the organoids back in the incubator for 24 h, 48 h, or 96 h.
   (b) 1.2 Change medium completely every other day.
   (c) Remove medium with IL1β and wash three times with 200 μL PBS.
2. Induction of injury and fibrosis with cisplatin. Cisplatin is a nephrotoxicant known for its ability to induce tubular cell damage in cancer patients as well as in laboratory animal models of kidney injury.
   (a) Remove medium and add 200 μL of medium with cisplatin. Place in incubator for 48 h.
   (b) Remove medium with cisplatin and wash three times with 200 μL PBS.

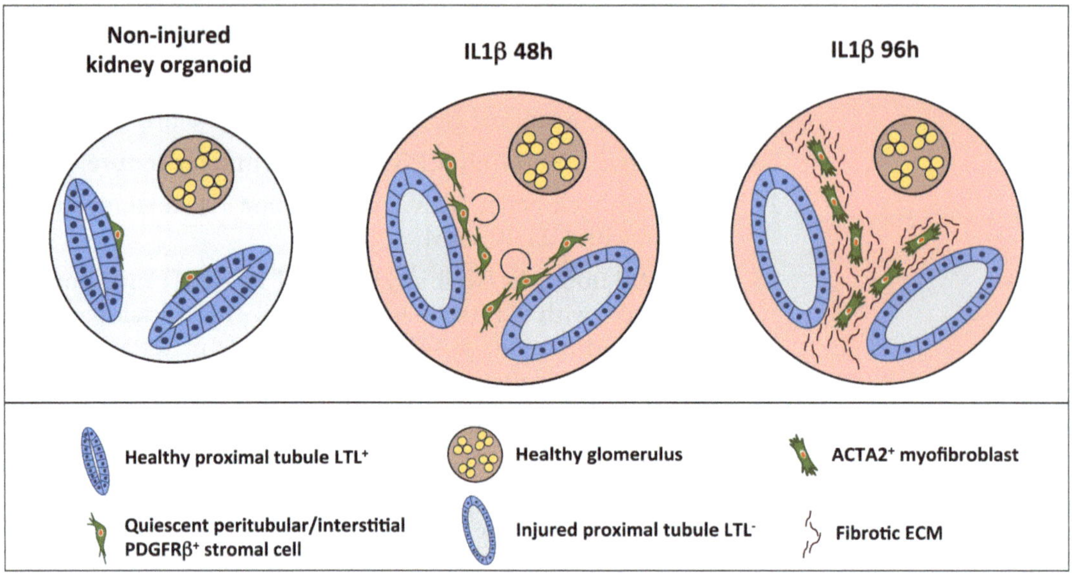

Fig. 2 Schematic representation of organoid fibrosis development during incubation with 100 ng/mL IL1β. (**a**) Non-damaged kidney organoids present healthy tubules that are labeled with LTL, and peritubular interstitial stromal cells identified by expression of the mesenchymal marker PDGFRβ. Incubation with IL1β triggers metabolic activation and proliferation of PDGFRβ$^+$ cells after 48 h, resulting in stroma growth and increased organoid size. Concomitantly, proximal tubule epithelial cells become damaged and the tubules dilate moderately. At 96 h PDGFRβ$^+$ cells adopt a myofibroblast phenotype, expressing ACTA2 and synthetizing excessive amount of ECM. The proximal tubules show severe dilation and loss of brush borders (LTL$^-$). Glomeruli show no sign of injury, indicated by lack of proliferative response or apoptosis

3.3 Assessment of Organoid Nephron Damage and Fibrosis Using Immunofluorescence Microscopy (Fig. 3)

These procedures do not need to be performed in a sterile laminar flow.

1. Organoid Fixation and Cryosectioning
   - (a) Add 200 μL PBS 4% PFA and incubate at room temperature for 20 min.
   - (b) Wash twice with 300 μL PBS.
   - (c) Add 300 μL of 20% sucrose. Incubate at 4 °C overnight.
   - (d) Next day, lift the organoids with a 200 μL pipette with a tip previously cut to ensure the whole organoid can be pipetted without damaging it. Deposit the organoid in a histology mold.
   - (e) Add OCT to the mold, embedding the organoids completely.
   - (f) Freeze the embedded organoids at −80 °C overnight, or until solid OCT blocks are formed.

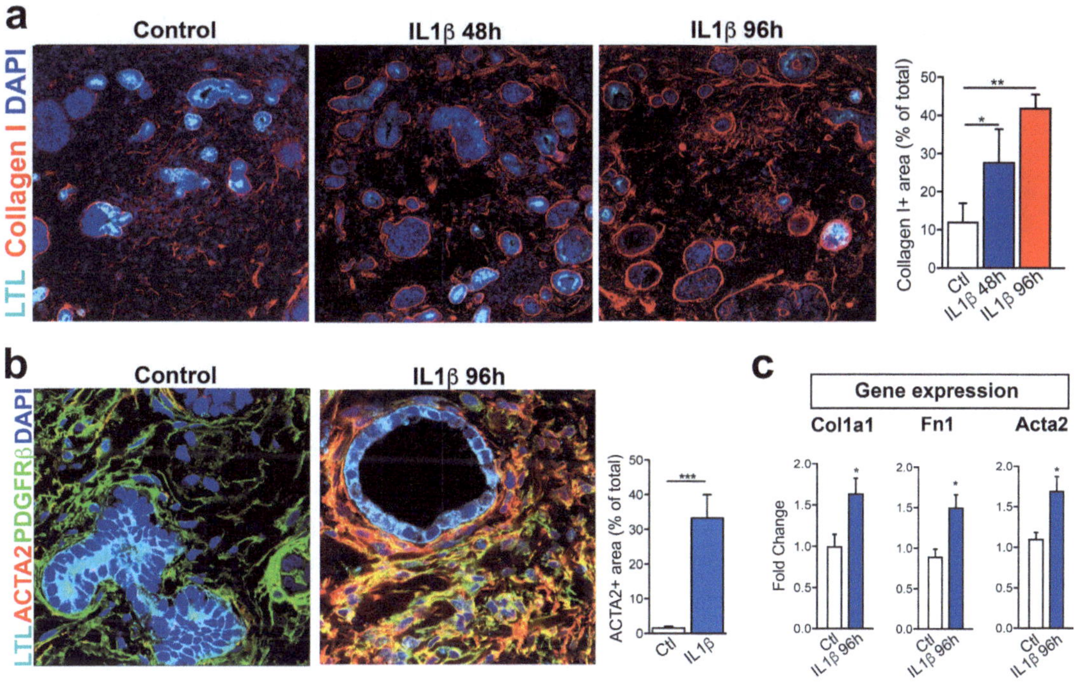

Fig. 3 Development of interstitial fibrosis in injured kidney **organoids**. (**a**) Immunofluorescence used to assess the progression of Collagen I deposition in basement membranes and interstitium, in kidney organoids treated with IL1β. Values on the bar graph represent mean ± SD. $^{**}P < 0.01$. Student's *t*-test. $n = 3$ organoids per treatment/time point. (**b**) Immunofluorescence for the detection of ACTA2 expression in interstitial PDGFRβ$^+$ stromal cells, after 96 h of stimulation with IL1β. Values on the bar graph represent mean ± SD. $^{**}P < 0.01$ *vs.* Ctl. One-way ANOVA. $n = 3$ organoids per treatment. (**c**) Assessment of Col1a1, Fn1, and Acta2 by qPCR. Values on the bar graph represent mean ± SD. $^{*}P < 0.05$; $^{**}P < 0.01$. Student's *t*-test. $n = 3$ organoids per treatment/time point. Figure adapted from [4] Lemos D *et al.* J Am Soc Nephrol. 2018 Jun;29(6):1690–1705, and reproduced with permission granted by the Journal of the American Society of Nephrology

(g) Using a cryostat, cut 15 μm sections, mounting the sections on positively charged glass slides.

(h) Store the slides at −80 °C until ready to use.

2. Immunolabeling for fluorescence microscopy and confocal microscopy

(a) Remove slides from −80 °C, let them sit at room temperature for 15 min.

(b) Wash slides three times for 5 min with PBS.

(c) Block samples with blocking buffer for 1 h.

(d) Wash slides three times with PBS for 5 min.

(e) Add primary antibodies diluted in dilution buffer and incubate overnight at 4 °C.

(f) Next day, wash slides three times for 5 min with PBS.

(g) Add secondary antibodies conjugated to fluorescent probes, incubate at room temperature for 1 h.
(h) Next day, wash slides three times with PBS for 5 min.
(i) Add vectashield with DAPI and place coverslip.
(j) Seal slides with nail polish, secure the coverslip.
(k) Examine samples in confocal or fluorescence microscope.

3. Measurement of injury and fibrosis markers with qPCR.
   (a) Wash organoids three times for 5 min with 200 μL PBS at room temperature.
   (b) Extract total RNA using RNeasy Plus Mini Kit.
   (c) Determine purity by measuring absorbance at 260 nm and 280 nm, and calculating the ratio A260/A280. Use RNA samples with a ratio of 2.0 or above.
   (d) Synthetize cDNA using iScript Reverse Transcription Supermix; using a combination of oligo(dT) and random primers.
   (e) Perform qPCR in Real-Time PCR System.

4 Notes

1. It is important to consider that the number of single cells obtained following treatment with Accutase will vary depending on the initial confluence of the 6-well plate well and the resulting cell viability. Under optimal conditions, one well at 80–90% confluence at 85–100% viability can yield approximately 2.5–3 $\times$ 10^6 live cells.
2. On day 5 of the differentiation protocol it is important to visually confirm that the cell monolayers contain regions were the cells appear more condensed. These condensations are the initial substrate for the subsequent formation of pre-tubular aggregates.
3. Following incubation in Phase III medium supplemented with 3 μM CHIR, the presence of renal vesicles (day 11) and subsequently of pre-tubular aggregates (day 14) can be visually assessed along the edges of the organoids using a bright-field microscope. The presence of these structures can also be confirmed by means of immunofluorescence for the detection of the two homeobox factors Paired Box 8 (PAX8) and LIM Homeobox 1 (LHX1).

4. Throughout the differentiation protocol, care must be taken when changing medium to not disturb the cell aggregates or the organoids mechanically with the pipette tips.
5. On day 21 nephron formation can be verified by means of immunofluorescence microscopy on sampled organoids through detection of nephron structures labeled with anti-Podocalyxin 1 (PDXL1); Cadherin 1 (CDH1); Lotus Tetranoglobus Lectin (LTL), antibodies.

References

1. Lam AQ, Freedman BS, Morizane R, Lerou PH, Valerius MT, Bonventre JV (2014) Rapid and efficient differentiation of human pluripotent stem cells into intermediate mesoderm that forms tubules expressing kidney proximal tubular markers. J Am Soc Nephrol 25 (6):1211–1225. https://doi.org/10.1681/ASN.2013080831
2. Morizane R, Lam AQ, Freedman BS, Kishi S, Valerius MT, Bonventre JV (2015) Nephron organoids derived from human pluripotent stem cells model kidney development and injury. Nat Biotechnol 33(11):1193–1200. https://doi.org/10.1038/nbt.3392
3. Morizane R, Bonventre JV (2017) Generation of nephron progenitor cells and kidney organoids from human pluripotent stem cells. Nat Protoc 12(1):195–207. https://doi.org/10.1038/nprot.2016.170
4. Lemos DR, McMurdo M, Karaca G, Wilflingseder J, Leaf IA, Gupta N, Miyoshi T, Susa K, Johnson BG, Soliman K, Wang G, Morizane R, Bonventre JV, Duffield JS (2018) Interleukin-1 beta activates a MYC-dependent metabolic switch in kidney stromal cells necessary for progressive tubulointerstitial fibrosis. J Am Soc Nephrol 29(6):1690–1705. https://doi.org/10.1681/ASN.2017121283
5. Freedman BS, Brooks CR, Lam AQ, Fu H, Morizane R, Agrawal V, Saad AF, Li MK, Hughes MR, Werff RV, Peters DT, Lu J, Baccei A, Siedlecki AM, Valerius MT, Musunuru K, McNagny KM, Steinman TI, Zhou J, Lerou PH, Bonventre JV (2015) Modelling kidney disease with CRISPR-mutant kidney organoids derived from human pluripotent epiblast spheroids. Nat Commun 6:8715. https://doi.org/10.1038/ncomms9715
6. Takasato M, Er PX, Chiu HS, Maier B, Baillie GJ, Ferguson C, Parton RG, Wolvetang EJ, Roost MS, Chuva de Sousa Lopes SM, Little MH (2015) Kidney organoids from human iPS cells contain multiple lineages and model human nephrogenesis. Nature 526(7574):564–568. https://doi.org/10.1038/nature15695
7. Kramann R, Fleig SV, Schneider RK, Fabian SL, DiRocco DP, Maarouf O, Wongboonsin J, Ikeda Y, Heckl D, Chang SL, Rennke HG, Waikar SS, Humphreys BD (2015) Pharmacological GLI2 inhibition prevents myofibroblast cell-cycle progression and reduces kidney fibrosis. J Clin Invest 125(8):2935–2951. https://doi.org/10.1172/JCI74929
8. Campanholle G, Ligresti G, Gharib SA, Duffield JS (2013) Cellular mechanisms of tissue fibrosis. 3. Novel mechanisms of kidney fibrosis. Am J Physiol Cell Physiol 304(7):C591–C603. https://doi.org/10.1152/ajpcell.00414.2012

Chapter 30

Decellularized Human Lung Scaffolds as Complex Three-Dimensional Tissue Culture Models to Study Functional Behavior of Fibroblasts

Gerald Burgstaller, Michael Gerckens, Oliver Eickelberg, and Melanie Königshoff

Abstract

In vitro culturing of cells in two-dimensional (2D) environments is a widespread used methodology in biomedical research. Most commonly, cells are cultured on artificial plastic dish surfaces, which lead to abnormal functional behaviors, as plastic does not reflect the native microenvironment found in vivo or in situ. Therefore, a multitude of three-dimensional (3D) cell culture systems were developed in the past years, which aim to bridge the gap between 2D cell culture dishes and the in vivo situation. One of the more recent development in the field, the generation of viable precision-cut tissue slices from various organs emerged as an exciting approach to study complex interactions and biological processes ex vivo in 3D. Decellularization of such tissue slices leads to the removal of all functional cells, and leaves behind a scaffold of extracellular matrix (ECM), which closely recapitulates the molecular composition, mechanical properties, topology, and microarchitecture of native ECM. Subsequently, decellularized precision-cut lung slices (PCLS), also called 3D lung tissue culture (3D-LTCs), can be successfully reseeded with a variety of cell types, including fibroblasts, which attach to and engraft into the matrix. Here, we describe the generation of PCLS from resected human lung tissue and their decellularization and recellularization with primary human fibroblasts. This novel 3D tissue culture model allows for various functional studies of fibroblast behavior on native ECM composition and topology.

Key words 3D cell culture models, Tissue engineering, Myofibroblasts, Fibroblasts, Human ex vivo models, Precision-cut lung slices, PCLS, 3D-LTCs, Decellularization, Recellularization, Pharmacological testing, Transdifferentiation

1 Introduction

Ideally, functional behavior of cells is studied in their native in situ or in vivo microenvironment. Signals from the extracellular matrix (ECM) instruct cells and fundamentally affect their biological processes, such as gene expression, adhesion, migration, proliferation, survival, apoptosis, and differentiation [1–3]. However, in vivo functional studies are largely restricted to animal models and it is

Boris Hinz and David Lagares (eds.), *Myofibroblasts: Methods and Protocols*, Methods in Molecular Biology, vol. 2299, https://doi.org/10.1007/978-1-0716-1382-5_30, © Springer Science+Business Media, LLC, part of Springer Nature 2021

well described that these findings are often poorly predictive for human pathophysiology. In contrast, in vitro studies investigating fundamental cellular processes are mostly restricted to 2D models utilizing rigid surfaces of cell culture dishes, all of which do not replicate the inherent intricacy of in vivo tissues in respect of cellular and molecular composition, as well as topology, architecture, and biomechanics. Moreover, cells cultured on 2D plastic surfaces might display abnormal cellular behavior as well as altered responses in pharmaceutical testing [4, 5]. Therefore, human ex vivo tissue cultures obtained from diverse healthy and diseased human tissues emerged as an exciting and highly relevant tool to bridge the experimental gap between in vitro and in vivo studies [3, 6–12]. Standardized slicing with vibratomes or similar devices yields reproducible cuts of human tissues with constant thicknesses as so-called precision-cut tissue slices (PCLS).

Within this chapter, we first provide a detailed and optimized protocol for the generation of precision-cut lung slices from human lungs, which are obtained from surgically resected soft-elastic patient lung tissue. Usually, these lung samples require special preparation, in contrast to whole lobe or entire lung explants. Furthermore, proper slicing entails the filling of the bronchoalveolar space by means of low-melting-point agarose, which provides sufficient stiffness for cutting and largely preserves the native lung structure and its architecture [11]. Human PCLS are subsequently generated with a vibratome to produce 500 μm thick slices, which can be cultured in using either regular cell culture dishes or transwell setups. Furthermore, we describe the process of decellularization and recellularization of human PCLS. Decellularization of human ex vivo tissue aims to remove all resident cells, but leaving an intact acellular extracellular matrix scaffold with integrated growth factors as well as an overall preserved tissue architecture. Decellularized PCLS scaffolds can be successfully recellularized with primary human lung fibroblasts. These recellularized human PCLS can be applied for (a) functional studies investigating the instructive nature of ECM, transdifferentiation of fibroblasts to myofibroblasts, migratory and proliferative behavior of fibroblasts and myofibroblasts, cell-matrix adhesions and subcellular structures, and (b) pharmacological testing of novel compounds interfering with pathological relevant aberrant fibroblast behavior [2, 3, 12].

2 Materials

The human tissue was provided by the CPC-M bioArchive and its partners at the Asklepios Biobank Gauting, the Klinikum der Universität München and the Ludwig-Maximilians-Universität München. The study was approved by the ethics committee of the Ludwig-Maximillian University [Munich, Germany (project number 455-12)] and by the Asklepios Biobank for Lung Diseases (Gauting, Germany, project number 333-10). Written consent was obtained from all patients. We gratefully acknowledge the provision of human biomaterial and clinical data from the CPC-M bioArchive and its partners at the Asklepios Biobank Gauting, the Klinikum der Universität München and the Ludwig-Maximilians-Universität München.

All solutions and consumables must be sterile and handled under sterile conditions.

2.1 Generation of Human Precision-Cut Lung Slices (PCLS)

1. Equipment: Hyrax V50 Vibratome (Zeiss).
2. Cyanoacrylate glue.
3. Tissue culture petri dishes: 15 cm dishes, 12-well and 96-well plates.
4. Laminar flow hood.
5. Water bath.
6. Scalpels.
7. Tweezers.
8. Forceps.
9. Threads.
10. Surgical clamps.
11. 30 mL Syringes.
12. Peripheral venous catheter.
13. Metal tray.
14. Agarose solution: low-melting 3% (w/v) agarose is dissolved in a minimum of 30 mL cultivation medium. The solution is heated in a microwave oven until boiling and then cooled to 42 °C in a water bath. 50 mL conical Falcon tubes (Corning) for aliquoting the agarose solution.
15. Culture medium: Dulbecco's Modified Eagle Medium (DMEM) F-12 supplemented with L-Glutamine, HEPES, 10,000 IE Penicillin, 10,000 μg Streptomycin, 2.5 μg/mL amphotericin B, and 0.1% (v/v) fetal bovine serum. Cell culture medium is used at 37 °C.
16. Resected lung tissue: fresh and tumor-free lung tissue from lobectomy, which must be kept in culture medium at 4 °C (*see* **Note 1**).
17. 4 mm Biopsy puncher.

2.2 Decellularization of PCLS

1. Deionized water.
2. Phosphate-Buffered Saline (PBS): 138 mM NaCl, 26 mM KCl, 84 mM Na_2HPO_4, 14 mM KH_2PO_4, pH = 7.4.

3. Decel-Solution A: 0.1% SDS dissolved in deionized water and warmed to 37 °C.
4. Decel-Solution B: 1 M NaCl in deionized water.
5. DNAse solution: 30 μg/mL DNAse (Roche) dissolved in 5 mM $MgCl_2$ in PBS.
6. Storage solution: 100 IE/mL Penicillin and 100 μg/mL Streptomycin in PBS.
7. Tube roller, Thermomixer for 50 mL tubes.

2.3 Recellularization of PCLS with Primary Human Fibroblasts

1. Primary human lung fibroblasts.
2. 0.25% Trypsin-EDTA.
3. PBS: 138 mM NaCl, 26 mM KCl, 84 mM Na_2HPO_4, 14 mM KH_2PO_4, pH = 7.4.
4. Decellularized or punched PCLS.
5. Culture medium.
6. Cell culture incubator equipped with a tube roller.

3 Methods

All procedures are performed under sterile conditions using laminar flow hoods as well as sterile solutions and consumables.

3.1 Generation of Human Precision-Cut Lung Slices (PCLS)

3.1.1 Human Lung Tissue Inflation by Agarose Filling

1. The lung tissue stored at 4 °C in culture media is transferred into a 15 cm cell culture dish.
2. A 30 mL syringe is filled with 42 °C warmed up low-melting agarose solution.
3. Remove the obturator of a peripheral venous catheter and attach it to a 30 mL syringe.
4. Within the piece of lung tissue identify a bronchus (0.5–3 mm in diameter) that ventilates an intact section (Fig. 1a).
5. The cannula of the venous catheter is inserted into the selected bronchus and gently pushed forward as far as possible.
6. The bronchus is sealed around the cannula by compressing the surrounding bronchial wall with forceps.
7. Lift the tissue with the forceps from the culture dish and manually start the filling at an approximate speed of 0.3 mL/s (*see* **Note 2**) (Fig. 1b).
8. If either high resistance or high agarose leakage during filling is experienced, the whole procedure can be repeated using a different bronchus (*see* **Notes 3** and **4**).
9. Apply enough agarose until the secondary pulmonary lobuli connected to the cannulated bronchus are entirely filled (*see* **Note 5**).

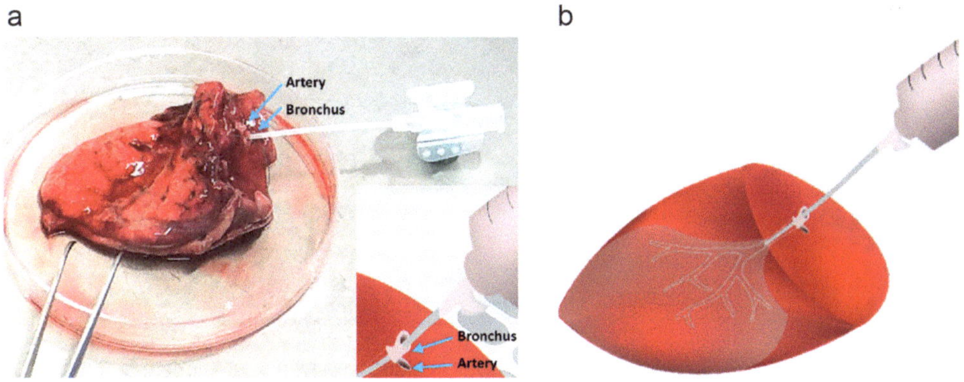

Fig. 1 Agarose filling of resected lung tissue. (**a**) The lung tissue is cannulated with a peripheral venous catheter which is inserted into a bronchus adjacent to the pulmonary artery. (**b**) Liquid agarose is poured into the lung tissue by a 30 mL syringe

10. After successful filling the cannula of the venous, catheter has to be removed and the bronchus immediately closed with a surgical clamp.
11. In order to facilitate solidification of the agarose, the tissue is incubated at 4 °C in culture medium.
12. The filling procedure with liquid agarose can be repeated if the resected piece of lung is large and consists of several bronchial entries.
13. The agarose-filled lung tissue is stored in culture medium at 4 °C until slicing.

3.1.2 Precision-Cut Lung Slicing of Agarose-Filled Human Lung Tissue

1. Identification of sections firmly filled with agarose within the lung tissue (*see* **Note 6**).
2. By using a scalpel excise a block of tissue with the approximate size of 1–1.5 cm^3. Ideally, one side of the block should be covered with pleura.
3. Individual tissue blocks are attached to the vibratome holder by their broadest base and preferably via their pleural side with a cyanoacrylate glue (*see* **Note** 7) (Fig. 2a).
4. Slicing of the lung tissue blocks is accomplished with a vibratome using the following parameter settings: thickness: 500 μm, frequency: 100 Hz, amplitude of the knife: 1.2 mm, forward speed of the blade of 3–12 μm/s (*see* **Note 8**) (Fig. 2b).
5. Single 500 μm thick precision-cut lung slices (Fig. 2c) are gently transferred with forceps from the vibratome tray to wells of 12-well plate which are filled with culture medium.

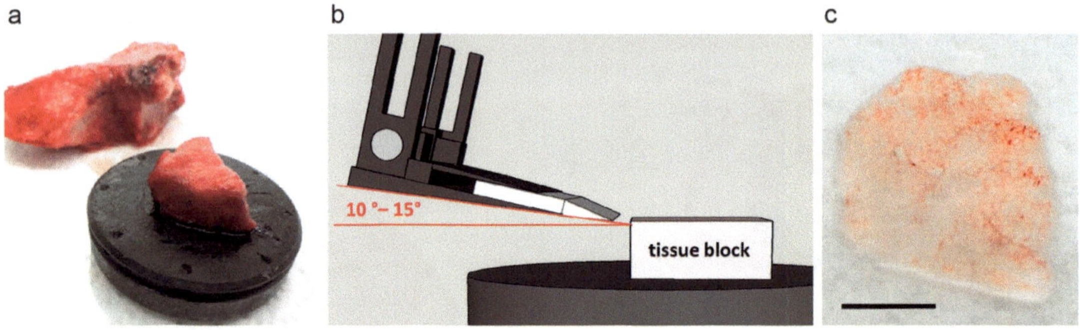

Fig. 2 Vibratome cutting of PCLS. (**a**) Preferably, the tissue is glued with its pleural surface to the surface of the tissue holder. (**b**) 500 μm thick slices are cut by the vibratome with a sapphire knife in a 10°–15° angle relative to the tissue. (**c**) The cutting procedure results in 2–3 cm^3 large intact precision-cut lung slices. Scale bar = 5 mm

6. The well plates containing the PCLS are incubated in a conventional cell culture incubator under standard culturing conditions (5% CO_2, 95% humidity, and 37 °C) (*see* **Note 9**).
7. Additionally, to get a higher number of tissue samples from the same block, a 4 mm biopsy puncher can be used to cut out various 4 mm punches from one PCLS. Punches can be incubated in 96-well plates in culture medium and standard culturing conditions.

3.2 Decellularization of PCLS

1. Transfer native human PCLS (Fig. 3a) or punches with forceps from the 12-well plate into a 50 mL Falcon tube filled with 50 mL deionized water.
2. Incubate the Falcon tube slowly rolling on a tube roller at room temperature for 10 min.
3. Remove the water from the Falcon tube and refill with fresh 50 mL deionized water.
4. Incubate the Falcon tube slowly rolling on a tube roller for 16 h at 4 °C.
5. Remove deionized water from the Falcon tube and refill with 50 mL Decel-Solution A and incubate the Falcon tube slowly rolling on a tube roller at room temperature for 2 h.
6. Remove Decel-Solution A and refill with 50 mL fresh Decel-Solution A and incubate the Falcon tube slowly rolling on a tube roller at room temperature for 2 h.
7. Remove Decel-Solution A and wash twice with 50 mL deionized water for 10 min.
8. Remove deionized water from the Falcon tube and refill with 50 mL Decel-Solution B and incubate the Falcon tube slowly rolling on a tube roller for 16 h at 4 °C.

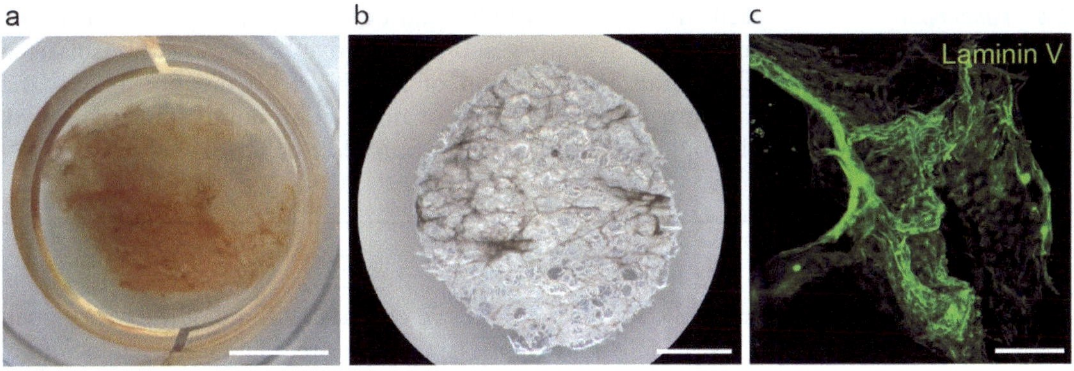

Fig. 3 Decellularization of PCLS. (**a**) 500 μm thick and 2–3 cm^3 large PCLS cultured in growth medium. Scale bar = 5 mm. (**b**) Decellularized and punched PCLS. Scale bar = 1 mm. (**c**) Immunofluorescence staining of Laminin V demonstrating intact microstructures (here: basal lamina) after decellularization. Scale bar = 100 μm

9. Remove Decel-Solution B and wash with 50 mL deionized water for 10 min.
10. Wash with 50 mL PBS for 10 min.
11. Remove PBS and leave 7.5 mL PBS in the Falcon Tube, add DNAse in a final concentration of 100 μg/mL and incubate slowly shaking for 2 h at 37 °C.
12. Wash twice with 50 mL deionized water for 10 min.
13. Store decellularized PCLS or punches (Fig. 3b) in storage solution in 12-well plates at 4 °C (*see* **Note 10**).

3.3 Recellularization of PCLS with Primary Human Lung Fibroblasts

1. Use a primary human fibroblast cell culture from a 10 cm cell culture dish, trypsinize, and generate a cell suspension of 3×10^6 cells/mL of primary human fibroblasts in cultivation medium and put 1.5 mL of this cell suspension in a 2 mL Eppendorf tube (*see* **Note 11**).
2. Add a maximum of 2–3 decellularized PCLS or 10–15 punches to the same 2 mL Falcon tube.
3. Seal the 2 mL Eppendorf tube with parafilm to allow a better gas exchange.
4. Incubate the 2 mL Eppendorf tube rolling at 10–15 rpm in a cell culture incubator at standard cell culture conditions (5% CO_2 and 37 °C) for 16 h.
5. Remove the recellularized PCLS from the cell suspension with a pair of tweezers, transfer PCLS one by one into a 12-well plate, and incubate them at standard cell culture conditions (5% CO_2 and 37 °C) until used for experimental work (*see* **Note 12**).

3.4 Functional Studies

Recellularized human PCLS can be used for (a) functional studies investigating migratory and proliferative behavior of fibroblasts as well as transdifferentiation of fibroblasts to myofibroblasts, and (b) pharmacological testing of novel compounds interfering with pathological relevant aberrant fibroblast behavior.

4 Notes

In the following paragraph, we emphasize some of the major challenges, which might appear and require adaptation when implementing these protocols. We also refer the reader also to the following referenced resources for further description and suggestions [10, 11, 13, 14].

1. Fresh and tumor-free lung tissue from lobectomy must be stored immediately after resection in chilled culture media at 4 °C. Prior to processing, the cold ischemia time of 4–8 h should not be exceeded.
2. The speed of filling with liquid agarose needs to be adapted depending on the state of the lung tissue in respect of airway resistance and/or atelectasis.
3. If high resistance is experienced various positioning of the venous catheter might help to properly fill the tissue. Additionally, plugs of already solidified agarose or other obstructions might block proximal bronchi and leads to incomplete filling. Here, using too much force on the syringe can lead to tissue damage in the obstructed parts.
4. The respiratory tree from the cannulated bronchus might be damaged due to the resection. This will lead to constant agarose leakage. Sealing the damaged peripheral airways with surgical clamps, sewing as well as a more peripheral repositioning of the catheter might help to fill at least a smaller portion of the tissue.
5. Inflation of the lung tissue with too much liquid agarose will lead to damage of lung structures.
6. Agarose solidification can be tested by softly squeezing the lung tissue with tweezers against the bottom of the cell culture dish. Sections which are solidly filled with agarose will not collapse.
7. The pleura provides a crucial barrier between the cyanoacrylate glue and the parenchymal tissue.
8. The protrusion speed depends on the tissue stiffness and has to be reduced if the slicing does not work properly, or if the tissue block itself starts to vibrate.

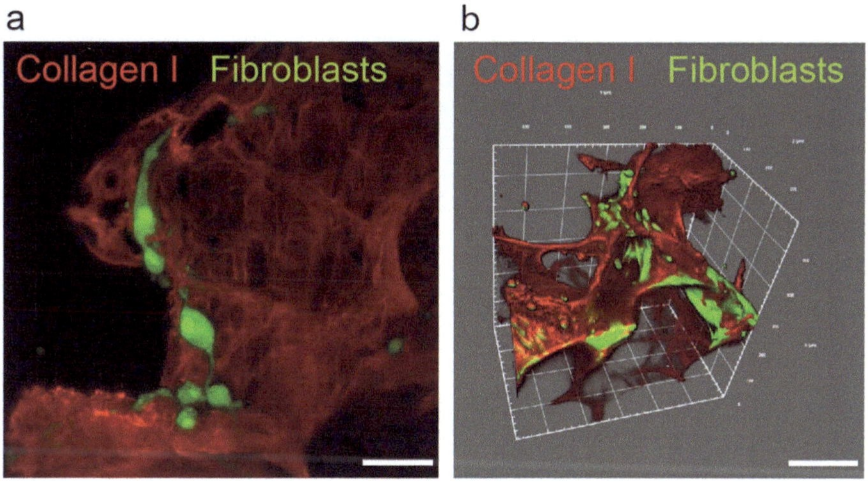

Fig. 4 Recellularization of acellular PCLS with primary human lung fibroblasts. (**a**) PCLS were recellularized with primary human fibroblasts, which were stained with a cell tracking dye (green). Collagen-I was live-immunostained with a Collagen-I antibody. Z-stacks were acquired by light-sheet fluorescence microscopy and are shown here as maximum intensity projection. Scale bar = 50 μm. (**b**) Software-based surface rendering of z-stacks acquired as described in (a). Scale bar = 200 μm

9. To avoid detrimental effects on tissue integrity due to the gluing process, stop slicing if only 2–3 mm of the tissue block are left.
10. Quality control of the decellularization process can be accomplished by histological and immunofluorescence stainings (Fig. 3c), protein content measurements (BCA), Western blotting, and qPCR [2, 12]. Ideally, no cellular components are left.
11. Depending on the requirements and the amount of cells available, the recellularization process can be upscaled on demand by using larger containers like 15 mL or 50 mL Falcon tubes.
12. Living cells, subcellular structures, as well as components of the extracellular matrix can be labeled either with numerous commercially available live tracking dyes or antibodies. Alternatively, recellularized PCLS can be fixed by standard fixing procedures and accordingly stained by immunofluorescent, immunohistological, and histological protocols. Living as well as fixed recellularized PCLS can be applied to widefield, brightfield, and confocal microscopy, lightsheet microscopy (Fig. 4a, b), as well as 2-photon microscopy [2, 9, 15, 16].

References

1. Hynes RO (2009) The extracellular matrix: not just pretty fibrils. Science 326 (5957):1216–1219. https://doi.org/10.1126/science.1176009
2. Burgstaller G, Sengupta A, Vierkotten S, Preissler G, Lindner M, Behr J, Konigshoff M, Eickelberg O (2018) Distinct niches within the extracellular matrix dictate fibroblast function in (cell free) 3D lung tissue cultures. Am J Physiol Lung Cell Mol Physiol 314(5):L708–L723. https://doi.org/10.1152/ajplung.00408.2017
3. Parker MW, Rossi D, Peterson M, Smith K, Sikstrom K, White ES, Connett JE, Henke CA, Larsson O, Bitterman PB (2014) Fibrotic extracellular matrix activates a profibrotic positive feedback loop. J Clin Invest 124 (4):1622–1635. https://doi.org/10.1172/JCI71386
4. Baker BM, Chen CS (2012) Deconstructing the third dimension: how 3D culture microenvironments alter cellular cues. J Cell Sci 125 (Pt 13):3015–3024. https://doi.org/10.1242/jcs.079509
5. Caliari SR, Burdick JA (2016) A practical guide to hydrogels for cell culture. Nat Methods 13 (5):405–414. https://doi.org/10.1038/nmeth.3839
6. Fisher RL, Vickers AE (2013) Preparation and culture of precision-cut organ slices from human and animal. Xenobiotica 43(1):8–14. https://doi.org/10.3109/00498254.2012.728013
7. Liberati TA, Randle MR, Toth LA (2010) In vitro lung slices: a powerful approach for assessment of lung pathophysiology. Expert Rev Mol Diagn 10(4):501–508. https://doi.org/10.1586/erm.10.21
8. Parrish AR, Gandolfi AJ, Brendel K (1995) Precision-cut tissue slices: applications in pharmacology and toxicology. Life Sci 57 (21):1887–1901. https://doi.org/10.1016/0024-3205(95)02176-j
9. Uhl FE, Vierkotten S, Wagner DE, Burgstaller G, Costa R, Koch I, Lindner M, Meiners S, Eickelberg O, Konigshoff M (2015) Preclinical validation and imaging of Wnt-induced repair in human 3D lung tissue cultures. Eur Respir J 46(4):1150–1166. https://doi.org/10.1183/09031936.00183214
10. Alsafadi HN, Uhl FE, Pineda RH, Bailey KE, Rojas M, Wagner DE, Konigshoff M (2020) Applications and approaches for 3D precision-cut lung slices: disease modeling and drug discovery. Am J Respir Cell Mol Biol. https://doi.org/10.1165/rcmb.2019-0276TR
11. Gerckens M, Alsafadi HN, Wagner DE, Lindner M, Burgstaller G, Konigshoff M (2019) Generation of human 3D lung tissue cultures (3D-LTCs) for disease modeling. J Vis Exp (144). https://doi.org/10.3791/58437
12. Booth AJ, Hadley R, Cornett AM, Dreffs AA, Matthes SA, Tsui JL, Weiss K, Horowitz JC, Fiore VF, Barker TH, Moore BB, Martinez FJ, Niklason LE, White ES (2012) Acellular normal and fibrotic human lung matrices as a culture system for in vitro investigation. Am J Respir Crit Care Med 186(9):866–876. https://doi.org/10.1164/rccm.201204-0754OC
13. Liu G, Betts C, Cunoosamy DM, Aberg PM, Hornberg JJ, Sivars KB, Cohen TS (2019) Use of precision cut lung slices as a translational model for the study of lung biology. Respir Res 20(1):162. https://doi.org/10.1186/s12931-019-1131-x
14. Zscheppang K, Berg J, Hedtrich S, Verheyen L, Wagner DE, Suttorp N, Hippenstiel S, Hocke AC (2018) Human pulmonary 3D models for translational research. Biotechnol J 13(1). https://doi.org/10.1002/biot.201700341
15. Burgstaller G, Vierkotten S, Lindner M, Konigshoff M, Eickelberg O (2015) Multidimensional immunolabeling and 4D time-lapse imaging of vital ex vivo lung tissue. Am J Physiol Lung Cell Mol Physiol 309(4):L323–L332. https://doi.org/10.1152/ajplung.00061.2015
16. Akram KM, Yates LL, Mongey R, Rothery S, Gaboriau DCA, Sanderson J, Hind M, Griffiths M, Dean CH (2019) Live imaging of alveologenesis in precision-cut lung slices reveals dynamic epithelial cell behaviour. Nat Commun 10(1):1178. https://doi.org/10.1038/s41467-019-09067-3

INDEX

Boris Hinz and David Lagares (eds.), *Myofibroblasts: Methods and Protocols*, Methods in Molecular Biology, vol. 2299, https://doi.org/10.1007/978-1-0716-1382-5, © Springer Science+Business Media, LLC, part of Springer Nature 2021

E

F

G

H

I

K

L

M

N

O

P

Q

R

S

MIX
Papier aus verantwortungsvollen Quellen
Paper from responsible sources
FSC® C105338

If you have any concerns about our products,
you can contact us on
ProductSafety@springernature.com

In case Publisher is established outside the EU,
the EU authorized representative is:
Springer Nature Customer Service Center GmbH
Europaplatz 3, 69115 Heidelberg, Germany

Printed by Libri Plureos GmbH
in Hamburg, Germany